Gut Microbiome and Environmental Toxicants

This book comprehensively reviews the intricate relationship between environmental toxicants and the gut microbiome. It explores the role of dietary choices and lifestyle in shaping and modulating the gut microbiome's response to environmental toxicants. It examines the intricate relationship between these toxic substances and the composition, function, and overall health implications of the gut microbiome. The chapters provide in-depth insights into the impacts of various toxicants, such as phthalates, pesticides, organic pollutants, bisphenols, and heavy metals, on the delicate microbial balance within our digestive systems. Specific chapters address the impact of lead, mercury, cadmium, and arsenic on the composition and function of the gut microbiome. The book concludes by addressing future prospects and challenges in understanding and mitigating the impacts of environmental toxicants on the gut microbiome and highlighting the importance of these efforts.

Key Features:

- Provides a comprehensive examination of the intricate relationship between environmental toxicants and the gut microbiome
- Reviews the possible mechanisms underlying bidirectional interactions between environmental pollutants and GI
- Examines the role of dietary choices and lifestyle factors in modulating the gut microbiome's response to environmental toxicants
- Covers the impact of toxic substances, phthalates, pesticides, and heavy metals on the gut microbiome
- Explores the practical implications of toxicant exposure on human health

This book is intended for researchers and scientists working in the fields of environmental toxicology, microbiology, pharmacology, and related disciplines.

Gut Microbiome and Environmental Toxicants

Impact on Human Health

Edited by
Gaurav Gupta, Md Sadique Hussain, Kamal Dua,
Ritu Gilhotra, and Murali Dhanasekaran

CRC Press
Taylor & Francis Group
Boca Raton London New York

CRC Press is an imprint of the
Taylor & Francis Group, an **informa** business

Designed cover image: Shutterstock

First edition published 2025
by CRC Press
2385 NW Executive Center Drive, Suite 320, Boca Raton FL 33431

and by CRC Press
4 Park Square, Milton Park, Abingdon, Oxon, OX14 4RN

CRC Press is an imprint of Taylor & Francis Group, LLC

ISBN: 9781032787343 (hbk)
ISBN: 9781032787350 (pbk)
ISBN: 9781003489221 (ebk)

DOI: 10.1201/9781003489221

Typeset in Times
by Deanta Global Publishing Services, Chennai, India

Contents

Preface

Environmental toxicants represent a significant and underexplored risk factor contributing to the global disease burden. These toxicants, ranging from industrial chemicals and pollutants to everyday household items, profoundly influence human health directly and through complex interactions with the gut microbiome. The gut microbiome, an intricate ecosystem within the human gastrointestinal tract, plays a crucial role in modulating immune responses, metabolism, and even the neurological functions of its host. Disturbing this delicate balance through exposure to environmental toxicants can lead to many health issues.

This book delves into the emerging field that intersects the gut microbiome study with the pervasive influence of environmental toxicants. It is designed to show how these interactions contribute to the aetiology of various diseases, from chronic gastrointestinal disorders to systemic conditions such as cancer, diabetes, and neurodegenerative diseases. By exploring the pathways through which toxicants impact the microbiome and human health, this book aims to establish a critical foundation for developing more effective diagnostic tools and therapeutic strategies.

Moreover, this book will serve as an essential resource for a diverse audience, including researchers in microbiology, environmental health, and pharmacology; health-care professionals; and policymakers looking to understand better and mitigate the impacts of environmental toxicants. It compiles contributions from leading experts worldwide, offering multidisciplinary perspectives on the challenges and opportunities at the intersection of microbiology, toxicology, and public health.

In crafting this volume, we have endeavored to create a comprehensive resource that informs and inspires further research and policy action to address the critical issues at the heart of environmental toxicants and the gut microbiome. We are immensely grateful to all contributors whose invaluable insights and dedication have made this publication possible.

Acknowledgement

We acknowledge the HEAL (Healthy Environments And Lives) National Research Network, which receives funding from the National Health and Medical Research Council (NHMRC) Special Initiative in Human Health and Environmental Change (Grant No. 2008937)

About the Editors

Gaurav Gupta is Adjunct Research Professor at the Centre for Global Health Research at Ajman University, UAE, and the Centre for Research Impact & Outcome at Chitkara College of Pharmacy, Chitkara University, Rajpura. Dr. Gupta's research interests lie in molecular mechanisms of receptor biology, epigenetics, and behavioral pharmacology. He has a distinguished record of over 500 publications in peer-reviewed international journals, amassing a cumulative impact factor of over 5,000. His scholarly work has garnered more than 17,000 citations, reflecting his significant contributions to the field. With an h-index of 58 and an i10-index of 340, Dr. Gupta's research has been widely recognized and respected within the scientific community. His academic and research achievements have earned him a place among the world's top 2% of scientists, a distinction he has held since 2020, as per the Stanford University database. This recognition underscores his global impact and influence in the scientific community. Dr. Gupta has also authored and edited several influential books, including titles such as *Targeting Epigenetics in Inflammatory Lung Diseases* and *Microbiome in Inflammatory Lung Diseases*, published by Springer. Throughout his career, Dr. Gupta has demonstrated a solid commitment to innovation in education and research. He has been actively involved in designing new experiments in pharmacology to enhance student learning and has embraced e-learning tools to foster student engagement. His teaching methodology incorporates a blend of traditional and modern techniques, ensuring that his students are well prepared for the challenges of the pharmaceutical sciences. Dr. Gupta's achievements have been recognized with several awards, including the Young Scientist Award 2019 and the Innovative Researcher Award from the Association of Pharmacy Professionals in 2017.

Md Sadique Hussain is Assistant Professor at Uttaranchal University, recognized for his extensive contributions to academia and literature. With over 50 research and review articles to his name, as well as three published book chapters, he has established a diverse and impactful portfolio. His work encompasses a range of peer-reviewed manuscripts and innovative research papers, showcasing his commitment to advancing knowledge across various disciplines. Md Sadique Hussain possesses a unique ability to explore complex subjects with clarity and insight, leaving a lasting impression in the field. His dedication to fostering intellectual curiosity and producing thought-provoking content has garnered respect and admiration from both colleagues and readers. As he continues to make significant contributions to the literary and academic communities, his influence on the world of writing and research is increasingly recognized.

Kamal Dua is Associate Professor in the discipline of pharmacy at the Graduate School of Health, University of Technology, Sydney (UTS). With over 17 years of research experience, he has specialized in drug delivery targeting inflammatory diseases. Additionally, Dr. Dua serves as the Node Leader of Drug Delivery Research in the Center for Inflammation at the Centenary Institute/UTS. In this role, he focuses on advancing targets identified in research projects to develop innovative formulations, taking the first steps toward clinical translation. Dr. Dua's research encompasses two complementary areas: drug delivery and immunology. His work explores how these disciplines can mutually benefit, contributing to the goal of promoting longer and healthier lives for the community. His commitment to synergy is reflected in his extensive publication record in reputable journals. Dr Dua's research interests are centered on harnessing the pharmaceutical potential of modulating critical regulators, such as interleukins and microRNAs. He also specializes in developing new and effective drug delivery formulations for managing inflammation in chronic airway diseases and cancer.

Ritu Gilhotra currently holds the position of ProPresident at SGVU and serves as a Professor and Dean at Gyan Vihar School of Pharmacy. She earned her Master's and Doctorate degrees in Pharmaceutics from the Faculty of Pharmacy at GJUST, Hisar, with a notable research output in the area of ocular drug delivery systems for the management of glaucoma. Dr. Gilhotra's primary research interest revolves around advanced drug delivery systems (DDSs) and drug targeting, in addition to pharmaceutical formulation and dosage form design. Within the realm of DDSs, her work focuses on formulating and evaluating innovative nanoparticles, microparticulates, and vesicular drug carriers aimed at targeting specific biological sites. Dr. Gilhotra has a substantial body of work to her credit, including book chapters and over 100 research articles published in indexed and peer-reviewed journals across the fields of pharmaceutics, pharmacy education, and pharmacy practice. Furthermore, she has obtained ten patents and two industrial designs, and she has facilitated one technology transfer.

Murali Dhanasekaran currently holds the position of Director of International Programs at Auburn University's Harrison College of Pharmacy. As a distinguished educator and scientist, he embarked on his academic journey with a B.Pharm from Annamalai University, furthering his education with an M.Pharm from Jadavpur University, and ultimately obtaining a PhD from the Indian Institute of Chemical Biology. His post-doctoral research led him through various institutions, including the University of Nebraska Medical Center, the University of North Dakota, and Texas A&M. During his tenure at Auburn, Murali has played a transformative role in pharmaceutical education. He not only teaches a diverse range of courses but also initiated an undergraduate program in drugs and biopharmaceutical sciences. His dedication to nurturing global academic relationships is evident in his collaborations with institutions in more than 15 countries. Dr. Murali's laboratory has been at the forefront of neuroscientific research, pioneering the development of various models to comprehend neurological disorders and exploring the therapeutic potential of both synthetic and natural bioactives. With a remarkable portfolio of over 145 research articles, 70 book chapters, and 8 books, his work has gained significant recognition, particularly his research on the neuroprotective effects of *Centella asiatica* in Alzheimer's disease. Presently, his research ambitiously delves into groundbreaking drug development for major neurological ailments, the effects of designer drugs, and the intricate interplay between chemotherapy and neurological health.

List of Contributors

Muhammad Afzal
Department of Pharmaceutical Sciences
Batterjee Medical College
Jeddah, Saudi Arabia

Neetu Agrawal
Institute of Pharmaceutical Research
GLA University
Mathura, U.P., India

Sattam Khulaif Alenezi
Department of Pharmacology and Toxicology,
College of Pharmacy, Qassim University,
Buraydah, Al Qassim,
Saudi Arabia

Taghreed Mohammad Saad Alhaidan
Department of Public Health,
College of Public Health and Health Informatics,
University of Ha'il, Ha'il,
Saudi Arabia
and
Scientific office manager in Advanz Pharma,
Riyadh, Saudi Arabia

Khalid Saad Alharbi
Department of Pharmacology and Toxicology
Unaizah College of Pharmacy
Qassim University
Qassim, Saudi Arabia

Waleed Hassan Almalki
Department of Pharmacology
College of Pharmacy
Umm Al-Qura University
Makkah, Saudi Arabia

Sana Saeed Alqarni
Department of Clinical Laboratory Sciences,
College of Applied Medical Science,
King Saud University,
Riyadh, Saudi Arabia

Samiyah Alshehri
Department of Pharmacology and Toxicology
College of Pharmacy
King Saud University
Riyadh, Saudi Arabia

Ahmed E. Altyar
Department of Pharmacy Practice
Faculty of Pharmacy
King Abdulaziz University
Jeddah, Saudi Arabia

Preeti Arya
Shri Ram College of Pharmacy
Ramba, Karnal, Haryana, India

Gurusha Bahl
School of Pharmaceutical Sciences
Jaipur National University
Jaipur, Rajasthan, India

Asif Ahmad Bhat
School of Pharmacy
Suresh Gyan Vihar University
Jaipur, Rajasthan, India

Mohammed Khaled Bin Break
Department of Pharmaceutical Chemistry
College of Pharmacy
University of Ha'il
Ha'il, Saudi Arabia

Benu Chaudhary
Shri Ram College of Pharmacy
Ramba, Karnal, Haryana, India

Dinesh Kumar Chellappan
School of Pharmacy
International Medical University
Bukit Jalil, Kuala Lumpur, Malaysia

Rajiv Dahiya
School of Pharmacy
Faculty of Medical Sciences
The University of the West Indies
St. Augustine, Trinidad and Tobago

Sunita Dahiya
Department of Pharmaceutical Sciences
School of Pharmacy
University of Puerto Rico
Medical Sciences Campus
San Juan, Puerto Rico, USA

Neeraj Kumar Fuloria
Faculty of Pharmacy
AIMST University
Kedah, Malaysia

Shivkanya Fuloria
Faculty of Pharmacy
AIMST University
Kedah, Malaysia

Fulvia di Pillo
Centre for Medicinal Chemistry (CQMED)
Campinas, SP, Brazil

Nehmat Ghaboura
Department of Pharmacy Practice
Batterjee Medical College
Jeddah, Saudi Arabia

Gowthamarajan K
Department of Pharmaceutics
JSS College of Pharmacy
JSS Academy of Higher Education and Research
Ooty, Nilgiris, Tamil Nadu, India

Ahsas Goyal
Institute of Pharmaceutical Research
GLA University
Mathura, U.P., India

Madan Mohan Gupta
School of Pharmacy
Faculty of Medical Sciences
The University of the West Indies
St. Augustine, Trinidad and Tobago
and
NIMS Institute of Pharmacy,
NIMS University Rajasthan,
Jaipur, Rajasthan, India

Saurabh Gupta
Department of Pharmacology
Chameli Devi Institute of Pharmacy
Indore, Madhya Pradesh, India

Mohd Imran
Department of Pharmaceutical Chemistry
College of Pharmacy
Northern Border University
Rafha, Saudi Arabia

Sheza Javaid
Forman Christian College
Lahore, Pakistan

Imran Kazmi
Department of Biochemistry
Faculty of Science
King Abdulaziz University
Jeddah, Saudi Arabia

Abida Khan
Department of Pharmaceutical Chemistry
College of Pharmacy
Northern Border University
Rafha, Saudi Arabia

Rashi Kulshrestha
School of Pharmacy
Suresh Gyan Vihar University
Jaipur, Rajasthan, India

Arvind Kumar
Department of Chemistry
Faculty of Science and Technology
The University of the West Indies
St. Augustine, Trinidad and Tobago

Parveen Kumar
NIMS Institute of Pharmacy,
NIMS University Rajasthan,
Jaipur, Rajasthan, India

Sokindra Kumar
Swami Vivekanand Subharti University
Subhartipuram, Meerut, India

Mohammad Jaffar Sadiq Mantargi
Department of Pharmaceutical Sciences
Pharmacy Program
Batterjee Medical College
Jeddah, Saudi Arabia

Mudasir Maqbool
Department of Pharmaceutical Sciences
University of Kashmir
Srinagar, Jammu and Kashmir, India

Riya Mishra
School of Pharmacy
Suresh Gyan Vihar University
Jaipur, Rajasthan, India

Fatima Mohsin
Forman Christian College
Lahore, Pakistan

Bibi Nazia Murtaza
Department of Zoology
Abbottabad University of Science and Technology (AUST)
Abbottabad, Pakistan

Muhammad Mustafa
Forman Christian College
Lahore, Pakistan

Muhammad Shahid Nadeem
Department of Biochemistry
Faculty of Science
King Abdulaziz University
Jeddah, Saudi Arabia

Kamal Narain
Faculty of Medicine
AIMST University
Kedah, Malaysia

Neeraj Patel
School of Pharmacy
Suresh Gyan Vihar University
Jaipur, Rajasthan, India

Zulfkar Qadrie
Department of Pharmacology
Government Medical College
Baramulla, Jammu and Kashmir, India

Sarita Rawat
Faculty of Pharmacy and Sciences,
Amrapali University,
Nainital, Uttarakhand, India

V V S Narayana Reddy
Department of Pharmaceutics
JSS College of Pharmacy
JSS University
Mysore, India

Roshan S
Deccan School of Pharmacy
Hyderabad, India

Mahendran Sekar
School of Pharmacy
Monash University
Subang Jaya, Selangor, Malaysia

Ayeh Bani Saeid
Faculty of Health
Australian Research Centre in Complementary and Integrative Medicine
University of Technology Sydney
Ultimo, Australia
and
Discipline of Pharmacy
Graduate School of Health
University of Technology
Sydney, NSW, Australia

Rahul Sharma
School of Pharmaceutical Sciences
Jaipur National University
Jaipur, Rajasthan, India

Somya Sharma
School of Pharmaceutical Sciences
Jaipur National University
Jaipur, Rajasthan, India

Yati Sharma
Institute of Pharmaceutical Research
GLA University
Mathura, U.P., India

Raihan Siddiqui
School of Pharmacy
Suresh Gyan Vihar University
Jaipur, Rajasthan, India

Mahaveer Singh
School of Pharmacy and Technology Management
SVKMs, NMIMS University, Shirpur Campus
Maharashtra, India

Sachin Singh
Faculty of Health
Australian Research Centre in Complementary and Integrative Medicine
University of Technology Sydney
Ultimo, Australia
and
School of Pharmaceutical Sciences
Lovely Professional University
Phagwara, Punjab, India

Neelam Singla
School of Pharmacy
Suresh Gyan Vihar University
Jaipur, Rajasthan, India

Vetriselvan Subramaniyan
Department of Medical Sciences
School of Medical and Life Sciences, Sunway University,
Selangor, Malaysia

Rahamat Unissa Syed
Department of Pharmaceutics
College of Pharmacy
University of Ha'il
Ha'il, Saudi Arabia

Riya Thapa
Uttaranchal Institute of Pharmaceutical Sciences,
Uttaranchal University
Dehradun, Uttarakhand, India

Vibhav Varshney
Institute of Pharmaceutical Research
GLA University
Mathura, U.P., India

1 Introduction to Environmental Toxicants

Neeraj Kumar Fuloria[1], Shivkanya Fuloria[1], Mahendran Sekar[2], Kamal Narain[3], and Sokindra Kumar[4]*

[1]Faculty of Pharmacy, AIMST University, Kedah, Malaysia
[2]School of Pharmacy, MONASH University Malaysia, Subang Jaya, Selangor, Malaysia
[3]Faculty of Medicine, AIMST University, Kedah, Malaysia
[4]KSCP, Swami Vivekanand Subharti University, Subhartipuram, Meerut, India
[*]Corresponding Author: Neeraj Kumar Fuloria.
Email id: neerajkumar@aimst.edu.my

1.1 INTRODUCTION

A widely recognized statement attributed to Paracelsus (1493–1541) articulates the foundational tenet of contemporary toxicology: "Solely the dose determines that a thing is not a poison" [1]. This principle emphasizes the critical importance of dosage in assessing the toxicity of substances. Another quote attributed to Paracelsus, "Die wichtigsten Dinge des Lebens spielen sich zwischen Anfang und Ende des Verdauungskanals ab" [The most important things in life take place between the beginning and end of the digestive tract], further underscores the significance of the digestive system in our understanding of toxicology [2, 3]. These excerpts establish the groundwork for a concise investigation into the interplay between metals and the gastrointestinal microbiota, emphasizing the crucial impact of dosage on metal toxicology [4]. Additionally, they underscore the advancing comprehension of the essential contribution made by trillions of microorganisms within the gastrointestinal tract to human health and disease [5]. The gastrointestinal system (GIT), being the human body's most densely inhabited region, accommodates a diverse array of microorganisms, with concentrations varying from 10^{1}–10^{3} CFU/ml in the stomach to 10^{11}–10^{12} CFU/ml in the colon. Recent studies have identified over 1,100 bacterial species in the gut, emphasizing the complexity of the gut microbiota [6–8]. Bacterial predominance characterizes the microorganisms within the intestinal tract, wherein archaea, fungi, and protozoa are regarded as subordinate constituents. Recent investigations into the intestinal virome indicate that the quantity of bacteriophage entities in the gastrointestinal system might equal or exceed the bacterial count [9]. This unveils novel prospects for comprehending the potential influence of the virome on interactions between the microbiome and its host [10]. The microbiota assumes a pivotal role in the formation, operation, and equilibrium of the gastrointestinal tract (GIT), synergizing with the immune and nervous systems of the host [11, 12]. This specialized component within an organ prompts thought-provoking inquiries regarding its engagement with consumed xenobiotics and the potential impact of such interactions on the host [13, 14]. While similarities can be observed in the composition and metabolic functionality of human intestinal microbiotas, the distinctiveness of individual microbiomes, particularly at the species and strain tiers, emerges as a pivotal consideration in evaluating the toxicological impact of xenobiotics in pharmaceutical research [15]. Given that intestinal microbiomes encounter diverse xenobiotics, encompassing oral therapeutic medications, substances subject to misuse, and environmental compounds, this examination will center on contemporary understandings of the impacts of specific heavy metals on the intestinal microbiome and, conversely, the potential influence of gut

DOI: 10.1201/9781003489221-1

microbes on the toxicity of metals [16, 17]. It is crucial to highlight that a substantial portion of the existing understanding originates from experimental studies involving animals. This is because the literature concerning the connections between toxic metals and the intestinal microbiome in humans remains relatively restricted [18].

Heavy metals, like mercury, lead, cadmium, and arsenic, are prevalent environmental contaminants with known toxicity to humans and ecosystems [19]. Their ubiquity in air, water, soil, and food sources raises concerns about potential health risks. The gastrointestinal tract is a primary route of exposure to heavy metals, as ingested food and water can be significant sources [20–22]. Understanding the influence of trace elements on the gastrointestinal microbiota is essential, considering the microbiome's pivotal role in preserving gastrointestinal well-being and impacting overall health [23]. Research suggests that trace elements have the potential to modify the structure and operation of the gastrointestinal microbiota, disturbing the intricate equilibrium necessary for optimal physiological performance [24, 25]. Figure 1.1 represents the diverse environmental metals causing toxicity.

1.2 EFFECTS OF HEAVY METALS ON THE GUT MICROBIOME

Research has revealed that exposure to heavy metals can lead to significant changes in the gut microbiome. Exposure to cadmium has been linked to a reduction in microbial diversity and modifications in the prevalence of particular bacterial taxa [26]. Likewise, the impact of mercury exposure has been demonstrated to influence the constitution of the intestinal microbiome, potentially contributing to the overall toxic effects of this metal [27, 28]. These changes in the gut microbiome induced by heavy metals can have far-reaching consequences. The microbiota significantly contributes to nutrient metabolism, modulation of the immune system, and safeguarding against pathogens in the body. Disruptions in this delicate ecosystem may compromise these functions, impacting the host's health [29].

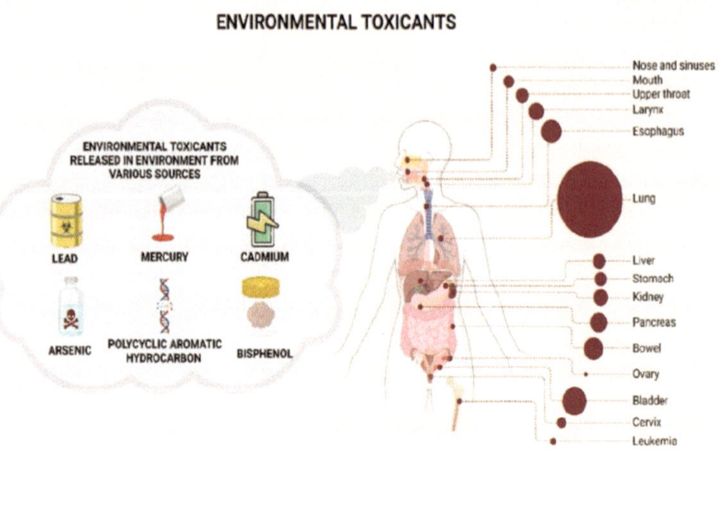

FIGURE 1.1 This image illustrates the diverse environmental metals causing toxicity, including arsenic, cadmium, lead, and mercury. These metals, present in different forms, pose significant health risks. Each metal exhibits unique toxic properties, impacting human health. The figure visually represents the complexity of metal pollution, emphasizing the need for comprehensive understanding and mitigation strategies.

1.2.1 Interactions between Gut Microbes and Metal Toxicity

Conversely, the gut microbiota can influence the toxicity of heavy metals through various mechanisms. Microbes can alter the chemical form of metals, affecting their absorption and bioavailability [30]. Additionally, certain bacteria possess the ability to detoxify or sequester metals, potentially mitigating their harmful effects on the host [31, 32]. The gut microbiome's involvement in modulating inflammation and immune responses further complicates the relationship between heavy metals and host toxicity [33]. Understanding the intricate interplay between gut microbes and metal toxicity is essential for comprehending the full spectrum of health outcomes associated with environmental exposures [34, 35]. The intricate relationship between heavy metals and the gut microbiome underscores the multifaceted nature of environmental toxicology [36]. Paracelsus's principle that "the dose determines that a thing is not a poison" takes on new dimensions when considering the dynamic interactions occurring within the gastrointestinal tract [37]. As our understanding of the gut microbiome expands, so does our awareness of its pivotal role in mediating the effects of environmental toxicants [38].

The effects of heavy metals on the gut microbiome and vice versa represent an evolving field of research with implications for human health and environmental stewardship. Recognizing the limitations of current knowledge and the challenges in translating findings from animal studies to human health emphasizes the need for continued research in this critical area [39]. Ultimately, a thorough understanding of the interplay between environmental pollutants and the gastrointestinal microbiota is imperative for formulating efficacious approaches to alleviate health hazards and ensure the welfare of both individuals and ecosystems [40, 41].

1.3 METALS CHANGE THE MICROBIOTA

Metals, both essential and non-essential, play a significant role in shaping the dynamics of the microbiota, particularly within the intricate ecosystems of the human gut [42]. Physiologically essential metals like cobalt (Co), manganese (Mn), iron (Fe), magnesium (Mg), nickel (Ni), and copper (Cu), when present at elevated concentrations, can exhibit antimicrobial properties [43]. Conversely, non-essential metals, such as silver, mercury, and lead, even at lower concentrations, pose toxicity to various microorganisms. The impact of metal-induced toxicity on microorganisms is multifaceted [44, 45]. The antibacterial properties, although beneficial for inhibiting the proliferation or eradicating pathogens, also present a risk to symbiotic and beneficial microorganisms within intricate microbial populations, such as the gut microbiota [46]. The non-selective characteristics of metal toxicity lead to oxidative stress through the production of ROS, reduction of antioxidants, protein malfunction due to oxidation, structural impairment, disruption of catalytic processes, and harm to biological membranes caused by metals [47]. Additionally, the antimicrobial activity may extend to disturbances in gene expression and genotoxicity, causing DNA damage. The gut microbiota, being exposed to toxic metals, experiences diverse effects on resident species [48, 49]. The impact varies based on factors such as intestinal location, the prevalence of vulnerable/immune strains, comprehensive variety, and metabolic functionality of the indigenous microbial populace [50]. Factors related to the host such as nutritional status, gender, age, and immune condition additionally enhance the intricacy of the microbiome-metal interplay [51]. Among the array of toxic metals, arsenic takes center stage in studies related to the gut microbiome. Arsenic compounds are extensively studied for their profound effects on the gut microbiota [52]. The complexity of these interactions is highlighted by the multitude of factors influencing outcomes, from the physiological state of the host to the specific microenvironment within the gut. Understanding the differential effects of toxic metals on various components of the gut microbiota is crucial [53]. The stomach, jejunum, ileum, cecum, and colon each host distinct microbial communities, and their response to metal exposure may differ significantly. The lumen-mucus interface also plays a pivotal role, as it presents a dynamic environment where metals interact with the microbiota [54]. Microenvironmental conditions further modulate the impact of metals on the gut microbiota. Factors like oxygen availability, pH, and redox potential influence the bioavailability and reactivity of metals, ultimately shaping their effects on microbial communities [55]. The prevalence of susceptible and resistant strains within the microbiota introduces an additional stratum of intricacy to these interactions, paralleled by the overall diversity and metabolic functionality of the adjacent microbial community [56]. Host-specific factors significantly contribute to the variability in microbiome-metal interactions. The nutritional status of the host, along with sex, age, and immune status, can influence the response of the gut microbiota to metal exposure [57, 58]. Nutrient availability and the overall health of the host may either exacerbate or mitigate the effects of metal toxicity on the microbiome [59].

While essential metals contribute to microbial homeostasis, their elevated concentrations, along with non-essential metals, can disrupt this delicate balance [60]. The resulting effects encompass oxidative stress, genotoxicity, and structural damage, with consequences varying based on the diverse factors influencing the microbiome-metal interaction [61]. Understanding these complexities is essential for unraveling the intricate relationships between metals and the gut microbiota, paving the way for targeted interventions and therapeutic strategies [62].

1.3.1 Arsenic

Arsenic, a potent environmental toxicant, is ubiquitously present in food sources such as rice, fish, seafood, and drinking water, raising significant concerns due to its potential health implications [63]. Exposure to arsenic has been linked to carcinogenicity and various adverse health outcomes across multiple organ systems [64]. Understanding the speciation of arsenic is crucial for accurate health risk assessment, especially considering its interactions with gut microbes and the potential for altering bioavailability [65]. In a study conducted by Richardson et al., the acute effects of toxic metal exposures on the rat gut microbiota were investigated. The study involved the administration of five different metals, each at specific doses, over five consecutive days [66, 67]. The metals included sodium arsenite, sodium dichromate, cobalt chloride, cadmium chloride, and nickel chloride. Stool specimens were obtained for microbiome assessments prior to the initial dosing and 24 hours following the fifth dosage [68]. The examination, utilizing 16S rRNA gene sequencing and PICRUSt, unveiled metal-induced disruptions in the intestinal microbiota of rats [69]. In a dose-dependent manner, cadmium, arsenic, and nickel showed significant changes in the variety and composition of bacteria. While still having an impact on the microbiome, cobalt and chromium showed less pronounced impacts [70]. Crucially, the microbiome's reaction to these metals varied, indicating that the substance delivered might cause different changes in the microbiota. For instance, the Bacteroidetes group S24–7 showed a significant decline in response to nickel, whereas other bacteria and proteobacteria showed an increase [71]. PICRUSt studies revealed that specimens containing nickel and arsenic had increased gene activities related to iron import, suggesting a potential strategy for reducing the harmful effects of these metals [72]. Contrasting these results with those of other investigations employing diverse metal exposure protocols in different animal models highlights the significant impact of experimental variables on research conclusions [73]. Lu et al. documented that exposure to arsenic in drinking water led to modifications in the configuration and metabolic patterns of the intestinal microbiota in female C57BL/6 mice. Changes in bile acid profiles and indole metabolites were observed in the experiment, along with decreases in some families of Firmicutes [74]. Another investigation using smaller, ecologically relevant concentrations of arsenic revealed dysbiosis with changes in the diversity and composition of the microbiota along with metagenomic modifications in the systems responsible for starch utilization, short-chain fatty acid production, and glucose metabolism [75]. Moreover, exposure to arsenic increased genes linked to DNA repair and oxidative stress markers. The preponderance of conjugative transposons and multidrug resistance genes that was seen was especially noteworthy, suggesting that heavy metals aid in the horizontal gene transfer that spreads multidrug resistance throughout the gastrointestinal tract [76]. The investigation conducted by Chi and colleagues delved into the gender-specific reactions to arsenic exposure. Variances in the microbial compositions of feces between male and female mice were identified, alongside distinct functional profiles determined through metagenomic sequencing [77]. This gender-specific impact was also witnessed in six-week-old human infants, where heightened urinary arsenic levels were significantly correlated with alterations in the composition of stool microbiota, particularly in male infants fed with formula [78]. Additional insights into the influence of arsenic on gut microbiota arise from a study involving Bangladeshi children exposed to varying levels of arsenic during prenatal development and early life. Children with elevated arsenic exposure displayed a higher abundance of proteobacteria, particularly Gammaproteobacteria, coupled with the enrichment of genes related to virulence and multidrug resistance. Specifically, there was an increase in strains of E. coli carrying arsenic resistance genes (ArsB, ArsC) [79]. Gokulan et al. examined the effects of a single exposure to arsenite as well as short recurring exposures on the immunological state and gut microbiota composition in adult and adolescent CD-1 mice. The study showed that different changes in bacterial survival and composition of the microbiota were dependent on the dose, length of contact, and stage of development of the animals [80]. Frequent exposure increased the number of bacteria with arsenic-resistant genes and caused the host to synthesize arsenite in order to detoxify. Furthermore, indicators of immunological modulation caused by exposure to arsenic were found, such as a decrease in the number of bacteria that convert protein to butyrate [81]. Coryell et al. used antibiotic-treated, mutant (lacking the detoxifying enzymes arsenite methyltransferase As3mt), free of germs, and gnotobiotic mice to study the mitigation of acute arsenic toxicity using many strategies [82]. Antibiotic-induced disruption of the microbiome resulted in increased arsenic bioaccumulation, whereas germ-free environments combined with As3mt deficiency led to higher post-acute exposure mortality [83]. Notably, transgenic mice that were hypersensitive and devoid of germs were protected from arsenic-induced death by the transplanting of human stool microbiota. Fecal bacteria *prausnitzii*, a bacterium frequently discovered in normal human microbiomes, was shown to provide at least some defense from arsenic poisoning in gnotobiotic rat trials [84]. These results imply that specific microbiome modification may be a viable treatment and prevention option for arsenic poisoning. The observed alterations in bacterial composition, diversity, and functional profiles underscore the complexity of the relationship between arsenic exposure and the gut microbiota [85]. Insights from these studies may inform strategies for mitigating arsenic toxicity through targeted microbiome manipulation, opening new avenues for research in environmental toxicology and human health [86]. Table 1.1 lists the potential mitigation approaches of arsenic.

1.3.2 Cadmium

Cadmium, a highly toxic metal, has emerged as a significant environmental concern due to its adverse impact on both human health and ecosystems [87]. The general population faces a considerable risk of exposure to cadmium primarily through the accumulation of this metal in food sources. However, it is the gastrointestinal tract (GIT) that serves as a primary target for cadmium toxicity. The deleterious effects of cadmium on the GIT are multifaceted [88]. One prominent aspect is the impairment of gut barrier function. Cadmium-induced changes in the viability of various components of the gut microbiota contribute to this impairment, creating a cascade effect. The repercussions include elevated levels of proinflammatory molecules such as LPS, potentially culminating in systemic inflammation [89]. A pivotal study conducted by Richardson et al. delved into the intricate relationship between cadmium toxicity and the gut microbiome. Several other research groups have subsequently explored this connection, yielding valuable insights [90]. However, it is essential to note that the observed alterations in bacterial communities vary significantly across studies, sometimes even displaying opposite trends. This variability can be attributed to differences in dosing regimens, animal models employed, and variations in sequencing methodologies [91]. The impact of cadmium on the intestinal microbiome of mice has been a subject of intense investigation. These studies have consistently reported substantial changes in bacterial communities, highlighting the

TABLE 1.1
Arsenic's impact and mitigation approaches.

Exposure Details	Microbiota Changes	Implications
Metal-induced disruptions in rat gut microbiota, 5 metals over 5 days.	Cadmium, arsenic, nickel show significant microbiota changes.	Potential strategy for reducing harmful effects via iron import.
Arsenic exposure in drinking water alters female mouse intestinal microbiota.	Changes in bile acid profiles, Firmicutes decrease.	Arsenic linked to DNA repair genes and multidrug resistance.
Gender-specific reactions to arsenic in mice, infants with altered microbiota.	Variances in microbial compositions, gender-specific impact.	Distinct functional profiles observed through metagenomic sequencing.
Children with elevated arsenic exposure show Proteobacteria abundance.	Enrichment of virulence and multidrug resistance genes.	Increase in E. coli strains with arsenic resistance genes.
Arsenite exposure effects on immunological state and gut microbiota.	Dose-dependent changes in bacterial survival, microbiota composition.	Frequent exposure increases bacteria with arsenic-resistant genes.
Mitigation strategies for acute arsenic toxicity using different mice models.	Antibiotic-induced microbiome disruption increases arsenic bioaccumulation.	Germ-free mice with As3mt deficiency show higher mortality.
Arsenic exposure's complex impact on microbiota, potential treatment avenue.	Specific microbiome modification may mitigate arsenic poisoning.	Insights for research in environmental toxicology and human health.

sensitivity of the gut microbiota to cadmium exposure [92]. The diversity and composition of these microbial populations undergo discernible shifts, reflecting the disruptive influence of cadmium on the delicate balance within the gut environment. Furthermore, the disruption of the gut barrier function, coupled with altered gut microbiota, contributes to the escalation of proinflammatory molecules. Lipopolysaccharide, a key player in inflammatory responses, sees an upsurge in concentration due to cadmium-induced changes [93]. This, in turn, sets the stage for systemic inflammation, extending the impact beyond the gastrointestinal tract. Understanding the nuances of cadmium toxicity requires consideration of the intricacies involved in various research findings [94]. The disparities in outcomes underscore the importance of standardizing experimental protocols and methodologies to draw more conclusive and generalizable results. Additionally, efforts to elucidate the impact of cadmium on human health should extend beyond murine models, accounting for the complexity introduced by inter-species variations [95]. As the global community grapples with environmental challenges, cadmium's pervasive presence and toxic consequences necessitate comprehensive strategies for mitigation. Efforts to minimize cadmium contamination in food sources, coupled with rigorous monitoring of industrial discharges, are crucial steps toward safeguarding both the environment and human health [96]. The evolving landscape of cadmium research underscores the need for continued exploration and collaboration to unravel the intricate dynamics of this toxic metal and its far-reaching implications [97]. Table 1.2 lists the potential mitigation approaches of cadmium.

TABLE 1.2
Cadmium's impact and mitigation approaches.

Exposure	Microbiota	Implications
Cadmium, highly toxic metal, poses significant environmental concern.	Impairs gut barrier function, alters microbiota, elevates proinflammation.	Cadmium toxicity's intricate relationship with gut microbiome.
Cadmium impact on GIT multifaceted; primary target is the gastrointestinal tract.	Observed varying bacterial community alterations due to cadmium.	Disparities in outcomes emphasize the need for standardized methodologies in research.
GIT vulnerability to cadmium toxicity; cascade effect on gut barrier function.	Studies on mice consistently report substantial shifts in microbial populations.	Cadmium-induced changes escalate proinflammatory molecules, fostering systemic inflammation.
Nuanced understanding of cadmium toxicity requires standardized methodologies.	Human health impact exploration should consider inter-species variations.	Mitigation strategies crucial; minimize cadmium contamination, monitor industrial discharges.

1.3.3 Lead

Lead, a ubiquitous metal, has long been a cause for concern due to its detrimental effects on public health. Despite increasing awareness and efforts to curb lead exposure, it continues to pose a global threat, as exemplified by the infamous water crisis in Flint, Michigan [98]. This pervasive issue necessitates a thorough examination of the various facets of lead exposure, ranging from its impact on gut microbiota to broader implications on metabolic pathways [99]. One significant study shedding light on the intricate relationship between lead and human health is the multi-omics investigation conducted by Gao and colleagues. This extensive investigation explored the impact of lead on the composition, diversity, and metabolic activity of the gastrointestinal microbiota [100]. Employing sophisticated methodologies such as comprehensive metagenomics sequencing of the entire genome and metabolomics using gas chromatography-mass spectrometry, the research provided a nuanced comprehension of how exposure to lead shapes the gastrointestinal microbiome and modulates essential metabolic pathways [101]. The findings of the study revealed that lead exposure triggers substantial alterations in the development of the gut microbiota. These changes, in turn, have cascading effects on various metabolic pathways within the body. Notably, some of these pathways are linked to oxidative stress and detoxification, elucidating the multipronged impact of lead on physiological processes [102]. Understanding these intricate connections is crucial for formulating effective strategies to mitigate the health risks associated with lead exposure. The gastrointestinal microbiome, an intricate community of microorganisms inhabiting the alimentary canal, assumes a crucial function in sustaining holistic well-being [103]. Any disruption in its composition and function can have far-reaching consequences for the host organism. The Gao et al. study highlights how lead, as a toxic metal, disrupts the delicate balance of the gut microbiome. This disruption not only affects the microbial diversity but also has implications for metabolic activities, contributing to the overall health challenges associated with lead exposure [104]. The study's use of advanced techniques, such as whole-genome metagenomics sequencing, allowed for a detailed analysis of the genetic material within the gut microbiota. This provided insights into how lead exposure influences specific genes and pathways, contributing to a more comprehensive understanding of the molecular mechanisms involved [105]. Simultaneously, gas chromatography–mass spectrometry metabolomics enabled the identification and quantification of various metabolites, offering a snapshot of the metabolic changes induced by lead. In light of these findings, it becomes evident that addressing the issue of lead exposure requires a holistic approach [106]. Beyond the immediate concerns of toxicity, understanding the intricate interplay between lead and the gut microbiome opens avenues for targeted interventions. Developing strategies to modulate the gut microbiota in the face of lead exposure could offer a novel and effective means of mitigating its health impacts [107]. The implications of lead exposure extend beyond individual health to broader environmental and societal concerns. The water crisis in Flint serves as a stark reminder of the far-reaching consequences of inadequate lead management. Therefore, a comprehensive understanding of lead's impact on gut microbiota and metabolic pathways is crucial not only for individual well-being but also for formulating policies and interventions that address this public health challenge on a larger scale [108]. The study by Gao and colleagues provides a valuable contribution to the ongoing discourse on lead exposure and its multifaceted impact on human health [109]. By employing a multi-omics approach, the research sheds light on the intricate connections between lead, gut microbiota, and metabolic pathways. This knowledge serves as a foundation for developing targeted interventions and policies that can effectively address the persistent global issue of lead exposure [110]. Table 1.3 lists the potential mitigation approaches of lead.

1.3.4 Mercury

Mercury, a versatile chemical element, exists in various forms, each with distinct properties and implications for environmental and human health. Its elemental, inorganic, and organic forms serve as global pollutants, exhibiting varying degrees of toxicity [111]. In the elemental form, mercury exists as Hg^0, while inorganic mercury is predominantly present as Hg^2+. The organic form, primarily CH3Hg or methylmercury, poses significant concerns due to its potent neurotoxicity and bioaccumulative nature [112]. Methylmercury, in particular, is a neurotoxicant of utmost concern. Its affinity for bioaccumulation in fish, which are often part of human diets, raises the stakes for potential health risks. The risks linked to methylmercury were underscored in a recent investigation conducted by Rothenberg et al., delving into the connection between the gut microbiota composition and metabolic functions in expectant mothers subjected to fetal methylmercury [113]. The research unveiled a noteworthy association of 17 bacterial genera with mercury biomarkers, underscoring the complex interaction between environmental exposure and the human microbiome. Dietary intake of methylmercury has been shown to induce changes in the gut microbiome and metabolome, as evidenced by studies on mice and larval fish [114]. These findings underscore the far-reaching impacts of mercury on both environmental ecosystems and human health. The interaction between mercury and gut microbiota not only sheds light on the intricate dynamics of toxic exposure but also emphasizes the need for comprehensive research to understand the broader implications on public health [115]. Mercury pollution is a global issue with disparate sources and pathways. Anthropogenic

TABLE 1.3
Lead's impact and mitigation approaches.

Exposure	Gut Microbiota Impact	Metabolic Pathways Implications
Lead, ubiquitous; poses global threat, Flint crisis.	Alters gut microbiota, affects diversity.	Lead exposure cascades effects on oxidative stress, detox pathways.
multi-omics investigation on lead impact.	Lead triggers substantial alterations in gut microbiota.	Disruption affects metabolic activities, links to health challenges.
Gastrointestinal microbiome crucial for well-being.	Lead disrupts balance, impacting microbial diversity, metabolism.	Advanced techniques provide insights into genetic and metabolic changes.
Lead exposure requires holistic approach; societal impact.	Foundation for interventions and policies.	Targeted interventions needed for mitigating health impacts.

TABLE 1.4
Mercury's impact and mitigation approaches.

Exposure	Gut Microbiota Impact	Metabolic Pathways Implications
Mercury, versatile element, forms with varying toxicity.	Gut microbiota linked to fetal methylmercury.	Methylmercury, a neurotoxicant, associates with 17 bacterial genera.
Methylmercury bioaccumulates in fish, raising health risks.	Dietary intake alters gut microbiome, metabolome in mice, larval fish.	Far-reaching impacts on environmental ecosystems and human health.
Global mercury pollution: elemental, inorganic, and organic forms.	Anthropogenic activities contribute to release, transformation.	Bioaccumulation in fish, direct human exposure through diet.
Mitigating mercury pollution requires interdisciplinary efforts.	Regulate industrial emissions, promote sustainable practices.	Gut microbiota insights open avenues for targeted health interventions.

activities such as industrial processes, mining, and combustion of fossil fuels contribute to the release of elemental and inorganic mercury into the environment. These pollutants can undergo complex transformations, eventually leading to the formation of methylmercury through microbial methylation processes [116]. The bioaccumulation of methylmercury in aquatic ecosystems, especially in fish, creates a direct pathway for human exposure through dietary consumption. Efforts to mitigate mercury pollution necessitate a multifaceted approach, encompassing both environmental and human health perspectives [117]. Monitoring and regulating industrial emissions, as well as implementing sustainable practices in mining and energy production, are crucial steps in reducing mercury release. Additionally, understanding the role of gut microbiota in mercury metabolism opens avenues for targeted interventions to mitigate the health impacts of exposure [118].

Mercury, in its elemental, inorganic, and organic forms, poses a significant threat to global ecosystems and human health. Methylmercury, with its neurotoxic properties and bioaccumulative nature, stands out as a particularly concerning aspect of mercury pollution [119]. The intricate relationship between mercury exposure, gut microbiota, and metabolic activities underscores the need for interdisciplinary research to develop comprehensive strategies for mitigating the environmental and health risks associated with mercury contamination [120]. Addressing mercury pollution requires a concerted effort to regulate industrial activities, promote sustainable practices, and advance our understanding of the complex interactions between pollutants and biological systems [121]. Table 1.4 lists the potential mitigation approaches of mercury.

1.4 MICROBIOTAS MODULATING METAL TOXICITY

The intricate interplay between gut microbiotas and host metabolism has emerged as a fascinating area of research, revealing the potential of microbiotas to modulate the toxicity of various metals [122]. Beyond their well-known roles in digestion and nutrient absorption, gut bacteria contribute to biotransformation processes, influencing the bioavailability and toxicity of metals within the gastrointestinal tract (GIT). This chapter explores the mechanisms through which gut microbiotas alter the fate of metals, focusing on processes like reduction, oxidation, methylation, and demethylation [123].

1.4.1 Biotransformation Dynamics

A prominent illustration of microbiota-facilitated metal biotransformation pertains to arsenic. Trivalent arsenite and its methylated derivatives, such as monomethylarsonous acid, manifest elevated toxicity compared to their corresponding pentavalent forms. Both host cells and gut microbes play pivotal roles in transforming arsenic into less or more toxic forms [124]. This process not only highlights the intricate relationship between the host and its microbiota but also emphasizes the potential impact on an individual's susceptibility to toxic metal exposure [125]. Similarly, gut bacteria participate in the reduction of highly toxic chromate [Cr(VI)] to its less-toxic form. The transformative capabilities of gut microbiotas extend to mercury as well, where certain microorganisms can methylate mercury, potentially increasing its bioavailability [126]. The identification of genetic markers, like the hgcAB gene pair, has facilitated the screening of orthologous methylation genes in bacterial and archaeal genomes, shedding light on the mechanisms involved in metal biotransformations [127].

1.4.2 Detoxification Mechanisms

Studies have shown that the gut microbiota may act as a detoxifying force by demethylating methylmercury or reducing inorganic mercury to its least toxic elemental form. The investigation of bacterial mer operons, encoding activities involved in these detoxification reactions, provides valuable insights into the potential protective role of gut bacteria against metal toxicity [128]. Another detoxification mechanism involves the binding of metals by intestinal microorganisms. This process aids in the elimination of toxicants from the GIT, potentially reducing their bioaccumulation. A notable study involving pregnant women and children demonstrated the beneficial effects of specific bacteria. Probiotic-supplemented yogurt was found to reduce the bioaccumulation of

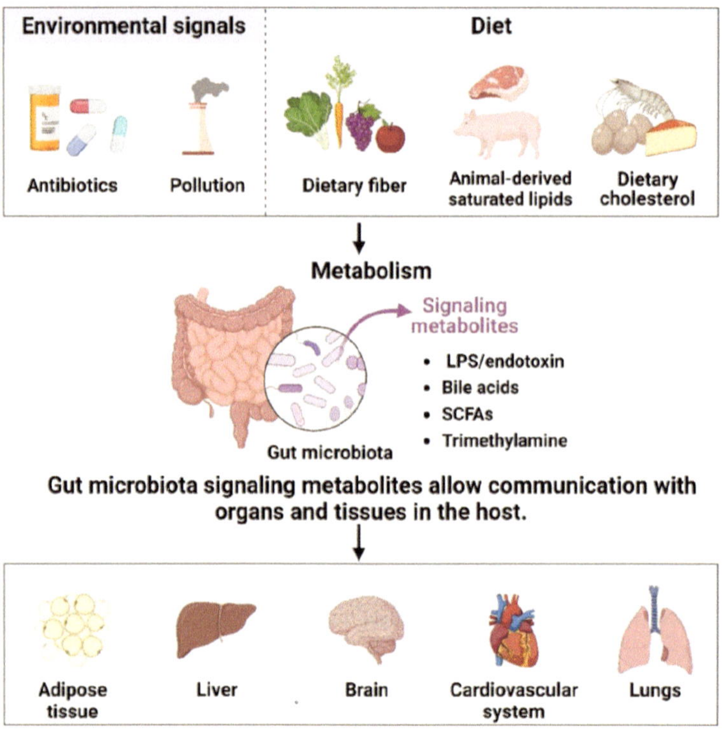

FIGURE 1.2 This figure illustrates the pivotal role of the resident gut microbiota in modulating the bioavailability and toxicity of metals within the host. The dynamic interplay between host cells and gut bacteria influences crucial processes like reduction, oxidation, methylation, and demethylation. The figure highlights detoxification mechanisms emphasizing the potential for leveraging the gut microbiota as a therapeutic avenue in managing metal toxicity.

arsenic and mercury, underscoring the potential application of microbiota-based interventions in mitigating metal toxicity [129]. The resident gut microbiota plays a crucial role in shaping the bioavailability and toxicity of metals within the host's body. The dynamic interplay between host cells and gut bacteria influences processes such as reduction, oxidation, methylation, and demethylation, ultimately impacting an individual's susceptibility to toxic metal exposure [130]. The identification of key genetic markers and ongoing research into detoxification mechanisms underscore the potential for harnessing the gut microbiota as a therapeutic avenue in managing metal toxicity [131]. As we continue to unravel the intricate relationship between microbiotas and metal metabolism, new opportunities for targeted interventions and personalized approaches to metal detoxification may emerge, offering novel solutions to mitigate the health risks associated with metal exposure [132]. Figure 1.2 shows the vital role of the gut microbiota in modulating the bioavailability and toxicity of metals within the host

1.5 CHALLENGES, LIMITATIONS, AND OPPORTUNITIES

The intricate relationship between the gut microbiome, metals, and their impact on human health is a subject of increasing significance in scientific research. Undoubtedly, the gut microbiome plays a crucial role in modulating the toxicity of metals and influencing their health effects on both humans and animals. However, as we delve into this complex interplay, we encounter various challenges, limitations, and exciting opportunities that shape the trajectory of our understanding.

1.5.1 Diverse Challenges in Experimental Design and Data Analytics

Understanding the microbiome-metal interrelationship poses a significant challenge in scientific research, primarily due to the diverse array of challenges embedded in experimental design and data analytics. The multifaceted nature of these challenges is rooted in several key factors that contribute to the variability in research outcomes [133]. One pivotal factor influencing the comprehensibility of studies is the choice of animal model. The selection of an appropriate model is critical as different species exhibit distinct microbiome compositions and responses to metal exposure. Researchers face the daunting task of identifying a model that adequately represents the system under investigation, adding a layer of complexity to experimental design. Metal dosing, another critical variable, introduces variability in the exposure levels across studies [134]. The absence of standardized dosing regimens makes it challenging to compare results directly. Efforts to establish dosing guidelines and ensure consistency in metal exposure are underway to enhance the reliability and reproducibility of findings. The mode of exposure further compounds the challenges in experimental design. Variations in exposure routes, such as ingestion, inhalation, or dermal exposure, can yield divergent outcomes [135]. Researchers grapple with standardizing exposure methods to establish a common ground for comparison. Advancements in sequencing technologies play a pivotal role in shaping the quality of data generated [136]. The diversity of sequencing platforms and methodologies employed across studies introduces a layer of intricacy in data interpretation. Standardization initiatives include the use of mock communities and spike-in controls to facilitate quality control and enhance data robustness [137]. Equally significant is the diversity in data analytics methodologies, which adds another layer of complexity [138]. Differences in statistical approaches, bioinformatics tools, and data processing pipelines contribute to the variability in research outcomes. Harmonizing these methodologies is essential to ensure the reproducibility and reliability of results. To address these challenges, the scientific community is actively engaged in standardization efforts [139]. Incorporating mock communities, employing spike-in controls for quality assurance, and adopting standardized sampling procedures are steps toward establishing a common framework [140]. Overcoming the diverse challenges in experimental design and data analytics is crucial for advancing our understanding of the intricate microbiome-metal interrelationship and fostering a more robust foundation for future research endeavors [141].

1.5.2 Overwhelming Complexity of Microbiome-Host Interactions

The complex interaction between the microbiota and its host poses a formidable obstacle for investigators exploring the realm of microbial ecosystems within the human organism. The microbiota, a varied assemblage of microorganisms inhabiting the human body, experiences dynamic fluctuations shaped by a multitude of elements, including environmental circumstances, dietary preferences, and specific biological traits [142]. This diversity and dynamism contribute to the complexity of microbiome-host interactions, rendering universal conclusions elusive. To navigate this intricate web, scientists are turning to cutting-edge omics approaches, including metagenomics, transcriptomics, proteomics, and metabolomics [143]. Metagenomics allows researchers to study the collective genetic material of microbial communities, shedding light on the functional potential of the microbiome [144]. Transcriptomics explores the expression of genes in response to various stimuli, providing insights into the molecular dynamics of host-microbiome interactions. Proteomics and metabolomics delve into the intricate world of proteins and metabolites, uncovering the biochemical signatures that define these relationships [145]. However, unraveling the intricacies of microbiome-host interactions goes beyond just identifying molecular components. Mechanistic confirmation is vital for a comprehensive understanding of the metal-microbiome-host interface [146]. Researchers employ these advanced techniques not only to observe changes but also to decipher the underlying mechanisms driving these changes [147]. This mechanistic insight is pivotal in discerning the impact of the microbiome on host physiology and vice versa. As researchers delve deeper into this complex realm, they grapple with the challenge of reconciling individual variations [148]. The uniqueness of each person's microbiome, shaped by their distinct lifestyle, genetics, and environment, adds an additional layer of complexity. Therefore, the quest for universal conclusions becomes an ongoing journey, prompting the scientific community to continually refine and expand their methodologies [149].

The overwhelming complexity of microbiome-host interactions necessitates a multidimensional approach [150]. Advanced omics techniques offer a glimpse into the intricate molecular landscape, while the pursuit of mechanistic understanding ensures a more profound comprehension of the dynamic interplay shaping human health [151]. The ongoing exploration of this complexity holds promise for unlocking new therapeutic avenues and enhancing our ability to harness the potential of the microbiome for personalized health interventions [152].

1.5.3 Integrating Gut Virome into the Equation

While progress has been made in understanding the bacterial composition of the gut microbiome, the role of the gut virome remains a largely unexplored territory [153]. The gut virome comprises a diverse array of viruses that infect bacteria (bacteriophages) and plays a crucial yet understudied role in shaping the dynamics of the microbiome. To comprehensively grasp the intricacies of the gut ecosystem, it is imperative to delve into the viral component [154].

Evaluating the involvement of the gut virome in the interplay between metals and the microbiome is an avenue ripe for exploration [155]. Recent studies suggest that certain metals can modulate virome composition, potentially influencing the balance of bacterial communities. This interplay introduces a novel dimension to our understanding of the gut environment, where metal-microbiome-virome interactions create a complex web of relationships [156].

This aspect adds another layer of complexity, requiring further research to unravel its significance in the broader context of human health. Investigating how the gut virome responds to environmental factors, including dietary metals, and understanding its potential impact on host health is essential [157]. Bridging this gap in knowledge not only enhances our understanding of the intricate gut ecosystem but also opens avenues for developing targeted interventions to manipulate the virome for therapeutic purposes [158]. As we unravel the mysteries of the gut virome, we pave the way for a more holistic comprehension of the factors shaping our internal microbial world and their implications for human well-being [159].

1.5.4 Pitfalls in Experimental Models and Design

The exploration of experimental models in scientific research demands meticulous attention due to inherent pitfalls in their design and the transferability of results. While it is imperative to choose models that mirror real-world scenarios, challenges arise in ensuring the reliability and applicability of findings [160]. Careful consideration must be given to the limitations of single exposure studies, urging researchers to delve into scenarios involving multiple metals or pollutants for a more comprehensive understanding of environmental exposures [161]. One intriguing avenue for investigation lies in alternative preclinical models, exemplified by research in highly social species like prairie voles [162]. By studying these unique models, researchers can gain novel insights into the intricate interplay of the microbiome-gut-brain behavior axis [163]. However, the adoption of such models introduces its own set of challenges, necessitating a balanced approach to address potential confounding factors [164].

As the scientific community navigates the complexities of experimental models, acknowledging and mitigating these pitfalls is crucial [165]. A thoughtful integration of diverse models and experimental designs, coupled with an awareness of their limitations, paves the way for robust and translatable research outcomes in the dynamic realm of scientific inquiry [166].

1.5.5 Global Health Threat and the Need for Surveillance Studies

Environmental pollution caused by toxic metals presents a pervasive global health threat, necessitating comprehensive surveillance studies to assess, address, and mitigate human exposure to these hazardous substances [167]. Current research underscores the urgency of understanding the intricate dynamics between environmental contaminants and human health. As a pivotal component of this effort, well-designed surveillance studies play a critical role in unraveling the complexities of toxic metal exposure. Microbiome research emerges as a key tool in combating this global health menace [168]. By delving into the microbial communities inhabiting the human body, scientists gain valuable insights that are instrumental in advancing the fields of toxicology and precision medicine. Integrating microbiome data into risk assessment models and treatment strategies becomes imperative for tailoring interventions to individual health profiles [169]. This personalized approach not only enhances the efficacy of health interventions but also contributes to the development of targeted preventive measures [170]. The global health threat posed by environmental pollution demands a proactive stance through robust surveillance studies. Incorporating microbiome research into these studies not only enriches our understanding of toxic metal exposure but also paves the way for personalized and effective health interventions on a global scale [171].

1.5.6 Opportunities for Advancement and Precision Medicine

In the realm of toxicology and precision medicine, the exploration of the intricate relationship between the gut microbiome, metals, and human health presents both challenges and exciting opportunities. Despite the complexities involved, the persistent efforts to standardize research methodologies and incorporate advanced omics techniques fuel optimism for groundbreaking advancements [172]. The ongoing efforts to understand the microbiome-metal interrelationship have led to collaborative initiatives that emphasize interdisciplinary research. These collaborative efforts hold the potential to unveil novel insights into the intricate dynamics at play, contributing to a more comprehensive comprehension of how metals interact with the microbiome and impact human health [173]. By standardizing research methodologies, the scientific community aims to establish a robust foundation for consistent and reliable findings across studies. One promising avenue for advancing personalized health care lies in precision medicine. Integrating individual microbiome profiles into treatment strategies offers a tailored approach to mitigating the health impacts of metal exposure [174]. This innovative approach holds the promise of more effective and targeted interventions, addressing the unique physiological responses of individuals to metal-induced stressors [175].

However, the challenges in unraveling this complex interplay are formidable. The multifaceted nature of the gut microbiome, coupled with the diverse array of metals and their varying effects on human health, poses significant obstacles [176]. Despite these challenges, the integration of technological advancements and a multidisciplinary approach provides a pathway forward. The continuous evolution of research methodologies and the application of cutting-edge omics techniques contribute to the refinement of our understanding [177]. In light of the global significance of environmental pollution, sustained research and surveillance studies are imperative. The integration of microbiome research emerges as a pivotal component in shaping the future of toxicology and precision medicine. By leveraging collaborative efforts, technological innovations, and a holistic approach to research, the scientific community can unlock new horizons, paving the way for more effective strategies to address the health implications of metal exposure on an individualized basis [178].

1.6 CONCLUSION

In conclusion, the study of environmental toxicants is crucial for understanding the pervasive impact of contaminants on ecosystems and human health. This introduction to environmental toxicants provides insights into the intricate relationship between pollutants, their sources, and the potential hazards they pose. By examining the pathways of exposure, evaluating ecological consequences, and assessing risks to public health, this field illuminates the need for robust regulatory frameworks and sustainable practices. As we delve deeper into the complexities of environmental toxicology, awareness and interdisciplinary efforts become imperative to mitigate the adverse effects of hazardous substances, fostering a healthier and more sustainable coexistence between humanity and the environment.

REFERENCES

1. R.M. Abdel-Megeed, Probiotics: a promising generation of heavy metal detoxification, *Biological Trace Element Research*, 199 (2021) 2406–2413.
2. M. Amde, J.F. Liu, L. Pang, Environmental application, fate, effects, and concerns of ionic liquids: a review, *Environmental Science & Technology*, 49 (2015) 12611–12627.
3. K.S. Alharbi, W.H. Almalki, M. Albratty, A.M. Meraya, A. Najmi, G. Vyas, S.K. Singh, K. Dua, G. Gupta, The therapeutic role of nutraceuticals targeting the Nrf2/HO-1 signaling pathway in liver cancer, *Journal of Food Biochemistry*, 46 (2022) e14357.
4. Z. Bai, N. Wang, M. Wang, Effects of microplastics on marine copepods, *Ecotoxicology and Environmental Safety*, 217 (2021) 112243.
5. P. Bhatt, S. Gangola, S. Ramola, M. Bilal, K. Bhatt, Y. Huang, Z. Zhou, S. Chen, Insights into the toxicity and biodegradation of fipronil in contaminated environment, *Microbiological Research*, 266 (2023) 127247.
6. A.M. Bolt, K.K. Mann, Tungsten: an emerging toxicant, alone or in combination, *Current Environmental Health Reports*, 3 (2016) 405–415.
7. P. Das, G. Gupta, V. Velu, R. Awasthi, K. Dua, H. Malipeddi, Formation of struvite urinary stones and approaches towards the inhibition—A review, *Biomedicine & Pharmacotherapy = Biomedecine & Pharmacotherapie*, 96 (2017) 361–370.
8. M.S. Hussain, A. Sharma, R. Kumar, Prebiotics and probiotics: a focused review of applications in respiratory disorders, *Carpathian Journal of Food Science and Technology*, 15 (2023) 183–207.
9. W. Brack, H.J. Klamer, M. López de Alda, D. Barceló, Effect-directed analysis of key toxicants in European river basins a review, *Environmental Science and Pollution Research International*, 14 (2007) 30–38.
10. E. Byeon, H.M. Kang, C. Yoon, J.S. Lee, Toxicity mechanisms of arsenic compounds in aquatic organisms, *Aquatic Toxicology*, 237 (2021) 105901.
11. S. Hussain, M.A. Khan, R. Rajan, J. Jyoti, S. Sharma, S.K. Sahu, Nanorobots: the future of healthcare, *AIP Conference Proceedings*, AIP Publishing, 2023.
12. M.R. Islam, A. Rauf, S. Akash, A. Kumer, M.S. Hussain, S. Akter, J.K. Gupta, L.T. Ansari, M.M.M.I. Raj, T.B. Emran, Recent perspective on the potential role of phytocompounds in the prevention of gastric cancer, *Process Biochemistry*, 2800 (2023) 020171-1–020171-12..

13. E. Calvo-Ochoa, C.A. Byrd-Jacobs, The olfactory system of zebrafish as a model for the study of neurotoxicity and injury: implications for neuroplasticity and disease, *International Journal of Molecular Sciences*, 20 (2019) 1639.
14. N.K. Jha, A. Sharma, S.K. Jha, S. Ojha, D.K. Chellappan, G. Gupta, K.K. Kesari, S. Bhardwaj, S.D. Shukla, M.M. Tambuwala, J. Ruokolainen, K. Dua, S.K. Singh, Alzheimer's disease-like perturbations in HIV-mediated neuronal dysfunctions: understanding mechanisms and developing therapeutic strategies, *Open Biology*, 10 (2020) 200286.
15. G. Cesarini, S. Secco, D. Taurozzi, I. Venditti, C. Battocchio, S. Marcheggiani, L. Mancini, I. Fratoddi, M. Scalici, C. Puccinelli, Teratogenic effects of environmental concentration of plastic particles on freshwater organisms, *The Science of the Total Environment*, 898 (2023) 165564.
16. Y.P. Chen, Q. Liu, Q.Y. Ma, L. Maltby, A.M. Ellison, Y. Zhao, Environmental toxicants impair liver and kidney function and sperm quality of captive pandas, *Ecotoxicology and Environmental Safety*, 162 (2018) 218–224.
17. J. Kaur, V. Mishra, S.K. Singh, M. Gulati, B. Kapoor, D.K. Chellappan, G. Gupta, H. Dureja, K. Anand, K. Dua, G.L. Khatik, K. Gowthamarajan, Harnessing amphiphilic polymeric micelles for diagnostic and therapeutic applications: Breakthroughs and bottlenecks, *Journal of Controlled Release: Official Journal of the Controlled Release Society*, 334 (2021) 64–95.
18. Y. Chisti, Environmental impact of toxic pollutants, *Biotechnology Advances*, 22 (2004) 431–432.
19. S. Dobaradaran, F. Soleimani, R. Akhbarizadeh, T.C. Schmidt, M. Marzban, R. BasirianJahromi, Environmental fate of cigarette butts and their toxicity in aquatic organisms: a comprehensive systematic review, *Environmental Research*, 195 (2021) 110881.
20. J. Flieger, M. Flieger, Ionic liquids toxicity-benefits and threats, *International Journal of Molecular Sciences*, 21 (2020) 6267.
21. S. Malhotra, N. Jain, J. Rathee, S. Kaul, U. Nagaich, M. Pandey, B. Gorain, G. Gupta, Nanoneuroscience: cutting-edge approach for disease management, *Recent Patents on Nanotechnology*, 18 (2024) 256–271.
22. M.S. Hussain, G. Gupta, R. Mishra, N. Patel, S. Gupta, S.I. Alzarea, I. Kazmi, P. Kumbhar, J. Disouza, H. Dureja, N. Kukreti, S.K. Singh, K. Dua, Unlocking the secrets: Volatile Organic Compounds (VOCs) and their devastating effects on lung cancer, *Pathology, Research and Practice*, 255 (2024) 155157.
23. S.J. Flora, Arsenic and dichlorvos: possible interaction between two environmental contaminants, *Journal of Trace Elements in Medicine and Biology: Organ of the Society for Minerals and Trace Elements (GMS)*, 35 (2016) 43–60.
24. S.J. Genuis, E. Kyrillos, The chemical disruption of human metabolism, *Toxicology Mechanisms and Methods*, 27 (2017) 477–500.
25. P. Prasher, M. Sharma, S.K. Singh, M. Gulati, D.K. Chellappan, F. Zacconi, G. De Rubis, G. Gupta, J. Sharifi-Rad, W.C. Cho, K. Dua, Luteolin: a flavonoid with a multifaceted anticancer potential, *Cancer Cell International*, 22 (2022) 386.
26. C.J. Gordon, Role of environmental stress in the physiological response to chemical toxicants, *Environmental Research*, 92 (2003) 1–7.
27. C.J. Gordon, A.F. Johnstone, C. Aydin, Thermal stress and toxicity, *Comprehensive Physiology*, 4 (2014) 995–1016.
28. M. Rahman, K. Alam, M.Z. Ahmad, G. Gupta, M. Afzal, S. Akhter, I. Kazmi, Jyoti, F.J. Ahmad, F. Anwar, Classical to current approach for treatment of psoriasis: a review, *Endocr Metab Immune Disord Drug Targets*, 12 (2012) 287–302.
29. L. Han, H. Gu, W. Lu, H. Li, W.X. Peng, N. Ling Ma, S.S. Lam, C. Sonne, Progress in phytoremediation of chromium from the environment, *Chemosphere*, 344 (2023) 140307.
30. T. Ivanković, J. Hrenović, Surfactants in the environment, *Arhiv za higijenu rada i toksikologiju*, 61 (2010) 95–110.
31. M. Jayapal, R.N. Bhattacharjee, A.J. Melendez, M.P. Hande, Environmental toxicogenomics: a post-genomic approach to analysing biological responses to environmental toxins, *The International Journal of Biochemistry & Cell Biology*, 42 (2010) 230–240.
32. A.K. Ramanunny, S. Wadhwa, M. Gulati, S. Vishwas, R. Khursheed, K.R. Paudel, S. Gupta, O. Porwal, S.M. Alshahrani, N.K. Jha, D.K. Chellappan, P. Prasher, G. Gupta, J. Adams, K. Dua, D. Tewari, S.K. Singh, Journey of Alpinia galanga from kitchen spice to nutraceutical to folk medicine to nanomedicine, *Journal of Ethnopharmacology*, 291 (2022) 115144.
33. S.W. Jeong, Y.J. Choi, Extremophilic microorganisms for the treatment of toxic pollutants in the environment, *Molecules*, 25 (2020) 4916.

34. O. Kaczerewska, R. Martins, J. Figueiredo, S. Loureiro, J. Tedim, Environmental behaviour and ecotoxicity of cationic surfactants towards marine organisms, *Journal of Hazardous Materials*, 392 (2020) 122299.
35. S. Rohilla, M. Singh, S. Priya, W.H. Almalki, S.M. Haniffa, V. Subramaniyan, S. Fuloria, N.K. Fuloria, M. Sekar, S.K. Singh, N.K. Jha, D.K. Chellappan, P. Negi, K. Dua, G. Gupta, Exploring the mechanical perspective of a new anti-tumor agent: melatonin, *Journal of Environmental Pathology, Toxicology and Oncology: Official Organ of the International Society for Environmental Toxicology and Cancer*, 42 (2023) 1–16.
36. M. Lechuga, M. Fernández-Serrano, F. Ríos, A. Fernández-Arteaga, R. Jiménez-Robles, Environmental impact assessment of nanofluids containing mixtures of surfactants and silica nanoparticles, *Environmental Science and Pollution Research International*, 29 (2022) 84125–84136.
37. C. Lei, Y. Sun, D.C.W. Tsang, D. Lin, Environmental transformations and ecological effects of iron-based nanoparticles, *Environmental Pollution*, 232 (2018) 10–30.
38. Z. Li, T. Lu, M. Li, M. Mortimer, L.H. Guo, Direct and gut microbiota-mediated toxicities of environmental antibiotics to fish and aquatic invertebrates, *Chemosphere*, 329 (2023) 138692.
39. A. Manciocco, G. Calamandrei, E. Alleva, Global warming and environmental contaminants in aquatic organisms: the need of the etho-toxicology approach, *Chemosphere*, 100 (2014) 1–7.
40. A. Manikandan, P. Suresh Babu, S. Shyamalagowri, M. Kamaraj, P. Muthukumaran, J. Aravind, Emerging role of microalgae in heavy metal bioremediation, *Journal of Basic Microbiology*, 62 (2022) 330–347.
41. P. Sharma, M. Mehta, D.S. Dhanjal, S. Kaur, G. Gupta, H. Singh, L. Thangavelu, S. Rajeshkumar, M. Tambuwala, H.A. Bakshi, D.K. Chellappan, K. Dua, S. Satija, Emerging trends in the novel drug delivery approaches for the treatment of lung cancer, *Chemico-Biological Interactions*, 309 (2019) 108720.
42. M. Narayanan, Y. Ma, Metal tolerance mechanisms in plants and microbe-mediated bioremediation, *Environmental Research*, 222 (2023) 115413.
43. D.W. Nebert, G. Zhang, E.S. Vesell, Genetic risk prediction: individualized variability in susceptibility to toxicants, *Annual Review of Pharmacology and Toxicology*, 53 (2013) 355-–375.
44. M.C. Petriello, B.J. Newsome, T.D. Dziubla, J.Z. Hilt, D. Bhattacharyya, B. Hennig, Modulation of persistent organic pollutant toxicity through nutritional intervention: emerging opportunities in biomedicine and environmental remediation, *The Science of the Total Environment*, 491–492 (2014) 11–16.
45. N. Singh, S. Vishwas, A. Kaur, H. Kaur, V. Kakoty, R. Khursheed, M. Chaitanya, M.R. Babu, A. Awasthi, L. Corrie, V. Harish, P. Yanadaiah, S. Gupta, A.A. Sayed, A. El-Sayed, I. Ali, O.A. Kensara, N. Ghaboura, G. Gupta, A.M. Dou, M. Algahtani, A.F. El-Kott, K. Dua, S.K. Singh, M.M. Abdel-Daim, Harnessing role of sesamol and its nanoformulations against neurodegenerative diseases, *Biomedicine & Pharmacotherapy = Biomedecine & Pharmacotherapie*, 167 (2023) 115512.
46. C. Roose-Amsaleg, A.M. Laverman, Do antibiotics have environmental side-effects? Impact of synthetic antibiotics on biogeochemical processes, *Environmental Science and Pollution Research International*, 23 (2016) 4000–4012.
47. D.A. Rossignol, S.J. Genuis, R.E. Frye, Environmental toxicants and autism spectrum disorders: a systematic review, *Translational Psychiatry*, 4 (2014) e360.
48. J.A. Rothman, K.A. Russell, L. Leger, Q.S. McFrederick, P. Graystock, The direct and indirect effects of environmental toxicants on the health of bumblebees and their microbiomes, *Proceedings Biological Sciences*, 287 (2020) 20200980.
49. D. Talukdar, P. Kumar, D. Sharma, V.M. Balaramnavar, O. Afzal, A.S.A. Altamimi, I. Kazmi, F.A. Al-Abbasi, S.I. Alzarea, G. Gupta, M.M. Gupta, Anticancer phytochemical-based nanoformulations: therapeutic intervention in cancer cell lines, *Journal of Environmental Pathology, Toxicology and Oncology: Official Organ of the International Society for Environmental Toxicology and Cancer*, 42 (2023) 79–93.
50. D.N. Roy, R. Goswami, A. Pal, The insect repellents: a silent environmental chemical toxicant to the health, *Environmental Toxicology and Pharmacology*, 50 (2017) 91–102.
51. A. Saravanan, P.S. Kumar, D.N. Vo, S. Jeevanantham, S. Karishma, P.R. Yaashikaa, A review on catalytic-enzyme degradation of toxic environmental pollutants: microbial enzymes, *Journal of Hazardous Materials*, 419 (2021) 126451.
52. S. Shaikh, J.S. Jagai, C. Ashley, S. Zhou, R.M. Sargis, Underutilized and under threat: environmental policy as a tool to address diabetes risk, *Current Diabetes Reports*, 18 (2018) 25.
53. R. Sharma, R. Garg, M. Bali, N.O. Eddy, Potential applications of green-synthesized iron oxide NPs for environmental remediation, *Environmental Monitoring and Assessment*, 195 (2023) 1397.

54. A. Singer, D. Nickisch, A. Gergs, Joint survival modelling for multiple species exposed to toxicants, *The Science of the Total Environment*, 857 (2023) 159266.
55. Z. Syed, M. Sogani, J. Rajvanshi, K. Sonu, Microbial biofilms for environmental bioremediation of heavy metals: a review, *Applied Biochemistry and Biotechnology*, 195 (2023) 5693–5711.
56. C. Tranchant, Introduction and classical environmental risk factors for Parkinson, *Revue Neurologique*, 175 (2019) 650–651.
57. C. Wang, L. Chen, J. Xu, L. Zhang, X. Yang, X. Zhang, C. Zhang, P. Gao, L. Zhu, Environmental behaviors and toxic mechanisms of engineered nanomaterials in soil, *Environmental Research*, 242 (2024) 117820.
58. Y.Y. Tan, P.K. Yap, G.L. Xin Lim, M. Mehta, Y. Chan, S.W. Ng, D.N. Kapoor, P. Negi, K. Anand, S.K. Singh, N.K. Jha, L.C. Lim, T. Madheswaran, S. Satija, G. Gupta, K. Dua, D.K. Chellappan, Perspectives and advancements in the design of nanomaterials for targeted cancer theranostics, *Chemico-Biological Interactions*, 329 (2020) 109221.
59. L. Wang, W.M. Wu, N.S. Bolan, D.C.W. Tsang, Y. Li, M. Qin, D. Hou, Environmental fate, toxicity and risk management strategies of nanoplastics in the environment: current status and future perspectives, *Journal of Hazardous Materials*, 401 (2021) 123415.
60. X. Wang, J. Zhu, Z. Xue, X. Jin, Y. Jin, Z. Fu, The environmental distribution and toxicity of short-chain chlorinated paraffins and underlying mechanisms: implications for further toxicological investigation, *The Science of the Total Environment*, 695 (2019) 133834.
61. J.F. Yi, Z.Z. Lin, X. Li, Y.Q. Zhou, Y. Guo, A short review on environmental distribution and toxicity of the environmentally persistent free radicals, *Chemosphere*, 340 (2023) 139922.
62. H.W. Yu, W.C. Cho, The role of microRNAs in toxicology, *Archives of Toxicology*, 89 (2015) 319–325.
63. International Agency for Research on Cancer, Arsenic and arsenic compounds, *IARC Monographs on the Evaluation of the Carcinogenic Risk of Chemicals to Humans*, 23 (1980) 39–141.
64. A. Castillejos Sepúlveda, L.M. Gatti, C.F. Kerl, A. Chennu, J.M. Klatt, Arsenic speciation analysis in porewater by a novel colorimetric assay, *The Science of the Total Environment*, 827 (2022) 154155.
65. H. Chen, H. Zhang, X. Wang, Y. Wu, Y. Zhang, S. Chen, W. Zhang, X. Sun, T. Zheng, W. Xia, S. Xu, Y. Li, Prenatal arsenic exposure, arsenic metabolism and neurocognitive development of 2-year-old children in low-arsenic areas, *Environment International*, 174 (2023) 107918.
66. K.A. Francesconi, D. Kuehnelt, Determination of arsenic species: a critical review of methods and applications, 2000–2003, *The Analyst*, 129 (2004) 373–395.
67. R. Wadhwa, K.R. Paudel, M. Mehta, S.D. Shukla, K. Sunkara, P. Prasher, N. Panth, R. Goyal, D.K. Chellappan, G. Gupta, P.M. Hansbro, A.A.A. Aljabali, M.M. Tambuwala, K. Dua, Beyond the obvious: smoking and respiratory infection implications on Alzheimer's disease, *CNS & Neurological Disorders Drug Targets*, 19 (2020) 698–708.
68. B. Gamboa-Loira, M.E. Cebrián, L. López-Carrillo, Arsenic exposure in northern Mexican women, *Salud publica de Mexico*, 62 (2020) 262–269.
69. A.H. Hall, Chronic arsenic poisoning, *Toxicology Letters*, 128 (2002) 69–72.
70. S. Kaur, M.R. Kamli, A. Ali, Role of arsenic and its resistance in nature, *Canadian Journal of Microbiology*, 57 (2011) 769–774.
71. D. Min, L. Cheng, D.F. Liu, J.Q. Liu, W.W. Li, H.Q. Yu, Single Strain-triggered biogeochemical cycle of arsenic, *Environmental Science & Technology*, 56 (2022) 16410–16418.
72. R.S. Oremland, C.W. Saltikov, J.F. Stolz, J.T. Hollibaugh, Autotrophic microbial arsenotrophy in arsenic-rich soda lakes, *FEMS Microbiology Letters*, 364 (2017) fnx146.
73. R.S. Oremland, J.F. Stolz, The ecology of arsenic, *Science*, 300 (2003) 939–944.
74. R.S. Oremland, J.F. Stolz, J.T. Hollibaugh, The microbial arsenic cycle in Mono Lake, California, *FEMS Microbiology Ecology*, 48 (2004) 15–27.
75. M.A. Rahman, C. Hassler, Is arsenic biotransformation a detoxification mechanism for microorganisms?, *Aquatic Toxicology*, 146 (2014) 212–219.
76. J.F. Reichard, A. Puga, Effects of arsenic exposure on DNA methylation and epigenetic gene regulation, *Epigenomics*, 2 (2010) 87–104.
77. A. Tariq, U. Ullah, M. Asif, I. Sadiq, Biosorption of arsenic through bacteria isolated from Pakistan, *International Microbiology: The Official Journal of the Spanish Society for Microbiology*, 22 (2019) 59–68.
78. J. Wu, J. Liang, L.O. Björn, J. Li, W. Shu, Y. Wang, Phosphorus-arsenic interaction in the 'soil-plant-microbe' system and its influence on arsenic pollution, *The Science of the Total Environment*, 802 (2022) 149796.

79. S. Wu, L. Wang, R. Gan, T. Tong, H. Bian, Z. Li, S. Du, Z. Deng, S. Chen, Signature arsenic detoxification pathways in halomonas sp. strain GFAJ-1, *mBio*, 9 (2018) 10–128.
80. L. Ye, H. Tian, C. Jing, Arsenic mobilization in nZVI residue by Alkaliphilus sp. IMB: Comparison between static and flowing incubation, *Environmental Pollution*, 319 (2023) 121019.
81. L. Zhou, F. Dong, X. Xi, L. Zhou, Q. Dai, M. Liu, Y. Han, G. Yang, Y. Zhang, Arsenic triggered nano-sized uranyl arsenate precipitation on the surface of Kocuria rosea, *Journal of Environmental Radioactivity*, 262 (2023) 107168.
82. S. Banerjee, S. Datta, D. Chattyopadhyay, P. Sarkar, Arsenic accumulating and transforming bacteria isolated from contaminated soil for potential use in bioremediation, *Journal of Environmental Science and Health. Part A, Toxic/Hazardous Substances & Environmental Engineering*, 46 (2011) 1736–1747.
83. D. Cai, S. Kong, Y. Shao, J. Liu, R. Liu, X. Wei, B. Bai, D. Werner, X. Gao, C. Li, Mobilization of arsenic from as-containing iron minerals under irrigation: effects of exogenous substances, redox condition, and intermittent flow, *Journal of Hazardous Materials*, 440 (2022) 129736.
84. A. Castillejos Sepúlveda, E. Metzger, S. Littmann, H. Taubner, A. Chennu, L. Gatti, D. de Beer, J.M. Klatt, Two-dimensional mapping of arsenic concentration and speciation with diffusive equilibrium in thin-film gels, *Environmental Science & Technology*, 57 (2023) 8107–8117.
85. Y. Chen, Z. Wang, Z. Luo, Y. Zhao, J. Yu, Decreasing arsenic accumulation but promoting arsenate biotransformation in Microcystis aeruginosa regulated by nano-Fe(2)O(3), *Environmental Science and Pollution Research International*, 29 (2022) 62423–62431.
86. R.L. Hsieh, Y.L. Huang, H.S. Shiue, S.R. Huang, M.I. Lin, S.C. Mu, C.J. Chung, Y.M. Hsueh, Arsenic methylation capacity and developmental delay in preschool children in Taiwan, *International Journal of Hygiene and Environmental Health*, 217 (2014) 678–686.
87. S.V. Adams, P.A. Newcomb, Cadmium blood and urine concentrations as measures of exposure: NHANES 1999–2010, *Journal of Exposure Science & Environmental Epidemiology*, 24 (2014) 163–170.
88. O. Andersen, J.B. Nielsen, G.F. Nordberg, Nutritional interactions in intestinal cadmium uptake--possibilities for risk reduction, *Biometals: An International Journal on the Role of Metal Ions in Biology, Biochemistry, and Medicine*, 17 (2004) 543–547.
89. O. Andersen, J.B. Nielsen, P. Svendsen, Oral cadmium toxicology, *Acta Pharmacologica et Toxicologica*, 59 Suppl 7 (1986) 44–47.
90. G. Bertin, D. Averbeck, Cadmium: cellular effects, modifications of biomolecules, modulation of DNA repair and genotoxic consequences (a review), *Biochimie*, 88 (2006) 1549–1559.
91. Deepika, A.K. Haritash, Cadmium uptake from soil by ornamental metallophytes: a meta-analytical approach, *Environmental Management*, 71 (2023) 1087–1097.
92. R. Fan, W. Xie, H. Ma, M. Zhu, K. Ma, X. Yan, Isolation of cadmium-resistant microbial strains and their immobilisation of cadmium in soil, *Biodegradation*, 34 (2023) 445–459.
93. L. Friberg, Cadmium, *Annual Review of Public Health*, 4 (1983) 367–373.
94. L. Friberg, Cadmium and the kidney, *Environmental Health Perspectives*, 54 (1984) 1–11.
95. R.A. Goyer, Transplacental transfer of cadmium and fetal effects, *Fundamental and Applied Toxicology: Official Journal of the Society of Toxicology*, 16 (1991) 22–23.
96. M.A. Gwon, M.J. Kim, H.G. Kang, Y.E. Joo, S.B. Jeon, P.S. Jeong, S.U. Kim, B.W. Sim, D.B. Koo, B.S. Song, Cadmium exposure impairs oocyte meiotic maturation by inducing endoplasmic reticulum stress in vitro maturation of porcine oocytes, *Toxicology In Vitro : An International Journal Published in Association with BIBRA*, 91 (2023) 105615.
97. H. Hong, Y. Xu, J. Xu, J. Zhang, Y. Xi, H. Pi, L. Yang, Z. Yu, Q. Wu, Z. Meng, W.S. Ruan, Y. Ren, S. Xu, Y.Q. Lu, Z. Zhou, Cadmium exposure impairs pancreatic β-cell function and exaggerates diabetes by disrupting lipid metabolism, *Environment International*, 149 (2021) 106406.
98. Lead and delinquency, *Environmental Health Perspectives*, 104 (1996) 600–601.
99. The lead treatment of cancer, *JAMA*, 313 (2015) 2084.
100. F. Barbosa, Jr., J.E. Tanus-Santos, R.F. Gerlach, P.J. Parsons, A critical review of biomarkers used for monitoring human exposure to lead: advantages, limitations, and future needs, *Environmental Health Perspectives*, 113 (2005) 1669–1674.
101. D. Bunch, A.L. Pyle-Eilola, Lead poisoning: clinical and laboratory considerations, *Advances in Clinical Chemistry*, 117 (2023) 209–222.
102. C.F. Delgado, M.A. Ullery, M. Jordan, C. Duclos, S. Rajagopalan, K. Scott, Lead exposure and developmental disabilities in preschool-aged children, *Journal of Public Health Management and Practice: JPHMP*, 24 (2018) e10–e17.

103. L.C. Dolan, B.M. Flannery, D. Hoffman-Pennesi, A. Gavelek, O.E. Jones, R. Kanwal, B. Wolpert, K. Gensheimer, S. Dennis, S. Fitzpatrick, A review of the evidence to support interim reference level for dietary lead exposure in adults, *Regulatory Toxicology and Pharmacology: RTP*, 111 (2020) 104579.
104. B. Gerson, Lead, *Clinics in Laboratory Medicine*, 10 (1990) 441–457.
105. P. Grandjean, Lead in danes. Historical and toxicological studies, *Environmental Quality and Safety*, 2 (1975) 6–75.
106. B. Gulson, Stable lead isotopes in environmental health with emphasis on human investigations, *The Science of the Total Environment*, 400 (2008) 75–92.
107. B.L. Gulson, K.R. Mahaffey, M. Vidal, C.W. Jameson, A.J. Law, K.J. Mizon, A.J. Smith, M.J. Korsch, Dietary lead intakes for mother/child pairs and relevance to pharmacokinetic models, *Environmental Health Perspectives*, 105 (1997) 1334–1342.
108. S.C. Hight, Lead migration from lead crystal wine glasses, *Food Additives and Contaminants*, 13 (1996) 747–765.
109. D.K. Houston, M.A. Johnson, Lead as a risk factor for hypertension in women, *Nutrition Reviews*, 57 (1999) 277–279.
110. D. Howarth, Lead exposure—implications for general practice, *Australian Family Physician*, 41 (2012) 311–315.
111. R.M. Aaronson, H.M. Spiro, Mercury and the gut, *The American Journal of Digestive Diseases*, 18 (1973) 583–594.
112. V. Andreoli, F. Sprovieri, Genetic aspects of susceptibility to mercury toxicity: an overview, *International Journal of Environmental Research and Public Health*, 14 (2017) 93.
113. A.V. Aravindan, S. Singh, Mercury embolism, *The Journal of the Association of Physicians of India*, 70 (2022) 11–12.
114. M.R. Hendzel, D. M. Jamieson, Determination of mercury in fish, *Analytical Chemistry*, 48 (1976) 926–928.
115. R.A. Bernhoft, Mercury toxicity and treatment: a review of the literature, *Journal of Environmental and Public Health*, 2012 (2012) 460508.
116. J.J. Berzas Nevado, R.C. Rodríguez Martín-Doimeadios, F.J. Guzmán Bernardo, M. Jiménez Moreno, A.M. Herculano, J.L. do Nascimento, M.E. Crespo-López, Mercury in the Tapajós River basin, Brazilian Amazon: a review, *Environment International*, 36 (2010) 593–608.
117. A. Bhan, N.N. Sarkar, Mercury in the environment: effect on health and reproduction, *Reviews on Environmental Health*, 20 (2005) 39–56.
118. T.Y. Chan, Inorganic mercury poisoning associated with skin-lightening cosmetic products, *Clinical Toxicology*, 49 (2011) 886–891.
119. S. Chandelia, V. Vohra, S. Prasad, M.K. Tiwari, K. Nallasamy, S.K. Angurana, Acute mercury poisoning in young children: when to treat?, *Indian Journal of Pediatrics*, 90 (2023) 629.
120. T.W. Clarkson, L. Magos, The toxicology of mercury and its chemical compounds, *Critical Reviews in Toxicology*, 36 (2006) 609–662.
121. S.A. Counter, L.H. Buchanan, Mercury exposure in children: a review, *Toxicology and Applied Pharmacology*, 198 (2004) 209–230.
122. S. Bolan, B. Seshadri, I. Grainge, N.J. Talley, R. Naidu, Gut microbes modulate bioaccessibility of lead in soil, *Chemosphere*, 270 (2021) 128657.
123. S. Bolan, B. Seshadri, A. Kunhikrishnan, I. Grainge, N.J. Talley, N. Bolan, R. Naidu, Differential toxicity of potentially toxic elements to human gut microbes, *Chemosphere*, 303 (2022) 134958.
124. M. Calatayud Arroyo, T. García Barrera, B. Callejón Leblic, A. Arias Borrego, M.C. Collado, A review of the impact of xenobiotics from dietary sources on infant health: early life exposures and the role of the microbiota, *Environmental Pollution*, 269 (2021) 115994.
125. X. Cao, J.P.M. van Putten, M. Wösten, Campylobacter jejuni benefits from the bile salt deoxycholate under low-oxygen condition in a PldA dependent manner, *Gut Microbes*, 15 (2023) 2262592.
126. X. Chang, M. Kang, Y. Shen, L. Yun, G. Yang, L. Zhu, X. Meng, J. Zhang, X. Su, Bacillus coagulans SCC-19 maintains intestinal health in cadmium-exposed common carp (Cyprinus carpio L.) by strengthening the gut barriers, relieving oxidative stress and modulating the intestinal microflora, *Ecotoxicology and Environmental Safety*, 228 (2021) 112977.
127. Z. Chen, S. Han, D. Zhou, S. Zhou, G. Jia, Effects of oral exposure to titanium dioxide nanoparticles on gut microbiota and gut-associated metabolism in vivo, *Nanoscale*, 11 (2019) 22398–22412.
128. T. Shoji, ATP-binding cassette and multidrug and toxic compound extrusion transporters in plants: a common theme among diverse detoxification mechanisms, *International Review of Cell and Molecular Biology*, 309 (2014) 303–346.

129. N. Ameen, M. Amjad, B. Murtaza, G. Abbas, M. Shahid, M. Imran, M.A. Naeem, N.K. Niazi, Biogeochemical behavior of nickel under different abiotic stresses: toxicity and detoxification mechanisms in plants, *Environmental Science and Pollution Research International*, 26 (2019) 10496–10514.
130. J.L. Hall, Cellular mechanisms for heavy metal detoxification and tolerance, *Journal of Experimental Botany*, 53 (2002) 1–11.
131. H. Huang, Q. Liu, T. Zhang, J. Zhang, J. Zhou, X. Jing, Q. Tang, C. Huang, Z. Zhang, Y. Zhao, G. Zhang, J. Yan, Y. Xia, Y. Xu, J. Li, Y. Li, J. He, Farnesylthiosalicylic acid-loaded albumin nanoparticle alleviates renal fibrosis by inhibiting Ras/Raf1/p38 signaling pathway, *International Journal of Nanomedicine*, 16 (2021) 6441–6453.
132. D.J. Liska, The detoxification enzyme systems, *Alternative Medicine Review: A Journal of Clinical Therapeutic*, 3 (1998) 187–198.
133. H. Itoh, K. Tago, M. Hayatsu, Y. Kikuchi, Detoxifying symbiosis: microbe-mediated detoxification of phytotoxins and pesticides in insects, *Natural Product Reports*, 35 (2018) 434–454.
134. I. Lyagin, E. Efremenko, Enzymes for detoxification of various mycotoxins: origins and mechanisms of catalytic action, *Molecules*, 24 (2019) 2362.
135. G. Yan, X. Chen, S. Du, Z. Deng, L. Wang, S. Chen, Genetic mechanisms of arsenic detoxification and metabolism in bacteria, *Current Genetics*, 65 (2019) 329–338.
136. R. Thapa, G. Gupta, P. Dave, S.K. Singh, A. Raizaday, W.H. Almalki, G. Vyas, S.K. Singh, K. Dua, Y. Singh, Current update on the protective effect of epicatechin in neurodegenerative diseases, *EXCLI Journal*, 21 (2022) 897–903.
137. P. Harnvoravongchai, M. Pipatthana, S. Chankhamhaengdecha, T. Janvilisri, Insights into drug resistance mechanisms in Clostridium difficile, *Essays in Biochemistry*, 61 (2017) 81–88.
138. R. Thapa, G. Gupta, A.A. Bhat, W.H. Almalki, S.I. Alzarea, I. Kazmi, S. Saleem, R. Khan, N. Altwaijry, H. Dureja, A review of Glycogen Synthase Kinase-3 (GSK3) inhibitors for cancers therapies, *International Journal of Biological Macromolecules*, 253 (2023) 127375.
139. G.A. Ahearn, P.K. Mandal, A. Mandal, Mechanisms of heavy-metal sequestration and detoxification in crustaceans: a review, *Journal of Comparative Physiology. B, Biochemical, Systemic, and Environmental Physiology*, 174 (2004) 439–452.
140. R. Thapa, A. Goyal, G. Gupta, A.A. Bhat, S.K. Singh, V. Subramaniyan, S. Sharma, P. Prasher, V. Jakhmola, S.K. Singh, Recent developments in the role of protocatechuic acid in neurodegenerative disorders, *EXCLI Journal*, 22 (2023) 595.
141. B. Mackowiak, H. Wang, Mechanisms of xenobiotic receptor activation: direct vs. indirect, *Biochimica et Biophysica Acta*, 1859 (2016) 1130–1140.
142. Y.J. Kim, M. Song, J.C. Ryu, Mechanisms underlying methotrexate-induced pulmonary toxicity, *Expert Opinion on Drug Safety*, 8 (2009) 451–458.
143. D. Li, T. He, M. Saleem, G. He, Metalloprotein-specific or critical amino acid residues: perspectives on plant-precise detoxification and recognition mechanisms under cadmium stress, *International Journal of Molecular Sciences*, 23 (2022) 1734.
144. R. Thapa, H. Ali, O. Afzal, A.A. Bhat, W.H. Almalki, S.I. Alzarea, I. Kazmi, A.S.A. Altamimi, N. Jain, M. Pandey, Unlocking the potential of mesoporous silica nanoparticles in breast cancer treatment, *Journal of Nanoparticle Research*, 25 (2023) 169.
145. R. Dringen, P.G. Pawlowski, J. Hirrlinger, Peroxide detoxification by brain cells, *Journal of Neuroscience Research*, 79 (2005) 157–165.
146. R. Thapa, O. Afzal, G. Kumar, A.A. Bhat, W.H. Almalki, S.I. Alzarea, I. Kazmi, A.S.A. Altamimi, V. Subramaniyan, L. Thangavelu, Unveiling the connection: long-chain non-coding RNAs and critical signaling pathways in breast cancer, *Pathology—Research and Practice*, 249 (2023) 154736.
147. H. Van Acker, T. Coenye, The role of reactive oxygen species in antibiotic-mediated killing of bacteria, *Trends in Microbiology*, 25 (2017) 456–466.
148. M.S. Hussain, G. Gupta, A. Goyal, R. Thapa, W.H. Almalki, I. Kazmi, S.I. Alzarea, S. Fuloria, D.U. Meenakshi, V. Jakhmola, From nature to therapy: Luteolin's potential as an immune system modulator in inflammatory disorders, *Journal of Biochemical and Molecular Toxicology*, 37 (2023) e23482.
149. Y.C. Yue, M.H. Li, H.B. Wang, B.L. Zhang, W. He, The toxicological mechanisms and detoxification of depleted uranium exposure, *Environmental Health and Preventive Medicine*, 23 (2018) 18.
150. M.M. Gupta, R. Gilhotra, D. Deopa, A.A. Bhat, R. Thapa, N. Singla, R. Kulshrestha, G. Gupta, Epigenetics of pulmonary tuberculosis, *Targeting Epigenetics in Inflammatory Lung Diseases*, Springer, 2023, pp. 127–144.

151. E. Pellegrini, A. Campanella, L. Cotrozzi, M. Tonelli, C. Nali, G. Lorenzini, What about the detoxification mechanisms underlying ozone sensitivity in Liriodendron tulipifera?, *Environmental Science and Pollution Research International*, 25 (2018) 8148–8160.
152. L. Avellaneda-Franco, S. Dahlman, J.J. Barr, The gut virome and the relevance of temperate phages in human health, *Frontiers in Cellular and Infection Microbiology*, 13 (2023) 1241058.
153. A.A. Bhat, R. Thapa, A. Goyal, V. Subramaniyan, D. Kumar, S. Gupta, S.K. Singh, K. Dua, G. Gupta, *Curcumin-Based Nanoformulations as an Emerging Therapeutic Strategy for Inflammatory Lung Diseases*, Newlands Press Ltd, 2023, pp. 583–586.
154. Z. Cao, N. Sugimura, E. Burgermeister, M.P. Ebert, T. Zuo, P. Lan, The gut virome: A new microbiome component in health and disease, *EBioMedicine*, 81 (2022) 104113.
155. R. Thapa, O. Afzal, A.S.A. Altamimi, A. Goyal, W.H. Almalki, S.I. Alzarea, I. Kazmi, V. Jakhmola, S.K. Singh, K. Dua, Galangin as an inflammatory response modulator: an updated overview and therapeutic potential, *Chemico–Biological Interactions*, 378 (2023) 110482.
156. S. Dagar, J. Singh, A. Saini, Y. Kumar, S. Chhabra, R.W. Minz, L. Rani, Gut bacteriome, mycobiome and virome alterations in rheumatoid arthritis, *Frontiers in Endocrinology*, 13 (2022) 1044673.
157. V. Fulci, L. Stronati, S. Cucchiara, I. Laudadio, C. Carissimi, Emerging roles of gut virome in pediatric diseases, *International Journal of Molecular Sciences*, 22 (2021) 4127.
158. A.C. Gregory, O. Zablocki, A.A. Zayed, A. Howell, B. Bolduc, M.B. Sullivan, The gut virome database reveals age-dependent patterns of virome diversity in the human gut, *Cell Host & Microbe*, 28 (2020) 724–740.e728.
159. J. Johansen, K. Atarashi, Y. Arai, N. Hirose, S.J. Sørensen, T. Vatanen, M. Knip, K. Honda, R.J. Xavier, S. Rasmussen, D.R. Plichta, Centenarians have a diverse gut virome with the potential to modulate metabolism and promote healthy lifespan, *Nature Microbiology*, 8 (2023) 1064–1078.
160. E.A. Kennedy, L.R. Holtz, Gut virome in early life: origins and implications, *Current Opinion in Virology*, 55 (2022) 101233.
161. S. Lam, X. Bai, A.N. Shkoporov, H. Park, X. Wu, P. Lan, T. Zuo, Roles of the gut virome and mycobiome in faecal microbiota transplantation, *The Lancet. Gastroenterology & Hepatology*, 7 (2022) 472–484.
162. A.A. Bhat, G. Gupta, A. Goyal, R. Thapa, W.H. Almalki, I. Kazmi, S.I. Alzarea, N. Kukreti, M. Sekar, D.U. Meenakshi, Unwinding circular RNA's role in inflammatory pulmonary diseases, *Naunyn-Schmiedeberg's Archives of Pharmacology*, 397 (2023) 1–22.
163. G. Liang, A.G. Cobián-Güemes, L. Albenberg, F. Bushman, The gut virome in inflammatory bowel diseases, *Current Opinion in Virology*, 51 (2021) 190–198.
164. N. Singla, R. Thapa, R. Kulshrestha, A.A. Bhat, S. Gupta, M. Purohit, S.K. Singh, G. Gupta, Introduction to epigenetics, *Targeting Epigenetics in Inflammatory Lung Diseases*, Springer, 2023, pp. 17–41.
165. L. Massimino, S. Lovisa, L. Antonio Lamparelli, S. Danese, F. Ungaro, Gut eukaryotic virome in colorectal carcinogenesis: is that a trigger?, *Computational and Structural Biotechnology Journal*, 19 (2021) 16–28.
166. I. Mukhopadhya, J.P. Segal, S.R. Carding, A.L. Hart, G.L. Hold, The gut virome: the 'missing link' between gut bacteria and host immunity?, *Therapeutic Advances in Gastroenterology*, 12 (2019) 1756284819836620.
167. E. Pargin, M.J. Roach, A. Skye, B. Papudeshi, L.K. Inglis, V. Mallawaarachchi, S.R. Grigson, C. Harker, R.A. Edwards, S.K. Giles, The human gut virome: composition, colonization, interactions, and impacts on human health, *Frontiers in Microbiology*, 14 (2023) 963173.
168. G. Pavia, N. Marascio, G. Matera, A. Quirino, Does the human gut virome contribute to host health or disease?, *Viruses*, 15 (2023) 2271.
169. A.A. Bhat, A. Goyal, R. Thapa, I. Kazmi, S.I. Alzarea, M. Singh, S. Rohilla, T.K. Saini, N. Kukreti, D.U. Meenakshi, Uncovering the complex role of interferon–gamma in suppressing type 2 immunity to cancer, *Cytokine*, 171 (2023) 156376.
170. H. Shareefdeen, C. Hill, The gut virome in health and disease: new insights and associations, *Current Opinion in Gastroenterology*, 38 (2022) 549–554.
171. A.N. Shkoporov, A.G. Clooney, T.D.S. Sutton, F.J. Ryan, K.M. Daly, J.A. Nolan, S.A. McDonnell, E.V. Khokhlova, L.A. Draper, A. Forde, E. Guerin, V. Velayudhan, R.P. Ross, C. Hill, The human gut virome is highly diverse, stable, and individual specific, *Cell Host & Microbe*, 26 (2019) 527–541.e525.
172. L. Spencer, B. Olawuni, P. Singh, Gut virome: role and distribution in health and gastrointestinal diseases, *Frontiers in Cellular and Infection Microbiology*, 12 (2022) 836706.
173. H.M. Tun, Y. Peng, L. Massimino, Z.Y. Sin, T.L. Parigi, A. Facoetti, S. Rahman, S. Danese, F. Ungaro, Gut virome in inflammatory bowel disease and beyond, *Gut*, 73 (2024) 350–360.

174. A.A. Bhat, O. Afzal, M. Afzal, G. Gupta, R. Thapa, H. Ali, I. Kazmi, S.I. Alzarea, S. Saleem, V.P. Samuel, MALAT1: a key regulator in lung cancer pathogenesis and therapeutic targeting, *Pathology-Research and Practice*, 253 (2023) 154991.
175. W.A. Walters, A.C. Granados, C. Ley, S. Federman, D. Stryke, Y. Santos, T. Haggerty, A. Sotomayor-Gonzalez, V. Servellita, R.E. Ley, J. Parsonnet, C.Y. Chiu, Longitudinal comparison of the developing gut virome in infants and their mothers, *Cell Host & Microbe*, 31 (2023) 187–198.e183.
176. M. Abubakar, E.R. Puppala, B.J. Dutta, K.C. Maharana, R. Thapa, S. Roshan, B. Tazneem, A. Khan, A.A. Bhat, Epigenetics of rhinovirus, *Targeting Epigenetics in Inflammatory Lung Diseases*, Springer Nature Singapore, 2023, pp. 185–197.
177. K. Yang, J. Niu, T. Zuo, Y. Sun, Z. Xu, W. Tang, Q. Liu, J. Zhang, E.K.W. Ng, S.K.H. Wong, Y.K. Yeoh, P.K.S. Chan, F.K.L. Chan, Y. Miao, S.C. Ng, Alterations in the gut virome in obesity and type 2 diabetes mellitus, *Gastroenterology*, 161 (2021) 1257–1269.e1213.
178. T. Zuo, X.J. Lu, Y. Zhang, C.P. Cheung, S. Lam, F. Zhang, W. Tang, J.Y.L. Ching, R. Zhao, P.K.S. Chan, J.J.Y. Sung, J. Yu, F.K.L. Chan, Q. Cao, J.Q. Sheng, S.C. Ng, Gut mucosal virome alterations in ulcerative colitis, *Gut*, 68 (2019) 1169–1179.

2 Introduction to Microbiome

Md Sadique Hussain[1], *Gurusha Bahl*[2], *Riya Mishra*[3], *Asif Ahmad Bhat*[3], *Riya Thapa*[1], *Raihan Siddiqui*[3], *Rahul Sharma*[2], *Rashi Kulshrestha*[3], *Neeraj Patel*[3], *and Gaurav Gupta*[4,5*]

[1]Uttaranchal Institute of Pharmaceutical Sciences, Uttaranchal University, Dehradun, Uttarakhand, India
[2]School of Pharmaceutical Sciences, Jaipur National University, Jaipur, Rajasthan, India
[3]School of Pharmacy, Suresh Gyan Vihar University, Jaipur, Rajasthan, India
[4]Centre of Medical and Bio-allied Health Sciences Research, Ajman University, Ajman, United Arab Emirates
[5]Centre for Research Impact & Outcome, Chitkara College of Pharmacy, Chitkara University, Rajpura, Punjab, India
*Corresponding Author: Gaurav Gupta.
Email id: gauravpharma25@gmail.com

2.1 INTRODUCTION

The human gastrointestinal tract (GIT) is a significant interface, spanning 250–400 m^2, that connects the individual, environmental variables, and antigens in the body. Throughout a typical lifespan, almost 60 metric tons of food traverse through the GIT, carrying many germs from the surroundings that might pose a substantial risk to gut health. The collection of bacteria, archaea, and eukarya living in the GIT is referred to as the "gut microbiota" (GM), and it has developed in parallel with the host, developing a symbiotic connection [1]. The term "microbiome" is frequently employed to depict the intricate assembly of microbes inhabiting a specific area of the body (Figure 2.1). This encompasses not only the microbes themselves but also the surrounding environment, including the host epithelium, immune elements, and various microbial and host-produced metabolites. Within the human body, the microbiome comprises an estimated 100 trillion microbes, predominantly residing in the gut. This ecosystem consists of diverse communities of bacteria, viruses, and fungi, exhibiting a complexity surpassing that of the human genome. A healthy gut typically harbors over 1000 different bacterial species, predominantly from the Bacteroidetes and Firmicutes phyla [2–4]. Emerging evidence underscores the profound interplay between the metabolic functions of the GM and human wellness and disease. Furthermore, the influence of the GM spreads to organs like the liver, muscles, and brain, reflecting its status as a pivotal component of human physiology. GM is indispensable for maintaining overall health, contingent upon the preservation of normal microbial equilibrium within the gut [5]. In the last few years, substantial attention has been given to the human GM, encompassing not only the microorganisms themselves but also their genomes and the surrounding gut environment [6]. This intricate ecosystem serves a multitude of vital functions, including the development of immunity [7], defense against pathogens [8], regulation of host nutrition such as the production of short-chain fatty acids crucial for energy metabolism [9], digestion of polysaccharides, synthesis of vitamins, and fat storage. Additionally, it plays roles in colonization resistance, modulation of the immune system, and even influences human behavior, rendering it an indispensable organ essential for proper bodily function [10–12]. Nonetheless, the possibility exists for these mechanisms to be disturbed due to modified

DOI: 10.1201/9781003489221-2

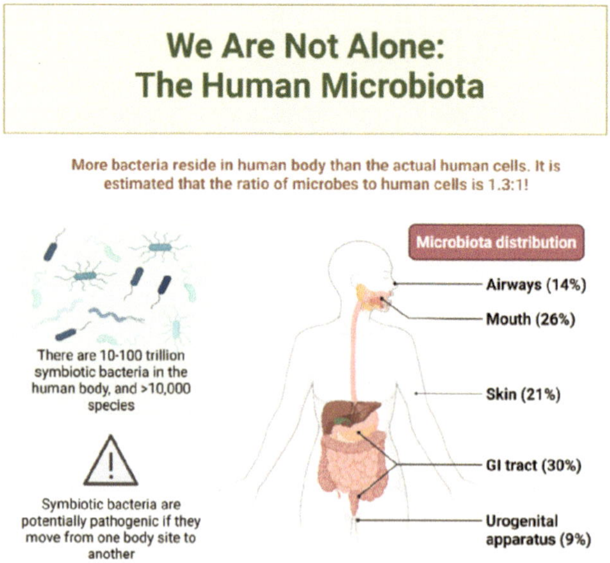

FIGURE 2.1 The human microbiome.

microbial makeup, a condition called dysbiosis [13]. The richness and variety of microbiomes within any environment, including the distinct habitats within a healthy body, can be characterized by both the total count and the proportional distribution of specific microorganisms in comparison to others coexisting within the same or different habitats. This microbial diversity, alongside the functional capabilities of the microbiota and the intricate interplays between these microorganisms and the human host, directly influence overall health and susceptibility to various diseases. For instance, a decrease in GM diversity may result in a reduction of metabolic functionalities, an elevation in pro-inflammatory compounds, and a diminished capacity for breaking down foreign substances. Such shifts can potentially lead to malnutrition, inflammatory bowel diseases (IBD), metabolic disorders, and may even compromise the effectiveness of certain treatments. Nevertheless, it is noteworthy that the relationship between diversity indices and health outcomes is not consistent across all bodily habitats; an increase in GM diversity, for instance, can sometimes precipitate conditions like bacterial vaginosis. Within specific bodily sites, the presence and activity of particular microbial taxa, along with their genetic, protein, and metabolic profiles, hold promise as potential indicators for early detection of diseases and for predicting the treatment efficacy of commonly utilized drugs in managing infectious or metabolic ailments [14, 15]. The makeup and functions of the GM are susceptible to various internal and external variables. For instance, exposure to different xenobiotics can disrupt the normal functioning of the GM [16]. Numerous research endeavors have confirmed notable disparities in both the content and the arrangement of microbial diversity among healthy people across various bodily regions, encompassing the GIT, respiratory system (RS), skin, reproductive organs (ROs), and urinary tract (UTs) [17]. Notably, the GIT harbors the greatest diversification in terms of community engagement, while the vaginal environment exhibits comparatively limited heterogenicity at the species level. Nevertheless, it is crucial to recognize that the richness and functional capabilities of indigenous microbial communities within bodily sites are not fixed entities. Over time, the diversity of these communities undergoes fluctuations. Notably, the oral cavity demonstrates the highest degree of variability, while the microbial taxa indigenous to the vagina and GIT tend to remain relatively stable [18]. These shifts in both structure and functionality of microbial communities across different bodily sites are influenced by a multitude of factors, including dietary patterns, environmental exposures, host genetics, age, gender, overall health, and lifestyle choices [19, 20].

2.2 DEVELOPMENT OF THE HUMAN MICROBIOTA (HM)

The establishment of the HM typically initiates shortly after birth, marking the rapid colonization of the GIT, while various life events such as illness, antibiotic use, and dietary changes can disrupt its balance. Delivery method also influences microbiota diversity, with vaginally born infants initially harboring a high lactobacilli reservoir, mirroring vaginal flora, whereas C-section delivered infants exhibit a delay in Bacteroides colonization but an influx of facultative anaerobes like *Clostridium* species. Early microbiota stages are characterized by low diversity dominated by Actinobacteria and Proteobacteria phyla. Throughout the first year, microbial diversity expands, eventually converging towards an adult-like profile unique to everyone [21]. By approximately 2.5 years old, infant microbiota closely resembles that of adults in composition, diversity, and function. However, in adulthood, GM, although relatively stable, remains susceptible to disruptions from life events. Notably, in individuals aged over 65, there is a shift in microbial community marked by increased Bacteroidetes and *Clostridium* cluster IV abundance, differing from younger individuals where cluster XIVa prevails [22, 23].

2.2.1 Differential Microbiome Composition across Human Body Regions

Within a healthy adult human, an astonishing 39 trillion bacterial cells reside, indicating a ratio of approximately 1.3 bacterial cells for every human cell. Nonetheless, this count solely encompasses bacteria and disregards other microbial components like viruses, fungi, and protists. Across various body regions, an estimated 1,000–1,500 microbial species coexist, each exhibiting a site-specific distribution. Remarkably, the number of subspecies boasting unique genomic makeup could potentially soar tenfold greater than the approximate species count. Furthermore, the collective genetic repertoire of the microbiome-mediated with our body surpasses our own genetic makeup by nearly 500 times. Despite their abundance, microbial biomass contributes only a minor fraction—approximately 1–3%—to overall body weight due to the diminutive size of these microorganisms [24, 25]. The composition of the microbiota in the GIT mirrors the physiological characteristics of each specific region, arranged both across and along the gut. Chemical, nutritional, and immunological variances create gradients that influence the density and makeup of the microbiota. For instance, the small intestine harbors elevated levels of acids, oxygen, and antimicrobials, alongside a brief transit period. These conditions restrict bacterial proliferation, allowing only fast-growing, facultative anaerobes capable of adhering to epithelial surfaces or mucus to thrive [26]. Unlike the relatively stable host genome, the HM is ever-changing and undergoes modifications throughout childhood, influenced by various environmental factors such as diet, antibiotic usage, and, notably, disease conditions. The most significant alterations in composition and functionality typically occur during infancy and early childhood. Factors affecting the GM of infants include gestational age; birth method; feeding choice; mother's nutritional condition; antibiotic administration; conditions affecting the GIT or leading to colorectal cancer (CRC); metabolic disorders like obesity and diabetes; immune-related disorders such as asthma, allergies, depression, and anxiety; as well as cardiometabolic diseases [27–30]. Furthermore, disruptions in the GM associated with metabolic diseases can have adverse effects on the gestational environment and may impact offspring in subsequent generations. Moreover, alterations in microbiome composition have been correlated with increased mortality risk [31]. The intricate and adaptable nature of the infant microbiota during these formative stages is considered to be important in establishing and maintaining balance with the host's immune system [32, 33].

2.2.1.1 The Microbiome of the GIT

The GIT's outer layer is a suitable habitat for anaerobic and aerobic bacteria to form populations. Neonatal gut colonization by microbes starts shortly following childbirth. Early in life, the GM is varied, mostly consisting of Proteobacteria, Actinobacteria, and Firmicutes [7, 34]. Longitudinal research has shown that the GM population in the adult gut is rather consistent, compared to the quick fluctuations seen in the initial couple of years of childhood. By age three, the GM stabilizes, reflecting the makeup and richness often observed in adulthood. The GM in healthy individuals is mostly made up of four bacterial phyla: Actinobacteria, Bacteroidetes, Firmicutes, and Proteobacteria, which together account for around 98% of the entire GM. The other 2% include uncommon bacterial phyla. The GM (Figure 2.2) in the GIT is mostly friendly or harmonious, and the natural presence of these indigenous bacterial species is crucial for preserving well-being [35, 36].

2.2.1.2 The Skin Microbiome

Our skin serves as a habitat for hundreds of microbial populations, including bacteria and fungi, coexisting in symbiosis over time. These skin-dwelling microorganisms play crucial roles in bolstering the body's defense mechanisms against pathogen colonization and in immunological development to discern between commensal and pathogenic strains within the microbial diversity. They achieve this through the recognition of specific microbial antigens or via microbial cascade molecules [37, 38]. The diversity of the skin microbiome is notably diverse and is primarily affected by the skin's physiological characteristics, including moisture levels, dryness, and sebaceous microenvironments. Sebaceous sites are often dominated by lipophilic *Propionibacterium* species, whereas humid skin regions, such as the feet and bends of the elbows, harbor an abundance of *Staphylococcus* and *Corynebacterium* species. Notably, *Propionibacterium acnes* and *Staphylococcus epidermidis* are prevalent and dominating varieties of bacteria on the skin of people [39, 40].

2.2.1.3 Respiratory Tract Microbiome

The human RS comprises two main parts: the upper and lower respiratory tract (RT). Both segments harbor distinct sets of bacterial communities, with the upper RT, particularly the oropharynx region, exhibiting the highest microbial load, while the lungs display minimal microbial presence. Common bacteria inhabiting the upper RT include *Corynebacterium spp.*, *Propionibacterium spp.*, *Staphylococcus spp.*, along with others. During early life, the lower RT microbiome is predominantly composed of Firmicutes and Proteobacteria, while in adults, it comprises Actinobacteria, Bacteroidetes, and Firmicutes. Additionally, healthy adult lungs harbor various fungal species, including *Eremothecium*, *Malassezia*, and *Systenostrema*. The indigenous microbial populations

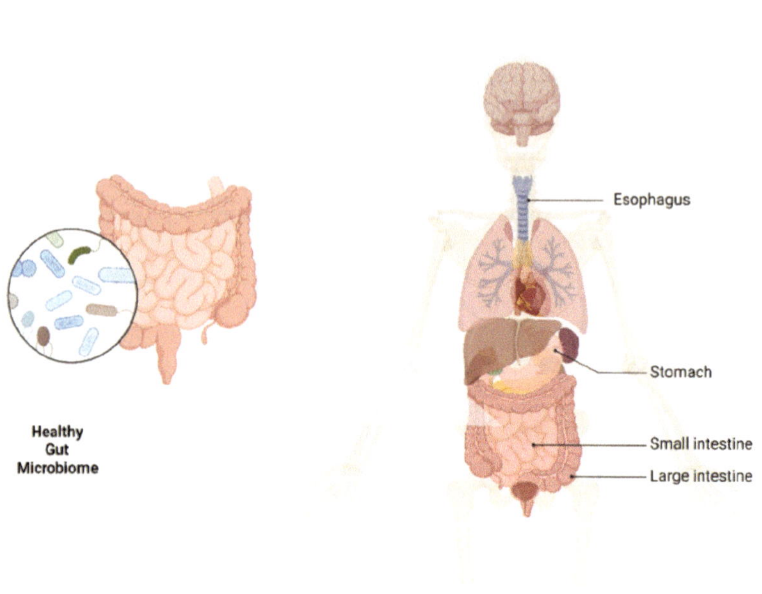

FIGURE 2.2 The gut microbiome.

within the RT serve defensive actions by preventing the spread of microorganisms responsible for infections [41–43].

2.2.1.4 The Microbiome of the Reproductive Organs

In women of reproductive age, the ROs houses stable yet intricate microbial communities that play pivotal roles in maintaining health and significantly contribute to birth outcomes [44, 45]. The vaginal microbiota (VM) produces and secretes substances to prevent the growth of invading microbes, reduce irritation, and maintain an acidic state. Additionally, microbial populations in the ROs of females are the primary origin of the first GM for kids. *Lactobacillus* is a prevalent organism in VM populations globally and makes lactic acid. The VM consists of about 35 distinct *Lactobacillus* species, including popular species such as *L. iners*, *L. crispatus*, *L. gasseri*, *L. jensenii*, and *L. johnsonii*. Metagenomic techniques have delineated three distinct clades within the VM: clade 1, composed exclusively of *Lactobacillus* species; clade 2, comprising a blend of *Lactobacillus* alongside additional microbial organisms; and clade 3, made up of non-*Lactobacillus* microbial taxa. Black women had higher pH values of around 4.7, whereas Hispanic women had pH levels of around 5.0, which promotes the presence of clade 3. The structure and operational characteristics of the ventromedial nucleus within all three groups have a substantial effect on birth results [46–48].

2.3 EXPLORING MICROBIOME DIVERSITY AND DYNAMICS

The development and widespread availability of Next-Generation Sequencing (NGS) approaches have been instrumental in enhancing our comprehension of the diversity, resilience, adaptability, and fluctuation of the HM over time and across different disease contexts. Based on current insights, it is estimated that each human body harbors approximately 10^14 bacterial cells, whereas the Earth is inhabited by roughly 10^30 bacterial and archaeal cells. The progression and accessibility of sequencing methodologies also facilitate investigations into the richness of species within intricate microbial communities and their resilience when exposed to foreign substances and various environmental influences. It is apparent that the balance between stability and adaptability of microbial community structures in diverse bodily habitats is notably distinct and influenced by a myriad of factors. Three pivotal elements—environmental conditions, host genetics, and the microbiome—interact intricately, collectively contributing to the maintenance of stability and equilibrium within the GIT and other bodily regions [32, 49].

2.3.1 Geographic Conditions

Geographical, socioeconomic, and cultural variables play crucial roles in shaping the composition, diversity, and genomic profiles of microbial communities across various regions of the body. The most pronounced variations in taxonomic makeup are observed in the GIT and skin of person's residing in various nations, especially among children and adults [50, 51]. For instance, individuals living in India exhibit notably different level of Firmicutes to Bacteroidetes compared to those in Japan. Further examination at the species level highlights the abundance of *P. copri*, *P. stercorea*, and other *Prevotella* species in Indian individuals, whereas these species are rare in Japanese subjects. Conversely, Japanese individuals exhibit a conserved presence of Bacteroides species such as *B. uniformis*, *B. ovatus*, and *B. fragilis* in their GM. In terms of fungal communities, both Indian and Japanese populations exhibit dominance by Ascomycota and Basidiomycota phyla, though Basidiomycota abundance is notably higher in healthy Japanese adults compared to Indians. At the genus level, *Candida* and *Saccharomyces* emerge as primary fungal genera in the GM of Indian and Japanese people, respectively. Common species within the *Candida* genus in Indian individuals include *C. albicans*, *C. tropicalis*, and *C. glabrata*, while *Saccharomyces cerevisiae* predominates in the guts of Japanese people [52, 53].

2.3.2 Genetic Influence in Host-Microbiome Interaction

While environmental factors exert significant influence over the structure of GM communities, there is undeniable evidence of a noteworthy connection between host genetics and GM composition [54]. Research comparing monozygotic and dizygotic twins has consistently shown that the GM of monozygotic twins exhibit greater similarity compared to dizygotic twins. Through genome-wide studies, researchers have identified at least nine loci in the human genome linked with variations in GM makeup. These genetic loci have been implicated in various complex diseases, cellular signaling pathways, inflammation regulation, defense against microbial invasion, and immune responses. The molecular effectors encoded by these genetic loci directly interact with the GM, influencing the release of pro-inflammatory cytokines upon activation. For instance, the functional single-nucleotide polymorphism (SNP) in the Lactase persistence gene (LCT) correlates with the relative abundance of *Bifidobacterium*. SNPs in the Nod2 gene are strongly associated with the presence of *Dorea* and *Subdoligranulum* species. Moreover, genes like CARD9 have been linked to IBD, resulting in a rise in pro-inflammatory microorganisms. Riboflavin, functioning as an oxidative mediator, stimulates the development of the common GM *Faecalibacterium prausnitzii*. SNPs in the CLEC4A–FAM90A1 gene impact riboflavin production and decrease the presence of *F. prausnitzii*. SNPs in the MEFV gene, which encodes pyrin, are linked to reduced bacterial diversity in the GIT, especially noticeable in individuals with hereditary Mediterranean illness. Lack of immunoglobulin IgA in humans is significantly associated with the onset of IBD [55, 56].

2.3.3 Age

Studies utilizing culture-independent DNA analysis have revealed the presence of bacteria in the placenta, as well as in the amniotic fluid and meconium of both full-term and preterm infants. During childbirth, the early colonization and makeup of the microbial populations in newborns' guts are influenced by various factors, including delivery mode, maternal physiology and health, prenatal antibiotic exposure, and birthing environment [57, 58]. In neonates, the GIT is primarily inhabited by facultative anaerobic species belonging to the Firmicutes and Proteobacteria phyla, which serves a crucial function in decreasing oxygen levels in the GIT and facilitating the constant colonization of strict anaerobes from the Actinobacteria, Bacteroidetes, and Firmicutes. Among newborns, the GM is predominantly composed of various species of Bifidobacterium, depending on their feeding habits. Notably, *B. longum subsp. infantis* is commonly found in the guts of breastfed child, while *B. longum subsp. longum* tends to dominate the GM of formula-fed infants. However, there exists significant interpersonal differences in complexity and variation among children, which is notably boosted compared to adult individuals [59, 60].

2.3.4 Dietary Influence

Diet stands out as the primary determinant in shaping the microbial makeup within the GIT and influencing the selection of GM. Research indicates that various dietary patterns, including veganism, consumption of specific plant species, gluten-free diets, dairy exclusion, fiber-rich vegetables, whole grains, as well as the intake of antibiotic-treated livestock, chickens, beef, and other high-fat meats, eggs, salty foods, sweets, packaged food, drinks with added sugar, and frequent seafood intake all drastically change the makeup and activity potential of the GM population [61–63]. Consistent consumption of animal-based diets is associated with increased levels of bile-tolerant microorganisms such as *Alistipes*, *Bilophila*, and *Bacteroides*, while reducing the abundance of Firmicutes species, which play crucial roles in metabolizing dietary plant polysaccharides. Moreover, there is a significant rise in the absolute abundance of the sulfite-reducing pathobiont, *Bilophila wadsworthia*, in response to an animal-based diet. Intake of saturated fats from milk, but not polyunsaturated fats from plants, also leads to elevated levels of *B. wadsworthia*. Furthermore, the host food preferences promote horizontal transmission of genes among microbial populations, increasing the microbe's genome potential for functioning [64–66].

2.3.5 Lifestyle Factors

Alcohol consumption, smoking habits, physical activity levels, vitamin intake, and probiotic usage play significant roles in shaping both the taxonomic and the functional composition, as well as the diversity, of the GM community. Consistent alcohol intake can distinctly differentiate the microbial profiles of individuals who drink from those who do not. Moreover, frequent alcohol consumption correlates with a notable increase in the alpha diversity of the GM. Specifically, the relative abundance of *Bifidobacterium ASV* at the genus level is substantially higher in the GIT of regular alcohol drinkers. Additional inquiry reveals consistent and robust increases in microbial population among consumers of red/white wine and beer/cider. Notably, daily alcohol consumption exerts the greatest influence on enhancing microbial diversity compared to infrequent or occasional consumption. Additionally, both vitamin and probiotic supplementation are associated with increased microbial diversity [19, 67, 68].

2.3.6 Therapeutic Interventions

Various treatments employed for preventing or managing infectious and metabolic diseases can induce transient disruptions in the richness, diversity, and functional capacity of the HM, with long-lasting ecological consequences. Among these interventions, antimicrobial drugs exhibit the most significant impact on the microbiome. While antibiotics are effective against pathogenic microbes, their non-specific action often results in the depletion of both pathogenic and beneficial bacteria lacking resistance mechanisms. This leads to acute disturbances in the microbiome, reducing the GM diversity. Broad-spectrum antibiotics can decrease the abundance of over 30% of GM, causing drastic declines in microbial populations. Antibiotic usage might lead to short-term consequences, such as *Clostridioides* difficile infection, diarrhea, and fungal colonization, while long-term consequences may involve the development of allergic conditions, malnutrition, and obesity. The extent of disruption based on factors such as the antibiotic, dosage, treatment period, route of delivery, excretion kinetics, and mode of action. For instance, a seven-day regimen of clindamycin can significantly reduce GM richness and diversity, with a lasting reduction in Bacteroides clonal diversity persisting for over two years. Macrolide administration via oral route diminishes the GM makeup and alters the abundance of *Bifidobacterium* and *Bacteroides.* In tuberculosis (TB) treatment, which typically lasts six to nine months and involves a combination of antibiotics have profound effects on microbial populations and makeup in the GIT are observed. Notably, the abundance of various *Firmicutes* species decreases significantly, while *Prevotella* abundance rises markedly. Although microbiome changes induced by antibiotics are usually reversible, the recovery period varies based on the antibiotic course. In TB, dysbiosis in the GM may persist long term even after treatment completion [53, 69–72].

2.4 CRUCIAL ROLE OF MICROBIOME IN HEALTH AND HUMAN PHYSIOLOGY

The composition and functional capabilities of microbial communities play pivotal roles in sustaining a healthy microbial ecosystem. Microbes residing in hosts utilize their cell surface receptors and internal sensory mechanisms to observe their surroundings, enabling them to acquire nourishment and energy for growth and adaptability, especially under challenging situations. Under homeostatic conditions, the functions performed by microbes and the byproducts of their metabolic activities significantly influence human physiology, reducing the likelihood of biochemical and viral illnesses. Additionally, Microbial metabolism may directly affect medications by changing their molecular makeup, permeation, bioavailability, and durability, thereby potentially enhancing or diminishing their effectiveness. Factors such as diet, exposure to xenobiotics, geographical location, age, host physiology, and lifestyle choices can modulate the diversity of GM over time, impacting overall GM composition. These intricate interactions underscore the dynamic interplay between the microbiome and human health [73]. Due to its expansive genomic and metabolic capabilities, the GM offers a multitude of advantageous features to the host. Among the significant functions of these bacteria are the preservation of mucosal barrier integrity, provision of essential nutrients like vitamins, and defense against pathogens. Moreover, the symbiotic relationship is essential for maintaining optimal immunity. The GM serves a crucial function in the synthesis of vitamins that the host organism cannot produce independently. Lactic acid bacteria are pivotal in vitamin B12 synthesis, a nutrient unattainable through animal, plant, or fungal synthesis. Bifidobacteria serve as primary producers of folate, a vitamin responsible for DNA synthesis and repair. Additionally, the GM has been demonstrated to synthesize other vitamins, including biotin, nicotinic acid, pantothenic acid, pyridoxine, riboflavin, thiamine, and vitamin K [74]. Also, the GIT serves as the primary reservoir of microbes within the human body, thus exerting a significant impact on human wellness. Gut bacteria, in particular, play a significant role in defending against invading pathogens and regulating essential host functions such as metabolism and immune system development [75–77]. This symbiotic relationship between GM and the host is crucial for maintaining health and combating diseases. GM contributes to the metabolism of various compounds, including polysaccharides, amino acids, xenobiotics, and micronutrients [78]. Additionally, they supply the host with vital vitamins B5 and B12, essential for neurotransmitter synthesis and stress hormone regulation, crucial for optimal nervous system function [79]. Another key function of the human GM is colonization tolerance, wherein non-pathogenic microbes outcompete and suppress the growth of potential pathogens, thereby bolstering immunity and conferring protection against infections like TB [80]. Furthermore, a balanced composition of GM is closely associated with optimal nervous system function, promoting good mental health [63, 66, 81]. However, any disruption in the GM can lead to complications, including infectious diseases. Variations in GM composition may allow potentially harmful bacteria to proliferate, leading to dysbiosis—a state characterized by an imbalance in microbial diversity [82]. Dysbiosis can result from various physiological factors such as diet, stress, age, antibiotic use, and environmental influences like radiation and geographical location.

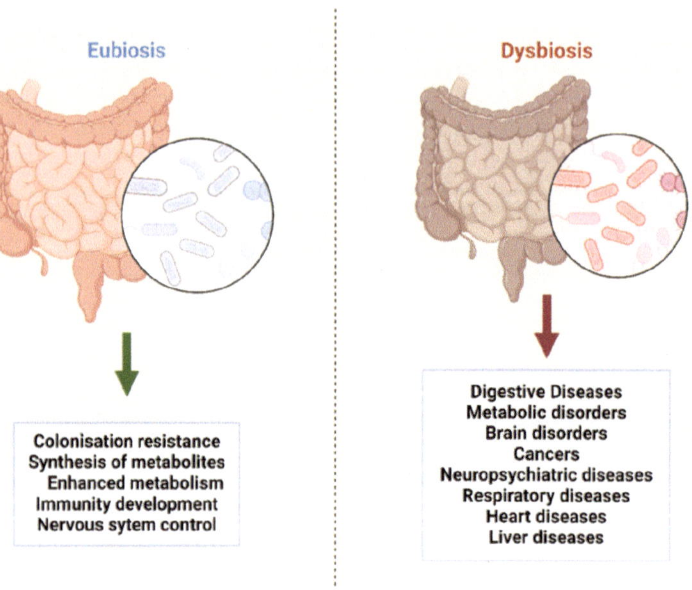

FIGURE 2.3 Gut microbiota in health and disease.

Figure 2.3 gives a summary of the impact of human GM on health and disease and how harmonious balance of microbes in the intestines, known as eubiosis, contributes to improved metabolic functions, resistance against pathogenic bacteria, and enhanced immunity. Conversely, an imbalance in GM diversity, termed dysbiosis, can lead to various disorders affecting digestion, metabolism, neural function, and potentially even cancer. For instance, certain *Lactobacillus* species decline during air travel, potentially contributing to flight-related anxiety. Similarly, exposure to heavy metals like cadmium can diminish beneficial bacteria such as *Bacteroides*, leading to reduced generation of metabolites like short-chain fatty acids (SCFAs) [83].

2.5 DISEASES LINKED TO DYSBIOSIS

It is highly likely that dysbiosis, or an inability to properly regulate the GM diversity, contributes to the onset and aggressiveness of various illnesses (Figure 2.4) [84, 85]. Dysbiosis can lead to diseases in three main ways: (a) acquisition or overgrowth of pathogens and their mechanisms on an opportunistic basis, (b) disruption or suppression of health-protective bacteria and their functions, and (c) a mixture involving reduction and boosts in functionality [86]. The diseases resulting from dysbiosis include the following.

2.5.1 *Clostridioides Difficile* Infection (CDI)

C. difficile, a member of the Firmicutes in the normal GM, is a Gram-positive, spore-forming bacterium that thrives in obligate anaerobic conditions. It may lead to various clinical signs, ranging from moderate diarrhea to serious colonic burst and deadly megacolon [87–89]. Interestingly, the leading cause of CDI is often the administration of broad-spectrum antibiotics, with diarrhea probably the most typical adverse effect [90]. Antibiotic use disturbs GM diversity and reduces their community, particularly affecting dominant microbes responsible for synthesizing secondary bile acids, which play a role in inhibiting the vegetative growth of C. difficile. This disruption allows C. difficile to proliferate and colonize the gut, leading to the secretion of toxins that interfere with intestinal ion absorption, ultimately causing diarrhea [91].

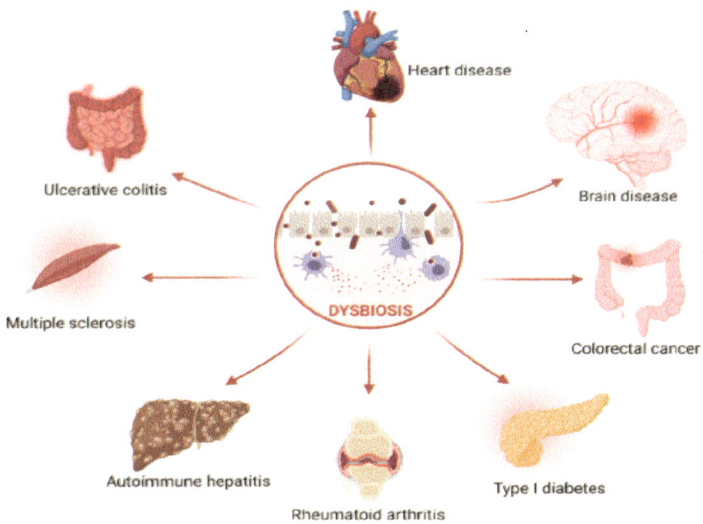

FIGURE 2.4 Various diseases associated with Dysbiosis.

2.5.2 Inflammatory Bowel Disease (IBD)

IBD is characterized by chronic and recurring inflammation of the GIT. There are two primary forms of IBD: Crohn's Disease (CD), involving persistent inflammation affecting any part of the GIT, and Ulcerative Colitis (UC), where inflammation is confined to the large intestine [92–94]. Clinical manifestations of both types include fever, abdominal discomfort, pain, and diarrhea. In individuals with genetic predisposition, antibodies may develop against commensal microbial antigens, leading to dysbiosis and compromising the integrity of the gut mucus barrier. This compromised barrier function stimulates local immunity, participating in severe gut inflammation. Dysbiotic features commonly observed in IBD patients include reduced levels of anti-inflammatory gut Firmicutes like *Faecalibacterium prausnitzii* and *Roseburia sp.* Additionally, there is an increase in pathogenic gut microorganisms such as *Enterobacteriaceae* species and *Bacteroides fragilis*, which produce high concentrations of endotoxic lipopolysaccharide (LPS) [95–98]. Treating GI inflammation in IBD individuals by using probiotics and prebiotics to promote the growth and activity of beneficial microbes may be a promising strategy. However, the effectiveness of therapy varies individually due to the intricate pathogenesis of IBD in each person, presenting a challenge in creating universally effective treatments [34, 99].

2.5.3 Cancer

Dysbiosis has been linked to the initiation and progression of various types of cancer, such as CRC, gastric cancer, pancreatic cancer, hepatocellular carcinoma (HCC), melanoma, and breast cancer. Through the activation of tumorigenic pathways, inflammation induction, and disruption of host DNA, dysbiosis and specific pathogenic bacteria in the gut may initiate or promote cancer development. Toll-like receptors (TLRs), integral to the innate immune system, become activated through pathogen-associated molecular patterns (PAMPs), resulting in heightened production of pro-inflammatory molecules within cells, thereby expediting carcinogenesis. Moreover, apart from inducing inflammation, certain bacteria have the ability to induce DNA damage by generating compounds that support tumorigenesis [100–102].

2.5.4 Obesity

A global health issue, obesity results from an accumulation of excess fat due to increased energy intake and decreased energy expenditure. It is associated with metabolic disruption, making obese individuals more susceptible to conditions like cardiovascular disease, type-II diabetes, and liver abnormalities [103–105]. The GM is now being studied for its potential role in obesity alongside genetic, behavioral, and environmental factors. Metagenomic studies have revealed a notable elevation in butyrate-generating Firmicutes and a decline in Bacteroidetes in the GM of obese individuals and genetically obese mice compared to healthy subjects. This Firmicutes to Bacteroidetes ratio correlates with body weight and fat deposition, suggesting a disproportionate balance in obese individuals. Given the association between gut dysbiosis and obesity, modifying microbiota composition through dietary interventions such as prebiotic or probiotic administration holds promise as therapy. In animal models, *Lactobacillus rhamnosus* has demonstrated reductions in plasma cholesterol, triacylglycerides, and white adipose tissue [23, 40, 106].

2.5.5 Periodontitis

This condition, which can lead to tooth loss, is an infectious disease affecting the tissues supporting the teeth. It is characterized by inflammation induced by bacteria and is closely associated with both the oral microbial community and the immune system of an individual. The microbial aspect plays a part in initiating the disease, while the host's immune response contributes to its progression. Recent studies suggest that oral dysbiosis may be a contributing factor to periodontitis, gradually worsening oral health over time until the disease manifests. During this process, specific microbial complexes emerge, starting with the orange complex, consisting of gram-negative anaerobic bacteria like *Prevotella intermedia*, *P. nigrescens*, *P. micros*, and F. *nucleatum*. As the condition progresses, the microbiota shifts to the red complex, including key periodontal pathogens like *Porphyromonas gingivalis*, *Treponema denticola*, and *Tannerella forsythia*. While microbial changes are recognized as playing a role in periodontitis development, it is essential for clinicians and investigators to also consider genetic, immunological, and environmental factors for a comprehensive understanding of the condition. New research indicates that the oral microbes is significantly involved in severe periodontitis but also serves as an indirect link between periodontitis and cardiovascular disease [3, 12, 107–109].

2.6 GUT MICROBIOME AND ENVIRONMENTAL TOXINS

The GM has surfaced as a fresh intermediary in human well-being. Conversely, exposure to various harmful environmental substances can readily disrupt the human GM. Disruption induced by environmental factors in the GM is closely linked with the risk of human diseases so exposure to environmental factors poses a significant risk for various human diseases, many of which intersect with those associated with the GM. The modification of the functional GM, which could detrimentally impact human health, is now recognized as a progressively acknowledged method through which environmental chemicals manifest their harmful influences. Tu et al. characterized the impairment caused by environmental exposure to the GM as GM toxicity, which serves as the bridge connecting environmental exposure to microbiome-related human ailments [20, 110, 111]. Specific environmental toxins have the potential to cause liver toxicity. Such exposure alters the liver's structure and function, ultimately resulting in liver diseases. Similarly, exposure to certain OPs induces alterations in the makeup of GM, likely resulting in adverse health consequences. For example, arsenic exposure has been shown to disrupt the diversity and makeup of the mouse GM, possibly contributing to its toxicity. Recognizing the significant role of the GM in human well-being, particularly its vulnerability to OPs, it becomes crucial to characterize GM toxicity. Given the emerging perception of the GM as a vital human organ, we propose defining environmentally induced functional damage in the GM as GM toxicity. The manifestation of diseases related to GM can be broadly understood as environmental elements leading to detrimental changes in the GM, which, in turn, impact human health through interactions between the host and GM. Essentially, GM toxicity resulting from environmental exposures contributes to adverse outcomes related to GM. By integrating GM toxicity into the spectrum of organ toxicities, we can now explore the interactions between the environment and diseases related to GM within the framework of toxicology [112]. Table 2.1 enlists the effects of ETs exposure on the microbiota and resulting poor effects related to GM toxicity [25, 113].

2.7 CONCLUSION

The HM, a complex and ever-changing ecosystem comprising diverse microbial communities inhabiting different regions of the body, emerges as a critical determinant of human health and overall well-being. By delving into the diversity, composition, and functionality of the microbiome across various bodily sites, this chapter has yielded profound insights into the intricate interplay among microbial communities, host factors, and environmental influences. We have observed the remarkable adaptability and resilience of the microbiome, from its establishment in infancy to its vulnerability to disturbances in adulthood. Through this exploration, we have uncovered the microbiome's crucial roles in preserving homeostasis, regulating immune responses, producing essential nutrients, and defending against pathogens. Additionally, we have acknowledged the microbiome's far-reaching influence beyond its immediate habitat, extending to distant organs and systems, highlighting its integral status within human physiology.

TABLE 2.1
Microbiome responses to environmental chemical exposure and adverse outcomes associated with gut microbiome toxicity.

Class	Chemical	Gut Bacteria	Adverse Outcome
Heavy metal	Arsenic	Catabacteriaceae ↓ Clostridiaceae ↓ Clostridiales Family XIII Incertae Sedis ↑ Erysipelotrichaceae ↓	Inflammation
	Manganese	Firmicutes↑ Bacterodetes ↓ in males Firmicutes ↓ in females	
	Cadmium Lead	γ-Proteobacteria ↓ Firmicutes/Bacteroidetes ratio ↓ Ruminococcus ↓ Oscillospira ↓	
Pesticide	Carbamate aldicarb	Clostridium ↑ Anaerostipes ↓	Inflammation
	Carbendazim	Firmicutes/Bacteroidetes ratio ↑	
	Diazinon	Lachnospiraceae Johnsonella ↑ in females, Lachnospiraceae Johnsonella ↓ in males	
	Imazalil	Clostridiales ↑ Lachnospiraceae ↑	
	Malathion	Mogibacteriaceae ↑	
Artificial sweetener	Ace-K	Bacteroides, Anaerostipes and Sutterella ↑ in males	Obesity in males
	Aspartame	Enterobacteriaceae ↑ Clostridium leptum ↑ Firmicutes/Bacteroidetes ratio	Diabetes
	Neotame	Firmicutes/Bacteroidetes ratio ↓	
	Saccharin	Corynebacterium ↑ Roseburia ↑ Ruminococcus ↓ Turicibacter ↑	Inflammation
	Sucralose	Christensenellaceae ↑ Clostridiaceae ↑ Erysipelotrichaceae ↓	
Others	Emulsifiers (P80 and CMC)	Akkermansia, Proteobacteria ↑	Colitis
	Nicotine	F16↓ in females, F16 ↑ in males	
	Triclosan	Turicibacteraceae ↓ Clostridiaceae ↓ Lactobacillaceae ↑ Streptococcaceae ↑ Christensenellaceae↓	

However, amidst the microbiome's beneficial contributions, we recognize the potential for dysbiosis to disrupt these intricate interactions and contribute to various health complications. Factors such as diet, environmental exposures, host genetics, age, and lifestyle choices significantly shape the diversity and functionality of the microbiome, underscoring the importance of maintaining a healthy microbial balance. As we continue to unravel the complexities of the HM, it becomes increasingly evident that this dynamic ecosystem holds immense potential for therapeutic interventions and personalized medicine. By leveraging microbiome-based biomarkers and interventions, we can strive to optimize human health and mitigate the risk of various diseases. The study of the HM represents a frontier of scientific exploration with profound implications for human health and disease. Through ongoing efforts to comprehend and manipulate this complex ecosystem, we pave the way for innovative health-care approaches that capitalize on the symbiotic relationship between humans and their microbial inhabitants.

REFERENCES

1. S. Bengmark, Ecological control of the gastrointestinal tract. The role of probiotic flora, *Gut*, 42 (1998) 2–7.
2. V.B. Young, The role of the microbiome in human health and disease: an introduction for clinicians, *BMJ*, 356 (2017) j831.
3. K.S. Alharbi, Y. Singh, W. Hassan Almalki, S. Rawat, O. Afzal, A.S. Alfawaz Altamimi, I. Kazmi, F.A. Al-Abbasi, S.I. Alzarea, S.K. Singh, S. Bhatt, D.K. Chellappan, K. Dua, G. Gupta, Gut microbiota disruption in COVID-19 or post-COVID illness association with severity biomarkers: a possible role of pre/pro-biotics in manipulating microflora, *Chemico–Biological Interactions*, 358 (2022) 109898.
4. S. Hussain, M.A. Khan, R. Rajan, J. Jyoti, S. Sharma, S.K. Sahu, Nanorobots: the future of healthcare, *AIP Conference Proceedings*, AIP Publishing, 2023.
5. J.K. Nicholson, E. Holmes, J. Kinross, R. Burcelin, G. Gibson, W. Jia, S. Pettersson, Host-gut microbiota metabolic interactions, *Science*, 336 (2012) 1262–1267.
6. J.R. Marchesi, J. Ravel, The vocabulary of microbiome research: a proposal, *Microbiome*, 3 (2015) 31.
7. T. Gensollen, S.S. Iyer, D.L. Kasper, R.S. Blumberg, How colonization by microbiota in early life shapes the immune system, *Science*, 352 (2016) 539–544.
8. A.J. Bäumler, V. Sperandio, Interactions between the microbiota and pathogenic bacteria in the gut, *Nature*, 535 (2016) 85–93.
9. G. den Besten, K. van Eunen, A.K. Groen, K. Venema, D.J. Reijngoud, B.M. Bakker, The role of short-chain fatty acids in the interplay between diet, gut microbiota, and host energy metabolism, *Journal of Lipid Research*, 54 (2013) 2325–2340.
10. Y. Moon, Microbiome-linked crosstalk in the gastrointestinal exposome towards host health and disease, *Pediatric Gastroenterology, Hepatology & Nutrition*, 19 (2016) 221–228.
11. G. Sharon, N. Garg, J. Debelius, R. Knight, P.C. Dorrestein, S.K. Mazmanian, Specialized metabolites from the microbiome in health and disease, *Cell Metabolism*, 20 (2014) 719–730.
12. S. Ashique, G. De Rubis, E. Sirohi, N. Mishra, M. Rihan, A. Garg, R.J. Reyes, B. Manandhar, S. Bhatt, N.K. Jha, T.G. Singh, G. Gupta, S.K. Singh, D.K. Chellappan, K.R. Paudel, P.M. Hansbro, B.G. Oliver, K. Dua, Short chain fatty acids: fundamental mediators of the gut-lung axis and their involvement in pulmonary diseases, *Chemico-Biological Interactions*, 368 (2022) 110231.
13. E. Thursby, N. Juge, Introduction to the human gut microbiota, *The Biochemical Journal*, 474 (2017) 1823–1836.
14. B. Das, G.B. Nair, Homeostasis and dysbiosis of the gut microbiome in health and disease, *Journal of Biosciences*, 44 (2019) 1–8.
15. B. Bashir, S. Alam, N. Khandale, D. Birla, S. Vishwas, N.K. Pandey, G. Gupta, K.R. Paudel, H. Dureja, P. Kumar, T.G. Singh, G. Kuppusamy, F.C. Zacconi, T.J.A. Pinto, M. Dhanasekaran, M. Gulati, K. Dua, S.K. Singh, Opening avenues for treatment of neurodegenerative disease using post-biotics: breakthroughs and bottlenecks in clinical translation, *Ageing Research Reviews*, 95 (2024) 102236.
16. T.S.B. Schmidt, J. Raes, P. Bork, The human gut microbiome: from association to modulation, *Cell*, 172 (2018) 1198–1215.
17. J. Lloyd-Price, A. Mahurkar, G. Rahnavard, J. Crabtree, J. Orvis, A.B. Hall, A. Brady, H.H. Creasy, C. McCracken, M.G. Giglio, D. McDonald, E.A. Franzosa, R. Knight, O. White, C. Huttenhower, Strains, functions and dynamics in the expanded human microbiome project, *Nature*, 550 (2017) 61–66.
18. I. Cho, S. Yamanishi, L. Cox, B.A. Methé, J. Zavadil, K. Li, Z. Gao, D. Mahana, K. Raju, I. Teitler, H. Li, A.V. Alekseyenko, M.J. Blaser, Antibiotics in early life alter the murine colonic microbiome and adiposity, *Nature*, 488 (2012) 621–626.
19. I. Vujkovic-Cvijin, J. Sklar, L. Jiang, L. Natarajan, R. Knight, Y. Belkaid, Host variables confound gut microbiota studies of human disease, *Nature*, 587 (2020) 448–454.
20. S. Bhatt, J. Kanoujia, S. Mohana Lakshmi, C.R. Patil, G. Gupta, D.K. Chellappan, K. Dua, Role of brain-gut-microbiota axis in depression: emerging therapeutic avenues, *CNS & Neurological Disorders Drug Targets*, 22 (2023) 276–288.
21. J.M. Rodríguez, K. Murphy, C. Stanton, R.P. Ross, O.I. Kober, N. Juge, E. Avershina, K. Rudi, A. Narbad, M.C. Jenmalm, J.R. Marchesi, M.C. Collado, The composition of the gut microbiota throughout life, with an emphasis on early life, *Microbial Ecology in Health and Disease*, 26 (2015) 26050.

22. M.J. Claesson, S. Cusack, O. O'Sullivan, R. Greene-Diniz, H. de Weerd, E. Flannery, J.R. Marchesi, D. Falush, T. Dinan, G. Fitzgerald, C. Stanton, D. van Sinderen, M. O'Connor, N. Harnedy, K. O'Connor, C. Henry, D. O'Mahony, A.P. Fitzgerald, F. Shanahan, C. Twomey, C. Hill, R.P. Ross, P.W. O'Toole, Composition, variability, and temporal stability of the intestinal microbiota of the elderly, *Proceedings of the National Academy of Sciences of the United States of America*, 108 Suppl 1 (2011) 4586–4591.
23. D.K. Chellappan, Q.L. Sze Ning, S.K. Su Min, S.Y. Bin, P.J. Chern, T.P. Shi, S.W. Ee Mei, T.H. Yee, O.J. Qi, L. Thangavelu, S. Rajeshkumar, P. Negi, J. Chellian, R. Wadhwa, G. Gupta, T. Collet, P.M. Hansbro, K. Dua, Interactions between microbiome and lungs: paving new paths for microbiome based bio-engineered drug delivery systems in chronic respiratory diseases, *Chemico-Biological Interactions*, 310 (2019) 108732.
24. C. Huttenhower, D. Gevers, R. Knight, S. Abubucker, J.H. Badger, A.T. Chinwalla, H.H. Creasy, A.M. Earl, M.G. FitzGerald, R.S. Fulton, M.G. Giglio, Structure, function and diversity of the healthy human microbiome, *Nature*, 486 (2012) 207–214.
25. L. Corrie, A. Awasthi, J. Kaur, S. Vishwas, M. Gulati, I.P. Kaur, G. Gupta, N. Kommineni, K. Dua, S.K. Singh, Interplay of gut microbiota in polycystic ovarian syndrome: role of gut microbiota, mechanistic pathways and potential treatment strategies, *Pharmaceuticals*, 16 (2023) 197.
26. G.P. Donaldson, S.M. Lee, S.K. Mazmanian, Gut biogeography of the bacterial microbiota, *Nature Reviews Microbiology*, 14 (2016) 20–32.
27. J. Ni, G.D. Wu, L. Albenberg, V.T. Tomov, Gut microbiota and IBD: causation or correlation?, *Nature Reviews Gastroenterology & Hepatology*, 14 (2017) 573–584.
28. E.A. Alhinai, G.E. Walton, D.M. Commane, The role of the gut microbiota in colorectal cancer causation, *International Journal of Molecular Sciences*, 20 (2019) 5295.
29. R.X. Ding, W.R. Goh, R.N. Wu, X.Q. Yue, X. Luo, W.W.T. Khine, J.R. Wu, Y.K. Lee, Revisit gut microbiota and its impact on human health and disease, *Journal of Food and Drug Analysis*, 27 (2019) 623–631.
30. L. Corrie, H. Singh, M. Gulati, S. Vishwas, D.K. Chellappan, G. Gupta, A.C. Paiva-Santos, F. Veiga, F. Alotaibi, A. Alam, R.D. Eri, P. Prasher, J. Adams, K.R. Paudel, K. Dua, S.K. Singh, Polysaccharide-fecal microbiota-based colon-targeted self-nanoemulsifying drug delivery system of curcumin for treating polycystic ovarian syndrome, *Naunyn-Schmiedeberg's Archives of Pharmacology*, 397 (2024) 6721–6743.
31. P. Ferretti, E. Pasolli, A. Tett, F. Asnicar, V. Gorfer, S. Fedi, F. Armanini, D.T. Truong, S. Manara, M. Zolfo, F. Beghini, R. Bertorelli, V. De Sanctis, I. Bariletti, R. Canto, R. Clementi, M. Cologna, T. Crifò, G. Cusumano, S. Gottardi, C. Innamorati, C. Masè, D. Postai, D. Savoi, S. Duranti, G.A. Lugli, L. Mancabelli, F. Turroni, C. Ferrario, C. Milani, M. Mangifesta, R. Anzalone, A. Viappiani, M. Yassour, H. Vlamakis, R. Xavier, C.M. Collado, O. Koren, S. Tateo, M. Soffiati, A. Pedrotti, M. Ventura, C. Huttenhower, P. Bork, N. Segata, Mother-to-infant microbial transmission from different body sites shapes the developing infant gut microbiome, *Cell Host & Microbe*, 24 (2018) 133–145.e135.
32. C.A. Lozupone, J.I. Stombaugh, J.I. Gordon, J.K. Jansson, R. Knight, Diversity, stability and resilience of the human gut microbiota, *Nature*, 489 (2012) 220–230.
33. S.B. Meropol, A. Edwards, Development of the infant intestinal microbiome: a bird's eye view of a complex process, *Birth Defects Research. Part C, Embryo Today: Reviews*, 105 (2015) 228–239.
34. K. Dua, G. Gupta, R. Awasthi, D.K. Chellappan, Why is there an emerging need to look for a suitable drug delivery platform in targeting and regulating microbiota?, *Panminerva Medica*, 60 (2018) 136–137.
35. B. Das, T.S. Ghosh, S. Kedia, R. Rampal, S. Saxena, S. Bag, R. Mitra, M. Dayal, O. Mehta, A. Surendranath, S.P.L. Travis, P. Tripathi, G.B. Nair, V. Ahuja, Analysis of the gut microbiome of rural and urban healthy indians living in sea level and high altitude areas, *Scientific Reports*, 8 (2018) 10104.
36. A. Kumar, P. Palit, S. Thomas, G. Gupta, P. Ghosh, R.P. Goswami, T. Kumar Maity, M. Dutta Choudhury, Osteoarthritis: prognosis and emerging therapeutic approach for disease management, *Drug Development Research*, 82 (2021) 49–58.
37. A.L. Byrd, Y. Belkaid, J.A. Segre, The human skin microbiome, *Nature Reviews Microbiology*, 16 (2018) 143–155.
38. R. Mikolajczyk, L.M. Roesner, [General aspects regarding the skin microbiome], Der Hautarzt; Zeitschrift fur Dermatologie, *Venerologie, und verwandte Gebiete*, 70 (2019) 400–406.
39. E.A. Grice, J.A. Segre, The skin microbiome, *Nature Reviews Microbiology*, 9 (2011) 244–253.
40. G. Singhvi, V. Girdhar, S. Patil, G. Gupta, P.M. Hansbro, K. Dua, Microbiome as therapeutics in vesicular delivery, *Biomedicine & Pharmacotherapy = Biomedecine & Pharmacotherapie*, 104 (2018) 738–741.

41. W.H. Man, W.A. de Steenhuijsen Piters, D. Bogaert, The microbiota of the respiratory tract: gatekeeper to respiratory health, *Nature Reviews Microbiology*, 15 (2017) 259–270.
42. W.A. de Steenhuijsen Piters, E.A. Sanders, D. Bogaert, The role of the local microbial ecosystem in respiratory health and disease, *Philosophical Transactions of the Royal Society of London. Series B, Biological Sciences*, 370 (2015) 20140294.
43. K.S. Alharbi, W.H. Almalki, M. Albratty, A.M. Meraya, A. Najmi, G. Vyas, S.K. Singh, K. Dua, G. Gupta, The therapeutic role of nutraceuticals targeting the Nrf2/HO–1 signaling pathway in liver cancer, *Journal of Food Biochemistry*, 46 (2022) e14357.
44. S. Kumar, N. Kumari, D. Talukdar, A. Kothidar, M. Sarkar, O. Mehta, P. Kshetrapal, N. Wadhwa, R. Thiruvengadam, B.K. Desiraju, G.B. Nair, S. Bhatnagar, S. Mukherjee, B. Das, The vaginal microbial signatures of preterm birth delivery in Indian women, *Frontiers in Cellular and Infection Microbiology*, 11 (2021) 622474.
45. A.A. Bhat, G. Gupta, K.S. Alharbi, O. Afzal, A.S.A. Altamimi, W.H. Almalki, I. Kazmi, F.A. Al-Abbasi, S.I. Alzarea, D.K. Chellappan, S.K. Singh, R. MacLoughlin, B.G. Oliver, K. Dua, Polysaccharide-based nanomedicines targeting lung cancer, *Pharmaceutics*, 14 (2022) 2788.
46. J. Ravel, P. Gajer, Z. Abdo, G.M. Schneider, S.S. Koenig, S.L. McCulle, S. Karlebach, R. Gorle, J. Russell, C.O. Tacket, R.M. Brotman, C.C. Davis, K. Ault, L. Peralta, L.J. Forney, Vaginal microbiome of reproductive-age women, *Proceedings of the National Academy of Sciences of the United States of America*, 108 Suppl 1 (2011) 4680–4687.
47. J. Xu, G. Bian, M. Zheng, G. Lu, W.Y. Chan, W. Li, K. Yang, Z.J. Chen, Y. Du, Fertility factors affect the vaginal microbiome in women of reproductive age, *American Journal of Reproductive Immunology*, 83 (2020) e13220.
48. A.A. Bhat, R. Thapa, O. Afzal, N. Agrawal, W.H. Almalki, I. Kazmi, S.I. Alzarea, A.S.A. Altamimi, P. Prasher, S.K. Singh, K. Dua, G. Gupta, The pyroptotic role of Caspase-3/GSDME signalling pathway among various cancer: a review, *International Journal of Biological Macromolecules*, 242 (2023) 124832.
49. G. Gupta, W.H. Al-Malki, I. Kazmi, L. Thangavelu, P.K. Gupta, N.K. Jha, P. Prasher, S.K. Singh, K. Dua, The role of HGF/MET in liver cancer, *Future Medicinal Chemistry*, 13 (2021) 1829–1832.
50. T. Yatsunenko, F.E. Rey, M.J. Manary, I. Trehan, M.G. Dominguez-Bello, M. Contreras, M. Magris, G. Hidalgo, R.N. Baldassano, A.P. Anokhin, A.C. Heath, B. Warner, J. Reeder, J. Kuczynski, J.G. Caporaso, C.A. Lozupone, C. Lauber, J.C. Clemente, D. Knights, R. Knight, J.I. Gordon, Human gut microbiome viewed across age and geography, *Nature*, 486 (2012) 222–227.
51. G. Gupta, D.K. Chellappan, T. de Jesus Andreoli Pinto, P.M. Hansbro, M. Bebawy, K. Dua, Tumor suppressor role of miR-503, *Panminerva Medica*, 60 (2018) 17–24.
52. S. Pareek, T. Kurakawa, B. Das, D. Motooka, S. Nakaya, T. Rongsen-Chandola, N. Goyal, H. Kayama, D. Dodd, R. Okumura, Y. Maeda, K. Fujimoto, T. Nii, T. Ogawa, T. Iida, N. Bhandari, T. Kida, S. Nakamura, G.B. Nair, K. Takeda, Comparison of Japanese and Indian intestinal microbiota shows diet-dependent interaction between bacteria and fungi, *NPJ Biofilms and Microbiomes*, 5 (2019) 37.
53. M.S. Hussain, O. Afzal, G. Gupta, A.S.A. Altamimi, W.H. Almalki, S.I. Alzarea, I. Kazmi, N.K. Fuloria, M. Sekar, D.U. Meenakshi, L. Thangavelu, A. Sharma, Long non-coding RNAs in lung cancer: unraveling the molecular modulators of MAPK signaling, *Pathology, Research and Practice*, 249 (2023) 154738.
54. M.J. Bonder, A. Kurilshikov, E.F. Tigchelaar, Z. Mujagic, F. Imhann, A.V. Vila, P. Deelen, T. Vatanen, M. Schirmer, S.P. Smeekens, D.V. Zhernakova, S.A. Jankipersadsing, M. Jaeger, M. Oosting, M.C. Cenit, A.A. Masclee, M.A. Swertz, Y. Li, V. Kumar, L. Joosten, H. Harmsen, R.K. Weersma, L. Franke, M.H. Hofker, R.J. Xavier, D. Jonkers, M.G. Netea, C. Wijmenga, J. Fu, A. Zhernakova, The effect of host genetics on the gut microbiome, *Nature Genetics*, 48 (2016) 1407–1412.
55. D. Knights, M.S. Silverberg, R.K. Weersma, D. Gevers, G. Dijkstra, H. Huang, A.D. Tyler, S. van Sommeren, F. Imhann, J.M. Stempak, H. Huang, P. Vangay, G.A. Al-Ghalith, C. Russell, J. Sauk, J. Knight, M.J. Daly, C. Huttenhower, R.J. Xavier, Complex host genetics influence the microbiome in inflammatory bowel disease, *Genome Medicine*, 6 (2014) 107.
56. P. Prasher, M. Sharma, S.K. Singh, M. Gulati, D.K. Chellappan, F. Zacconi, G. De Rubis, G. Gupta, J. Sharifi-Rad, W.C. Cho, K. Dua, Luteolin: a flavonoid with a multifaceted anticancer potential, *Cancer Cell International*, 22 (2022) 386.
57. M.G. Dominguez-Bello, E.K. Costello, M. Contreras, M. Magris, G. Hidalgo, N. Fierer, R. Knight, Delivery mode shapes the acquisition and structure of the initial microbiota across multiple body habitats in newborns, *Proceedings of the National Academy of Sciences of the United States of America*, 107 (2010) 11971–11975.

58. S. Rohilla, M. Singh, S.I. Alzarea, W.H. Almalki, F.A. Al-Abbasi, I. Kazmi, O. Afzal, A.S.A. Altamimi, S.K. Singh, D.K. Chellappan, K. Dua, G. Gupta, Recent developments and challenges in molecular-targeted therapy of non-small-cell lung cancer, *Journal of Environmental Pathology, Toxicology and Oncology: Official Organ of the International Society for Environmental Toxicology and Cancer*, 42 (2023) 27–50.
59. F. Sommer, F. Bäckhed, The gut microbiota--masters of host development and physiology, *Nature Reviews Microbiology*, 11 (2013) 227–238.
60. S. Rohilla, M. Singh, S. Priya, W.H. Almalki, S.M. Haniffa, V. Subramaniyan, S. Fuloria, N.K. Fuloria, M. Sekar, S.K. Singh, N.K. Jha, D.K. Chellappan, P. Negi, K. Dua, G. Gupta, Exploring the mechanical perspective of a new anti-tumor agent: melatonin, *Journal of Environmental Pathology, Toxicology and Oncology: Official Organ of the International Society for Environmental Toxicology and Cancer*, 42 (2023) 1–16.
61. M.J. Claesson, I.B. Jeffery, S. Conde, S.E. Power, E.M. O'Connor, S. Cusack, H.M. Harris, M. Coakley, B. Lakshminarayanan, O. O'Sullivan, G.F. Fitzgerald, J. Deane, M. O'Connor, N. Harnedy, K. O'Connor, D. O'Mahony, D. van Sinderen, M. Wallace, L. Brennan, C. Stanton, J.R. Marchesi, A.P. Fitzgerald, F. Shanahan, C. Hill, R.P. Ross, P.W. O'Toole, Gut microbiota composition correlates with diet and health in the elderly, *Nature*, 488 (2012) 178–184.
62. B.D. Muegge, J. Kuczynski, D. Knights, J.C. Clemente, A. González, L. Fontana, B. Henrissat, R. Knight, J.I. Gordon, Diet drives convergence in gut microbiome functions across mammalian phylogeny and within humans, *Science*, 332 (2011) 970–974.
63. R. Thapa, O. Afzal, A.A. Bhat, A. Goyal, A.S. Alfawaz Altamimi, W.H. Almalki, S.I. Alzarea, I. Kazmi, S.K. Singh, K. Dua, L. Thangavelu, G. Gupta, New horizons in lung cancer management through ATR/CHK1 pathway modulation, *Future Medicinal Chemistry*, 15 (2023) 1807–1818.
64. J.H. Hehemann, G. Correc, T. Barbeyron, W. Helbert, M. Czjzek, G. Michel, Transfer of carbohydrate-active enzymes from marine bacteria to Japanese gut microbiota, *Nature*, 464 (2010) 908–912.
65. A.A. Bhat, G. Gupta, A. Goyal, R. Thapa, W.H. Almalki, I. Kazmi, S.I. Alzarea, N. Kukreti, M. Sekar, D.U. Meenakshi, S.K. Singh, R. MacLoughlin, K. Dua, Unwinding circular RNA's role in inflammatory pulmonary diseases, *Naunyn-Schmiedeberg's Archives of Pharmacology*, 397 (2023) 2567–2588.
66. R. Thapa, G. Gupta, A.A. Bhat, W.H. Almalki, S.I. Alzarea, I. Kazmi, S. Saleem, R. Khan, N. Altwaijry, H. Dureja, S.K. Singh, K. Dua, A review of Glycogen Synthase Kinase-3 (GSK3) inhibitors for cancers therapies, *International Journal of Biological Macromolecules*, 253 (2023) 127375.
67. G. De Rubis, K.R. Paudel, L. Corrie, S. Mehndiratta, V.K. Patel, P.S. Kumbhar, A.S. Manjappa, J. Disouza, V. Patravale, G. Gupta, B. Manandhar, R. Rajput, A.K. Robinson, R.J. Reyes, A. Chakraborty, D.K. Chellappan, S.K. Singh, B.G.G. Oliver, P.M. Hansbro, K. Dua, Applications and advancements of nanoparticle-based drug delivery in alleviating lung cancer and chronic obstructive pulmonary disease, *Naunyn-Schmiedeberg's Archives of Pharmacology*, 397 (2023) 2793–2833.
68. A.A. Bhat, R. Thapa, A. Goyal, V. Subramaniyan, D. Kumar, S. Gupta, S.K. Singh, K. Dua, G. Gupta, Curcumin-based nanoformulations as an emerging therapeutic strategy for inflammatory lung diseases, *Future Medicinal Chemistry*, 15 (2023) 583–586.
69. L. Maier, M. Pruteanu, M. Kuhn, G. Zeller, A. Telzerow, E.E. Anderson, A.R. Brochado, K.C. Fernandez, H. Dose, H. Mori, K.R. Patil, P. Bork, A. Typas, Extensive impact of non-antibiotic drugs on human gut bacteria, *Nature*, 555 (2018) 623–628.
70. L. Maier, C.V. Goemans, J. Wirbel, M. Kuhn, C. Eberl, M. Pruteanu, P. Müller, S. Garcia-Santamarina, E. Cacace, B. Zhang, C. Gekeler, T. Banerjee, E.E. Anderson, A. Milanese, U. Löber, S.K. Forslund, K.R. Patil, M. Zimmermann, B. Stecher, G. Zeller, P. Bork, A. Typas, Unravelling the collateral damage of antibiotics on gut bacteria, *Nature*, 599 (2021) 120–124.
71. S. Becattini, Y. Taur, E.G. Pamer, Antibiotic-induced changes in the intestinal microbiota and disease, *Trends in Molecular Medicine*, 22 (2016) 458–478.
72. N. Gulati, D.K. Chellappan, R. MacLoughlin, G. Gupta, S.K. Singh, B.G. Oliver, K. Dua, H. Dureja, Advances in nano-based drug delivery systems for the management of cytokine influx-mediated inflammation in lung diseases, *Naunyn-Schmiedeberg's Archives of Pharmacology*, 397 (2023) 3695–3707.
73. Y. Fan, O. Pedersen, Gut microbiota in human metabolic health and disease, *Nature Reviews Microbiology*, 19 (2021) 55–71.
74. A.A. Abou-Zeid, Y.M. Shehata, R. el-Sherbeny, Microbial production of cobalamin (vitamin B 12), Zentralblatt fur Bakteriologie, Parasitenkunde, Infektionskrankheiten und Hygiene. Zweite naturwissenschaftliche Abt.: allgemeine, landwirtschaftliche und technische *Mikrobiologie*, 128 (1973) 285–296.
75. M. Kachrimanidou, E. Tsintarakis, Insights into the role of human gut microbiota in clostridioides difficile infection, *Microorganisms*, 8 (2020) 200.

76. K.R. Paudel, B. Manandhar, S.K. Singh, G. Gupta, P.M. Hansbro, D.K. Chellappan, K. Dua, Cytotoxic mechanisms of berberine-phytantriol liquid crystalline nanoparticles against non-small-cell lung cancer, *EXCLI Journal*, 22 (2023) 516–519.
77. M.S. Hussain, G. Gupta, M. Afzal, S.M. Alqahtani, V.P. Samuel, W. Hassan Almalki, I. Kazmi, S.I. Alzarea, S. Saleem, H. Dureja, S.K. Singh, K. Dua, L. Thangavelu, Exploring the role of lncrna neat1 knockdown in regulating apoptosis across multiple cancer types: a review, *Pathology, Research and Practice*, 252 (2023) 154908.
78. S.R. Gill, M. Pop, R.T. Deboy, P.B. Eckburg, P.J. Turnbaugh, B.S. Samuel, J.I. Gordon, D.A. Relman, C.M. Fraser-Liggett, K.E. Nelson, Metagenomic analysis of the human distal gut microbiome, *Science*, 312 (2006) 1355–1359.
79. E. Andrès, N.H. Loukili, E. Noel, G. Kaltenbach, M.B. Abdelgheni, A.E. Perrin, M. Noblet-Dick, F. Maloisel, J.L. Schlienger, J.F. Blicklé, Vitamin B12 (cobalamin) deficiency in elderly patients, *CMAJ: Canadian Medical Association Journal = journal de l'Association Medicale Canadienne*, 171 (2004) 251–259.
80. F. Yang, Y. Yang, L. Chen, Z. Zhang, L. Liu, C. Zhang, Q. Mai, Y. Chen, Z. Chen, T. Lin, L. Chen, H. Guo, L. Zhou, H. Shen, X. Chen, L. Liu, G. Zhang, H. Liao, L. Zeng, G. Zeng, The gut microbiota mediates protective immunity against tuberculosis via modulation of lncRNA, *Gut Microbes*, 14 (2022) 2029997.
81. M. Clapp, N. Aurora, L. Herrera, M. Bhatia, E. Wilen, S. Wakefield, Gut microbiota's effect on mental health: the gut-brain axis, *Clinics and Practice*, 7 (2017) 987.
82. J. Bien, V. Palagani, P. Bozko, The intestinal microbiota dysbiosis and Clostridium difficile infection: is there a relationship with inflammatory bowel disease?, *Therapeutic Advances in Gastroenterology*, 6 (2013) 53–68.
83. V.D. Appanna, V.D. Appanna, *Dysbiosis, Probiotics, and Prebiotics: In Diseases and Health*, Springer, 2018.
84. M.S. Hussain, A. Sharma, R. Kumar, Prebiotics and probiotics: a focused review of applications in respiratory disorders, *Carpathian Journal of Food Science and Technology*, 15 (2023) 183–207.
85. M. Kumar, M.S. Hussain, S. Raj, R. Verma, S. Sharma, S.K. Sahu, An overview of treatment modalities and management aspects for obesity, *Current Nutrition and Food Science*, 19 (2023) 105–113.
86. L.J. Wilkins, M. Monga, A.W. Miller, Defining dysbiosis for a cluster of chronic diseases, *Scientific Reports*, 9 (2019) 12918.
87. R.H. Sunenshine, L.C. McDonald, Clostridium difficile-associated disease: new challenges from an established pathogen, *Cleveland Clinic Journal of Medicine*, 73 (2006) 187–197.
88. M.A. Hakami, A. Hazazi, F.R. Khan, O. Abdulaziz, K. Alshaghdali, A. Abalkhail, S.A. Nassar, B.I.A. Omar, F. Almarshadi, G. Gupta, A.S. Binshaya, PVT1 lncRNA in lung cancer: a key player in tumorigenesis and therapeutic opportunities, *Pathology, Research and Practice*, 253 (2024) 155019.
89. A.A. Bhat, O. Afzal, M. Afzal, G. Gupta, R. Thapa, H. Ali, W. Hassan Almalki, I. Kazmi, S.I. Alzarea, S. Saleem, V.P. Samuel, S.K. Gubbiyappa, V. Subramaniyan, MALAT1: a key regulator in lung cancer pathogenesis and therapeutic targeting, *Pathology, Research and Practice*, 253 (2024) 154991.
90. S.M. Pear, T.H. Williamson, K.M. Bettin, D.N. Gerding, J.N. Galgiani, Decrease in nosocomial Clostridium difficile-associated diarrhea by restricting clindamycin use, *Annals of Internal Medicine*, 120 (1994) 272–277.
91. D.A. Antonopoulos, S.M. Huse, H.G. Morrison, T.M. Schmidt, M.L. Sogin, V.B. Young, Reproducible community dynamics of the gastrointestinal microbiota following antibiotic perturbation, *Infection and Immunity*, 77 (2009) 2367–2375.
92. J. Lennard-Jones, Classification of inflammatory bowel disease, *Scandinavian Journal of Gastroenterology*, 24 (1989) 2–6.
93. M.S. Hussain, A.S.A. Altamimi, M. Afzal, W.H. Almalki, I. Kazmi, S.I. Alzarea, S. Saleem, P. Prasher, B. Oliver, S.K. Singh, R. MacLoughlin, K. Dua, G. Gupta, From carcinogenesis to therapeutic avenues: lncRNAs and mTOR crosstalk in lung cancer, *Pathology, Research and Practice*, 253 (2024) 155015.
94. M.S. Hussain, O. Afzal, G. Gupta, A. Goyal, W.H. Almalki, I. Kazmi, S.I. Alzarea, A.S. Alfawaz Altamimi, N. Kukreti, A. Chakraborty, S.K. Singh, K. Dua, Unraveling NEAT1's complex role in lung cancer biology: a comprehensive review, *EXCLI Journal*, 23 (2024) 34–52.
95. K. Machiels, M. Joossens, J. Sabino, V. De Preter, I. Arijs, V. Eeckhaut, V. Ballet, K. Claes, F. Van Immerseel, K. Verbeke, M. Ferrante, J. Verhaegen, P. Rutgeerts, S. Vermeire, A decrease of the butyrate-producing species Roseburia hominis and Faecalibacterium prausnitzii defines dysbiosis in patients with ulcerative colitis, *Gut*, 63 (2014) 1275–1283.

96. J.G. Ruseler-van Embden, H.C. Both-Patoir, Anaerobic gram-negative faecal flora in patients with Crohn's disease and healthy subjects, *Antonie van Leeuwenhoek*, 49 (1983) 125–132.
97. R. Thapa, O. Afzal, M. Afzal, G. Gupta, A.A. Bhat, W. Hassan Almalki, I. Kazmi, S.I. Alzarea, S. Saleem, P. Arora, S.K. Singh, K. Dua, From LncRNA to metastasis: The MALAT1-EMT axis in cancer progression, *Pathology, Research and Practice*, 253 (2024) 154959.
98. M.S. Hussain, G. Gupta, R. Mishra, N. Patel, S. Gupta, S.I. Alzarea, I. Kazmi, P. Kumbhar, J. Disouza, H. Dureja, N. Kukreti, S.K. Singh, K. Dua, Unlocking the secrets: Volatile Organic Compounds (VOCs) and their devastating effects on lung cancer, *Pathology, Research and Practice*, 255 (2024) 155157.
99. M. Joossens, V. De Preter, V. Ballet, K. Verbeke, P. Rutgeerts, S. Vermeire, Effect of oligofructose-enriched inulin (OF-IN) on bacterial composition and disease activity of patients with Crohn's disease: results from a double-blinded randomised controlled trial, *Gut*, 61 (2012) 958.
100. D. Chen, J. Wu, D. Jin, B. Wang, H. Cao, Fecal microbiota transplantation in cancer management: current status and perspectives, *International Journal of Cancer*, 145 (2019) 2021–2031.
101. S.Y. Lam, J. Yu, S.H. Wong, M.P. Peppelenbosch, G.M. Fuhler, The gastrointestinal microbiota and its role in oncogenesis, *Best Practice & Research Clinical Gastroenterology*, 31 (2017) 607–618.
102. M.R. Islam, A. Rauf, S. Akash, A. Kumer, M.S. Hussain, S. Akter, J.K. Gupta, L.T. Ansari, M.M.M.I. Raj, T.B. Emran, Recent perspective on the potential role of phytocompounds in the prevention of gastric cancer, *Process Biochemistry*, 135 (2023) 83–101.
103. K.L. Donnelly, C.I. Smith, S.J. Schwarzenberg, J. Jessurun, M.D. Boldt, E.J. Parks, Sources of fatty acids stored in liver and secreted via lipoproteins in patients with nonalcoholic fatty liver disease, *The Journal of Clinical Investigation*, 115 (2005) 1343–1351.
104. P.D. Cani, J. Amar, M.A. Iglesias, M. Poggi, C. Knauf, D. Bastelica, A.M. Neyrinck, F. Fava, K.M. Tuohy, C. Chabo, A. Waget, E. Delmée, B. Cousin, T. Sulpice, B. Chamontin, J. Ferrières, J.F. Tanti, G.R. Gibson, L. Casteilla, N.M. Delzenne, M.C. Alessi, R. Burcelin, Metabolic endotoxemia initiates obesity and insulin resistance, *Diabetes*, 56 (2007) 1761–1772.
105. M.S. Hussain, G. Sharma, The burden of cardiovascular diseases due to COVID-19 pandemic, *The Thoracic and Cardiovascular Surgeon*, 72 (2024) 40–50.
106. H.Y. Lee, J.H. Park, S.H. Seok, M.W. Baek, D.J. Kim, K.E. Lee, K.S. Paek, Y. Lee, J.H. Park, Human originated bacteria, Lactobacillus rhamnosus PL60, produce conjugated linoleic acid and show anti-obesity effects in diet-induced obese mice, *Biochimica et Biophysica Acta*, 1761 (2006) 736–744.
107. S.G. Nath, R. Raveendran, Microbial dysbiosis in periodontitis, *Journal of Indian Society of Periodontology*, 17 (2013) 543.
108. M.A. Curtis, P.I. Diaz, T.E. Van Dyke, The role of the microbiota in periodontal disease, *Periodontology 2000*, 83 (2020) 14–25.
109. A.S. Plachokova, S. Andreu-Sánchez, M.P. Noz, J. Fu, N.P. Riksen, Oral microbiome in relation to periodontitis severity and systemic inflammation, *International Journal of Molecular Sciences*, 22 (2021) 5876.
110. R.J. Turesky, K. Lu, Biomarkers of environmental toxicants: exposure and biological effects, *Toxics*, 8 (2020) 37.
111. K.A. Thayer, J.J. Heindel, J.R. Bucher, M.A. Gallo, Role of environmental chemicals in diabetes and obesity: a national toxicology program workshop review, *Environmental Health Perspectives*, 120 (2012) 779–789.
112. K. Lu, R.P. Abo, K.A. Schlieper, M.E. Graffam, S. Levine, J.S. Wishnok, J.A. Swenberg, S.R. Tannenbaum, J.G. Fox, Arsenic exposure perturbs the gut microbiome and its metabolic profile in mice: an integrated metagenomics and metabolomics analysis, *Environmental Health Perspectives*, 122 (2014) 284–291.
113. P. Tu, L. Chi, W. Bodnar, Z. Zhang, B. Gao, X. Bian, J. Stewart, R. Fry, K. Lu, Gut Microbiome toxicity: connecting the environment and gut microbiome-associated diseases, *Toxics*, 8 (2020) 19.

3 The Gut Reaction
Modulating Microbial Response to Toxicants through Diet and Lifestyle

Vetriselvan Subramaniyan[1], Gaurav Gupta[2,3], Kamal Dua[4,5], Sachin Singh[5,6], Gowthamarajan K[7], and V V S Narayana Reddy[8]*

[1]Department of Medical Sciences School of Medical and Life Sciences, Sunway University, Selangor, Malaysia
[2]Centre for Research Impact & Outcome, Chitkara College of Pharmacy, Chitkara University, Punjab, India
[3]Centre of Medical and Bio-allied Health Sciences Research, Ajman University, Ajman, United Arab Emirates
[4]Discipline of Pharmacy, Graduate School of Health, University of Technology Sydney, NSW, Australia
[5]Faculty of Health, Australian Research Centre in Complementary and Integrative Medicine, University of Technology Sydney, Ultimo, Australia
[6]School of Pharmaceutical Sciences, Lovely Professional University, Phagwara, Punjab, India
[7]Department of Pharmaceutics, JSS College of Pharmacy, JSS Academy of Higher Education and Research, Ooty, Nilgiris, Tamil Nadu, India
[8]Department of Pharmaceutics, JSS College of Pharmacy, Ootacamund, JSS University, Mysore, India
*Corresponding Author: Vetriselvan Subramaniyan
Email id: vetris@sunway.edu.my

3.1 INTRODUCTION

In recent years, the exploration of the human microbiome and its profound impact on overall health has sparked a growing fascination within the scientific community. At the heart of this inquiry lies the gut, a vibrant hub teeming with diverse microbial communities that exert significant influence over a myriad of physiological processes. This burgeoning interest extends beyond understanding the mere existence of these microbial communities to unraveling their responses to environmental toxicants, particularly how lifestyle choices – primarily diet – can intricately shape this complex interplay. “The Gut Reaction: Modulating Microbial Response to Toxicants through Diet and Lifestyle” is designed to be a comprehensive exploration delving into the captivating links between the gut microbiome, exposure to toxicants, and the potential for dietary and lifestyle modifications to shape microbial responses [1, 2].

Our journey takes us deep into the intricate terrain of human health, where an increasing focus is directed toward the microscopic realm within – the human microbiome. The gut, a bustling epicenter hosting a multitude of diverse microbial communities, emerges as a crucial orchestrator of physiological processes that transcend the traditional scope of digestion. Recent years have witnessed a

DOI: 10.1201/9781003489221-3

surge in scientific curiosity surrounding the nuanced relationship between these microbial inhabitants and our overall well-being. The hidden ecosystem within the gut, brimming with microscopic life, holds the potential to profoundly shape our health [3]. Venturing beyond the confines of our bodies, this exploration takes us into the realm of environmental toxicants, probing their interaction with the gut microbiome. Specifically, we aim to unravel the intricate crossroads where lifestyle choices, particularly dietary habits, exert influence over the complex dance between microbial communities and the toxicants that permeate our surroundings [4].

"The Gut Reaction: Modulating Microbial Response to Toxicants through Diet and Lifestyle" strives to untangle the captivating connections between the gut microbiome, its responses to environmental toxicants, and the potential impact of dietary and lifestyle interventions on these microbial reactions. This journey leads us into the heart of emerging fields, where toxicology and microbiome research converge, shaping new paradigms for understanding the delicate balance between our internal microbial ecosystems and the challenges presented by the external environment [5]. As we embark on this scientific odyssey, our aim extends beyond merely illuminating these intricate relationships; we also seek to explore the practical implications of these findings for individuals looking to optimize their health. The choices made in daily life, particularly those related to diets and lifestyles, play a pivotal role in modulating the delicate interplay between the gut microbiome and environmental toxicants [6]. Through the microscopic landscapes of the gut, we can unravel the communication pathways between microbial communities and discern how environmental toxicants disrupt or harmonize this delicate balance. This exploration encompasses the potential for dietary choices and lifestyle modifications to serve as strategic tools in steering the microbial response to toxicants. Beyond being a scientific journey, it is an exploration into the practical applications of this knowledge, with the ultimate goal of empowering individuals to cultivate a resilient and thriving internal ecosystem [7]. Join us as we venture into the depths of the gut, where the intricate dance between microbial communities and environmental challenges unfolds, and where the choices made reverberate through the delicate tapestry of our health and well-being [8].

3.2 NATURAL DRUG MOLECULES AND MICROBIAL MODULATION

As an example of the practical applications discussed, consider the case of curcumin, a bioactive compound found in turmeric. Studies have indicated that curcumin possesses anti-inflammatory and antioxidant properties that can potentially influence the gut microbiome [9, 10].

Research has suggested that curcumin may modulate the composition and activity of gut microbes, promoting the growth of beneficial bacteria while inhibiting the proliferation of harmful ones. This microbial modulation is significant as it contributes to a balanced and diverse gut microbiome, which is essential for overall health [3]. In the context of the exploration presented in this chapter, this example demonstrates how dietary choices, in this case the incorporation of curcumin, can potentially influence the microbial response to toxicants. It underscores the practical application of knowledge about the gut microbiome in making informed dietary decisions for the promotion of a resilient and thriving internal ecosystem [11, 12]. One practical example of this application involves substances like karanjin, found in *Pongamia pinnata* seeds. Karanjin has demonstrated antimicrobial properties and could potentially influence the microbial response within the gut. If incorporated into the diet, it could represent a strategic tool in modulating microbial responses to toxicants, illustrating the practical applications discussed in this chapter. Join the exploration into the depths of the gut, where the intricate dance between microbial communities and environmental challenges unfolds and where the choices made reverberate through the delicate tapestry of health and well-being [13].

3.2.1 The Microscopic Universe Within

Understanding this microscopic universe forms the foundation for comprehending the interplay between the gut microbiome and environmental toxicants. This chapter embarks on a journey to demystify the intricate world of the human microbiome, portraying it as a microscopic universe thriving within our bodies. This foundational exploration sets the stage for a comprehensive understanding, where the dynamic interplay between the gut microbiome and environmental toxicants will be unravelled [14, 15].

3.2.2 Unveiling the Microbial Communities

The chapter begins by delving into the composition and diversity of microbial communities residing in the gut. It emphasizes the significance of this vibrant hub, highlighting its role as a critical orchestrator of various physiological processes that extend far beyond the traditional understanding of digestion. The microbial communities within this microscopic universe play pivotal roles in nutrient metabolism, immune system modulation, and even neurochemical signalling [16].

3.2.2.1 Curcumin as an Example

To illustrate the importance of this microbial universe, we may introduce examples such as curcumin, a bioactive compound found in turmeric. Studies suggest that curcumin interacts with the gut microbiome, influencing the composition and activity of microbial communities. This interaction not only showcases the dynamic nature of the microbial universe but also hints at the potential implications for overall health. Curcumin, with its anti-inflammatory and antioxidant properties, serves as an example of how dietary components can impact the delicate balance within the gut [9, 17].

3.2.2.2 Karanjin as an Example

Expanding the exploration, we introduce karanjin, a compound found in Pongamia pinnata seeds. Karanjin has demonstrated antimicrobial properties, indicating its potential to influence the gut microbiome. This example further underscores the diversity of compounds that can interact with the microbial universe, shaping its intricate dynamics. Through examples like karanjin, we aim to emphasize that the microbial universe is not only influenced by diet but also by a range of compounds present in our environment [18].

3.2.3 Roles in Maintaining Physiological Harmony

The study delves into the roles played by these microbial communities in maintaining physiological harmony. It explores how a balanced and diverse gut microbiome contributes to overall well-being, influencing everything from nutrient absorption to immune responses. The study aims to convey that the microscopic universe within us is not just a passive observer but also an active participant in the intricate symphony of physiological processes. Ultimately, understanding this microscopic universe serves as the foundation for the exploration of the gut microbiome's interaction with environmental toxicants [19].

3.2.3.1 Toxicants and Microbial Dialogue

This intricate interaction is explored, shedding light on how toxicants influence microbial responses within the gut. The study aims to delineate the complex web of communication between external toxicants and the internal microbial communities, elucidating the potential consequences for human health. This chapter guides readers through the crossroads where environmental toxicants and the gut microbiome engage in a dynamic dialogue. This exploration unveils the intricate interaction between external toxicants and the internal microbial communities within the gut. It illuminates the complex web of communication that occurs between these two entities, offering insights into the potential consequences for human health [20].

3.2.3.2 Understanding the Toxicant-Microbiome Nexus

The study establishes the importance of understanding the nexus between environmental toxicants and the gut microbiome. It explores how various toxicants from the external environment find their way into the gut and interact with the resident microbial communities. This interaction forms a dynamic dialogue, shaping the responses of the microbial universe within [21].

3.2.3.3 Impact on Microbial Responses

A significant focus is placed on elucidating how toxicants influence microbial responses within the gut. Examples such as curcumin and karanjin may be revisited to demonstrate the diverse ways in which external compounds can modulate the microbial landscape. For instance, curcumin's anti-inflammatory properties may counteract the effects of certain toxicants, highlighting the potential protective role of dietary components. On the other hand, karanjin's antimicrobial properties may influence the composition of the gut microbiome, showcasing the multifaceted nature of the toxicant-microbiome dialogue [22].

3.2.3.4 Dynamic Web of Communication

The chapter dives deep into the intricate web of communication that occurs between toxicants and the internal microbial communities. It explores how toxicants can directly or indirectly affect the abundance and diversity of microbial species. Furthermore, the dialogue between toxicants and the gut microbiome is not unidirectional; microbial communities can also metabolize or modify toxicants, influencing their toxicity or bioavailability [23].

3.2.3.5 Consequences for Human Health

As the toxicant-microbiome dialogue unfolds, it elucidates the potential consequences for human health. Disruptions in the balance of the gut microbiome induced by toxicants may contribute to various health issues, including inflammation, metabolic disorders, and compromised immune function. We connect the intricate interplay between toxicants and the gut microbiome with broader health outcomes, emphasizing the far-reaching implications of this dialogue [24]. Real-world examples and case studies may be incorporated to illustrate the toxicant-microbiome dialogue. Instances where environmental exposures have led to shifts in the gut microbiome, subsequently impacting health, can serve as poignant examples. This could include studies linking certain pollutants or heavy metals to alterations in gut microbial composition and associated health consequences [25, 26].

3.2.3.6 Setting the Stage for Intervention

The focus shifts to setting the stage for potential interventions. The understanding gained about the toxicant-microbiome dialogue becomes a crucial foundation for considering strategies to mitigate adverse effects. The chapter aims to instill a sense of agency, highlighting the potential for dietary and lifestyle interventions to modulate this dynamic dialogue for the benefit of human health. Through this exploration, readers are equipped with the knowledge to appreciate the ongoing dialogue between environmental toxicants and the gut microbiome, and its profound implications for our well-being [3].

3.3 DIETARY INFLUENCES ON MICROBIAL RESILIENCE

Moving forward, the chapter delves into the profound impact of dietary choices on the resilience of the gut microbiome. We explore how specific nutrients and dietary patterns nourish and fortify microbial communities, contributing to their adaptability and robustness. This chapter aims to emphasize the symbiotic relationship between our dietary habits and the health of the gut microbiome, offering insights into strategies for enhancing microbial resilience. This exploration delves into the intricate relationship between dietary choices and the resilience of the gut microbiome. Understanding how specific nutrients and dietary patterns contribute to the adaptability and robustness of microbial communities is crucial for maintaining optimal gut health. The highlights the symbiotic nature of this relationship, providing valuable insights into strategies for enhancing microbial resilience [27].

3.3.1 Nutrient Contributions

Certain nutrients play a pivotal role in fortifying microbial communities. Take curcumin, for instance, a bioactive compound found in turmeric. Research suggests that curcumin possesses anti-inflammatory and antioxidant properties, contributing to a balanced and resilient gut environment. Another example is karanjin, a compound from *Pongamia pinnata* seeds with antimicrobial effects that can influence the diversity and resilience of microbial species [14, 28].

3.3.2 Dietary Patterns

Beyond individual nutrients, the overall composition of one's diet significantly influences microbial resilience. A diet rich in fiber, commonly found in fruits, vegetables, and whole grains, acts as a prebiotic, promoting the growth of beneficial bacteria. On the other hand, diets high in processed foods and low in fiber may compromise microbial diversity and resilience [29].

3.3.3 Case Studies and Examples

Real-world examples further illustrate the impact of dietary choices on microbial resilience. Consider individuals adopting a Mediterranean diet, characterized by high fiber, plant-based foods and healthy fats. Studies show that such dietary patterns contribute to a more diverse and resilient gut microbiome. These case studies offer tangible evidence of how intentional dietary choices can positively shape microbial health [30].

3.3.4 Strategies for Enhancement

Practical strategies are explored to empower individuals to enhance microbial resilience through their dietary habits. Incorporating fermented foods, such as yogurt or kimchi, introduces beneficial probiotics. Increasing fiber intake supports the growth of microbes that thrive on dietary fibers. Choosing a variety of nutrient-rich foods ensures a diverse microbial ecosystem, fostering adaptability and resilience [31].

3.3.5 Balancing Act

The chapter navigates the delicate balance required in dietary choices for optimal microbial health. Extremes, such as excessively restrictive diets or overconsumption of certain nutrients, can disrupt the equilibrium. The emphasis is on promoting a balanced and varied diet that sustains a flourishing gut microbiome. Dietary influences emerge as a critical determinant in shaping microbial resilience, setting the stage for subsequent discussions on lifestyle choices. This interconnected narrative underscores the holistic approach needed to cultivate a thriving internal microbial ecosystem. In essence, the exploration into dietary influences on microbial resilience offers a nuanced perspective on the role of nutrition in maintaining a healthy gut microbiome. Through the examples of curcumin and karanjin and practical strategies, readers gain valuable insights into the symbiotic relationship between dietary habits and the resilience of their internal microbial communities [32].

3.3.6 Lifestyle Choices and Microbiome Harmony

Our perspective encompasses lifestyle choices beyond diet, recognizing their influence on achieving harmony within the microbiome. Factors such as physical activity, stress management, and sleep patterns are explored for their roles in shaping the delicate equilibrium of the gut microbiome. This chapter underscores the holistic nature of lifestyle choices and their collective impact on maintaining microbial balance [33]. The exploration beyond dietary considerations acknowledges the profound influence of lifestyle choices on achieving harmony within the microbiome. By delving into factors such as physical activity, stress management, and sleep patterns, the chapter aims to underscore the holistic nature of lifestyle choices and their collective impact on maintaining the delicate equilibrium of the gut microbiome [34].

3.3.6.1 Physical Activity

Physical activity emerges as a significant contributor to microbiome harmony. Regular exercise has been associated with increased microbial diversity and a more balanced composition of gut bacteria. For example, engaging in aerobic exercises or strength training can positively influence the abundance of beneficial microbes. This highlights the interconnectedness between an active lifestyle and fostering a resilient gut microbiome [35].

3.3.6.2 Stress Management

The impact of stress on the gut microbiome is explored, emphasizing the importance of effective stress management for microbial harmony. Chronic stress can lead to dysbiosis, an imbalance in gut bacteria. Strategies such as mindfulness, meditation, or yoga are examples of stress management techniques that have been linked to a more favorable microbial profile, promoting overall gut health [36].

3.3.6.3 Sleep Patterns

Quality sleep is recognized as a crucial factor in maintaining microbiome balance. Disruptions to sleep patterns can influence the composition of gut bacteria. For instance, irregular sleep schedules or insufficient sleep may contribute to microbial imbalances. Establishing consistent and adequate sleep patterns becomes integral in supporting the resilience and equilibrium of the gut microbiome [37].

3.3.6.4 Holistic Nature of Lifestyle Choices

The chapter weaves together the various lifestyle choices, emphasizing their collective impact on microbiome harmony. It illuminates how these factors do not operate in isolation but synergistically contribute to the overall well-being of the gut microbiome. For instance, a person with a balanced exercise routine, effective stress management practices, and healthy sleep patterns is more likely to maintain a harmonious gut microbiome [38].

3.3.6.5 Practical Implications

Practical insights are provided to empower individuals in making lifestyle choices conducive to microbiome harmony. Incorporating regular physical activity, adopting stress reduction techniques, and prioritizing quality sleep become actionable steps. Examples of individuals who have successfully integrated these lifestyle choices to positively impact their gut health may serve as inspiration [39].

3.3.6.6 Long-Term Impact

The study explores the long-term impact of lifestyle choices on microbiome harmony and, consequently, overall health. Scientific evidence suggests that sustaining a balanced lifestyle positively influences the stability and diversity of the gut microbiome over time. This underscores the notion that lifestyle choices are not transient interventions but enduring contributors to microbial equilibrium [3].

3.3.6.7 Interconnectedness

This chapter synthesizes the understanding of the gut's microbial universe, its dialogue with environmental toxicants, and the dietary influences explored. Lifestyle choices emerge as a dynamic force that can either enhance or disrupt the delicate balance within the gut microbiome. Through practical examples and the interconnected nature of various lifestyle factors, readers gain valuable insights into the collective influence of these choices on the delicate equilibrium of the gut microbiome [1].

3.4 BALANCING ACT – STRATEGIES FOR MODULATING MICROBIAL RESPONSE

The study explores various approaches for maintaining a delicate balance within the gut microbiome, especially in the face of environmental toxicants. It outlines strategies, encompassing dietary modifications and lifestyle adjustments, to strategically modulate microbial responses to toxicants. This study aims to provide practical insights for individuals seeking to navigate the challenges posed by environmental exposures [40].

3.4.1 Strategic Dietary Modifications

Dietary interventions as key strategies for modulating microbial responses. Examples such as curcumin, found in turmeric, may be highlighted for its anti-inflammatory properties, potentially mitigating the impact of certain toxicants on the gut microbiome. Dietary choices rich in antioxidants and phytochemicals become crucial components of a strategy aimed at promoting microbial resilience and counteracting toxicant-induced disruptions. Beyond diet, the exploration extends to lifestyle adjustments known to support a healthy gut microbiome. Regular exercise is emphasized, as it has been linked to enhanced microbial diversity and a more robust microbiome. Stress management techniques, such as meditation or deep breathing exercises, are presented as integral components of a holistic approach to modulating microbial responses, especially in the face of environmental challenges [41].

3.4.2 Probiotics and Prebiotics

Strategic use of probiotics and prebiotics. Probiotics, found in fermented foods like yogurt, introduce beneficial bacteria to the gut, aiding in the restoration of microbial balance. Prebiotics, often present in dietary fibers, serve as nourishment for these beneficial microbes, fostering their growth and promoting a resilient microbiome [42]. Recognizing the individualized nature of microbial responses, the exploration advocates for personalized approaches. Genetic factors, lifestyle variations, and the unique composition of one's gut microbiome necessitate tailored strategies. Examples of how individuals have successfully implemented personalized approaches to navigate toxicant exposures may serve as inspirations for readers seeking customized solutions [43].

3.4.3 Mitigating Environmental Exposures

The study offers insights into reducing environmental exposures as a proactive strategy. Recommendations may include minimizing exposure to pollutants, incorporating organic and locally sourced foods, and adopting eco-friendly practices. This approach aligns with the notion that modulating microbial responses involves not only internal strategies but also external considerations in creating a microbiome-friendly environment [22].

3.4.4 Long-Term Resilience Building

Emphasizing long-term resilience, the exploration discusses sustained strategies rather than short-term interventions. Consistent dietary choices, ongoing lifestyle adjustments, and a proactive approach to mitigating environmental exposures are positioned as elements contributing to the enduring health of the gut microbiome. This approach aims to empower individuals with tools for building sustained microbial resilience over time. Through examples, personalized approaches, and a focus on long-term resilience, readers are equipped with actionable insights for navigating the intricate interplay between their microbiomes and environmental toxicants [44].

3.5 FUTURE HORIZONS – IMPLICATIONS FOR PERSONALIZED NUTRITION AND ENVIRONMENTAL HEALTH

As we look ahead, we explore the future horizons in the intersection of microbiome research and toxicology. The implications of these findings for personalized nutrition and environmental health are contemplated, envisioning a future where individualized approaches are tailored based on one's unique microbial profile. This study stimulates thought on the potential advancements and applications of microbiome science in shaping individualized health strategies [45].

In contemplating the future of microbiome research and toxicology, the intersection of these fields holds promise with significant implications for personalized nutrition and environmental health. The envisioned future foresees individualized health approaches tailored to the unique microbial profile of each person. This forward-looking exploration stimulates thinking about the potential advancements and applications of microbiome science in shaping personalized health strategies [46].

3.5.1 Personalized Nutrition

In this future landscape, personalized nutrition emerges as a transformative concept. Instead of generic dietary recommendations, individuals could receive tailored advice based on their specific microbial composition. For instance, if someone's microbiome thrives with certain nutrients, personalized nutrition might optimize their intake for improved well-being. The use of examples like curcumin, known for its anti-inflammatory properties, may illustrate how knowledge of one's microbiome can guide dietary choices for enhanced health [47].

3.5.2 Microbiome as a Health Indicator

The future sees the microbiome evolving into a dynamic health indicator. Changes in microbial composition become a valuable tool to assess an individual's health status, vulnerabilities, and areas for improvement. Monitoring shifts in the microbiome could serve as an early warning system, prompting personalized interventions and health strategies. This approach may draw parallels with current practices of using biomarkers for health assessment [48].

3.5.3 Precision Medicine in Toxicology

Applying precision medicine principles to toxicology becomes a pivotal aspect of this future scenario. Understanding how an individual's unique microbiome responds to toxicants allows for targeted interventions. For instance, certain individuals might exhibit resilience to specific environmental toxins due to their microbiome, guiding tailored regulatory measures and interventions. This could be illustrated through examples like karanjin, showcasing how personalized toxicology considerations may evolve [49].

3.5.4 Environmental Health Interventions

The envisioned future highlights the interconnectedness of environmental health and the microbiome. Interventions may extend beyond personal habits to broader practices informed by microbiome knowledge. Urban planning, agriculture, and industry regulations could be shaped with consideration for their impact on human microbial ecosystems. This holistic approach aligns environmental practices with microbiome-friendly initiatives [50].

3.5.5 Technological Advancements

Technological advancements play a crucial role in realizing this future vision. Advanced sequencing technologies, artificial intelligence, and big data analytics become essential tools. They enable a deeper understanding of the complex relationships between the microbiome, toxicants, and personalized health. Examples of how technology facilitates personalized health insights may include breakthroughs in microbiome sequencing techniques [51].

3.5.6 Ethical Considerations

As this future unfolds, ethical considerations come to the forefront. Privacy concerns, consent for microbiome data usage, and the potential for genetic determinism are contemplated. Striking a balance between leveraging microbiome data for personalized health benefits while respecting individual autonomy and privacy becomes a critical aspect of ethical considerations [52].

3.5.7 Integration with Previous Insights

Building upon foundational knowledge, this exploration of future horizons integrates microbiome research, toxicology, and personalized health into a cohesive narrative. It positions the future as a dynamic space where microbiome science transforms how we approach nutrition, toxicology, and environmental health, bridging the current understanding with visionary possibilities. In essence, the envisioned future holds exciting prospects where microbiome science revolutionizes personalized health strategies. By tailoring nutrition, toxicology, and environmental considerations to individual microbiomes, this future scenario illustrates a paradigm shift toward precision and holistic approaches to well-being [53].

3.6 CASE STUDIES: SUCCESS STORIES IN HARNESSING DIET

This segment highlights a compilation of real-world case studies demonstrating the effectiveness of dietary interventions in addressing and mitigating the consequences of exposure to toxicants. Through these instances, the narrative unfolds, showcasing the pivotal role of tailored dietary strategies not only in supporting the gut microbiome but also in mitigating the harmful effects of environmental toxicants. The examples emphasize the instrumental nature of specific dietary approaches in fostering resilience and minimizing adverse impacts on the delicate balance within the gut microbiome [54].

The exploration unfolds as a collection of real-world case studies, serving as success stories in leveraging dietary interventions to mitigate the impact of toxicant exposure. Through these examples, we delve into instances where specific dietary strategies have proven effective in supporting the gut microbiome and minimizing the adverse effects of environmental toxicants [1].

3.6.1 Curcumin: A Dietary Triumph

One notable case study revolves around the use of curcumin, a compound found in turmeric known for its anti-inflammatory and antioxidant properties. Studies have demonstrated that curcumin possesses the potential to counteract the harmful effects of certain toxicants. For instance, it may mitigate oxidative stress caused by environmental pollutants, showcasing the capacity of dietary components to act as protective agents within the intricate dance of the gut microbiome and toxicants [55].

3.6.2 Karanjin: Nature's Defense

Another compelling case study involves karanjin, a natural compound derived from the seeds of the karanja tree. Research suggests that karanjin exhibits antimicrobial and antioxidant properties, potentially aiding in fortifying the gut microbiome against toxicant-induced disruptions. By exploring the utilization of karanjin in dietary strategies, we uncover nature's potential role in providing tools to navigate the challenges posed by environmental toxicants [56].

3.6.3 Probiotics and Prebiotics: Balancing Act

The case studies extend to the use of probiotics and prebiotics, showcasing how these dietary components contribute to the balance within the gut microbiome. Probiotics, such as specific strains of beneficial bacteria, have been employed to enhance microbial resilience and counteract the detrimental effects of certain toxicants. Prebiotics, on the other hand, serve as nourishment for beneficial microbes, fostering a harmonious environment within the gut. The exploration includes instances where plant-based diets have emerged as success stories. Rich in fiber and phytonutrients, plant-based diets contribute to microbial diversity and overall gut health. By examining cases where individuals have adopted plant-centric dietary patterns, we uncover how such choices have been instrumental in supporting the gut microbiome and mitigating the impact of toxicants [57].

3.6.4 Omega-3 Fatty Acids: Guardians of Gut Health

Omega-3 fatty acids present another intriguing case study. Found in fatty fish, flaxseeds, and walnuts, these essential fatty acids have demonstrated anti-inflammatory properties and may contribute to a resilient gut microbiome. Success stories involving the incorporation of omega-3-rich foods provide insights into the potential of dietary choices in modulating microbial responses to toxicants. Several case studies highlight the efficacy of combining multiple dietary interventions for a comprehensive approach. For example, integrating curcumin supplementation with a plant-based diet or probiotic support showcases the potential synergies in harnessing various dietary strategies to enhance the gut microbiome's resilience against toxicant exposure [58].

3.6.5 Realizing Dietary Potential

The case studies underscore the tangible successes achieved through strategic dietary interventions. Whether through the targeted use of specific compounds like curcumin and karanjin or the adoption of broader dietary patterns, these real-world examples illuminate the potential for harnessing the power of diet to navigate and mitigate the impact of environmental toxicants on the gut microbiome. Despite significant progress, there remain gaps in comprehending the full spectrum of interactions and their implications. The diverse and intricate responses of the gut microbiome to various toxicants pose a significant challenge. Understanding the nuanced ways in which microbial communities react to different environmental stressors is an ongoing area of exploration. Recognizing the individual variability in microbial compositions and responses adds a layer of complexity. Tailoring interventions to account for this diversity remains a challenge, requiring a more personalized approach. Continued advancements in technology, such as high-throughput sequencing and metabolomics, present opportunities to unravel previously inaccessible details of microbial responses. This can lead to a more comprehensive understanding of the dynamics at play. The integration of multi-omics approaches, combining data from genomics, metagenomics, and metabolomics, holds promise in deciphering the intricate web of interactions within the gut. This holistic approach could uncover novel insights. The growing field of precision medicine provides opportunities to tailor interventions based on an individual's unique microbiome profile. Personalized approaches may enhance the efficacy of strategies aimed at modulating microbial responses to toxicants [59].

Curcumin and Precision Medicine: The utilization of curcumin, with its anti-inflammatory properties, could be explored within the context of precision medicine. Tailoring curcumin interventions based on an individual's specific microbial profile may enhance its effectiveness in mitigating toxicant-induced stress. Karanjin and Multi-Omics Approaches: Investigating the effects of karanjin, a natural compound, using multi-omics approaches could reveal comprehensive insights into how it influences the gut microbiome. Integrating genomics, metagenomics, and metabolomics data may provide a holistic understanding, encapsulating the ongoing challenges in understanding the microbiome-toxicant interplay while presenting exciting opportunities for future exploration. It serves as a call to action, inspiring further research and technological advancements to unravel the complexities within our gut and optimize interventions for individual health. The journey through this exploration not only enlightens the scientific community but also empowers individuals to make informed choices for a resilient and thriving internal ecosystem [60].

3.7 UNRAVELING THE GUT MICROBIOME: A MICROSCOPIC WORLD WITHIN

This section of the exploration dives into the intricate world of the gut microbiome, portraying it as a microscopic universe within the human body. The emphasis is on unraveling the complexities of microbial communities residing in the gut and understanding their diverse roles. From exploring the composition of these microbial populations to deciphering their influence on various physiological processes, this segment offers a comprehensive journey into the hidden intricacies of the gut microbiome. This segment serves as an illuminating gateway into the microscopic universe thriving within the human body – the gut microbiome. Let us delve into the intricacies of this exploration, unveiling the mysteries of microbial communities and their profound impact on our physiological processes. The exploration begins by scrutinizing the diverse composition of microbial communities inhabiting the gut. In this intricate world, various species, such as bacteria, fungi, and viruses, coexist in a delicate balance. For instance, the spice-derived compound curcumin has drawn attention for its potential to influence microbial composition. Studies suggest curcumin's antimicrobial properties may modulate the abundance of specific bacterial strains, contributing to a balanced microbial ecosystem [61].

3.7.1 Roles in Physiological Processes

Moving beyond mere existence, this segment aims to unravel the dynamic roles played by these microbial communities. The gut microbiome, often considered as a symbiotic partner, significantly impacts physiological processes extending beyond digestion. For instance, karanjin, a natural compound found in the *Pongamia pinnata* tree, has exhibited anti-inflammatory properties. Understanding how karanjin interacts with the gut microbiome provides insights into its potential to modulate inflammatory responses, showcasing the intricate relationship between external compounds and internal microbial communities. The exploration accentuates the impact of microbial communities on overall health. Examples like curcumin and karanjin showcase how external compounds can modulate the gut microbiome, influencing factors like inflammation. Furthermore, understanding the gut microbiome's role in nutrient absorption and immune system regulation highlights its significance for maintaining holistic well-being. This segment delves into the complex web of interactions within the gut microbiome. It explores how microbial communities communicate and cooperate, contributing to the balance essential for health. For instance, curcumin's interactions with specific bacteria may influence microbial communication, potentially affecting the overall ecosystem. "A Microscopic World Within" is a captivating exploration into the hidden intricacies of the gut microbiome. By examining the composition, roles, and influence of microbial communities, this segment lays the foundation for understanding how external factors, such as curcumin and karanjin, can modulate this microscopic universe. As we embark on this journey, the goal is not only to uncover the scientific nuances but also to grasp the practical implications for individuals aiming to cultivate a resilient and thriving internal ecosystem [62].

3.7.2 Toxicants and Microbial Dialogue: The Crossroads in the Gut

This section delves into the dynamic interaction between environmental toxicants and the microbial communities within the gut. It serves as a focal point where toxicants and microbes engage in a dialogue, highlighting the pivotal crossroads within the gut. The exploration aims to unravel how these dialogues influence the delicate balance of the gut microbiome, shedding light on the intricate interplay between external toxicants and the internal microbial ecosystems. The emphasis is on understanding the complexities of this crossroads and its potential implications for overall health. This section navigates the intricate crossroads within the gut, where environmental toxicants and microbial communities engage in a dynamic dialogue, shaping the delicate balance of the gut microbiome. Let us delve into the complexities of this interaction, shedding light on the interplay between external toxicants and the internal microbial ecosystems. At the heart of this exploration is the dynamic interaction between environmental toxicants and the diverse microbial communities within the gut. The gut serves as a unique arena where these entities engage in a dialogue, creating a crossroads that play a pivotal role in shaping the overall health landscape. This interaction forms a nuanced tapestry of communication, with each toxicant presenting a unique script in the dialogue [63].

3.7.2.1 Influence on Microbial Balance

This section aims to unravel how these dialogues influence the delicate balance of the gut microbiome. Understanding the impact of toxicants on microbial communities is crucial for comprehending their role in disrupting or harmonizing the internal ecosystems. For instance, curcumin, known for its antioxidant properties, may influence the gut dialogue by mitigating oxidative stress caused by certain toxicants, potentially preserving the microbial balance. The exploration identifies key crossroads within the gut where toxicants and microbial communities intersect. These crossroads are pivotal points where external factors can exert influence, leading to potential shifts in microbial composition and function. Examining the specific molecular pathways involved in these interactions provides a deeper understanding of the mechanisms at play. The emphasis is on unraveling the potential implications of these toxicant-microbe dialogues for overall health. It is crucial to explore how disruptions in the gut microbiome, induced by toxicants, may contribute to various health conditions. For instance, karanjin, a bioactive compound, could engage in this dialogue by influencing the gut microbial balance, potentially contributing to anti-inflammatory effects and overall health benefits [64].

3.7.2.2 Complexities and Challenges

This exploration acknowledges the complexities inherent in deciphering the toxicant-microbe dialogue. Variability in individual responses, diverse toxicant types, and the multifaceted nature of microbial communities contribute to the challenges in understanding this intricate interplay. Robust methodologies, including advanced omics techniques, are essential for unraveling the complexities involved. "The Crossroads in the Gut" provides a profound journey into the dynamic interplay between environmental toxicants and gut microbial communities. By exploring the nuances of these dialogues and their potential implications for overall health, this section lays the groundwork for comprehending the impact of external factors on the delicate balance within our internal ecosystems. As we navigate through this exploration, the goal is not only to uncover scientific intricacies but also to glean practical insights for fostering a resilient and thriving gut microbiome in the face of environmental challenges. This segment explores the profound impact of diet on the resilience of the gut microbiome. It delves into how dietary choices play a crucial role in nourishing and sustaining the diverse microbial communities within the gut. The focus is on understanding how specific nutrients and dietary patterns contribute to the robustness and adaptability of the microbiome, ultimately influencing overall health. The exploration aims to unveil the intricate relationship between diet and microbial resilience, highlighting the significance of mindful dietary choices in fostering a healthy and thriving gut ecosystem. This segment is a journey into the profound impact of diet on the resilience of the gut microbiome. Let us explore how dietary choices contribute to nourishing and sustaining diverse microbial communities within the gut, emphasizing the intricate relationship between diet and microbial resilience [1].

3.7.2.3 Nourishing the Microbial Ecosystem

At the core of this exploration is the recognition of diet as a key determinant in the health and resilience of the gut microbiome. The gut, teeming with diverse microbial communities, relies on specific nutrients derived from our diet for sustenance. Just as a varied diet benefits human health, it plays a crucial role in nourishing the microbial ecosystem within.

This segment delves into the impact of specific nutrients and dietary patterns on microbial resilience. For instance, curcumin, a bioactive compound found in turmeric, has been associated with anti-inflammatory and antioxidant properties. Its inclusion in the diet may contribute to microbial resilience by mitigating inflammation and oxidative stress within the gut. Understanding how dietary choices contribute to the adaptability and robustness of the microbiome is key. Certain dietary components act as fuel for beneficial microbes, fostering their growth and enhancing their ability to adapt to changing conditions. The segment explores how a diverse and balanced diet promotes microbial diversity, a hallmark of a resilient gut ecosystem. The emphasis is on the significance of mindful dietary choices in fostering a healthy gut ecosystem. For instance, considering the impact of dietary fiber on promoting the growth of beneficial bacteria illustrates the importance of conscious food choices. A diet rich in fiber, found in fruits, vegetables, and whole grains, provides nourishment for microbes that contribute to a resilient gut [28].

3.7.2.4 Overall Health Implications

This exploration aims to unveil the intricate relationship between diet and microbial resilience and, in turn, its implications for overall health. A resilient gut microbiome is associated with improved digestion, immune function, and even mental well-being. Examining the role of dietary components, such as karanjin from *Pongamia pinnata* seeds, in promoting microbial balance adds depth to our understanding of the health benefits associated with specific dietary choices. While exploring the positive aspects, this segment acknowledges challenges and considerations in maintaining microbial resilience through diet. Factors such as individual variations in responses to specific diets and the influence of external factors underscore the complexities involved in this intricate relationship. "Dietary Influences on Microbial Resilience: Nourishing the Microbiome" provides a comprehensive exploration into the role of diet in shaping the resilience of the gut microbiome. By understanding how specific nutrients and dietary patterns contribute to the adaptability and robustness of microbial communities, this segment highlights the significance of mindful dietary choices in fostering a healthy and thriving gut ecosystem. As we delve into this exploration, the goal is not only to unravel scientific intricacies but also to offer practical insights for individuals seeking to optimize their health through dietary decisions that support a resilient gut microbiome. This section explores the broader impact of lifestyle choices on achieving harmony within the microbiome, extending beyond dietary considerations. It delves into how various aspects of lifestyle, such as physical activity, stress management, and sleep patterns, influence the equilibrium of the gut microbiome. The focus is on understanding the holistic nature of lifestyle choices and their role in fostering a harmonious relationship with the microbial communities residing within the gut. By going beyond dietary factors, the exploration aims to elucidate how overall lifestyle choices contribute to the well-being of the microbiome and, consequently, impact human health [65].

3.7.2.5 Holistic Influence of Lifestyle Choices

Beyond the dinner plate, various lifestyle factors significantly influence the equilibrium of the gut microbiome. This segment recognizes the holistic nature of lifestyle choices, considering how they extend beyond what we eat to encompass broader aspects such as physical activity, stress management, and sleep patterns. Physical activity emerges as a pivotal element in fostering microbiome harmony. Regular exercise has been associated with positive shifts in microbial diversity and composition. For example, studies have suggested that exercise may enhance the abundance of beneficial bacteria, contributing to a more balanced and resilient gut microbiome. Stress, a ubiquitous aspect of modern life, can impact the gut microbiome. Chronic stress may alter the composition of microbial communities, potentially leading to an imbalance. Conversely, effective stress management techniques, such as mindfulness and relaxation practices, may contribute to a more favorable microbial environment within the gut. The exploration extends to the realm of sleep patterns and their influence on the gut microbiome. Disruptions in sleep have been linked to changes in microbial composition. Adequate and quality sleep, on the other hand, is associated with a healthier gut microbiome. Understanding these connections emphasizes the importance of sleep-in supporting microbiome harmony. The focus of this segment is on understanding how lifestyle choices contribute to the equilibrium of the gut microbiome. By examining the impact of physical activity, stress management, and sleep patterns on microbial communities, we gain insights into the delicate balance within the gut. This equilibrium is crucial for overall health, influencing processes ranging from digestion to immune function [66].

3.7.2.6 Beyond Dietary Factors

While diet plays a fundamental role, this exploration goes beyond dietary factors, recognizing the multifaceted influence of lifestyle choices. For example, curcumin, a compound found in turmeric, is not only relevant in the context of diet but also in holistic health practices. Its anti-inflammatory properties may contribute to the overall well-being of individuals practicing stress management techniques and engaging in regular physical activity. Ultimately, this segment aims to elucidate how overall lifestyle choices contribute to the well-being of the microbiome and, consequently, impact human health. The practical implications extend to empowering individuals to adopt a holistic approach to health. By making mindful choices in various aspects of life, individuals can positively influence the harmonious relationship with their gut microbiome. "Lifestyle Choices and Microbiome Harmony: Beyond the Plate" offers comprehensive insights into the broader impact of lifestyle choices on achieving harmony within the microbiome. By exploring the connections between physical activity, stress management, sleep patterns, and the gut microbiome, this segment provides a holistic perspective. It emphasizes the intricate interplay beyond dietary factors and highlights the practical implications for individuals seeking to optimize their well-being through mindful lifestyle choices [67].

3.7.2.7 Balancing Act: Strategies for Modulating Microbial Response to Toxicants

This section focuses on the delicate equilibrium required to modulate the microbial response to environmental toxicants. It explores various strategies aimed at maintaining a balance within the gut microbiome when exposed to harmful substances. The emphasis is on understanding how interventions, ranging from dietary modifications to lifestyle adjustments, can be strategically employed to mitigate the impact of toxicants on the microbial communities in the gut. The exploration aims to provide insights into practical approaches for achieving a harmonious balance, ensuring the resilience and health of the gut microbiome in the face of external challenges. This section intricately explores the delicate equilibrium required to modulate the microbial response to environmental toxicants, emphasizing various strategies aimed at maintaining balance within the gut microbiome when exposed to harmful substances. The central theme revolves around understanding and implementing strategies to preserve balance within the gut microbiome amidst exposure to environmental toxicants. This balancing act is crucial for the resilience and health of the microbial communities residing within the gut [44]. One of the key strategies involves dietary modifications. For instance, curcumin, a polyphenol found in turmeric, has been studied for its potential to counteract the adverse effects of certain environmental toxins. Curcumin's anti-inflammatory and antioxidant properties make it a promising dietary component to support the gut microbiome in the presence of toxicants.

Beyond diet, lifestyle adjustments play a pivotal role in the balancing act. Regular physical activity, stress management practices, and sufficient sleep contribute to a resilient gut microbiome. These lifestyle factors can be regarded as proactive measures to counteract the disruptive effects of environmental toxicants. Another facet of the exploration involves the use of probiotics and prebiotics. Probiotics, such as those found in fermented foods like yogurt, introduce beneficial bacteria to the gut, potentially enhancing microbial diversity and resilience. Prebiotics, found in foods like garlic and onions, provide nourishment for existing beneficial bacteria, fostering a healthier microbial environment. This section aims to deepen our understanding of microbial resilience. Karanjin, a natural compound found in the seeds of the Pongamia pinnata tree, exemplifies a substance with potential benefits. Studies suggest karanjin possesses antimicrobial properties, showcasing its potential role in supporting the gut microbiome's resilience against harmful substances. The overarching goal is to strategically mitigate the impact of toxicants on the gut microbiome. By implementing a combination of dietary modifications, lifestyle adjustments, and the incorporation of substances like curcumin and karanjin, individuals can aim for a harmonious balance. This approach acknowledges the dynamic relationship between environmental challenges and the gut microbiome, offering practical strategies to navigate and counteract potential disruptions [3].

3.7.2.8 Ensuring Resilience and Health:

The exploration goes beyond theoretical considerations, providing practical approaches for achieving a harmonious balance. By emphasizing the resilience and health of the gut microbiome, individuals can adopt informed strategies to navigate the challenges posed by environmental toxicants. This holistic view considers the multifaceted nature of the balancing act, recognizing that interventions must address various aspects to ensure long-term microbial well-being. By exploring dietary modifications, lifestyle adjustments, and the potential benefits of substances like curcumin and karanjin, this section offers practical guidance for individuals seeking to proactively support their gut microbiome in the face of external challenges. This section explores the potential future developments in understanding the implications of the gut microbiome for personalized nutrition and environmental health. It delves into how advancements in research may lead to personalized approaches, tailoring nutritional recommendations based on an individual's unique microbial profile. Additionally, the exploration considers the broader environmental health implications, envisioning a future where insights from microbiome research contribute to strategies for mitigating environmental challenges. The focus is on the promising horizons that personalized nutrition and environmental health may offer as our understanding of the interplay between the gut microbiome and these factors continues to evolve. The central theme revolves around the prospect of personalized nutrition, a revolutionary concept that tailors dietary recommendations based on an individual's unique microbial profile. Curcumin, the active compound in turmeric, stands as an example. Studies suggest that curcumin's impact on the gut microbiome might vary among individuals due to their distinct microbial compositions. In the envisioned future, individuals could receive personalized advice on incorporating curcumin or similar compounds into their diets, optimizing the benefits based on their specific microbial landscape [9]. The exploration highlights how advancements in microbiome research could pave the way for a more nuanced understanding of the intricate relationships between dietary components, the gut microbiome, and individual health. By analyzing an individual's microbial profile, researchers may unravel personalized responses to substances like curcumin, offering targeted recommendations for optimal well-being. Beyond individual nutrition, the section delves into broader environmental health implications. Karanjin, a natural compound with potential antimicrobial properties found in Pongamia pinnata seeds, exemplifies the broader environmental considerations. As research progresses, insights into how environmental factors impact the gut microbiome may inform strategies for mitigating environmental challenges. The envisioned future sees a holistic approach where knowledge derived from microbiome research contributes to developing interventions to safeguard both individual and environmental health. The exploration considers a future where insights from the gut microbiome contribute to strategies for mitigating environmental challenges. This includes understanding how certain environmental toxins may disrupt the microbial balance within the gut and developing targeted interventions to counteract these disruptions. The potential for personalized approaches extends to environmental health, with tailored strategies to enhance microbial resilience against specific environmental stressors [68].

3.8 PROMISING HORIZONS

The section concludes by highlighting the promising horizons that personalized nutrition and environmental health may offer. It envisions a future where individuals receive highly customized nutritional guidance, leveraging insights from their unique gut microbiome. Simultaneously, these personalized approaches extend to environmental health, fostering a symbiotic relationship between personalized well-being and a sustainable environment. By considering examples like curcumin and karanjin, the section provides a glimpse into the exciting possibilities of tailoring dietary and environmental strategies based on individual microbial profiles, presenting a future where health recommendations are finely tuned to the intricacies of one's internal and external ecosystems.

This section presents a collection of case studies showcasing successful instances where dietary interventions have been employed to effectively navigate and mitigate the impact of toxicant exposure. These real-world examples illustrate how specific dietary strategies have been instrumental in supporting the gut microbiome and minimizing the adverse effects of environmental toxicants. The focus is on providing tangible success stories that demonstrate the potential efficacy of dietary approaches in promoting resilience and health in the face of toxicant challenges. By delving into these practical examples, the exploration illustrates how targeted dietary strategies have played a pivotal role in bolstering the gut microbiome and mitigating the harmful effects of environmental toxicants [14].

3.8.1 Curcumin as a Dietary Ally

One notable case study revolves around the dietary inclusion of curcumin, a bioactive compound found in turmeric. Numerous studies suggest that curcumin possesses anti-inflammatory and antioxidant properties, potentially mitigating the detrimental effects of certain environmental toxins. Individuals exposed to specific pollutants or contaminants may benefit from curcumin's protective attributes. These case studies shed light on how incorporating curcumin into the diet has, in certain instances, supported the gut microbiome's resilience, offering a tangible example of the positive impact of dietary choices on navigating toxicant exposure. Another compelling case study centers on the dietary integration of karanjin, a bioactive compound derived from *Pongamia pinnata* seeds. Research indicates that karanjin exhibits antimicrobial properties, suggesting its potential role in fortifying the gut microbiome against harmful invaders, including certain environmental toxins. Case studies showcase scenarios where individuals exposed to specific toxicants have experienced positive outcomes through the incorporation of karanjin into their diets. This highlights the versatility of dietary interventions, where natural compounds like karanjin contribute to the body's defense mechanisms against toxic insults. The focus of this section extends beyond individual compounds, emphasizing strategic dietary approaches tailored to unique toxicant challenges. Case studies examine instances where well-designed diets, rich in specific nutrients and bioactive compounds, have contributed to maintaining a resilient and healthy gut microbiome despite environmental stressors. These success stories serve as a testament to the efficacy of personalized dietary interventions in navigating diverse toxicant exposures. The overarching theme of the case studies is the effective minimization of adverse effects associated with toxicant exposure. Through dietary interventions, individuals have achieved not only a reduction in the negative impacts on the gut microbiome but also an enhancement of overall well-being. These tangible outcomes underscore the practical significance of informed dietary choices in promoting resilience and health amid environmental challenges [69].

3.8.2 Practical Insights for Individuals

This section serves not only as a compilation of success stories but also as a source of practical insights for individuals facing similar challenges. By examining these case studies, readers gain a deeper understanding of how specific dietary strategies can be tailored to mitigate the impact of toxicants on the gut microbiome. The goal is to empower individuals with knowledge that translates into actionable steps, fostering a resilient internal ecosystem in the face of environmental stressors.

Through the lens of curcumin, karanjin, and strategic dietary approaches, these cases illuminate the transformative potential of informed dietary choices in promoting health and resilience. This section explores the future path in the evolving field that intersects microbiome research and toxicology. It delves into the challenges faced in understanding the complex interplay between the gut microbiome and toxic substances, offering insights into potential obstacles on the journey ahead. Simultaneously, it highlights the exciting opportunities that lie within this intersection, suggesting avenues for further exploration and advancement. The focus is on navigating the uncharted territory where microbiome science meets toxicology, acknowledging both the difficulties and the prospects that shape the road ahead [70].

3.8.3 Challenges in Microbiome-Toxicology Integration

The first challenge lies in comprehending the intricate web of interactions between the gut microbiome and toxic substances. The microbiome's vast diversity and the multitude of toxicants present a complex puzzle that requires deciphering. Examples such as curcumin and karanjin demonstrate the need for a nuanced understanding of how specific compounds interact with microbial communities, adding layers of complexity to the research. The inherent variability in individuals' microbiomes poses another hurdle. Each person's microbial profile responds uniquely to toxic exposures, making it challenging to generalize findings. Overcoming this challenge involves unraveling the nuances of individual responses, considering genetic factors, lifestyle, and environmental influences. The microbiome's dynamic nature presents an ongoing challenge. Factors like diet, lifestyle, and environmental exposures continuously shape microbial communities. Understanding the temporal aspects of these changes becomes pivotal in deciphering how the microbiome responds to and influences toxicological events over time. The intersection of microbiome research and toxicology opens avenues for personalized interventions. By understanding an individual's unique microbiome profile, tailored strategies can be developed to mitigate the impact of specific toxicants. This personalized approach, exemplified by the potential use of curcumin and karanjin, holds promise in optimizing health outcomes. The evolving field provides opportunities for developing novel therapeutic interventions. Insights gained from microbiome-toxicology research may lead to the discovery of compounds, similar to curcumin and karanjin, with therapeutic potential in protecting the gut microbiome and mitigating toxic effects. These therapeutic approaches may become crucial in enhancing resilience against environmental challenges [71].

TABLE 3.1
Summarizing the studies and their effects.

Signaling Pathways	Effects	Target Study/Treatment	Type of Animal Study	Reference
Extrinsic pathway proteins FAS/FADD/Caspases 8/3/9, Downregulation of KRAS	Apoptosis, Inhibition of cell proliferation	Non-small cell lung cancer cells (A549 cell line)	Cell line	[73]
Lipoxygenase-1 (LOX-1)	Anti-inflammatory properties	Karanja (Pongamia pinnata) and its derivatives	Rats (paw and ear edema model)	[74]
IL-17A, IL-17F, IL-23, RORγt, TLR-78	Free radical quenching, molecular docking studies	Psoriasis management with Karanjin and Pongapin	Protein 3D structures	[75]
AMP-activated protein kinase (AMPK)	Increased glucose uptake, GLUT4 translocation	Type 2 diabetes management with Karanjin	L6-GLUT4myc myotubes	[49]
CYP1A1 inhibition	Inhibition of 7-ethoxyresorufin-O-deethylase activity	Identification of CYP1A1 inhibitors from edible plants	Human embryonic kidney (HEK293) cells	[76]
NF-κB, TNFα	Anti-inflammatory properties	Experimental arthritis model and in vitro macrophage study	Adjuvant induced arthritis model (AIA) in two doses	[77]
H(+), K(+)-ATPase activity	Anti-ulcerogenic properties; Inhibition of oxidative stress	Gastric ulcers induced by swim and ethanol stress	Adult male albino rats	[56]
Antioxidative, Anti-inflammatory	Reduction in BMI, body weight, waist circumference	Curcumin supplementation in randomized controlled trials (RCTs)	Clinical trial	[78]
Antioxidant, Anti-inflammatory	Decreased muscle pain intensity, increased antioxidant capacity, reduced CK activity, improved muscle performance, anti-inflammatory effect	Exercise-induced muscle damage (EIMD) in humans	Clinical trial	[79]
Immune-modulatory (cytokines, T-cell responses)	Positive effects on relieving COVID-19-related inflammatory response	Patients with COVID-19 (randomized clinical trials)	Clinical trial	[80]
Not specified in the provided abstract	Reduced exercise-induced muscle damage, muscle soreness, TNF-α concentration; Improved muscle strength and joint flexibility; Reduced exercise-induced muscle damage, muscle soreness, TNF-α concentration; Improved muscle strength and joint flexibility	Curcumin supplementation for exercise-induced muscle damage	Clinical trial	[6]

Understanding the interplay between the microbiome and toxicants contributes to the development of broader environmental health strategies. This involves envisioning ways to mitigate the impact of environmental toxicants not only on individual health but also on broader ecosystems. The potential benefits extend beyond personal well-being to encompass the health of communities and ecosystems. The section culminates in a call to navigate the uncharted territory where microbiome science meets toxicology. It encourages researchers and practitioners to overcome challenges through collaborative efforts, technological advancements, and innovative research methodologies. By doing so, the road ahead becomes a canvas for groundbreaking discoveries that will reshape our understanding of how the microbiome and toxicology intersect. Through examples like curcumin and karanjin, it paints a vision of personalized interventions, therapeutic applications, and holistic environmental health strategies that can potentially revolutionize how we navigate the complexities of the microbiome-toxicology interface [72]. Table 3.1 summarizes the recent studies.

3.9 CONCLUSION

In conclusion, this chapter not only contributes to the scientific understanding of the delicate balance between internal microbial ecosystems and external environmental challenges but also emphasizes the practical implications for individuals seeking to optimize their health. The incorporation of dietary and lifestyle interventions as strategic tools in modulating microbial responses to toxicants becomes a central theme, empowering readers to cultivate a resilient and thriving internal ecosystem. The exploration invites individuals to join a scientific odyssey into the depths of the gut, where the choices made resonate through the delicate tapestry of health and well-being.

REFERENCES

1. Conlon MA, Bird AR. The impact of diet and lifestyle on gut microbiota and human health. *Nutrients.* 2014;7(1):17–44. doi: 10.3390/nu7010017.
2. Balaguer-Trias J, Deepika D, Schuhmacher M, Kumar V. Impact of contaminants on microbiota: linking the gut-brain axis with neurotoxicity. *Int J Environ Res Public Health.* 2022;19(3):1368. doi: 10.3390/ijerph19031368.
3. Hills RD Jr, Pontefract BA, Mishcon HR, Black CA, Sutton SC, Theberge CR. Gut microbiome: profound implications for diet and disease. *Nutrients.* 2019;11(7):1613. doi: 10.3390/nu11071613.
4. Viljakainen J, Raju SC, Viljakainen H, Figueiredo RAO, Roos E, Weiderpass E, Rounge TB. Meal regularity plays a role in shaping the saliva microbiota. *Front Microbiol.* 2020;11:757. doi: 10.3389/fmicb.2020.00757.
5. Ohland CL, Jobin C. Microbial activities and intestinal homeostasis: a delicate balance between health and disease. *Cell Mol Gastroenterol Hepatol.* 2015;1(1):28–40. doi: 10.1016/j.jcmgh.2014.11.004.
6. Beba M, Mohammadi H, Clark CCT, Djafarian K. The effect of curcumin supplementation on delayed-onset muscle soreness, inflammation, muscle strength, and joint flexibility: a systematic review and dose-response meta-analysis of randomized controlled trials. *Phytother Res.* 2022;36(7):2767–2778. doi: 10.1002/ptr.7477.
7. Sisto A, Vicinanza F, Campanozzi LL, Ricci G, Tartaglini D, Tambone V. Towards a transversal definition of psychological resilience: a literature review. *Medicina.* 2019;55(11):745. doi: 10.3390/medicina55110745.
8. Modak A, Ronghe V, Gomase KP, Mahakalkar MG, Taksande V. A comprehensive review of motherhood and mental health: postpartum mood disorders in focus. *Cureus.* 2023;15(9):e46209. doi: 10.7759/cureus.46209.
9. Scazzocchio B, Minghetti L, D'Archivio M. Interaction between gut microbiota and curcumin: a new key of understanding for the health effects of curcumin. *Nutrients.* 2020;12(9):2499. doi: 10.3390/nu12092499.
10. Hewlings SJ, Kalman DS. Curcumin: a review of its effects on human health. *Foods.* 2017;6(10):92. doi: 10.3390/foods6100092..
11. Leeming ER, Johnson AJ, Spector TD, Le Roy CI. Effect of diet on the gut microbiota: rethinking intervention duration. *Nutrients.* 2019;11(12):2862. doi: 10.3390/nu11122862.

12. Al Bander Z, Nitert MD, Mousa A, Naderpoor N. The gut microbiota and inflammation: an overview. *Int J Environ Res Public Health.* 2020;17(20):7618. doi: 10.3390/ijerph17207618.
13. Berg G, Rybakova D, Fischer D, Cernava T, Vergès MC, Charles T, Chen X, Cocolin L, Eversole K, Corral GH, Kazou M, Kinkel L, Lange L, Lima N, Loy A, Macklin JA, Maguin E, Mauchline T, McClure R, Mitter B, Ryan M, Sarand I, Smidt H, Schelkle B, Roume H, Kiran GS, Selvin J, Souza RSC, van Overbeek L, Singh BK, Wagner M, Walsh A, Sessitsch A, Schloter M. Microbiome definition re-visited: old concepts and new challenges. *Microbiome.* 2020;8(1):103. doi: 10.1186/s40168-020-00875-0.
14. Chen Z, Liang N, Zhang H, Li H, Guo J, Zhang Y, Chen Y, Wang Y, Shi N. Resistant starch and the gut microbiome: exploring beneficial interactions and dietary impacts. *Food Chem X.* 2024;21:101118. doi: 10.1016/j.fochx.2024.101118.
15. Frame LA, Costa E, Jackson SA. Current explorations of nutrition and the gut microbiome: a comprehensive evaluation of the review literature. *Nutr Rev.* 2020;78(10):798–812. doi: 10.1093/nutrit/nuz106.
16. Lee C, Lee H, Park JC, Im SH. Microbial components and effector molecules in T helper cell differentiation and function. *Immune Netw.* 2023;23(1):e7. doi: 10.4110/in.2023.23.e7.
17. Peng Y, Ao M, Dong B, Jiang Y, Yu L, Chen Z, Hu C, Xu R. Anti-inflammatory effects of curcumin in the inflammatory diseases: status, limitations and countermeasures. *Drug Des Devel Ther.* 2021;15:4503–4525. doi: 10.2147/DDDT.S327378.
18. Zhang P. influence of foods and nutrition on the gut microbiome and implications for intestinal health. *Int J Mol Sci.* 2022;23(17):9588. doi: 10.3390/ijms23179588.
19. Chen CZ, Li P, Liu L, Li ZH. Exploring the interactions between the gut microbiome and the shifting surrounding aquatic environment in fisheries and aquaculture: a review. *Environ Res.* 2022;214(Pt 4):114202. doi: 10.1016/j.envres.2022.114202.
20. Carabotti M, Scirocco A, Maselli MA, Severi C. The gut-brain axis: interactions between enteric microbiota, central and enteric nervous systems. *Ann Gastroenterol.* 2015;28(2):203–209.
21. Lipka V, Panstruga R. Dynamic cellular responses in plant-microbe interactions. *Curr Opin Plant Biol.* 2005;8(6):625–631. doi: 10.1016/j.pbi.2005.09.006. Epub 2005 Sep 22.
22. Hasan N, Yang H. Factors affecting the composition of the gut microbiota, and its modulation. *PeerJ.* 2019;7:e7502. doi: 10.7717/peerj.7502.
23. Pant A, Maiti TK, Mahajan D, Das B. Human gut microbiota and drug metabolism. *Microb Ecol.* 2023;86(1):97–111. doi: 10.1007/s00248-022-02081-x.
24. Wiertsema SP, van Bergenhenegouwen J, Garssen J, Knippels LMJ. The interplay between the gut microbiome and the immune system in the context of infectious diseases throughout life and the role of nutrition in optimizing treatment strategies. *Nutrients.* 2021;13(3):886. doi: 10.3390/nu13030886.
25. Bist P, Choudhary S. Impact of heavy metal toxicity on the gut microbiota and its relationship with metabolites and future probiotics strategy: a review. *Biol Trace Elem Res.* 2022;200(12):5328–5350. doi: 10.1007/s12011-021-03092-4. Epub 2022 Jan 7.
26. Assefa S, Köhler G. Intestinal microbiome and metal toxicity. *Curr Opin Toxicol.* 2020;19:21–27. doi: 10.1016/j.cotox.2019.09.009.
27. Tsiaoussis J, Antoniou MN, Koliarakis I, Mesnage R, Vardavas CI, Izotov BN, Psaroulaki A, Tsatsakis A. Effects of single and combined toxic exposures on the gut microbiome: Current knowledge and future directions. *Toxicol Lett.* 2019;312:72–97. doi: 10.1016/j.toxlet.2019.04.014.
28. Fu J, Zheng Y, Gao Y, Xu W. Dietary fiber intake and gut microbiota in human health. *Microorganisms.* 2022;10(12):2507. doi: 10.3390/microorganisms10122507.
29. Fuloria S, Subramaniyan V, Karupiah S, Kumari U, Sathasivam K, Meenakshi DU, Wu YS, Sekar M, Chitranshi N, Malviya R, Sudhakar K, Bajaj S, Fuloria NK. Comprehensive Review of Methodology to Detect Reactive Oxygen Species (ROS) in mammalian species and establish its relationship with antioxidants and cancer. *Antioxidants.* 2021;10(1):128. doi: 10.3390/antiox10010128.
30. Fuloria S, Subramaniyan V, Karupiah S, Kumari U, Sathasivam K, Meenakshi DU, Wu YS, Guad RM, Udupa K, Fuloria NK. A comprehensive review on source, types, effects, nanotechnology, detection, and therapeutic management of reactive carbonyl species associated with various chronic diseases. *Antioxidants.* 2020;9(11):1075. doi: 10.3390/antiox9111075.
31. Watroly MN, Sekar M, Fuloria S, Gan SH, Jeyabalan S, Wu YS, Subramaniyan V, Sathasivam KV, Ravi S, Mat Rani NNI, Lum PT, Vaijanathappa J, Meenakshi DU, Mani S, Fuloria NK. Chemistry, biosynthesis, physicochemical and biological properties of Rubiadin: a promising natural anthraquinone for new drug discovery and development. *Drug Des Devel Ther.* 2021;15:4527–4549. doi: 10.2147/DDDT.S338548.

32. Yap KM, Sekar M, Seow LJ, Gan SH, Bonam SR, Mat Rani NNI, Lum PT, Subramaniyan V, Wu YS, Fuloria NK, Fuloria S. *Mangifera indica* (Mango): a promising medicinal plant for breast cancer therapy and understanding its potential mechanisms of action. *Breast Cancer.* 2021;13:471–503. doi: 10.2147/BCTT.S316667.
33. Malviya R, Raj S, Fuloria S, Subramaniyan V, Sathasivam K, Kumari U, Unnikrishnan Meenakshi D, Porwal O, Hari Kumar D, Singh A, Chakravarthi S, Kumar Fuloria N. Evaluation of antitumor efficacy of chitosan-tamarind gum polysaccharide polyelectrolyte complex stabilized nanoparticles of simvastatin. *Int J Nanomedicine.* 2021;16:2533–2553. doi: 10.2147/IJN.S300991.
34. Chinnasamy V, Subramaniyan V, Chandiran S, Kayarohanam S, Kanniyan DC, Velaga VSSR, Muhammad S. Antiarthritic activity of achyranthes aspera on formaldehyde - induced arthritis in rats. *Open Access Maced J Med Sci.* 2019;7(17):2709–2714. doi: 10.3889/oamjms.2019.559.
35. Akhlaq M, Azad AK, Fuloria S, Meenakshi DU, Raza S, Safdar M, Nawaz A, Subramaniyan V, Sekar M, Sathasivam KV, Wu YS, Miret MM, Fuloria NK. Fabrication of tizanidine loaded patches using flaxseed oil and coriander oil as a penetration enhancer for transdermal delivery. *Polymers.* 2021;13(23):4217. doi: 10.3390/polym13234217.
36. Bhat AA, Thapa R, Goyal A, Subramaniyan V, Kumar D, Gupta S, Singh SK, Dua K, Gupta G. Curcumin-based nanoformulations as an emerging therapeutic strategy for inflammatory lung diseases. *Future Med Chem.* 2023;15(7):583–586. doi: 10.4155/fmc-2023-0048.
37. Thayumanavan G, Jeyabalan S, Fuloria S, Sekar M, Ravi M, Selvaraj LK, Bala L, Chidambaram K, Gan SH, Rani NNIM, Begum MY, Subramaniyan V, Sathasivam KV, Meenakshi DU, Fuloria NK. Silibinin and naringenin against bisphenol a-induced neurotoxicity in zebrafish model-potential flavonoid molecules for new drug design, development, and therapy for neurological disorders. *Molecules.* 2022;27(8):2572. doi: 10.3390/molecules27082572.
38. Reshma A, Tamilanban T, Chitra V, Subramaniyan V, Gupta G, Fuloria NK, Sekar M, Fuloria S, Sahu R, Narayanan J, Chakravarthy S, Selvaraj S. Anti-obesity effects of olivetol in adult zebrafish model induced by short-term high-fat diet. *Sci Rep.* 2023;13(1):18449.
39. Rarokar NR, Saoji SD, Deole NV, Gaikwad M, Pandey A, Kamaraj C, Chinni SV,Subramaniyan V, Ramachawolran G, Dharashivkar S. Preparation and formula optimization of cephalexin loaded transferosomal gel by QbD to enhance the transdermal delivery: in vitro, ex vivo and in vivo study. *J Drug Del Sci Technol.* 2023;89:104968.
40. Muteeb G, Rehman MT, Shahwan M, Aatif M. Origin of antibiotics and antibiotic resistance, and their impacts on drug development: a narrative review. *Pharmaceuticals.* 2023;16(11):1615. doi: 10.3390/ph16111615.
41. Toussaint L, Nguyen QA, Roettger C, Dixon K, Offenbächer M, Kohls N, Hirsch J, Sirois F. Effectiveness of progressive muscle relaxation, deep breathing, and guided imagery in promoting psychological and physiological states of relaxation. *Evid Based Complement Alternat Med.* 2021;2021:5924040. doi: 10.1155/2021/5924040.
42. Davani-Davari D, Negahdaripour M, Karimzadeh I, Seifan M, Mohkam M, Masoumi SJ, Berenjian A, Ghasemi Y. Prebiotics: definition, types, sources, mechanisms, and clinical applications. *Foods.* 2019;8(3):92. doi: 10.3390/foods8030092.
43. Weis BK, Balshaw D, Barr JR, Brown D, Ellisman M, Lioy P, Omenn G, Potter JD, Smith MT, Sohn L, Suk WA, Sumner S, Swenberg J, Walt DR, Watkins S, Thompson C, Wilson SH. Personalized exposure assessment: promising approaches for human environmental health research. *Environ Health Perspect.* 2005;113(7):840–848. doi: 10.1289/ehp.7651.
44. Fassarella M, Blaak EE, Penders J, Nauta A, Smidt H, Zoetendal EG. Gut microbiome stability and resilience: elucidating the response to perturbations in order to modulate gut health. *Gut.* 2021;70(3):595–605. doi: 10.1136/gutjnl-2020-321747.
45. Cullen CM, Aneja KK, Beyhan S, Cho CE, Woloszynek S, Convertino M, McCoy SJ, Zhang Y, Anderson MZ, Alvarez-Ponce D, Smirnova E, Karstens L, Dorrestein PC, Li H, Sen Gupta A, Cheung K, Powers JG, Zhao Z, Rosen GL. Emerging priorities for microbiome research. *Front Microbiol.* 2020;11:136. doi: 10.3389/fmicb.2020.00136.
46. Gupta G, Hussain MS, Thapa R, Dahiya R, Mahapatra DK, Bhat AA, Singla N, Subramaniyan V, Rawat S, Jakhmola V, Roshan S, Dua K. Hope on the horizon: Wharton's jelly mesenchymal stem cells in the fight against COVID-19. *Regen Med.* 2023;18(9):675–678.
47. Mujafarkani N, Ahamed FMM, Babu KS, Debnath S, Sayed AA, Albadrani GM, Al-Ghadi MQ, Kumarasamy V, Subramaniyan V, Kamaraj C, Abdel-Daim MM. Unveiling a novel terpolymer-metal complex: a detailed exploration of synthesis, characterization, and its potential as an antimicrobial and antioxidant agent. *Heliyon.* 2023;9(10):e20459.

48. Pletcher MJ, Pignone M. Evaluating the clinical utility of a biomarker: a review of methods for estimating health impact. *Circulation.* 2011;123(10):1116–1124. doi: 10.1161/CIRCULATIONAHA.110.943860.
49. Jaiswal N, Yadav PP, Maurya R, Srivastava AK, Tamrakar AK. Karanjin from Pongamia pinnata induces GLUT4 translocation in skeletal muscle cells in a phosphatidylinositol-3-kinase-independent manner. *Eur J Pharmacol.* 2011;670(1):22–28. doi: 10.1016/j.ejphar.2011.08.049.
50. Armstrong R. Towards the microbial home: an overview of developments in next-generation sustainable architecture. *Microb Biotechnol.* 2023;16(6):1112–1130. doi: 10.1111/1751-7915.14256.
51. Batko K, Ślęzak A. The use of big data Analytics in healthcare. *J Big Data.* 2022;9(1):3. doi: 10.1186/s40537-021-00553-4.
52. Howe Iii EG, Elenberg F. Ethical challenges posed by big data. *Innov Clin Neurosci.* 2020;17(10–12):24–30.
53. Arafah A, Khatoon S, Rasool I, Khan A, Rather MA, Abujabal KA, Faqih YAH, Rashid H, Rashid SM, Bilal Ahmad S, Alexiou A, Rehman MU. The future of precision medicine in the cure of alzheimer's disease. *Biomedicines.* 2023;11(2):335. doi: 10.3390/biomedicines11020335.
54. Singh RK, Chang HW, Yan D, Lee KM, Ucmak D, Wong K, Abrouk M, Farahnik B, Nakamura M, Zhu TH, Bhutani T, Liao W. Influence of diet on the gut microbiome and implications for human health. *J Transl Med.* 2017;15(1):73. doi: 10.1186/s12967-017-1175-y.
55. Poljšak B, Fink R. The protective role of antioxidants in the defence against ROS/RNS-mediated environmental pollution. *Oxid Med Cell Longev.* 2014;2014:671539. doi: 10.1155/2014/671539.
56. Vismaya, Belagihally SM, Rajashekhar S, Jayaram VB, Dharmesh SM, Thirumakudalu SK. Gastroprotective properties of Karanjin from Karanja (Pongamia pinnata) seeds; role as antioxidant and H, K-ATPase inhibitor. *Evid Based Complement Alternat Med.* 2011;2011:747246. doi: 10.1093/ecam/neq027.
57. Ji J, Jin W, Liu SJ, Jiao Z, Li X. Probiotics, prebiotics, and postbiotics in health and disease. *MedComm.* 2023;4(6):e420. doi: 10.1002/mco2.420.
58. Arafah A, Khatoon S, Rasool I, Khan A, Rather MA, Abujabal KA, Faqih YAH, Rashid H, Rashid SM, Bilal Ahmad S, Alexiou A, Rehman MU. The future of precision medicine in the cure of Alzheimer's disease. *Biomedicines.* 2023;11(2):335. doi: 10.3390/biomedicines11020335..
59. Zhao Q, Chen Y, Huang W, Zhou H, Zhang W. Drug-microbiota interactions: an emerging priority for precision medicine. *Signal Transduct Target Ther.* 2023;8(1):386. doi: 10.1038/s41392-023-01619-w.
60. Brown ED, Williams BK. Resilience and resource management. *Environ Manage.* 2015;56(6):1416–1427. doi: 10.1007/s00267-015-0582-1.
61. Adamczak A, Ożarowski M, Karpiński TM. Curcumin, a natural antimicrobial agent with strain-specific activity. *Pharmaceuticals.* 2020;13(7):153. doi: 10.3390/ph13070153.
62. Hanson K, Brikci N, Erlangga D, Alebachew A, De Allegri M, Balabanova D, Blecher M, Cashin C, Esperato A, Hipgrave D, Kalisa I, Kurowski C, Meng Q, Morgan D, Mtei G, Nolte E, Onoka C, Powell-Jackson T, Roland M, Sadanandan R, Stenberg K, Vega Morales J, Wang H, Wurie H. The lancet global health commission on financing primary health care: putting people at the centre. *Lancet Glob Health.* 2022 May;10(5):e715–e772. doi: 10.1016/S2214-109X(22)00005-5. Epub 2022 Apr 4. Erratum in: Lancet Glob Health. 2023;11(4):e504.
63. Pennington L, McConachie H. Mother-child interaction revisited: communication with non-speaking physically disabled children. *Int J Lang Commun Disord.* 1999;34(4):391–416. doi: 10.1080/136828299247351.
64. Bhatt P, Kumar V, Subramaniyan V, Nagarajan K, Sekar M, Chinni SV, Ramachawolran G. Plasma modification techniques for natural polymer-based drug delivery systems. *Pharmaceutics.* 2023;15(8):2066.
65. Williams GM, Tapsell LC, Beck EJ. Gut health, the microbiome and dietary choices: An exploration of consumer perspectives. *Nutr Diet.* 2023;80(1):85–94. doi: 10.1111/1747-0080.12769.
66. Childs CE, Calder PC, Miles EA. Diet and immune function. *Nutrients.* 2019;11(8):1933. doi: 10.3390/nu11081933.
67. Begdache L, Patrissy CM. Customization of diet may promote exercise and improve mental wellbeing in mature adults: the role of exercise as a mediator. *J Pers Med.* 2021;11(5):435. doi: 10.3390/jpm11050435.
68. Karl JP, Hatch AM, Arcidiacono SM, Pearce SC, Pantoja-Feliciano IG, Doherty LA, Soares JW. Effects of psychological, environmental and physical stressors on the gut microbiota. *Front Microbiol.* 2018;9:2013. doi: 10.3389/fmicb.2018.02013.
69. Walker C, Gibney ER, Mathers JC, Hellweg S. Comparing environmental and personal health impacts of individual food choices. *Sci Total Environ.* 2019;685:609–620. doi: 10.1016/j.scitotenv.2019.05.404
70. Lozupone CA, Stombaugh JI, Gordon JI, Jansson JK, Knight R. Diversity, stability and resilience of the human gut microbiota. *Nature.* 2012;489(7415):220–230. doi: 10.1038/nature11550.

71. Zam W. Gut microbiota as a prospective therapeutic target for curcumin: a review of mutual influence. *J Nutr Metab.* 2018;2018:1367984. doi: 10.1155/2018/1367984.
72. Hsu KY, Ho CT, Pan MH. The therapeutic potential of curcumin and its related substances in turmeric: from raw material selection to application strategies. *J Food Drug Anal.* 2023;31(2):194–211. doi: 10.38212/2224-6614.3454.
73. Kumar G, Pandey DM, Ghosh M, Dall'Acqua S, Gupta R, Tiwari NP, Mohd Siddique U, Vishwakarma L, Guleri SK, Lal UR, Dubey S. Karanjin, A promising bioactive compound possessing anti-cancer activity against experimental model of non-small cell lung cancer cells. *Anticancer Agents Med Chem.* 2023;24:317–333. doi: 10.2174/0118715206255557231024095245. Epub ahead of print.
74. Rekha MJ, Bettadaiah BK, Muthukumar SP, Govindaraju K. Synthesis, characterization and anti-inflammatory properties of karanjin (Pongamia pinnata seed) and its derivatives. *Bioorg Chem.* 2021;106:104471. doi: 10.1016/j.bioorg.2020.104471.
75. Ghosh A, Tiwari GJ. Role of nitric oxide-scavenging activity of Karanjin and Pongapin in the treatment of Psoriasis. *3 Biotech.* 2018;8(8):338. doi: 10.1007/s13205-018-1337-5.
76. Joshi P, Sonawane VR, Williams IS, McCann GJP, Gatchie L, Sharma R, Satti N, Chaudhuri B, Bharate SB. Identification of karanjin isolated from the Indian beech tree as a potent CYP1 enzyme inhibitor with cellular efficacy via screening of a natural product repository. *Medchemcomm.* 2018;9(2):371–382. doi: 10.1039/c7md00388a.
77. Bose M, Chakraborty M, Bhattacharya S, Mukherjee D, Mandal S, Mishra R. Prevention of arthritis markers in experimental animal and inflammation signalling in macrophage by Karanjin isolated from Pongamia pinnata seed extract. *Phytother Res.* 2014;28(8):1188–1195. doi: 10.1002/ptr.5113.
78. Unhapipatpong C, Polruang N, Shantavasinkul PC, Julanon N, Numthavaj P, Thakkinstian A. The effect of curcumin supplementation on weight loss and anthropometric indices: an umbrella review and updated meta-analyses of randomized controlled trials. *Am J Clin Nutr.* 2023;117(5):1005–1016. doi: 10.1016/j.ajcnut.2023.03.006. Epub 2023 Mar 9. PMID: 36898635.
79. Nanavati K, Rutherfurd-Markwick K, Lee SJ, Bishop NC, Ali A. Effect of curcumin supplementation on exercise-induced muscle damage: a narrative review. *Eur J Nutr.* 2022;61(8):3835–3855. doi: 10.1007/s00394-022-02943-7.
80. Abdelazeem B, Awad AK, Elbadawy MA, Manasrah N, Malik B, Yousaf A, Alqasem S, Banour S, Abdelmohsen SM. The effects of curcumin as dietary supplement for patients with COVID-19: a systematic review of randomized clinical trials. *Drug Discov Ther.* 2022;16(1):14–22. doi: 10.5582/ddt.2022.01017.

4 Role of Gut Microbiome on Human Health

Waleed Hassan Almalki[1]*

[1]Department of Pharmacology, College of Pharmacy, Umm Al-Qura University, Makkah, Saudi Arabia

*Corresponding Author: Waleed Hassan Almalki. Email id: Whmalki@uqu.edu

4.1 INTRODUCTION

The complex, dynamic, and regionally varied ecosystem known as the human gut microbiome (GM) is home to a broad variety of microorganisms, such as viruses, bacteria, fungus, and archae. Both the microorganisms and their human host can communicate with one another [1]. Ten times more genetic material than the human genome is included in the collection of genes that are common to all intestinal bacteria. In some respects, it is also regarded as the "essential organ" of the human body [2]. The largest microenvironment in humans, the GM works in unison with its host to maintain normal physiological processes in a state of dynamic equilibrium. Bacteroides, Proteus, Firmicutes, and Actinomycetes are the four main groups that make up GM [3, 4]. One of the most important markers for determining GM dysfunction is the Firmicutes to Bacteroidetes ratio. Furthermore, the location, homogeneity, and variety of GM serve as significant indicators of the makeup of the gut flora [5]. The gut flora will alter in composition and function as soon as the GM problem arises, and it may even be the underlying cause of some diseases. This is the result of several complex regulatory processes [6, 7]. GM research has advanced quickly in recent years because of the fast expansion of technologies, including high-throughput sequencing, genomics, molecular biology, and bioinformatics analysis [8]. Numerous studies have shown that GM and its metabolites have a notable effect on preserving intestinal equilibrium in the host and affecting the start of a variety of illnesses, including metabolic, neurological, and cardiovascular diseases (CVDs) as well as gastrointestinal problems (Figure 4.1). The host's health will be impacted by the imbalance of GM in a number of ways, including bile acids, choline, the gastrointestinal-cerebral axis, and short chain fatty acids (SCFAs) [9, 10]. It is yet unknown, however, just how GM affects human health. This first part of this chapter summarizes the most recent research on the relationship between GM and human disorders, as well as how such relationships are managed [11, 12]. The features of altered GM linked with several conditions are then described as well as possible mechanisms involving microbial metabolism and byproducts that may be in charge of the onset and growth of illness, hence, providing a possibility center of attention for sickness management [13].

4.2 GM AND NEURODEGENERATIVE DISEASES (NDS)

In the medical world, there has been a lot of attention lately to the connection between GM and NDs. GM may have an impact on the immune system, hormone control, and nervous system communication via the gut-brain axis, according to an increasing amount of research [14]. This may result in the onset and exacerbation of illnesses that impact the central nervous system (CNS), including Parkinson's disease (PD) and Alzheimer's disease (AD) [15, 16].

DOI: 10.1201/9781003489221-4

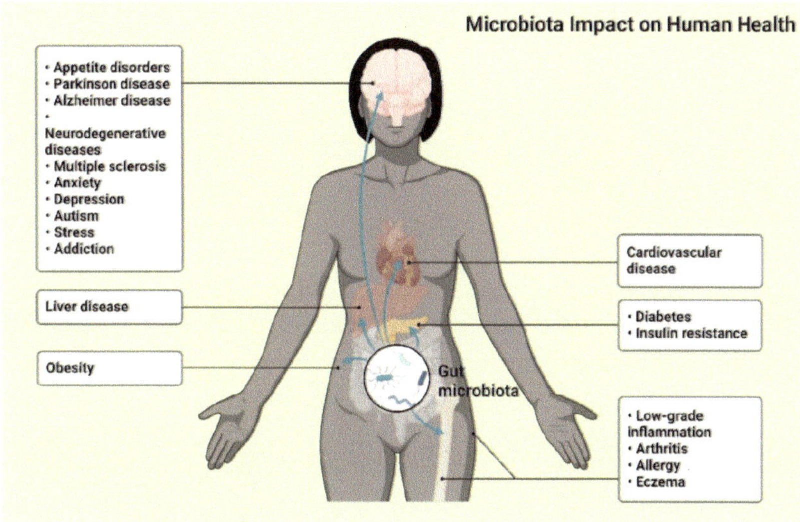

FIGURE 4.1 The figure illustrates the intricate interplay between microbiota and human health. Microbiota, comprising diverse microbial communities residing in the body, significantly influence various physiological processes. Through symbiotic interactions, microbiota contribute to digestion, immune system regulation, and nutrient absorption.

4.2.1 Parkinson's Disease (PD)

PD is a prevalent NDs that affects the CNS. Its pathogenic features include abnormal synuclein aggregation in Lewy bodies and dopaminergic neuron degeneration and death in the substantia nigra [17]. Numerous studies suggest that there are substantial differences in the GM composition between those with PD and those in good health, notwithstanding the wide range of findings. Several studies have shown a decrease in the amount of *Brautella*, *Prevotella*, and *Faecococcus* in the GM of patients with PD, whereas *Bifidobacterium*, *Pasteurella*, and *Enterococcus* were found to be much more prevalent [18, 19]. Although *Bifidobacterium* was shown to be more common in people with PD, a subsequent decline in *Bifidobacterium* may indicate whether PD will advance. This suggests that the existence of such alterations, such as an increase in *Bifidobacterium* in people with PD, may represent a defense process against the worsening of NDs [20]. It suggests that a probiotic treatment with *Bifidobacterium* may stop PD from progressing into more severe stages [21, 22]. The quantity of *Lactobacillus* presents in individuals with PD, however, differed greatly across studies. The results are not entirely unexpected, given that several variables, including eating choices, age, geography, body mass index, and ethnicity, have an influence on genetic variation [23]. The composition of *Lactobacillus* varies greatly among the world's nations due to differences in dietary patterns. Even within a single nation, there are variations in *Lactobacillus* strains across areas, such as Southeast and Southwest China [24]. This discrepancy might be due to variations in regional eating traditions. Furthermore, as *Lactobacillus* may increase IBS symptoms that are prevalent in diarrhea and decrease those that are predominant in constipation, the rise in *Lactobacillus* may be related to the frequent constipation that PD patients experience [25, 26]. Figure 4.2 shows the intricate disruption of microbiota resulting from bacterial infection. The intestinal-cerebral axis is a two-way communication channel between the gut bacteria and CNS that regulates brain growth, aging, and neuronal function. The interaction between the GM and brain axis occurs via three basic routes [27]. Substance cues come first. Microbes may affect the neurological system directly or indirectly by the production of other metabolites and SCFAs. They may also have an effect on the neuroendocrine system, regulate GABA, 5-HT, and other neurotransmitter levels, and directly

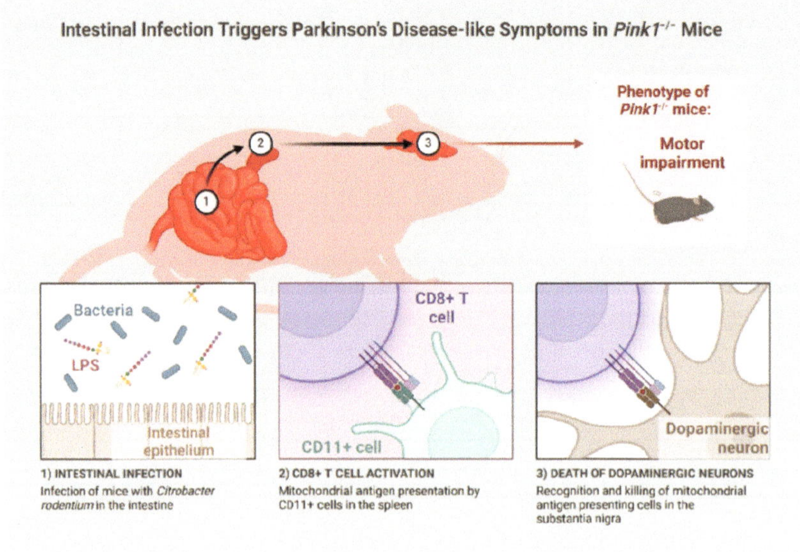

FIGURE 4.2 The figure illustrates the intricate disruption of microbiota resulting from bacterial infection. Within this symbiotic ecosystem, bacterial pathogens infiltrate and alter the composition of the microbiota, leading to dysbiosis. This disruption can induce inflammation, compromise host immune responses, and contribute to the pathogenesis of infectious diseases.

affect the neurological system [28, 29]. The brain pathways are involved in the second. Bacteria and their metabolites may impact the vagus and intestinal nerve networks, which, in turn, can impact the brain and behavior. The immune system is the last system in place [30]. The two main mediating mechanisms are microglia and cytokines in the systemic circulation, which is a measure of inflammation. These pathways might be involved in the development of nervous system problems, such NDs [31, 32].

Many investigations have shown that the intestinal-cerebral axis has a role in PD. PD patients may have neurons with an excess of microglia activation and an accumulation of a-synuclein due to altered gut flora (Table 4.1) [33]. Mice with non-eradicated microbial communities showed greater a-synuclein aggregations in their brains than germ-free animals without gut flora, according to an investigation using mice overexpressing a-synuclein. In the same research, fecal microflora from PD patients was transferred into microorganism-free mice, exacerbating the motor symptoms brought on by alpha-synuclein, rather than using healthy donor feces [34, 35]. Recent research indicates that a significant part of the pathogenesis of PD may include the reduction of ghrelin and SCFAs that occurs in PD patients as a consequence of GM change. Despite inconsistent results among studies, several of them show a significant difference in the GM makeup between those with PD and those who are healthy [36]. In PD, microbes may affect the neurological system directly or indirectly via chemical signals, neuronal connections, or the immune system. Therefore, past research suggests that intestinal lesions associated with the beginning of PD may be significantly influenced by the GM [37, 38].

4.2.2 Alzheimer's Disease (AD)

The most common NDs affecting the CNS is AD. Characterized by hyperphosphorylation of tau, beta-amyloid protein accumulation, beta-amyloid plaques, and neurofibrillary tangles; these factors set off an inflammatory cascade that eventually results in irreversible brain damage due to neuronal death or necrosis [39]. Similarities exist between the beginning of PD and AD, including intestinal

TABLE 4.1
A concise overview of the associations between specific gut microbes, altered gut flora, and their implications for Parkinson's disease, encompassing neurological impact, mediating systems, and potential interventions.

Gut Microbe	Parkinson's Link	Neurological Impact	Mediating System	Implications for PD
Bifidobacterium	Increase, Potential Defense	Direct/Indirect Effects	Substance Cues	Potential Probiotic Intervention
Lactobacillus	Varied, Influenced by Factors	Influence on IBS	Gut-Brain Pathways	Association with Constipation
Microbial Flora	Altered in PD Patients	Impact on a-Synuclein	Immune System	Significance in Parkinson's Pathology
Short-Chain Fatty Acids	Reduction in PD Pathogenesis	Regulation of Ghrelin	Gut-Brain Axis	Implications for Parkinson's Disease
Neurotransmitters	Altered Levels in PD	Impact on Neuroendocrine System	Gut-Brain Communication	Implications for Parkinson's Symptoms
Proinflammatory Mediators	Potential Involvement in PD	Influence on Inflammation	Immune Response	Connection to Neurodegenerative Process

bacteria and an unbalanced gut-brain axis. A growing amount of research has shown a link between AD and GM (Table 4.2) [40, 41]. Clinical data suggests that GM has a role in the early pathophysiology of AD in humans. When compared to those in excellent health, the diversity of fecal flora was much lower, and the changes in intestinal flora at the stage of moderate cognitive impairment were shown to be comparable to those in the early stages of AD [42]. This suggests that dysbacteriosis may aid in the early detection of AD. Several research studies have shown evidence of a close connection between AD and inflammation caused by GM. The imbalance in GM seen in AD individuals may lead to a decline in pro-inflammatory bacteria such as *Shigella* and *Escherichia* and an increase in anti-inflammatory bacteria [43]. Pro-inflammatory compounds generated by various Gut bacterial taxa have the capacity to increase peripheral immune cell invasion, CNS, and plasma inflammation. Mice with AD have altered GM makeup, reduced epithelial barrier integrity, and prolonged inflammation in the gut and throughout the body, in contrast to wild-type mice in normal health [44, 45].

In recent times, there has been a proposition indicating that microbial infection accelerates the progression of AD [46]. Abbott and colleagues have reported the presence of herpes simplex virus type 1 infection within the brains of individuals afflicted with AD. Subsequent investigations have revealed that amyloid beta plaques serve as a form of defensive protein, potentially directly enveloping and enclosing the virus that penetrates the brain. These so-called protective proteins accumulate in the brain, become difficult to get rid of, and eventually cause significant damage to neural tissue [47, 48]. This study raises the possibility that microbial infections are the underlying cause of AD. One of the main pathological features of AD, apart from microbial infection, is the buildup of amyloid beta protein in the brain, which causes plaques and subsequently nerve damage [49]. Growing evidence has shown that in models of AD, GM affects the degree of beta-amyloid protein pathology and cognitive decline. For example, compared to mice grown normally, APPPS1 animals maintained in a germ-free environment showed a substantial decline in both neuronal beta-amyloid protein buildup and Iba-1-positive microglia, which ultimately resulted in an overall reduction in neuroinflammation [50]. Similar effects were shown when the GM of healthy mice was added to models of AD, which decreased glial activation, cognitive loss, and the formation of neurofibrillary tangles and beta-amyloid protein plaques. One possible cause of AD is an imbalance in gut flora brought on by aging [51]. This imbalance may initiate and exacerbate AD by

TABLE 4.2
A comprehensive overview of Alzheimer's disease (AD), highlighting key features, the impact of gut microbiota, microbial infections, and therapeutic implications, aiding understanding and potential treatment strategies.

Alzheimer's Features	Gut Microbiota Impact	Microbial Infections in AD	Therapeutic Implications
Amyloid, Tau, Neurofibrillary Tangles	Dysbiosis, Inflammation	Herpes Simplex Virus Type 1 Role	Gut Flora Control for Treatment
Similarities with PD	Altered Gut-Brain Axis	Amyloid Beta Plaques in AD	Prebiotics, GV-971, Neuroprotection
Fecal Flora Diversity	GM Influence on AD	AD Pathogenesis and Gut-Brain Axis	Phenylalanine, Isoleucine Effects
Early AD Detection	Inflammatory Bacterial Changes	Beta-Amyloid in Alzheimer's	GM Dysbiosis, Neuroinflammation
AD and GM Connection	GM Dysbiosis Effects	Aging and Gut Flora Imbalance	Therapeutic Approaches for AD

activating the CCAAT/enhancer binding protein beta/asparagine endopeptidase pathway in the stomach and brain. Further analysis has demonstrated that the prebiotic R13 may prevent this process by encouraging certain bacteria and reducing amyloid plaques in the digestive system [52, 53]. Furthermore, recent research found that the changed GM composition causes a buildup of phenylalanine and isoleucine in the blood, which promotes the maturation and proliferation of T helper 1 cells that support inflammation. Additionally, GV-971 reduced neuroinflammation, enhanced cognitive impairment, and blocked peripheral phenylalanine and isoleucine accumulation, according to experimental validation in the same study [54]. AD patients' pathogenesis includes a disturbance in the gut-brain axis, which is the first step toward intestinal microbiota involvement. A growing body of research has shown a clear link between AD and inflammation caused by GM [55]. The aforementioned studies unequivocally demonstrate that GM dysbiosis may alter the dynamic balance of GM and result in neuroinflammation during AD, indicating that this may represent a unique therapeutic approach for AD. Treatment for neurological diseases may include controlling gut flora [56, 57].

4.3 GM AND CARDIOVASCULAR DISEASES

Atherosclerosis, hypertension, and heart failure are examples of cardiovascular diseases (CVDs). Globally, CVDs are the leading cause of mortality [58]. A growing body of research has shown that GM and its metabolic byproducts interact with the host in several ways, influencing the growth and progression of CVDs [59]. Numerous studies have connected the metabolic byproducts of the GM, which include bile acids, SCFAs, and trimethylamine oxide (TMAO), to CVDs [60, 61].

4.3.1 Hypertension

One of the most common risk factors for CVDs is high blood pressure. Changes in gut bacterial compositions, gut function, and the gut-brain system relationship are all associated with hypertension [62]. Microbial diversity, abundance, and homogeneity were all significantly lower in hypertensive people, whereas the ratio of Firmicutes to Bacteroidetes was noticeably greater. Blood pressure regulation is mostly regulated by changes in the gut flora, and the production of GM byproducts may be a key mechanism behind these changes [63]. The concentration of circulating TMAO in serum or plasma has been shown to have a significant dose-dependent link with both the risk of hypertension and CV metabolic risk markers such as HDL cholesterol. This conclusion was drawn from a meta-analysis of eighteen observational studies [64, 65].

Genetic modification has implications that extend far beyond intestinal health. The blood-brain barrier may be crossed by metabolites from the GM, which can then alter brain function by entering the circulation. The disturbance of SCFAs balance brought on by GM dysbiosis may increase the generation of 5-hydroxytryptamine (5-HT) by intestinal chromaffin cells [66]. This compound may interact with the 5-HT3 receptor on the intestinal vagus nerve, hence impairing sensory activation of the vagus nerve from the gut to the brain. Vasoconstriction may result from the release of 5-HT into the circulation. Both 5-HT and SCFA may affect blood pressure via the circulation due to the blood-brain barrier being compromised by high blood pressure [67, 68]. Moreover, the production of hydrogen sulfide (H2S) by genetically modified crops has a detrimental effect on human health. It has been shown that H2S can control a number of physiological functions, such as angiogenesis, hypotension, and vasorelaxation [69]. The absence of H2S occurs prior to the onset of hypertension. Exogenous H2S donor may prevent hypertension in spontaneously hypertensive rats, suggesting a significant role for H2S deficiency in blood pressure control. via a several methods, such as the suppression of oxidative stress or inflammation and the induction of vascular smooth muscle relaxation via ion channel mediation, H2S may be able to lower blood pressure [70].

A growing body of research indicates that intestinal dysbacteriosis might be related to vascular permeability, blood pressure, and the control of inflammatory processes. Through the induction of inflammation and vascular immune cell infiltration, GM exacerbates the hypertension and vascular dysfunction caused by angiotensin II [71]. Ex vivo studies have shown that the addition of human feces elevated the mice's blood pressure, pointing to a possible link between genetically modified organisms (genMO) and the onset of hypertension. According to a different study, administering antibiotics to rats lowers the Firmicutes to Bacteroidetes ratio, which, in turn, lowers raised blood pressure and helps balance the dysbiotic hypertension gene [72]. Furthermore, eradicating the whole gut flora with a cocktail of antibiotics markedly decreased the frequency of hypertension-associated aneurysms. This suggests that the immune system and inflammation are regulated by the GM, which serves in the pathogenesis of aneurysms [73]. A high-polyphenol diet has been shown in a randomized controlled experiment to significantly increase intestinal permeability in older adults. It also led to a drop in blood pressure and an elevation in the percentage of butyrate- and cellulose-producing bacteria in the GM [74]. The GM's diversity, abundance, and homogeneity significantly decreased in those with high blood pressure diagnoses. Genetic modification affects hypertension by changing the way the digestive system is controlled and how its metabolites are produced [75]. These results demonstrate a strong correlation between GM and hypertension, suggesting that bringing GM back to balance might be a useful strategy for controlling high blood pressure [76].

4.3.2 Atherosclerosis

Atherosclerosis is the primary cause of cerebral infarction, coronary heart disease, and peripheral vascular disease, accounting for around 50% of all CVDs. Atherosclerotic plaques were shown to include both live oral bacteria and bacterial DNA, suggesting a possible role for genMO in the onset and progression of atherosclerosis [77]. These results point to a possible link between CVDs and genetic engineering (GM). There is a discernible difference in the makeup of GM between healthy persons and those who have atherosclerosis, according to metagenomic-wide association research [78]. Beneficial bacterial growth was restricted by the considerably higher numbers of Enterobacteriaceae and Enterobacter aerogenes in patient samples with atherosclerosis than in control samples [79]. Numerous research on humans and animals has shown that dysbiosis in the gut linked to atherosclerosis may also increase intestinal permeability, which, in turn, raises the quantity of lipopolysaccharide taken up into circulation. Thus, the formation of atherosclerotic plaques may be influenced by an infection, either direct or indirect, of the wall cells of the artery [80].

Furthermore, the development of atherosclerosis may be influenced by the impacts of GM and its metabolites. Acetate, propionate, and butyrate are a few of the SCFAs produced by human genetic modification via the fermentation of food fibers [81]. Two instances of G-protein coupled receptors that are triggered by signaling molecules known as SCFAs are GPR41 and Olfr78. These receptors have the power to elevate the production of peptide YY (PYY) and glucagon-like peptide 1 (GLP-1), which may lower blood pressure and prevent atherosclerosis [82]. Bile acids have anti-atherogenic properties and are a result of the metabolism of cholesterol. One of the main kinds of substances that genetically alters is a basophil activator (BA) [83]. It works by activating the Farnesoid X receptor (FXR), which suppresses the expression of inflammatory markers on monocytes, macrophages, and dendritic cells, hence preventing the development of atherosclerosis. Another important GM molecule linked to the advancement of atherosclerosis is TMAO [84]. According to epidemiological studies, TMAO is a new marker that may be used to assess the risk of CVDs. Research has shown that there is a positive correlation between high levels of TMAO and atherosclerosis [85]. Choline, betaine, and carnitine were degraded by the GM to produce trimethylamine [86]. This was then transferred to the liver, where flavin monooxygenase converted it into TMAO. A higher level of TMEA accelerates the migration of macrophages and their conversion into foam cells, which, in turn, stimulates the production of pro-inflammatory cytokines and atherosclerosis [87]. Moreover, TMAO causes oxidative stress and reduces vascular reactivity, impairing endothelial function and eventually advancing atherosclerosis. Moreover, TMAO may activate platelets, which promotes thrombosis and the rupture of atherosclerotic plaque [88]. Recently, Nemet et al. observed phenylalanine as phenylacetylglutamine (pagln) by non-targeted metabonomics. The metabolite pagln, which is generated from the GM, has the capability to elevate characteristics linked to platelet activation and thrombosis potential. Pamphln triggers physiological effects downstream of its interaction with G protein-coupled receptors. Subsequent research showed that pagln raises the risk of thrombosis by acting via a G protein coupled receptor [89].

People with atherosclerosis have a genetic composition that is quite different from that of healthy people. Following the identification of bacterial DNA in atherosclerotic plaques, other investigations revealed a link between the GM and the development of atherosclerosis [90]. Atherosclerosis may be impacted by the growth of gut microbes and the chemicals they produce. Some alkaloids have been transformed by microorganisms into dangerous compounds that encourage the synthesis of inflammatory cytokines, which, in turn, stimulate inflammation and hasten the onset of atherosclerosis [91]. Taken together, these studies demonstrate that genMO and their offspring possess the capacity to regulate the onset and advancement of atherosclerosis via many mechanisms. Reducing GM may be one way to address CVDs [92].

4.4 GM AND METABOLIC DISEASES

Although both hereditary and environmental variables impact the development of metabolic diseases, in people with comparable caloric consumption and genetic profiles, the makeup of the GM is linked to an increased vulnerability to metabolic disorders [93]. An increasing body of recent research indicates that aberrations in GM are closely related to a number of metabolic illnesses, including diabetes, obesity, and non-alcoholic fatty liver disease (NAFLD) [94].

4.4.1 Obesity

Over 1.9 billion people are overweight, and over 650 million people are obese. Public health is severely hampered by excess weight and its metabolic effects. Studies reveal that genMO have a significant impact on the rise in obesity [95]. Less diversity and richness in the GM of obese individuals is suggestive of GM dysbiosis, which is associated with obesity. *Bacteroides*, *Faecalibacterium prausnitzii*, and *Akkermansia muciniphila* were less common, but the Firmicutes phylum showed a significant increase in frequency [96]. In 2006, Turnbaugh and colleagues discovered variations in the intestinal microecology between individuals classified as lean and those classified as obese. Additionally, they observed that germ-free mice receiving gut microbiota transplants from obese mice exhibited increased fat accumulation and weight gain compared to those receiving transplants from lean mice. These findings imply that dysbiosis of gut microbiota might play a role in the development of obesity [97]. An examination of metagenomic-wide association conducted on a Chinese teenage population that was both lean and obese shown a significant reduction in the prevalence of *Bacteroides thetaiotaomicron* in obese people. A negative association between this and the blood glutamate levels was discovered [98]. Animal studies revealed that *B. thetaiotaomicron gavage* reduced blood concentrations of branched chain amino acids and alleviated diet-induced obesity and weight gain in mice [99]. Moreover, two additional advantages of bariatric surgery point to targeting the GM as a viable therapy for obesity: lowering the quantity of *B. thetaiotaomicron* and altering the GM in ways that may last for ten years [100].

Furthermore, nondigestible carbohydrates may be fermented by the GM, which yields significant metabolites such as succinate and SCFAs. Recent research found that these byproducts are significantly involved in the development of obesity and the issues that accompany it [101]. SCFAs increase energy expenditure and decrease appetite to preserve energy homeostasis and delay the development of obesity [102]. Increased sensations of fullness and reduced food intake are linked to enteroendocrine L-cell-secreted GLP-1 and PYY [103]. Through the receptors GPR41 and GPR43, SCFAs modify immune cell activity, hence controlling GLP-1 and PYY. Moreover, by augmenting lipid oxidation, SCFAs may promote the upregulation of heat-generating proteins (PPARg, PGC1a, and UCP1) as well as proteins associated with lipid oxidation (CPT-I, UCP2) [104]. By doing this, you will consume more calories and prevent obesity. The phylum Succinicella produces succinat via GM. It is possible for bacteria to increase intestinal gluconeogenesis because of their ability to attach to the GPR91 receptor [105]. This might then initiate a gut-brain glucose pathway that affects body mass and fat tissue [106]. In an investigation involving experimental animals, De Vadder and colleagues uncovered the significance of succinate, an essential molecule synthesized internally by *Prevotella*, in the modulation of glucose metabolism. Following oligofructose supplementation, mice showed an increase in cecal succinate content. This resulted in the stimulation of gluconeogenesis and prevented the mice from developing fat and impaired glucose tolerance [107]. Obese individuals often have abnormalities in their gut flora, including decreased intestinal flora richness and variety [108]. Some goods induce obesity by elevating energy consumption and promoting fat oxidation; other products made from genMO indigestible carbohydrates prevent obesity by boosting calorie intake and lowering hunger [109]. To prevent and treat obesity, the aforementioned studies have shown the role that genMO and their byproducts play in the beginning and progression of the disease.

4.4.2 Diabetes

An extensive metabolic disease featured by high blood glucose levels is called diabetes. Gestational diabetes mellitus (GDM), type 2 diabetes mellitus (T2DM), and type 1 diabetes mellitus (T1DM) are all included. An increasing body of research indicates that genMO increase the chance of developing diabetes [110]. T1DM is one kind of diabetes that requires insulin. Although the precise etiology of T1DM is still unknown, it is known that both environmental and genetic factors involve in the disease's formation [111]. When combined, genMO pose a serious risk to the environment. Numerous studies have shown that individuals with T1DM have a GM very different from healthy individuals [112]. Patients with T1DM have lower levels of GM variety, as seen by lower concentrations of Clostridium and Prevotella. T1DM may develop spontaneously in diabetic mice that are not fat [113]. It is possible to increase inflammation in the pancreas of anti-diabetic mice by introducing the fecal microbiota of non-obese diabetic mice into these animals. This suggests that GM has little effect on the pathogenic process of T1DM [114]. Insulin is not necessary for T2DM; instead, it is characterized by decreased insulin output and resistance to insulin. Patients with T2DM exhibited considerably decreased in GM, according to a Swedish cohort research [115]. When compared to healthy persons, the quantity of Dallella in the stomachs of those with T2DM is greater than that of *Bifidobacteria* and *Akkermansia* [116]. According to epidemiological research, those who have undergone a colectomy are more likely than those who have not to develop T2DM [117]. This finding suggests that glucose metabolism may be regulated by the GM and hormone synthesis at the distal end of the colon. In a large prospective cohort study that looked at the richness and diversity of GM in several subcategories, a significant negative correlation was discovered between the baseline g-linolenic acid level and the diversity and composition of GM during the follow-up period [118]. Furthermore, a significant link was discovered between the risk of T2DM and an increase in g-linolenic acid levels [119]. The findings suggests that GM influences the interaction between g-linolenic acid and T2DM. The findings mentioned earlier point to a link between T2DM and GM alterations [120]. Pregnancy complications are prevalent, and there is a higher chance of adverse pregnancy outcomes in women with GDM. Several investigations have shown that gut microbial problems in GDM patients are characterized by an increase in Rumenococcus, Desulfovibrio, Enterobacter, Bacteroides, and Prevotella, along with a decrease in Bifidobacterium and Fischeri [121]. It is interesting that alterations in the GM signals may still be seen eight months after conception. GM dysbiosis coexists with obesity, reduced glucose tolerance, and inflammation in patients with GDM [122]. The organization of metagenomic linkage groups in people with GDM is different from that in the control group, according to a metagenomic investigation [123]. Furthermore, a functional analysis shows that, in comparison to the microbial community in control subjects, the microbial community in patients with GDM has higher levels of membrane transport, phosphotransferase system (PTS), lipopolysaccharide (LPS), energy metabolism pathways, and PTS [124]. Moreover, there is a connection between PTS and the lipopolysaccharide signaling pathways and the level of glucose tolerance [125]. Furthermore, the most recent comprehensive investigation revealed that probiotic supplementation may control GM to stop GDM [126]. Numerous studies have connected genMO byproducts to the development and progression of diabetes. Disruptors of intestinal epithelial cell tight junctions, such as lipopolysaccharide, flagellin, lipoteichoic acid, and peptidoglycan, may cause inflammation by transmitting signals related to TRL2 and TRL4. Insulin resistance and inflammation may both rise in response to elevated levels of branched-chain amino acids and TMAO [127]. Increased glucagon-like peptide-2 production may result in inflammation, endotoxemia, and impaired intestinal barrier function when SCFA levels are raised. By secreting GLP-1 and PYY, intestinal passage and insulin insensitivity may be inhibited. Intestinal cells produce GLP-1 when bile acids (BAs) connect to the TGR5 signaling receptor, improving glucose absorption and insulin sensitivity [128]. A team of Chinese researchers has made significant strides in understanding how small-molecule metabolites regulate blood glucose levels [129]. It has been discovered that hycholic acid (HCA) may successfully cure T2DM by regulating the blood glucose balance [130]. By inducing intestinal endocrine cells to produce and release GLP-1, HCA regulates blood glucose levels by inhibiting the FXR signal and activating TGR5. A clinical data analysis also showed an inverse correlation between blood glucose levels and diabetes and serum HCA [131]. With less emphasis on bile acids, HCA is expected to become a viable treatment drug for diabetes. Additionally, it may be used in combination with other hypoglycemic medications for optimal clinical efficacy [132]. Moreover, because intestinal flora controls the metabolism of cholic acid, research and development projects aimed at developing new drugs will concentrate on determining how to increase cholic acid levels in diabetics by influencing their GM. However, enteric neurons allow the GM to independently regulate blood glucose levels. The distribution of GM throughout the digestive system is not uniform [133]. Innate enteric-associated neurons control intestinal motility and secretion by keeping an eye on dietary cues and signals from commensal microorganisms [134]. Research has shown that gut bacteria control CART neurons, which control blood glucose levels, as opposed to the CNS. Moreover, studies have shown that the GM is essential for the function of intrinsic enteric-associated neurons in different regions of the mouse gut [135]. Environmental factors and genetics work together to influence T1DM. Patients with T1DM and healthy individuals have very different GM. Patients with T1DM have less genetic diversity [136]. Transplanting the feces of healthy individuals might one day lead to a novel approach to the treatment of T1DM. Additionally, GM uses intestinal neurons to sense nutritional cues and signals from commensal microorganisms to regulate secretory processes and intestinal motility [137]. This allows GM to independently manage blood glucose levels. The peripheral neural circuit's finding provides new insights into the roles of GM in metabolism, tissue physiology, immunology, and nervous system activities. In the future, controlling GM or their species' chemical production may be part of treating diabetes [138]. Figure 4.3 illustrates the mechanism by which autoimmune disorders precipitate diabetes.

4.4.3 Non-Alcoholic Fatty Liver Disease

Hepatocellular carcinoma and liver cirrhosis are possible outcomes of untreated non-alcoholic fatty liver disease (NAFLD). It is a severe, ongoing illness. Alongside the obesity crisis, NAFLD is becoming more common [139]. Environmental and genetic factors may have an impact on the complicated disorder known as NAFLD. Although there is still much to learn about the molecular processes behind the development of NAFLD, investigations have demonstrated a strong association between GM and the illness [140].

To prevent intestinal microorganisms from infiltrating the liver, the integrity of the gut barrier must be maintained. In people with NAFLD, intestinal permeability was higher than in healthy

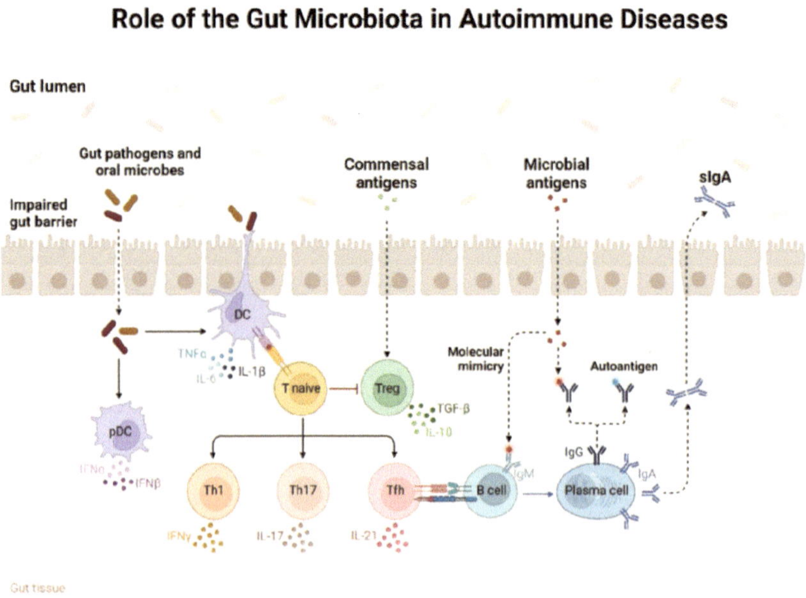

FIGURE 4.3 The figure illustrates the intricate mechanism by which autoimmune disorders precipitate diabetes. At its core, autoimmune dysfunction triggers the immune system to erroneously attack pancreatic beta cells, responsible for insulin production and regulation. This assault leads to a gradual decline in insulin levels, disrupting glucose metabolism and resulting in diabetes onset.

persons, and this was associated with the degree of hepatic steatosis, as per a meta-analysis [141]. Furthermore, NAFLD patients' GM are not like those of healthy individuals. Rumenococcus, Prevotella, and Flavobacterium were less common than Lactobacillus, Dorea, and Streptococcus in the intestines of NAFLD patients [142]. Because of the increased lipid content in their livers, mice that received liver transplants containing the feces of NAFLD patients developed hepatic steatosis [143]. This suggests that alterations in the gut flora might be one of the primary causes of NAFLD [144]. genMO, often referred to as byproducts of saccharolytic and proteolytic fermentation, may influence the liver-gut relationship via a variety of pathways and eventually engage in the growth of non-alcoholic fatty liver disease (NAFLD). SCFAs have a beneficial effect on the liver's function and metabolic processes [145]. In particular, butyrate promotes the production of mucosal secretions and tight junction proteins, which shields the liver from harmful substances and strengthens the intestinal barrier. Some of the possible underlying processes include decreased expression of tumor necrosis factor, increased glycogen storage, decreased activity of liver fatty acid synthase, and enhanced hepatic lipid oxidation via the AMPK–acetyl–CoA carboxylase pathway [146]. Moreover, by controlling BA metabolism and FXR signaling pathways, the GM may prevent the production of hepatic lipids. Studies indicate that glycine and the beginning of non-alcoholic fatty liver disease may be directly related [147]. Test rat non-alcoholic fatty liver disease (NAFLD) is significantly reduced by glycine via changes in blood metabolite concentrations, GM composition, fat oxidation, and glutathione generation [148]. Moreover, during the late fed phase, hepatic lipogenesis is suppressed by fibroblast growth factor-15/19, a gastrointestinal hormone generated by bile acid [149]. Fibroblast growth factor-15/19 suppresses hepatic lipogenesis by binding to its downstream nuclear receptor small heterodimer partner via an epigenetic mechanism. Therefore, DNA methyltransferase DNA methyltransferase-3a recruits lipogenic gene sites such fatty acid synthase. This research might provide fresh insights into the management of metabolic disorders [150]. Furthermore, endogenous ethanol generated by GM may cause direct harm to liver cells, hastening the development of non-alcoholic fatty liver disease. Because endogenous ethanol may alter the function of the gut barrier and increase peripheral cell activation of the nuclear factor-kB pathway, it can produce elevated portal endotoxemia [151]. Studies like the one cited above have shown how important the GM and liver are to the development and course of non-alcoholic fatty liver disease. NAFLD patients have distinct GM from healthy individuals in addition to genetic differences [152]. The gut-liver axis may be impacted by certain metabolites from the GM in a number of ways, which may contribute to the development of non-alcoholic fatty liver disease [153].

4.5 GM AND GASTROINTESTINAL DISEASES

During a protracted evolutionary process, interactions and limitations with the host are how intestinal flora manages intestinal homeostasis. Additionally, it constantly adapts to both the process of natural selection and individual acclimation [154]. Microbes in the intestines help break down food remnants that the host is unable to metabolize, process nutrients in the gastrointestinal tract, and take up nutrients from the host. A growing amount of research is connecting disorders of the digestive tract, such as colorectal cancer (CRC) and inflammatory bowel disease (IBD), to GM dysbiosis [155].

4.5.1 Inflammatory Bowel Disease (IBD)

IBD, which includes Crohn's disease and ulcerative colitis, is caused by immune system dysregulation that results in microbial dysbiosis and intestinal inflammation [156]. Many pieces of evidence suggested that genMO may have a role in promoting intestinal inflammation [157]. Reduced GM diversity, one of the primary indicators of dysbiosis, is often seen in patients with IBD. In contrast to individuals without health complications, individuals with IBD exhibit a notable decline in gut microbiota diversity. This reduction is characterized by a significant decrease in Firmicutes and a concurrent increase in Enterobacter and Proteobacteria [158]. Furthermore, alterations in intestinal fungi have been observed in patients with IBD. Patients with Crohn's disease had higher concentrations of Aspergillus albicans, Candida albicans, and Cryptococcus neoformans in their colons, ilea, and feces, suggesting a significant increase in the variety of fungus present. More evidence confirming the role of GM in the development of IBD has come from investigations conducted on experimental animals [159]. After ingesting the gut bacteria from mice with IBD, animals without germs showed clinical signs of the condition [160]. Similarly, C Eftychi et al. demonstrated that a mix of specific symbiotic flora can stimulate colitis in mice with impaired intestinal barrier functioning. According to the findings above, some gut microorganisms may be associated with inflammatory enteritis; however, further research is required to pinpoint the precise microbial community in question [161]. Furthermore, the substances that GM generate have an impact on the course of IBD. One of the many impacts that SCFAs have on mucosal immunity is that they promote B cell proliferation. producing IL-18 and maintaining the integrity of the mucosa by activating the inflammasome. By directly activating FXR, which has anti-inflammatory properties and offers a defense against chemically induced colitis, BAs have an immunomodulatory impact [162]. GM synthesizes indoles from tryptophan, which modulate mucosal immunity by activating polycyclic aromatic hydrocarbon receptors. Furthermore, by using epigenetics, gut microorganisms may control how host cells behave. The most current research indicates that by encouraging the development of intestinal epithelial cells and the repair of intestinal lesions, the GM metabolite inositol triphosphate may exacerbate IBD [163]. Additionally, it has the ability to reverse the suppressive impact of butyric acid on histone deacetylase 3 and activate protein deacetylase 3. The aforementioned research suggests that genMO may induce or exacerbate IBD; the precise mechanism behind this effect remains unknown. Therefore, to develop a new treatment approach for IBD, additional study on the relationship between GM and IBD is required [164].

4.5.2 Colorectal Cancer (CRC)

CRC is among the most common cancers in the digestive tract [165]. GM dysbiosis and CRC have been linked in studies. The intestinal microbiota of individuals with CRC exhibits dysbiosis, characterized by a decrease in helpful gut bacteria and an increase in dangerous bacteria [166]. The pathogenic bacteria release toxins that harm intestinal lining cells and cause a protracted inflammatory response, both of which are factors in the progression of CRC. Studies conducted in vitro and in vivo have shed further light on the process by which GM promotes the development and metastasis of cancers [167]. Tsoi H et al. observed that Peptostreptococcus anaerobius (Pa) is substantially enriched in feces samples and colon cancer tissues of individuals with CRC, which can induce the abnormal growth of colon cells. Experiments on cells and animals provided further details on the mechanism via which Pa develops CRC. Researchers found that Pa may bind to the PCWBR2 surface protein on CRC cells, which would then cause the PI3K-Akt-NF kB signaling cascade to be triggered, which would eventually cause CRC [168]. Studies conducted by Garrett WS and colleagues have revealed that *Fusobacterium nucleatum* possesses the ability to attach to cadherin and the inhibitory immune receptor TIGIT on the surface of CRC cells via adhesin FadA and Fap2, respectively. This interaction activates the oncogenic Wnt/beta-catenin signaling pathway while simultaneously suppressing the activity of tumor-infiltrating lymphocytes and natural killer cells. Consequently, this mechanism contributes to the advancement of CRC. Furthermore, studies have shown that the GM may change the way many host genes work biologically [169]. The p53 gene alterations in the animal model of CRC affect distinct areas of the gut in different ways. The p53 mutant has the predicted carcinogenic impact in the distal gut, but it may greatly lower the occurrence and development of cancers in the proximal intestine [170]. Additional investigation revealed that the metabolite gallic acid, which is generated by gut bacteria, has the ability to change the tumor-suppressive qualities of mutant p53 into carcinogenic ones [171]. It suggests that altered cancer genes may interact with GM. According to a recently released study, the gastrointestinal tract's microbiota may produce memory T cells that are unique to symbiotic relationships and show cross-reactivity with tumor antigens [172]. Additionally, certain *Enterococci* phages make use of this molecular pathway to boost anti-tumor immune treatment efficacy. According to the previously cited research, genMO may cause cancer directly or indirectly by suppressing the immune system or causing inflammation. Moreover, GM infections are a significant contributor to the incidence of CRC [173].

4.6 CONCLUSION AND FUTURE PERSPECTIVE

Because genMO are closely related to human health and illnesses, they provide enormous prospects for disease prevention, diagnosis, and therapy. This chapter highlights the pathophysiology and mechanism of genMO in multi-system illnesses as an area of ongoing research. In recent years, there has been a growing concern in using metagenomic investigation to investigate the relationship between gut microorganisms and illnesses. Obtaining high-quality reference genome collections might substantially elevate the accuracy and clarity of metagenomic studies and provide the groundwork for closely examining the connection between the GM that lives in the human gut and phenotype. The most recent studies found that the human GM is a gene collection consisting of 200,000 genomes and 171 million protein sequences. The fact that 71% of the strains among 4,644 species are uncultivable and that strain diversity varies greatly across continents indicates that most microorganisms still need experimental study. However, there are still a lot of unanswered questions about the connection between illnesses and gut microorganisms. First, opinions on what constitutes "good" gut flora vary. How may GM disorders be clinically described and robust microbiota screening approaches created for precise diagnosis to identify individuals with intestinal microbial problems? Second, the exact mechanism by which the gut bacterial issue relates to the disease is yet unknown. Moreover, it is unclear whether a single strain of bacteria or the whole GM is to blame for the illness. The exact mechanism by which genMO cause and progress through illness is yet unknown. Moreover, may the gut microorganisms that aid in medicinal treatment be used for illness prevention and enhancing general immune function? Finally, since host immune systems, gut bacteria, and illnesses are all intricately linked, a patient may need a variety of GM therapies. Currently, there are few practical approaches for treating clinical disorders by regulating gut flora. Therefore, for forward-looking studies requiring large sample numbers, it is crucial to include state-of-the-art techniques using approaches like metagenomics, transcriptomics, proteomics, and metabolomics. Individually customized genMO may be used in the future in precision medicine to treat and prevent certain illnesses. Focusing on the composition and metabolic functions of GM may be a novel approach to illness prevention and treatment. To sum up, there are many prospects for fundamental and translational research as well as epidemiological and clinical assessments that may help us better comprehend this intricate gut ecology. This necessitates more meticulous high-throughput sequencing, multi-omics analysis, dynamic long-term monitoring of alterations in gut flora and the onset of disease, and rational experiment design. Look more closely at specific strains to identify potentially harmful bacteria that are conditionally associated with illness. It may realize the promise of precision medicine and provide fresh perspectives on how to treat illness.

REFERENCES

1. E. Thursby, N. Juge, Introduction to the human gut microbiota, *The Biochemical Journal*, 474 (2017) 1823–1836.
2. M.S. Hussain, A. Sharma, R. Kumar, Prebiotics and probiotics: a focused review of applications in respiratory disorders, *Carpathian Journal of Food Science and Technology*, 15 (2023) 183–207.
3. Q. Liu, Q. Luo, Y. Ju, G. Song, Role of the mechanical microenvironment in cancer development and progression, *Cancer Biology & Medicine*, 17 (2020) 282–292.
4. K.S. Alharbi, N.K. Fuloria, S. Fuloria, S.B. Rahman, W.H. Al-Malki, M.A. Javed Shaikh, L. Thangavelu, S.K. Singh, V.S. Rama Raju Allam, N.K. Jha, D.K. Chellappan, K. Dua, G. Gupta, Nuclear factor-kappa B and its role in inflammatory lung disease, *Chemico–Biological Interactions*, 345 (2021) 109568.
5. F. Magne, M. Gotteland, L. Gauthier, A. Zazueta, S. Pesoa, P. Navarrete, R. Balamurugan, The firmicutes/bacteroidetes ratio: a relevant marker of gut dysbiosis in obese patients?, *Nutrients*, 12 (2020) 1474.
6. N. Hasan, H. Yang, Factors affecting the composition of the gut microbiota, and its modulation, *PeerJ*, 7 (2019) e7502.
7. A.A. Bhat, O. Afzal, M. Afzal, G. Gupta, R. Thapa, H. Ali, W. Hassan Almalki, I. Kazmi, S.I. Alzarea, S. Saleem, V.P. Samuel, S.K. Gubbiyappa, V. Subramaniyan, MALAT1: A key regulator in lung cancer pathogenesis and therapeutic targeting, *Pathology, Research and Practice*, 253 (2024) 154991.
8. D. Park, S.H. Park, Y.W. Ban, Y.S. Kim, K.C. Park, N.S. Kim, J.K. Kim, I.Y. Choi, A bioinformatics approach for identifying transgene insertion sites using whole genome sequencing data, *BMC Biotechnology*, 17 (2017) 67.
9. H. Xu, X. Wang, W. Feng, Q. Liu, S. Zhou, Q. Liu, L. Cai, The gut microbiota and its interactions with cardiovascular disease, *Microbial Biotechnology*, 13 (2020) 637–656.
10. A.A. Bhat, G. Gupta, R. Dahiya, R. Thapa, A. Gahtori, M. Shahwan, V. Jakhmola, A. Tiwari, M. Kumar, H. Dureja, S.K. Singh, K. Dua, V. Kumarasamy, V. Subramaniyan, CircRNAs: Pivotal modulators of TGF-β signalling in cancer pathogenesis, *Noncoding RNA Research*, 9 (2024) 277–287.
11. A.S. Bawa, K.R. Anilakumar, Genetically modified foods: safety, risks and public concerns-a review, *Journal of Food Science and Technology*, 50 (2013) 1035–1046.
12. N.B. Charbe, N.D. Amnerkar, B. Ramesh, M.M. Tambuwala, H.A. Bakshi, A.A.A. Aljabali, S.C. Khadse, R. Satheeshkumar, S. Satija, M. Metha, D.K. Chellappan, G. Shrivastava, G. Gupta, P. Negi, K. Dua, F.C. Zacconi, Small interfering RNA for cancer treatment: overcoming hurdles in delivery, *Acta Pharmaceutica Sinica B*, 10 (2020) 2075–2109.
13. R.D. Hills, Jr., B.A. Pontefract, H.R. Mishcon, C.A. Black, S.C. Sutton, C.R. Theberge, Gut microbiome: profound implications for diet and disease, *Nutrients*, 11 (2019) 1613.
14. Y. Chen, J. Zhou, L. Wang, Role and mechanism of gut microbiota in human disease, *Frontiers in Cellular and Infection Microbiology*, 11 (2021) 625913.

15. E. De Pablo-Fernández, A.J. Lees, J.L. Holton, T.T. Warner, Prognosis and neuropathologic correlation of clinical subtypes of parkinson disease, *JAMA Neurology*, 76 (2019) 470–479.
16. G. Gupta, M. Bebawy, T.J.A. Pinto, D.K. Chellappan, A. Mishra, K. Dua, Role of the tristetraprolin (Zinc Finger Protein 36 Homolog) gene in cancer, *Critical Reviews in Eukaryotic Gene Expression*, 28 (2018) 217–221.
17. L. Stefanis, α-Synuclein in Parkinson's disease, *Cold Spring Harbor Perspectives in Medicine*, 2 (2012) a009399.
18. B. Zapała, T. Stefura, M. Wójcik-Pędziwiatr, R. Kabut, M. Bałajewicz-Nowak, T. Milewicz, A. Dudek, A. Stój, M. Rudzińska-Bar, Differences in the composition of gut microbiota between patients with parkinson's disease and healthy controls: a cohort study, *Journal of Clinical Medicine*, 10 (2021) 5698.
19. G. Gupta, D.K. Chellappan, T. de Jesus Andreoli Pinto, P.M. Hansbro, M. Bebawy, K. Dua, Tumor suppressor role of miR-503, *Panminerva Medica*, 60 (2018) 17–24.
20. S. Gerhardt, M.H. Mohajeri, Changes of colonic bacterial composition in parkinson's disease and other neurodegenerative diseases, *Nutrients*, 10 (2018) 708.
21. C.H. Lin, H.C. Lai, M.S. Wu, Gut-oriented disease modifying therapy for Parkinson's disease, *Journal of the Formosan Medical Association = Taiwan yi zhi*, 122 (2023) 9–18.
22. M.A. Hakami, A. Hazazi, O. Abdulaziz, H.H. Almasoudi, A.Y.M. Alhazmi, S.S. Alkhalil, N.S. Alharthi, H.M. Alhuthali, W.H. Almalki, G. Gupta, F.R. Khan, HOTAIR: A key regulator of the Wnt/β-catenin signaling cascade in cancer progression and treatment, *Pathology, Research and Practice*, 253 (2024) 154957.
23. C. Li, L. Cui, Y. Yang, J. Miao, X. Zhao, J. Zhang, G. Cui, Y. Zhang, Gut microbiota differs between parkinson's disease patients and healthy controls in Northeast China, *Frontiers in Molecular Neuroscience*, 12 (2019) 171.
24. M. Zommiti, M.G.J. Feuilloley, N. Connil, Update of probiotics in human world: a nonstop source of benefactions till the end of time, *Microorganisms*, 8 (2020) 1907.
25. M.N.M. Aziz, J. Kumar, K.N. Muhammad Nawawi, R.A. Raja Ali, N.M. Mokhtar, irritable bowel syndrome, depression, and neurodegeneration: a bidirectional communication from gut to brain, *Nutrients*, 13 (2021) 3061.
26. M.A. Hakami, A. Hazazi, F.R. Khan, O. Abdulaziz, K. Alshaghdali, A. Abalkhail, S.A. Nassar, B.I.A. Omar, F. Almarshadi, G. Gupta, A.S. Binshaya, PVT1 lncRNA in lung cancer: A key player in tumorigenesis and therapeutic opportunities, *Pathology, Research and Practice*, 253 (2024) 155019.
27. M. Carabotti, A. Scirocco, M.A. Maselli, C. Severi, The gut-brain axis: interactions between enteric microbiota, central and enteric nervous systems, *Annals of Gastroenterology*, 28 (2015) 203–209.
28. Y. Gong, A. Chen, G. Zhang, Q. Shen, L. Zou, J. Li, Y.B. Miao, W. Liu, Cracking brain diseases from gut microbes-mediated metabolites for precise treatment, *International Journal of Biological Sciences*, 19 (2023) 2974–2998.
29. M.S. Hussain, O. Afzal, G. Gupta, A.S.A. Altamimi, W.H. Almalki, S.I. Alzarea, I. Kazmi, N.K. Fuloria, M. Sekar, D.U. Meenakshi, L. Thangavelu, A. Sharma, Long non-coding RNAs in lung cancer: unraveling the molecular modulators of MAPK signaling, *Pathology, Research and Practice*, 249 (2023) 154738.
30. J. Appleton, The gut-brain axis: influence of microbiota on mood and mental health, *Integrative Medicine*, 17 (2018) 28–32.
31. W.Y. Wang, M.S. Tan, J.T. Yu, L. Tan, Role of pro-inflammatory cytokines released from microglia in Alzheimer's disease, *Annals of Translational Medicine*, 3 (2015) 136.
32. M.S. Hussain, O. Afzal, G. Gupta, A. Goyal, W.H. Almalki, I. Kazmi, S.I. Alzarea, A.S. Alfawaz Altamimi, N. Kukreti, A. Chakraborty, S.K. Singh, K. Dua, Unraveling NEAT1's complex role in lung cancer biology: a comprehensive review, *EXCLI Journal*, 23 (2024) 34–52.
33. E.M. Klann, U. Dissanayake, A. Gurrala, M. Farrer, A.W. Shukla, A. Ramirez-Zamora, V. Mai, V. Vedam-Mai, The gut-brain axis and its relation to Parkinson's disease: a review, *Frontiers in Aging Neuroscience*, 13 (2021) 782082.
34. Y. Bhattarai, P.C. Kashyap, Parkinson's disease: are gut microbes involved?, *American Journal of Physiology. Gastrointestinal and Liver Physiology*, 319 (2020) G529–G540.
35. M.S. Hussain, A.S.A. Altamimi, M. Afzal, W.H. Almalki, I. Kazmi, S.I. Alzarea, S. Saleem, P. Prasher, B. Oliver, S.K. Singh, R. MacLoughlin, K. Dua, G. Gupta, From carcinogenesis to therapeutic avenues: lncRNAs and mTOR crosstalk in lung cancer, *Pathology, Research and Practice*, 253 (2024) 155015.
36. N. Yemula, C. Dietrich, V. Dostal, M. Hornberger, Parkinson's disease and the gut: symptoms, nutrition, and microbiota, *Journal of Parkinson's Disease*, 11 (2021) 1491–1505.

37. D. Yang, D. Zhao, S.Z. Ali Shah, W. Wu, M. Lai, X. Zhang, J. Li, Z. Guan, H. Zhao, W. Li, H. Gao, X. Zhou, L. Yang, The role of the gut microbiota in the pathogenesis of Parkinson's disease, *Frontiers in Neurology*, 10 (2019) 1155.
38. M.S. Hussain, G. Gupta, M. Afzal, S.M. Alqahtani, V.P. Samuel, W. Hassan Almalki, I. Kazmi, S.I. Alzarea, S. Saleem, H. Dureja, S.K. Singh, K. Dua, L. Thangavelu, Exploring the role of lncrna neat1 knockdown in regulating apoptosis across multiple cancer types: a review, *Pathology, Research and Practice*, 252 (2023) 154908.
39. R. Rajmohan, P.H. Reddy, Amyloid-beta and phosphorylated tau accumulations cause abnormalities at synapses of Alzheimer's disease neurons, *Journal of Alzheimer's Disease*, 57 (2017) 975–999.
40. M. Szandruk-Bender, B. Wiatrak, A. Szeląg, The risk of developing Alzheimer's disease and Parkinson's disease in patients with inflammatory bowel disease: a meta-analysis, *Journal of Clinical Medicine*, 11 (2022).
41. M.S. Hussain, A.A. Majami, H. Ali, G. Gupta, W.H. Almalki, S.I. Alzarea, I. Kazmi, R.U. Syed, N.E. Khalifa, M.K. Bin Break, R. Khan, N. Altwaijry, R. Sharma, The complex role of MEG3: an emerging long non-coding RNA in breast cancer, *Pathology, Research and Practice*, 251 (2023) 154850.
42. A. Varesi, E. Pierella, M. Romeo, G.B. Piccini, C. Alfano, G. Bjørklund, A. Oppong, G. Ricevuti, C. Esposito, S. Chirumbolo, A. Pascale, The potential role of gut microbiota in Alzheimer's disease: from diagnosis to treatment, *Nutrients*, 14 (2022) 668.
43. W. Xiao, J. Su, X. Gao, H. Yang, R. Weng, W. Ni, Y. Gu, The microbiota-gut-brain axis participates in chronic cerebral hypoperfusion by disrupting the metabolism of short-chain fatty acids, *Microbiome*, 10 (2022) 62.
44. M. Sochocka, K. Donskow-Łysoniewska, B.S. Diniz, D. Kurpas, E. Brzozowska, J. Leszek, The gut microbiome alterations and inflammation-driven pathogenesis of Alzheimer's disease-a critical review, *Molecular Neurobiology*, 56 (2019) 1841–1851.
45. I. Kazmi, K.S. Alharbi, F.A. Al-Abbasi, W.H. Almalki, S.K. G, A. Yasmeen, A. Khan, G. Gupta, Role of epithelial-to-mesenchymal transition markers in different stages of endometriosis: expression of the snail, slug, Zeb1, and twist genes, *Critical Reviews in Eukaryotic Gene Expression*, 31 (2021) 89–95.
46. D.-o. Seo, D.M. Holtzman, Current understanding of the Alzheimer's disease-associated microbiome and therapeutic strategies, *Experimental & Molecular Medicine*, 56 (2024) 86–94.
47. Z.H. Guo, S. Khattak, M.A. Rauf, M.A. Ansari, M.N. Alomary, S. Razak, C.Y. Yang, D.D. Wu, X.Y. Ji, Role of nanomedicine-based therapeutics in the treatment of CNS disorders, *Molecules*, 28 (2023) 1283.
48. M.A.J. Shaikh, A.S.A. Altamimi, M. Afzal, G. Gupta, N. Singla, R. Gilhotra, W.H. Almalki, I. Kazmi, S.I. Alzarea, P. Prasher, S.K. Singh, K. Dua, Unraveling the impact of miR-21 on apoptosis regulation in glioblastoma, *Pathology, Research and Practice*, 254 (2024) 155121.
49. I. Vojtechova, T. Machacek, Z. Kristofikova, A. Stuchlik, T. Petrasek, Infectious origin of Alzheimer's disease: amyloid beta as a component of brain antimicrobial immunity, *PLoS Pathogens*, 18 (2022) e1010929.
50. S. Chandra, S.S. Sisodia, R.J. Vassar, The gut microbiome in Alzheimer's disease: what we know and what remains to be explored, *Molecular Neurodegeneration*, 18 (2023) 9.
51. X.H. Qian, R.Y. Xie, X.L. Liu, S.D. Chen, H.D. Tang, Mechanisms of short-chain fatty acids derived from gut microbiota in Alzheimer's disease, *Aging and Disease*, 13 (2022) 1252–1266.
52. C. Chen, E.H. Ahn, S.S. Kang, X. Liu, A. Alam, K. Ye, Gut dysbiosis contributes to amyloid pathology, associated with C/EBPβ/AEP signaling activation in Alzheimer's disease mouse model, *Science Advances*, 6 (2020) eaba0466.
53. G. Shrivastava, A.A. Aljabali, S.H. Shahcheraghi, M. Lotfi, M.D. Shastri, S.D. Shukla, D.K. Chellappan, N.K. Jha, K. Anand, H. Dureja, R.M. Pabari, V. Mishra, A.G. Almutary, A.M. Alnuqaydan, N. Charbe, P. Prasher, P. Negi, R. Goyal, K. Dua, G. Gupta, Á. Serrano-Aroca, B. Bahar, D. Barh, P.K. Panda, K. Takayama, K. Lundstorm, P. McCarron, H. Bakshi, M.M. Tambuwala, Targeting LIN28: a new hope in prostate cancer theranostics, *Future Oncology*, 17 (2021) 3873–3880.
54. X. Wang, G. Sun, T. Feng, J. Zhang, X. Huang, T. Wang, Z. Xie, X. Chu, J. Yang, H. Wang, S. Chang, Y. Gong, L. Ruan, G. Zhang, S. Yan, W. Lian, C. Du, D. Yang, Q. Zhang, F. Lin, J. Liu, H. Zhang, C. Ge, S. Xiao, J. Ding, M. Geng, Sodium oligomannate therapeutically remodels gut microbiota and suppresses gut bacterial amino acids-shaped neuroinflammation to inhibit Alzheimer's disease progression, *Cell Research*, 29 (2019) 787–803.
55. Y. Zhang, R. Geng, Q. Tu, Gut microbial involvement in Alzheimer's disease pathogenesis, *Aging*, 13 (2021) 13359–13371.
56. R. Solanki, A. Karande, P. Ranganathan, Emerging role of gut microbiota dysbiosis in neuroinflammation and neurodegeneration, *Frontiers in Neurology*, 14 (2023) 1149618.

57. G. Singhvi, P. Manchanda, V. Krishna Rapalli, S. Kumar Dubey, G. Gupta, K. Dua, MicroRNAs as biological regulators in skin disorders, *Biomedicine & Pharmacotherapy = Biomedecine & Pharmacotherapie*, 108 (2018) 996–1004.
58. K.A. Birdwell, M. Park, Post-transplant cardiovascular disease, *Clinical journal of the American Society of Nephrology*, 16 (2021) 1878–1889.
59. H. Xue, X. Chen, C. Yu, Y. Deng, Y. Zhang, S. Chen, X. Chen, K. Chen, Y. Yang, W. Ling, Gut microbially produced Indole-3-propionic acid inhibits atherosclerosis by promoting reverse cholesterol transport and its deficiency is causally related to atherosclerotic cardiovascular disease, *Circulation Research*, 131 (2022) 404–420.
60. A.K. Duttaroy, Role of gut microbiota and their metabolites on atherosclerosis, hypertension and human blood platelet function: a review, *Nutrients*, 13 (2021) 144.
61. C.L. Tan, Y. Chan, M. Candasamy, J. Chellian, T. Madheswaran, L.P. Sakthivel, V.K. Patel, A. Chakraborty, R. MacLoughlin, D. Kumar, N. Verma, V. Malyla, P.K. Gupta, N.K. Jha, L. Thangavelu, H.P. Devkota, S. Bhatt, P. Prasher, G. Gupta, M. Gulati, S.K. Singh, K.R. Paudel, P.M. Hansbro, B.G. Oliver, K. Dua, D.K. Chellappan, Unravelling the molecular mechanisms underlying chronic respiratory diseases for the development of novel therapeutics via in vitro experimental models, *European Journal of Pharmacology*, 919 (2022) 174821.
62. W.H. Tang, T. Kitai, S.L. Hazen, Gut microbiota in cardiovascular health and disease, *Circulation Research*, 120 (2017) 1183–1196.
63. T. Yang, M.M. Santisteban, V. Rodriguez, E. Li, N. Ahmari, J.M. Carvajal, M. Zadeh, M. Gong, Y. Qi, J. Zubcevic, B. Sahay, C.J. Pepine, M.K. Raizada, M. Mohamadzadeh, Gut dysbiosis is linked to hypertension, *Hypertension*, 65 (2015) 1331–1340.
64. D. Li, Y. Lu, S. Yuan, X. Cai, Y. He, J. Chen, Q. Wu, D. He, A. Fang, Y. Bo, P. Song, D. Bogaert, K. Tsilidis, S.C. Larsson, H. Yu, H. Zhu, E. Theodoratou, Y. Zhu, X. Li, Gut microbiota-derived metabolite trimethylamine-N-oxide and multiple health outcomes: an umbrella review and updated meta-analysis, *The American Journal of Clinical Nutrition*, 116 (2022) 230–243.
65. R. Thapa, O. Afzal, M. Afzal, G. Gupta, A.A. Bhat, W. Hassan Almalki, I. Kazmi, S.I. Alzarea, S. Saleem, P. Arora, S.K. Singh, K. Dua, From LncRNA to metastasis: the MALAT1-EMT axis in cancer progression, *Pathology, Research and Practice*, 253 (2024) 154959.
66. Y.P. Silva, A. Bernardi, R.L. Frozza, The role of short-chain fatty acids from gut microbiota in gut-brain communication, *Frontiers in Endocrinology*, 11 (2020) 25.
67. K.N. Browning, Role of central vagal 5-HT3 receptors in gastrointestinal physiology and pathophysiology, *Frontiers in Neuroscience*, 9 (2015) 413.
68. R. Thapa, O. Afzal, G. Gupta, A.A. Bhat, W.H. Almalki, S.I. Alzarea, I. Kazmi, A.S.A. Altamimi, V. Subramaniyan, L. Thangavelu, S.K. Singh, K. Dua, Unveiling the connection: Long-chain non-coding RNAs and critical signaling pathways in breast cancer, *Pathology, Research and Practice*, 249 (2023) 154736.
69. H. Kimura, Production and physiological effects of hydrogen sulfide, *Antioxidants & Redox Signaling*, 20 (2014) 783–793.
70. F.U. Ahmad, M.A. Sattar, H.A. Rathore, M.H. Abdullah, S. Tan, N.A. Abdullah, E.J. Johns, Exogenous hydrogen sulfide (H2S) reduces blood pressure and prevents the progression of diabetic nephropathy in spontaneously hypertensive rats, *Renal Failure*, 34 (2012) 203–210.
71. J. Maiuolo, C. Carresi, M. Gliozzi, R. Mollace, F. Scarano, M. Scicchitano, R. Macrì, S. Nucera, F. Bosco, F. Oppedisano, S. Ruga, A.R. Coppoletta, L. Guarnieri, A. Cardamone, I. Bava, V. Musolino, S. Paone, E. Palma, V. Mollace, The contribution of gut microbiota and endothelial dysfunction in the development of arterial hypertension in animal models and in humans, *International Journal of Molecular Sciences*, 23 (2022) 3698.
72. N. de la Visitación, I. Robles-Vera, J. Moleón, C. González-Correa, N. Aguilera-Sánchez, M. Toral, M. Gómez-Guzmán, M. Sánchez, R. Jiménez, N. Martin-Morales, F. O'Valle, M. Romero, J. Duarte, Gut Microbiota has a crucial role in the development of hypertension and vascular dysfunction in toll-like receptor 7-driven lupus autoimmunity, *Antioxidants*, 10 (2021) 1426.
73. J. Jordan, R. Moeller, S. Chakraborty, M. Vijay-Kumar, B. Joe, Pressure from the bugs within, *Hypertension*, 73 (2019) 977–979.
74. A. Kumar Singh, C. Cabral, R. Kumar, R. Ganguly, H. Kumar Rana, A. Gupta, M. Rosaria Lauro, C. Carbone, F. Reis, A.K. Pandey, Beneficial effects of dietary polyphenols on gut microbiota and strategies to improve delivery efficiency, *Nutrients*, 11 (2019) 2216.
75. P.A. Jose, D. Raj, Gut microbiota in hypertension, *Current Opinion in Nephrology and Hypertension*, 24 (2015) 403–409.

76. M. Cai, L. Lin, F. Jiang, Y. Peng, S. Li, L. Chen, Y. Lin, Gut microbiota changes in patients with hypertension: a systematic review and meta-analysis, *Journal of Clinical Hypertension*, 25 (2023) 1053–1068.
77. J. Frostegård, Immunity, atherosclerosis and cardiovascular disease, *BMC Medicine*, 11 (2013) 117.
78. B. Deng, L. Tao, Y. Wang, Natural products against inflammation and atherosclerosis: targeting on gut microbiota, *Frontiers in Microbiology*, 13 (2022) 997056.
79. A. Davin-Regli, J.M. Pagès, Enterobacter aerogenes and Enterobacter cloacae; versatile bacterial pathogens confronting antibiotic treatment, *Frontiers in Microbiology*, 6 (2015) 392.
80. J.Y. Yoo, S. Sniffen, K.C. McGill Percy, V.B. Pallaval, B. Chidipi, Gut dysbiosis and immune system in atherosclerotic cardiovascular disease (ACVD), *Microorganisms*, 10 (2022) 108.
81. X. Shen, L. Li, Z. Sun, G. Zang, L. Zhang, C. Shao, Z. Wang, Gut microbiota and atherosclerosis-focusing on the plaque stability, *Frontiers in Cardiovascular Medicine*, 8 (2021) 668532.
82. M. Priyadarshini, K.U. Kotlo, P.K. Dudeja, B.T. Layden, Role of short chain fatty acid receptors in intestinal physiology and pathophysiology, *Comprehensive Physiology*, 8 (2018) 1091–1115.
83. J.Y. Chiang, Bile acid metabolism and signaling, *Comprehensive Physiology*, 3 (2013) 1191–1212.
84. H. Herrema, M. Nieuwdorp, A.K. Groen, Microbiome and cardiovascular disease, *Handbook of Experimental Pharmacology*, 270 (2022) 311–334.
85. Y. Zhu, Q. Li, H. Jiang, Gut microbiota in atherosclerosis: focus on trimethylamine N-oxide, *Acta Pathologica, Microbiologica, et Immunologica Scandinavica*, 128 (2020) 353–366.
86. R. Thapa, H. Ali, O. Afzal, A.A. Bhat, W.H. Almalki, S.I. Alzarea, I. Kazmi, A.S.A. Altamimi, N. Jain, M. Pandey, Unlocking the potential of mesoporous silica nanoparticles in breast cancer treatment, *Journal of Nanoparticle Research*, 25 (2023) 169.
87. E.E. Blaak, E.E. Canfora, Increased circulating choline, L-carnitine and TMAO levels are related to changes in adiposity during weight loss: role of the gut microbiota?, *Annals of Translational Medicine*, 6 (2018) S92.
88. Y.H. Oktaviono, A. Dyah Lamara, P.B.T. Saputra, J.N. Arnindita, D. Pasahari, M.E. Saputra, N.M.A. Suasti, The roles of trimethylamine-N-oxide in atherosclerosis and its potential therapeutic aspect: a literature review, *Biomolecules & Biomedicine*, 23 (2023) 936–948.
89. I. Nemet, P.P. Saha, N. Gupta, W. Zhu, K.A. Romano, S.M. Skye, T. Cajka, M.L. Mohan, L. Li, Y. Wu, M. Funabashi, A.E. Ramer-Tait, S.V. Naga Prasad, O. Fiehn, F.E. Rey, W.H.W. Tang, M.A. Fischbach, J.A. DiDonato, S.L. Hazen, A cardiovascular disease-linked gut microbial metabolite acts via adrenergic receptors, *Cell*, 180 (2020) 862–877.e822.
90. E.E. Ziganshina, D.M. Sharifullina, A.P. Lozhkin, R.N. Khayrullin, I.M. Ignatyev, A.M. Ziganshin, Bacterial communities associated with atherosclerotic plaques from Russian individuals with atherosclerosis, *PLoS One*, 11 (2016) e0164836.
91. M.D. Pieczynska, Y. Yang, S. Petrykowski, O.K. Horbanczuk, A.G. Atanasov, J.O. Horbanczuk, Gut microbiota and its metabolites in atherosclerosis development, *Molecules*, 25 (2020) 594.
92. C. Bárcena, R. Valdés-Mas, P. Mayoral, C. Garabaya, S. Durand, F. Rodríguez, M.T. Fernández-García, N. Salazar, A.M. Nogacka, N. Garatachea, N. Bossut, F. Aprahamian, A. Lucia, G. Kroemer, J.M.P. Freije, P.M. Quirós, C. López-Otín, Healthspan and lifespan extension by fecal microbiota transplantation into progeroid mice, *Nature Medicine*, 25 (2019) 1234–1242.
93. C.D. Davis, The gut microbiome and its role in obesity, *Nutrition Today*, 51 (2016) 167–174.
94. Z. Safari, P. Gérard, The links between the gut microbiome and non-alcoholic fatty liver disease (NAFLD), *Cellular and Molecular Life Sciences*, 76 (2019) 1541–1558.
95. A.M. Mahmoud, An overview of epigenetics in obesity: the role of lifestyle and therapeutic interventions, *International Journal of Molecular Sciences*, 23 (2022) 1341.
96. E. Rinninella, P. Raoul, M. Cintoni, F. Franceschi, G.A.D. Miggiano, A. Gasbarrini, M.C. Mele, What is the healthy gut microbiota composition? a changing ecosystem across age, environment, diet, and diseases, *Microorganisms*, 7 (2019) 14.
97. R. Thapa, O. Afzal, A.A. Bhat, A. Goyal, A.S. Alfawaz Altamimi, W.H. Almalki, S.I. Alzarea, I. Kazmi, S.K. Singh, K. Dua, New horizons in lung cancer management through ATR/CHK1 pathway modulation, *Future Medicinal Chemistry*, 15 (2023) 1807–1818.
98. X. Nie, J. Chen, X. Ma, Y. Ni, Y. Shen, H. Yu, G. Panagiotou, Y. Bao, A metagenome-wide association study of gut microbiome and visceral fat accumulation, *Computational and Structural Biotechnology Journal*, 18 (2020) 2596–2609.

99. N. Yoshida, T. Yamashita, T. Osone, T. Hosooka, M. Shinohara, S. Kitahama, K. Sasaki, D. Sasaki, T. Yoneshiro, T. Suzuki, T. Emoto, Y. Saito, G. Ozawa, Y. Hirota, Y. Kitaura, Y. Shimomura, Y. Okamatsu-Ogura, M. Saito, A. Kondo, S. Kajimura, T. Inagaki, W. Ogawa, T. Yamada, K.I. Hirata, Bacteroides spp. promotes branched-chain amino acid catabolism in brown fat and inhibits obesity, *iScience*, 24 (2021) 103342.
100. D. Ciobârcă, A.F. Cătoi, C. Copăescu, D. Miere, G. Criş an, Bariatric Surgery in obesity: effects on gut microbiota and micronutrient status, *Nutrients*, 12 (2020) 235.
101. G. den Besten, K. van Eunen, A.K. Groen, K. Venema, D.J. Reijngoud, B.M. Bakker, The role of short-chain fatty acids in the interplay between diet, gut microbiota, and host energy metabolism, *Journal of Lipid Research*, 54 (2013) 2325–2340.
102. R. Thapa, O. Afzal, A.S.A. Altamimi, A. Goyal, W.H. Almalki, S.I. Alzarea, I. Kazmi, V. Jakhmola, S.K. Singh, K. Dua, Galangin as an inflammatory response modulator: an updated overview and therapeutic potential, *Chemico-Biological Interactions*, 378 (2023) 110482.
103. M.L. Ricardo-Silgado, A. McRae, A. Acosta, Role of enteroendocrine hormones in appetite and glycemia, *Obesity Medicine*, 23 (2021) 100332.
104. M.H. Kim, S.G. Kang, J.H. Park, M. Yanagisawa, C.H. Kim, Short-chain fatty acids activate GPR41 and GPR43 on intestinal epithelial cells to promote inflammatory responses in mice, *Gastroenterology*, 145 (2013) 396–406.e391–310.
105. S. Fernández-Veledo, J. Vendrell, Gut microbiota-derived succinate: Friend or foe in human metabolic diseases?, *Reviews in Endocrine & Metabolic Disorders*, 20 (2019) 439–447.
106. H.R. Wachsmuth, S.N. Weninger, F.A. Duca, Role of the gut-brain axis in energy and glucose metabolism, *Experimental & Molecular Medicine*, 54 (2022) 377–392.
107. X. Liao, L. Song, B. Zeng, B. Liu, Y. Qiu, H. Qu, Y. Zheng, M. Long, H. Zhou, Y. Wang, Y. Du, J. Xu, R. Shen, Q. Tong, L. Cai, X. Li, S. Guo, G. Yang, Z. Zhu, X. Pu, H. Wei, H. Zheng, Alteration of gut microbiota induced by DPP-4i treatment improves glucose homeostasis, *EBioMedicine*, 44 (2019) 665–674.
108. N. Singla, R. Thapa, R. Kulshrestha, A.A. Bhat, S. Gupta, M. Purohit, S.K. Singh, G. Gupta, Introduction to epigenetics, *Targeting Epigenetics in Inflammatory Lung Diseases*, Springer, 2023, pp. 17–41.
109. C. Sanmiguel, A. Gupta, E.A. Mayer, Gut Microbiome and obesity: a plausible explanation for obesity, *Current Obesity Reports*, 4 (2015) 250–261.
110. American Diabetes Association, Diagnosis and classification of diabetes mellitus, *Diabetes Care*, 32 Suppl 1 (2009) S62–67.
111. S. Marucci, G. De Iaco, G. Lisco, F. Mariano, V.A. Giagulli, E. Guastamacchia, G. De Pergola, V. Triggiani, Eating disorders and type 1 diabetes: a perspective, *Endocrine, Metabolic & Immune Disorders Drug Targets*, 22 (2022) 1245–1251.
112. W.Z. Li, K. Stirling, J.J. Yang, L. Zhang, Gut microbiota and diabetes: From correlation to causality and mechanism, *World Journal of Diabetes*, 11 (2020) 293–308.
113. H. Zhou, X. Zhao, L. Sun, Y. Liu, Y. Lv, X. Gang, G. Wang, Gut microbiota profile in patients with type 1 diabetes based on 16S rRNA gene sequencing: a systematic review, *Disease Markers*, 2020 (2020) 3936247.
114. W. Bielka, A. Przezak, A. Pawlik, The role of the gut microbiota in the pathogenesis of diabetes, *International Journal of Molecular Sciences*, 23 (2022) 480.
115. L. Zhang, J. Chu, W. Hao, J. Zhang, H. Li, C. Yang, J. Yang, X. Chen, H. Wang, Gut microbiota and type 2 diabetes mellitus: association, mechanism, and translational applications, *Mediators of Inflammation*, 2021 (2021) 5110276.
116. M.S. Hussain, G. Gupta, A. Goyal, R. Thapa, W.H. Almalki, I. Kazmi, S.I. Alzarea, S. Fuloria, D.U. Meenakshi, V. Jakhmola, From nature to therapy: Luteolin's potential as an immune system modulator in inflammatory disorders, *Journal of Biochemical and Molecular Toxicology*, 37 (2023) e23482.
117. L. Liu, J. Zhang, Y. Cheng, M. Zhu, Z. Xiao, G. Ruan, Y. Wei, Gut microbiota: a new target for T2DM prevention and treatment, *Frontiers in Endocrinology*, 13 (2022) 958218.
118. M.S.A. Palmnäs-Bédard, G. Costabile, C. Vetrani, S. Åberg, Y. Hjalmarsson, J. Dicksved, G. Riccardi, R. Landberg, The human gut microbiota and glucose metabolism: a scoping review of key bacteria and the potential role of SCFAs, *The American Journal of Clinical Nutrition*, 116 (2022) 862–874.
119. M.M. Gupta, R. Gilhotra, D. Deopa, A.A. Bhat, R. Thapa, N. Singla, R. Kulshrestha, G. Gupta, Epigenetics of pulmonary tuberculosis, *Targeting Epigenetics in Inflammatory Lung Diseases*, Springer, 2023, pp. 127–144.
120. I.S.S. A, A.B. C, J.S. A, Changes in plasma free fatty acids associated with type-2 diabetes, *Nutrients*, 11 (2019) 2022.

121. C. Neri, E. Serafino, M. Morlando, A. Familiari, Microbiome and gestational diabetes: interactions with pregnancy outcome and long-term infant health, *Journal of Diabetes Research*, 2021 (2021) 9994734.
122. R.F. Ionescu, R.M. Enache, S.M. Cretoiu, B.S. Gaspar, Gut microbiome changes in gestational diabetes, *International Journal of Molecular Sciences*, 23 (2022) 12839.
123. R. Dahiya, R. Thapa, N.K. Kumawat, M. Singh, S. Jakhotiya, D. Deopa, Y. Singh, N. Singla, G. Gupta, Epigenetic mechanisms in inflammation, *Targeting Epigenetics in Inflammatory Lung Diseases*, Springer, 2023, pp. 43–55.
124. Y.S. Kuang, J.H. Lu, S.H. Li, J.H. Li, M.Y. Yuan, J.R. He, N.N. Chen, W.Q. Xiao, S.Y. Shen, L. Qiu, Y.F. Wu, C.Y. Hu, Y.Y. Wu, W.D. Li, Q.Z. Chen, H.W. Deng, C.J. Papasian, H.M. Xia, X. Qiu, Connections between the human gut microbiome and gestational diabetes mellitus, *GigaScience*, 6 (2017) 1–12.
125. A.A. Bhat, R. Thapa, A. Goyal, V. Subramaniyan, D. Kumar, S. Gupta, S.K. Singh, K. Dua, G. Gupta, Curcumin-based nanoformulations as an emerging therapeutic strategy for inflammatory lung diseases, Newlands Press Ltd London, UK, 2023, pp. 583–586.
126. K. Kamińska, D. Stenclik, W. Błażejewska, P. Bogdański, M. Moszak, Probiotics in the prevention and treatment of gestational diabetes mellitus (GDM): a review, *Nutrients*, 14 (2022) 4303.
127. R. Jastrząb, D. Graczyk, P. Siedlecki, Molecular and cellular mechanisms influenced by postbiotics, *International Journal of Molecular Sciences*, 22 (2021) 13475.
128. Y. Fu, S. Li, Y. Xiao, G. Liu, J. Fang, A metabolite perspective on the involvement of the gut microbiota in type 2 diabetes, *International Journal of Molecular Sciences*, 24 (2023) 14991.
129. A.A. Bhat, R. Thapa, O. Afzal, N. Agrawal, W.H. Almalki, I. Kazmi, S.I. Alzarea, A.S.A. Altamimi, P. Prasher, S.K. Singh, The pyroptotic role of Caspase-3/GSDME signalling pathway among various cancer: a review, *International Journal of Biological Macromolecules*, 242 (2023) 124832.
130. J. Yin, H. Xing, J. Ye, Efficacy of berberine in patients with type 2 diabetes mellitus, *Metabolism: Clinical and Experimental*, 57 (2008) 712–717.
131. X. Zheng, T. Chen, R. Jiang, A. Zhao, Q. Wu, J. Kuang, D. Sun, Z. Ren, M. Li, M. Zhao, S. Wang, Y. Bao, H. Li, C. Hu, B. Dong, D. Li, J. Wu, J. Xia, X. Wang, K. Lan, C. Rajani, G. Xie, A. Lu, W. Jia, C. Jiang, W. Jia, Hyocholic acid species improve glucose homeostasis through a distinct TGR5 and FXR signaling mechanism, *Cell Metabolism*, 33 (2021) 791–803.e797.
132. K. Takebayashi, Y. Aso, T. Inukai, Role of bile acid sequestrants in the treatment of type 2 diabetes, *World Journal of Diabetes*, 1 (2010) 146–152.
133. X. Zhang, Y. Han, W. Huang, M. Jin, Z. Gao, The influence of the gut microbiota on the bioavailability of oral drugs, *Acta Pharmaceutica Sinica. B*, 11 (2021) 1789–1812.
134. A.A. Bhat, G. Gupta, R. Dahiya, R. Thapa, A. Gahtori, M. Shahwan, V. Jakhmola, A. Tiwari, M. Kumar, H. Dureja, CircRNAs: pivotal modulators of TGF-β signalling in cancer pathogenesis, *Non-Coding RNA Research*, 9 (2024) 277–287.
135. Z.H. Geng, Y. Zhu, Q.L. Li, C. Zhao, P.H. Zhou, Enteric nervous system: the bridge between the gut microbiota and neurological disorders, *Frontiers in Aging Neuroscience*, 14 (2022) 810483.
136. P. Zheng, Z. Li, Z. Zhou, Gut microbiome in type 1 diabetes: a comprehensive review, *Diabetes/Metabolism Research and Reviews*, 34 (2018) e3043.
137. L. He, R. Chen, B. Zhang, S. Zhang, B.A. Khan, D. Zhu, Z. Wu, C. Xiao, B. Chen, F. Chen, K. Hou, Fecal microbiota transplantation treatment of autoimmune-mediated type 1 diabetes mellitus, *Frontiers in Immunology*, 13 (2022) 930872.
138. S. Calabrò, S. Kankowski, M. Cescon, G. Gambarotta, S. Raimondo, K. Haastert-Talini, G. Ronchi, Impact of gut microbiota on the peripheral nervous system in physiological, regenerative and pathological conditions, *International Journal of Molecular Sciences*, 24 (2023) 8061.
139. Y. Chen, W. Wang, M.P. Morgan, T. Robson, S. Annett, Obesity, non-alcoholic fatty liver disease and hepatocellular carcinoma: current status and therapeutic targets, *Frontiers in Endocrinology*, 14 (2023) 1148934.
140. N.B. Wajsbrot, N.C. Leite, G.F. Salles, C.A. Villela-Nogueira, Non-alcoholic fatty liver disease and the impact of genetic, epigenetic and environmental factors in the offspring, *World Journal of Gastroenterology*, 28 (2022) 2890–2899.
141. T.J.I. De Munck, P. Xu, H.J.A. Verwijs, A.A.M. Masclee, D. Jonkers, J. Verbeek, G.H. Koek, Intestinal permeability in human nonalcoholic fatty liver disease: a systematic review and meta-analysis, *Liver International: Official Journal of the International Association for the Study of the Liver*, 40 (2020) 2906–2916.
142. Y. Xia, M. Ren, J. Yang, C. Cai, W. Cheng, X. Zhou, D. Lu, F. Ji, Gut microbiome and microbial metabolites in NAFLD and after bariatric surgery: correlation and causality, *Frontiers in Microbiology*, 13 (2022) 1003755.

143. A.A. Bhat, A. Goyal, R. Thapa, I. Kazmi, S.I. Alzarea, M. Singh, S. Rohilla, T.K. Saini, N. Kukreti, D.U. Meenakshi, Uncovering the complex role of interferon-gamma in suppressing type 2 immunity to cancer, *Cytokine*, 171 (2023) 156376.
144. K. Brandl, B. Schnabl, Intestinal microbiota and nonalcoholic steatohepatitis, *Current Opinion in Gastroenterology*, 33 (2017) 128–133.
145. P. Iruzubieta, J.M. Medina, R. Fernández-López, J. Crespo, F. de la Cruz, A role for gut microbiome fermentative pathways in fatty liver disease progression, *Journal of Clinical Medicine*, 9 (2020) 1369.
146. L. Peng, Z.R. Li, R.S. Green, I.R. Holzman, J. Lin, Butyrate enhances the intestinal barrier by facilitating tight junction assembly via activation of AMP-activated protein kinase in Caco-2 cell monolayers, *The Journal of Nutrition*, 139 (2009) 1619–1625.
147. R. Radun, M. Trauner, Role of FXR in bile acid and metabolic homeostasis in NASH: pathogenetic concepts and therapeutic opportunities, *Seminars in Liver Disease*, 41 (2021) 461–475.
148. A.A. Bhat, O. Afzal, N. Agrawal, R. Thapa, W.H. Almalki, I. Kazmi, S.I. Alzarea, A.S.A. Altamimi, N. Kukreti, A. Chakraborty, A comprehensive review on the emerging role of long non-coding RNAs in the regulation of NF-κB signaling in inflammatory lung diseases, *International Journal of Biological Macromolecules*, 253 (2023) 126951.
149. O. Kucera, Z. Cervinkova, Experimental models of non-alcoholic fatty liver disease in rats, *World Journal of Gastroenterology*, 20 (2014) 8364–8376.
150. G. Guthrie, C. Vonderohe, D. Burrin, Fibroblast growth factor 15/19 expression, regulation, and function: an overview, *Molecular and Cellular Endocrinology*, 548 (2022) 111617.
151. X. Chen, Z. Zhang, H. Li, J. Zhao, X. Wei, W. Lin, X. Zhao, A. Jiang, J. Yuan, Endogenous ethanol produced by intestinal bacteria induces mitochondrial dysfunction in non-alcoholic fatty liver disease, *Journal of Gastroenterology and Hepatology*, 35 (2020) 2009–2019.
152. S. Bashiardes, H. Shapiro, S. Rozin, O. Shibolet, E. Elinav, Non-alcoholic fatty liver and the gut microbiota, *Molecular Metabolism*, 5 (2016) 782–794.
153. S. Pezzino, M. Sofia, G. Faletra, C. Mazzone, G. Litrico, G. La Greca, S. Latteri, Gut-liver axis and non-alcoholic fatty liver disease: a vicious circle of dysfunctions orchestrated by the gut microbiome, *Biology*, 11 (2022) 1622.
154. X. Hu, J. Guo, C. Zhao, P. Jiang, T. Maimai, L. Yanyi, Y. Cao, Y. Fu, N. Zhang, The gut microbiota contributes to the development of Staphylococcus aureus-induced mastitis in mice, *The ISME Journal*, 14 (2020) 1897–1910.
155. C.M. Guinane, P.D. Cotter, Role of the gut microbiota in health and chronic gastrointestinal disease: understanding a hidden metabolic organ, *Therapeutic Advances in Gastroenterology*, 6 (2013) 295–308.
156. V.K.R. Balaga, A. Pradhan, R. Thapa, N. Patel, R. Mishra, N. Singla, Morin: a comprehensive review on its versatile biological activity and associated therapeutic potential in treating cancers, *Pharmacological Research*, 7 (2023) 100264.
157. P.T. Santana, S.L.B. Rosas, B.E. Ribeiro, Y. Marinho, H.S.P. de Souza, Dysbiosis in inflammatory bowel disease: pathogenic role and potential therapeutic targets, *International Journal of Molecular Sciences*, 23 (2022).
158. S. Sultan, M. El-Mowafy, A. Elgaml, T.A.E. Ahmed, H. Hassan, W. Mottawea, Metabolic influences of gut microbiota dysbiosis on inflammatory bowel disease, *Frontiers in Physiology*, 12 (2021) 715506.
159. M. Olaisen, M.L. Richard, V. Beisvåg, A.V.B. Granlund, E.S. Røyset, O. Rué, T.C. Martinsen, A.K. Sandvik, H. Sokol, R. Fossmark, The ileal fungal microbiota is altered in Crohn's disease and is associated with the disease course, *Frontiers in Medicine*, 9 (2022) 868812.
160. Y.J. Zhang, S. Li, R.Y. Gan, T. Zhou, D.P. Xu, H.B. Li, Impacts of gut bacteria on human health and diseases, *International Journal of Molecular Sciences*, 16 (2015) 7493–7519.
161. L. Zheng, X.L. Wen, Gut microbiota and inflammatory bowel disease: the current status and perspectives, *World Journal of Clinical Cases*, 9 (2021) 321–333.
162. H.J. Wu, E. Wu, The role of gut microbiota in immune homeostasis and autoimmunity, *Gut Microbes*, 3 (2012) 4–14.
163. M. Wyatt, K.L. Greathouse, Targeting dietary and microbial tryptophan-indole metabolism as therapeutic approaches to colon cancer, *Nutrients*, 13 (2021) 1189.
164. J.R. Davie, Inhibition of histone deacetylase activity by butyrate, *The Journal of Nutrition*, 133 (2003) 2485s–2493s.
165. M. Tufail, C. Wu, M.S. Hussain, Dietary, addictive and habitual factors, and risk of colorectal cancer, *Nutrition*, 120 (2024) 112334.
166. J. Gagnière, J. Raisch, J. Veziant, N. Barnich, R. Bonnet, E. Buc, M.A. Bringer, D. Pezet, M. Bonnet, Gut microbiota imbalance and colorectal cancer, *World Journal of Gastroenterology*, 22 (2016) 501–518.

167. J. Sun, Impact of bacterial infection and intestinal microbiome on colorectal cancer development, *Chinese Medical Journal*, 135 (2022) 400–408.
168. M.H. Lee, Harness the functions of gut microbiome in tumorigenesis for cancer treatment, *Cancer Communications*, 41 (2021) 937–967.
169. R. Thapa, G. Gupta, A.A. Bhat, W.H. Almalki, S.I. Alzarea, I. Kazmi, S. Saleem, R. Khan, N. Altwaijry, H. Dureja, A review of Glycogen Synthase Kinase-3 (GSK3) inhibitors for cancers therapies, *International Journal of Biological Macromolecules*, 253 (2023) 127375.
170. E. Kadosh, I. Snir-Alkalay, A. Venkatachalam, S. May, A. Lasry, E. Elyada, A. Zinger, M. Shaham, G. Vaalani, M. Mernberger, T. Stiewe, E. Pikarsky, M. Oren, Y. Ben-Neriah, The gut microbiome switches mutant p53 from tumour-suppressive to oncogenic, *Nature*, 586 (2020) 133–138.
171. M. Abubakar, E.R. Puppala, B.J. Dutta, K.C. Maharana, R. Thapa, S. Roshan, B. Tazneem, A. Khan, A.A. Bhat, Epigenetics of rhinovirus, *Targeting Epigenetics in Inflammatory Lung Diseases*, Springer Nature Singapore, 2023, pp. 185–197.
172. M. Wang, G. Yang, Y. Tian, Q. Zhang, Z. Liu, Y. Xin, The role of the gut microbiota in gastric cancer: the immunoregulation and immunotherapy, *Frontiers in Immunology*, 14 (2023) 1183331.
173. M.S. Islam, J. Fan, F. Pan, The power of phages: revolutionizing cancer treatment, *Frontiers in Oncology*, 13 (2023) 1290296.

5 Phthalates and the Gut Microbiome

Gaurav Gupta[1,2], *Md Sadique Hussain*[3*], *Mudasir Maqbool*[4], *Asif Ahmad Bhat*[5], *Riya Thapa*[6], *Sarita Rawat*[7], *Saurabh Gupta*[8], *Roshan S*[9], *Dinesh Kumar Chellappan*[10], *Sachin Kumar Singh*[11,12], *Kamal Dua*[12,13], *and Ayeh Bani Saeid*[12,13]

[1]Centre for Research Impact & Outcome, Chitkara College of Pharmacy, Chitkara University, Rajpura, Punjab, India
[2]Centre of Medical and Bio-allied Health Sciences Research, Ajman University, Ajman, United Arab Emirates
[3]Uttaranchal Institute of Pharmaceutical Sciences, Uttaranchal University, Dehradun, Uttarakhand, India
[4]Department of Pharmaceutical Sciences, University of Kashmir, Srinagar, Jammu and Kashmir, India
[5]School of Pharmacy, Suresh Gyan Vihar University, Jaipur, Rajasthan, India
[6]Uttaranchal Institute of Pharmaceutical Sciences, Uttaranchal University, Dehradun, Uttarakhand, India
[7]Faculty of Pharmacy and Sciences, Amrapali University, Nainital, Uttarakhand, India
[8]Department of Pharmacology Chameli Devi Institute of Pharmacy Indore, Madhya Pradesh, India
[9]Deccan School of Pharmacy, Hyderabad, India
[10]School of Pharmacy, International Medical University, Bukit Jalil, Kuala Lumpur, Malaysia
[11]School of Pharmaceutical Sciences, Lovely Professional University, Phagwara, Punjab, India
[12]Faculty of Health, Australian Research Centre in Complementary and Integrative Medicine, University of Technology Sydney, Ultimo, Australia
[13]Discipline of Pharmacy, Graduate School of Health, University of Technology Sydney, Australia
*Corresponding Author: Md Sadique Hussain.
Email id: sadiquehussain007@gmail.com

5.1 INTRODUCTION

5.1.1 Definition and Background of Phthalates

The persistent exposure to various chemical compounds found in the environment, encountered by humans in their day-to-day endeavors, can have detrimental effects on human health, thus posing a worldwide concern. Phthalates (PAEs) represent a group of chemicals extensively utilized across various industrial sectors, such as plastics, food packaging, playthings, and medical apparatuses. They function as additives in plastics to confer flexibility to the final products [1–3]. PAEs can escape into the surroundings through multiple pathways such as ingestion, inhalation, and skin contact, resulting in their widespread presence as contaminants [4]. Investigations have associated

DOI: 10.1201/9781003489221-5

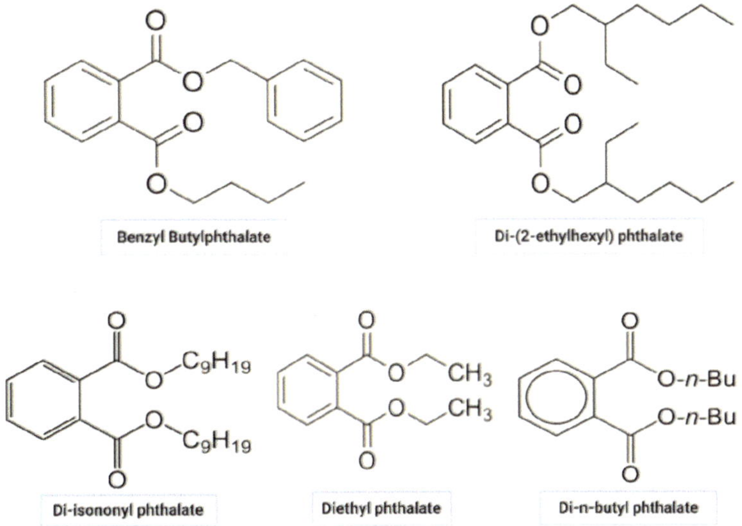

FIGURE 5.1 Most commonly used PAEs.

PAEs with diverse health repercussions, including allergic reactions, respiratory issues like asthma, reproductive harm, and disturbances in the endocrine system. Products aimed at pregnant women and children are particularly worrisome due to potential health ramifications associated with PAEs. In response to reproductive concerns, the European Union has imposed limitations on specific PAEs in consumer goods [5, 6]. Standard analytical methods like gas and liquid chromatography combined with mass spectrometry are frequently employed to identify PAEs in food items and biological specimens. Ongoing research endeavors are aimed at comprehending the repercussions of PAEs on both human health and the environment [7–9]. Figure 5.1 depicts the chemical structure of the frequently utilized PAEs.

5.1.2 Significance of the Gut Microbiome (GM)

The human gastrointestinal tract (GIT) contains an immense number of bacterial cells, estimated to be in a ratio of around 1:1 with our cells [10]; consequently, a dynamic biological community with metabolic processes akin to the liver. There is a growing inclination to designate the microbiome as a novel human organ. Given their close link with the highly biologically active intestinal mucosa, characterized by a high cellular turnover rate, their impact on human health, and their role in disease prevention or precipitation, can be significant. The GM establishes a symbiotic relationship with the host, contributing to the maintenance of wellness and metabolic equilibrium through the generation of various metabolites. The collective genetic material of these microbes inhabiting the intestines is referred to as the metagenome. The establishment of the GM commences at conception and is close to completion within three years in healthy individuals; however, it is susceptible to modification by environmental variables, particularly the composition of the diet and antibiotic therapy [11, 12]. The GM is integral to functions like food breakdown, fermentation, metabolic processes, and synthesis of vitamins. Additionally, it aids in the advancement and regulation of the immune system and ensures a beneficial equilibrium of intestinal microorganisms [13–16].

Research on GM has witnessed rapid global expansion, offering new perspectives in disorders and actions aimed at treating those [17]. Bacterial diversity inhabits the whole GIT, from the mouth to the rectum. The abundance and makeup of Indigenous microbial populations vary greatly across anatomical locations and are influenced by movement rates, host secretions, surrounding factors, substrate accessibility, and the strength of the intestinal barrier [18]. Gram-positive Firmicutes and gram-negative Bacteroidetes are the predominant bacterial groups in the overall microbiome [19]. Bacteria residing in the distal regions of the GIT promote host well-being by synthesizing vitamins and essential amino acids and producing vital metabolites from indigestible food components [20]. Metabolites, particularly short-chain fatty acids (SCFA) like butyrate, propionate, and acetate, result from fermentation of indigestible carbohydrates, predominantly fiber. These SCFAs functions as the primary energy source for colonic epithelial cells, thereby fortifying the gut barrier and defensive mechanisms [21–23].

5.1.3 Rationale for Studying the Interaction between PAEs and the Gut Microbiome

Investigating the interplay between PAEs and GM is crucial given their pervasive presence in the environment and potential effects on human well-being. PAEs, widely utilized as plasticizers in numerous consumer goods, result in extensive exposure through ingestion, inhalation, and skin contact. Increasing research suggests that PAEs can disturb the structure and operation of the GM, leading to dysbiosis and adverse health consequences. Understanding the mechanisms by which PAEs influence the composition and function of the GM is essential for comprehending PAEs-induced toxicity and identifying potential therapeutic strategies to mitigate their detrimental effects on human health. This chapter has the potential to inform regulatory measures, public health initiatives, and consumer decisions geared toward minimizing PAE exposure and preserving GM integrity. Consequently, the primary objective of this chapter is to evaluate and consolidate current knowledge concerning how PAEs influence the composition and operation of the GM, alongside the underlying mechanisms. Additionally, the chapter seeks to explore the implications of PAEs-induced alterations in the GM for human health.

5.2 PHTHALATES: SOURCES, EXPOSURE, AND HEALTH EFFECTS

Exposure of humans to PAEs can be categorized into direct and indirect exposures [24]. Direct exposure involves the direct utilization of PAEs or products having PAEs, while indirect contact occurs when PAEs contaminate the surrounding or certain products via leaching or infiltration. Moreover, exposure pathways can be meticulously classified into ingestion, respiration, cutaneous, and medicinal exposure, respectively [1, 25]. Research indicates that worldwide output of PAE exceeded over 6 million tons annually by 2017 [26]. The rising annual consumption of PAEs corroborates that metabolites of PAEs are frequently detected in human fluid samples.

Food constitutes a fundamental aspect of human life, yet certain studies have detected PAEs in specific food items such as alcoholic drinks, cooking oils, dairy products, drinking water, meat, poultry, and vegetables. The physical characteristics of PAEs, including their fat solubility, low melting point, high boiling point, and solubility in ethanol, facilitate their migration into food upon exposure, subsequently leading to ingestion by the body [27, 28].

In the medical field, di(2-ethylhexyl) phthalate (DEHP) is commonly utilized as a plasticizer in PVC medical devices [29, 30]. Specifically, certain medical devices rely on the flexibility provided by DEHP for indwelling purposes [31]. Since DEHP does not form chemical bonds with plastics, it may be extracted from them upon use, leading to accumulation in the body. Common impurities in drug formulations arise from the leaching, migration, or degradation of plasticizers. Both tablet and capsule medications are coated with polymers to regulate when and where they operate within the GIT [32, 33]. Additionally, semi-volatile PAEs like Diethyl phthalate (DEP) may transfer from primary and secondary packing materials into medicinal goods, as well as ancillary components due to their chemical properties. For instance, research by Singh et al. found that DEP moved from scotch tape into a medicinal product, even with main (LDPE vessel closing mechanism) and secondary external obstacles such as packing materials (carton and labeling). PAEs are capable of migrating and building up in medicinal formulations from auxiliary components [34]. Moreover, earlier studies have reported the negative impacts of DEP on the GIT, cardiovascular, and reproductive systems [35–37].

In the case of children's products, plastics hold a significant market value due to their affordability and durability, including toys, teething items, tableware, etc. [38]. Consequently, children may ingest PAEs via hand-to-mouth and toy-to-mouth contact during usage. Studies have measured PAE levels in toys marketed in Malaysia and observed that DEHP was the most commonly identified PAE [39], which can lead to toxicity and endocrine disruption (ED). Many products, such as cleaners, nail polish, perfumes, and personal care products, contain PAEs [40, 41]. DEP, in particular, is frequently used in fragrances products to prolong their scent [42]. Additionally, research has shown that PAEs in the environment can deposit on cotton and polyester fabrics, with higher concentrations observed in cotton [43]. These items primarily introduce people to PAEs via skin contact [44–46]. PAEs can enter the atmosphere from various sources, including household products, apparel, construction materials, and automotive components, leading to human exposure through inhalation. Studies have quantified the dispersion of PAEs in the atmosphere of Asan Lake, South Korea, and identified DEHP as the most prevalent PAE detected [47].

5.2.1 Health Effects Associated with Phthalate Exposure

The extensive presence of PAEs is raising significant apprehension regarding their potential effects on human health. Over the past 20 years, mounting evidence suggests that these substances, upon conversion into primary and secondary metabolites, may function as suspected ED chemicals (EDCs) by linking with various signaling systems within the endocrine. Utilizing various techniques of biomonitoring that enable the identification of indicators, including both parent compounds and their metabolites in living samples, numerous investigators have endeavored to elucidate the purported function of PAEs in a broad spectrum of pathophysiological illnesses affecting humans (Figure 5.2).

Thus far, exposure to PAEs has been associated with a variety of health concerns, including disturbances in endocrine and reproductive function [48], early onset of puberty, endometriosis, anomalies in sexual development, infertility, alterations in fetal development, breast and skin cancer, obesity, type 2 diabetes [49, 50], attention-deficit hyperactivity disorder, autism spectrum disorders, cardiotoxicity [51], hepatotoxicity, nephrotoxicity [52], asthma, and allergies [53, 54]. Upon absorption, PAEs undergo metabolic conversion through hydrolysis by esterases or lipases to form their respective monoesters or phthalic acid, followed by further transformation via sulfonation or glucuronidation before elimination [55]. To assess the risk associated with PAEs exposure, the US Environmental Protection Agency (EPA) and other scientific organizations have established reference doses, expressed as micrograms (μg) per kilogram (kg) of body weight per day (d), for various PAEs: 3,500 for mono-methyl phthalate (MMP), 800 for DEP, 100 for dibutyl phthalate (DBP), 200 for butyl benzyl phthalate (BBP), 80 for the sum of DEHP metabolites, 120 for diisononyl phthalate (DINP), and 3,500 for di-n-octyl phthalate (DnOP). Specifically, exposure levels to DEHP metabolites (mono(2-ethyl-5-oxohexyl) phthalate (MEOHP), mono(2-ethyl-5-hydroxyhexyl) phthalate (MEHHP), and mono(2-ethylhexyl) phthalate (MEHP)) should not exceed 20 μg/kg bw/d [56]. It is estimated that individuals are exposed to at least 1.0 gram per day of PAEs in their daily lives. Given that PAEs are rapidly metabolized and excreted, urinary analysis is considered a suitable method for assessing these compounds. Notably, children and adults may respond differently to PAE exposure, with children being particularly susceptible due to their hand-to-mouth behavior, which increases the likelihood of DEHP ingestion [57–60].

5.3 FACTORS INFLUENCING GUT MICROBIOME COMPOSITION

The GM refers to the collection of microorganisms residing in the human GIT. This dynamic community is predominantly composed of species from the prokaryotic domain, encompassing bacteria

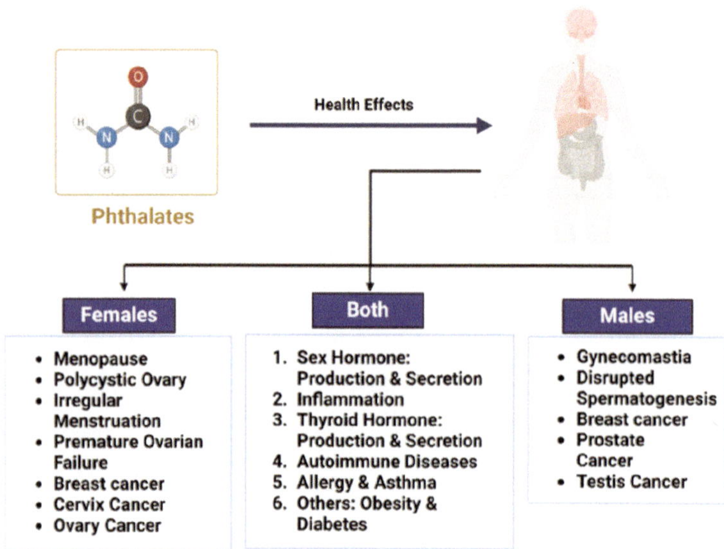

FIGURE 5.2 Health effects in humans caused by Phthalates.

and other unicellular microbes devoid of specialized organelles. Additionally, fungi, parasites, archaea, and viruses contribute to the complexity of this environment. The aggregate genetic and functional profile of these microbial species is termed the GM [61, 62]. From a broader perspective, humans can be categorized based on their enterotypes, representing distinct gut microbial compositions observed in specific populations [63]. Within a single individual, the collective gut microbial species have been demonstrated to possess approximately 3.3 million genes. This vast gene repertoire, when contrasted with the human genome's roughly 23,000 genes, underscores the extensive magnitude and potential impact of these species on human health [64]. Furthermore, the human body harbors nearly an equivalent number of bacterial cells as human cells [65, 66].

The diversity of the GM is intricately molded by various determinants, encompassing geographical distance, host evolutionary lineage, dietary patterns, SES, age, and potentially gender (Figure 5.3). These factors collectively influence the configuration of the GM, thereby holding potential implications for the understanding of human wellness.

5.3.1 Geographical Distance

Geographical distance emerges as a notable determinant influencing the diversity of GM, as demonstrated by studies across diverse animal species. For instance, an inquiry into fish species originating from the Yellow River underscored geographical distance as the primary factor delineating the composition of the GM. The microbial community structure demonstrated an augmented dissimilarity concomitant with the widening geographical span [67]. Likewise, an investigation involving Mediterranean scorpionfishes underscored the significance of geographical provenance in influencing the diversity and constitution of GM communities. This influence surpassed the impact of host phylogenetic relatedness within the studied fish group [68, 69]. These cumulative observations emphasize the influence of geographical distance in molding the composition of GM across diverse animal species.

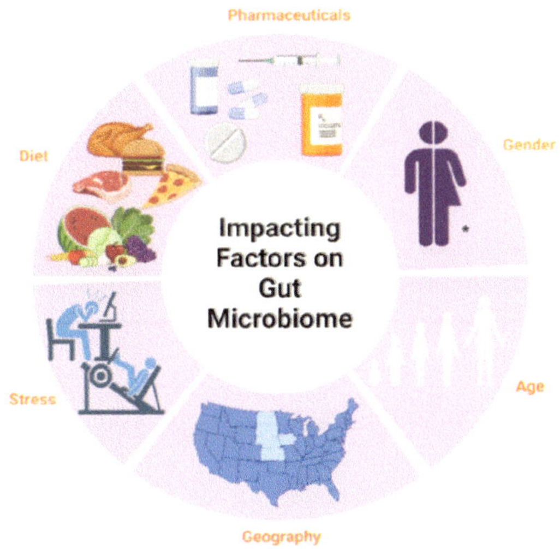

FIGURE 5.3 Factors influencing the composition of Gut microbiota.

5.3.2 Host Evolutionary History

Numerous investigations have probed the impact of host evolutionary lineage on the diversity of the GM, elucidating distinctive insights within diverse vertebrate clades. In a study concentrating on fish species from the Yellow River in China, host phylogeny emerged as a notable determinant influencing the diversity of the GM community. While geographical distance predominantly governed the GM, resulting in an escalated dissimilarity in microbial community structure with greater geographical separation, the microbial communities demonstrated heightened uniformity at higher taxonomic resolutions of host classification, indicative of the predominant influence of homogeneous selection in facilitating community convergence. Although the host diet exerted a substantial impact on GM diversity, it elucidated only a constrained variation across the assessed metrics [67]. In a thorough examination of an extensive dataset encompassing the distal gut 16S rRNA microbiomes of a diverse array of animals, including 80% wild specimens across Mammalia, Aves, Amphibia, Reptilia, and Actinopterygii, the study methodically unraveled the influences of host evolutionary lineage and diet on GM diversity. The results emphasized that each factor distinctly shaped facets of diversity, with evolutionary history emerging as the primary determinant predicting GM differentiation among the investigated vertebrate clades [70, 71]. Additional revelations were gleaned from an examination of two phylogenetically linked fish, namely the three-spined stickleback and the nine-spined stickleback. In this context, habitat emerged as a potent influencer of GM diversity in both host species, surpassing the explanatory efficacy of host species in the variability of GM composition. The investigation underscored that robust, habitat-specific effects took precedence over considerations such as geographical distance and historical freshwater colonization, emphasizing the principal association of GM with localized environmental conditions [72, 73]. These collective inquiries collectively attest that host evolutionary lineage, as manifested in the host's phylogenetic relationships, profoundly molds the diversity of the GM across a spectrum of vertebrate clades, encompassing fish and other wildlife. The amalgamated results imply that the configuration of the GM in diverse animal species is intricately modulated by the interplay of both host evolutionary history and localized environmental conditions.

5.3.3 Diet

Diet emerges as a central determinant influencing the diversity of the GM, and numerous studies have scrutinized the intricate relationship between dietary patterns and the composition of the GM [74]. For instance, an investigation involving a substantial cohort of asymptomatic Chinese adults unveiled that a general dietary pattern characterized by diminished refined grain consumption exhibited a correlation with heightened abundances of the genus Anaerostipes, along with a specific species within it. The study further delineated 17 associations elucidating the linkages between the intake of various food groups and specific genera and species. This was exemplified by adverse associations between red meat consumption and the relative abundances of the genus *Weissella*, including an uncultured species within it [75]. Comparably, an investigation conducted within an extensive population-based cohort in the Netherlands established a correlation between elevated diet quality and enhanced alpha diversity, elucidating a portion of the variability in beta diversity. The overarching quality of the diet demonstrated associations with the relative abundance of four families, specifically Ruminococcaceae, Erysipelotrichaceae, Lachnospiraceae, and Christensenellaceae [76, 77]. A separate investigation concentrating on urban Chinese adults showcased that sustained diet quality, spanning eight food groups, exhibited a positive correlation with GM alpha-diversity and the prevalence of Firmicutes, Tenericutes, and Actinobacteria, along with specific genera/species within these phyla. Notably, this included *Coprococcus*, *Bifidobacterium* (specifically *Bifidobacterium adolescentis*), *Faecalibacterium* (specifically *Faecalibacterium prausnitzii*), and the order RF39 [78]. These collective findings underscore the influence of dietary patterns in shaping the composition of the GM, suggesting potential implications for human health and disease.

5.3.4 Socioeconomic Status (SES)

Several investigations have explored the correlation between SES and the diversity of the GM, yielding noteworthy insights. One study detected a link between the SES of neighborhoods and the diversity of the GM. Notably, it revealed that lower diversity in GM and the presence of multi-drug resistant microorganisms were associated with neighborhoods characterized by lower SES. This research underscores the conceivable influence of the socioeconomic environment as a determinant affecting the diversity of GM [79]. An additional investigation explored the relationship between family SES and host genetics in conjunction with the diversity of GM in infants and children. The findings suggested that family SES played a substantial role in predicting a latent variable reflecting the abundance of six genera identified through whole-genome shotgun sequencing. While SES did not predict alpha diversity, it did exert an influence on beta diversity, underscoring the impact of malleable environmental factors, including family SES, on the initial composition of GM [80]. In a study conducted on a population of school-aged children in Latin America, associations between 14 clinical and environmental covariates and inter-individual variability in GM composition were discerned. Significantly, indicators of SES were identified as influential factors affecting the prevalence of the most abundant taxa and the distribution of enterotypes. Furthermore, age exhibited a positive correlation with heightened diversity in children of normal weight but manifested a negative correlation in overweight and obese children, thereby emphasizing the role of SES in molding the composition of GM in children [81]. In aggregate, these results imply an association between SES, encompassing both neighborhood and family levels, and the diversity of GM. This highlights the impact of social and environmental determinants on the intricate configuration of the GM.

5.3.5 Age

Age emerges as a crucial determinant significantly shaping the diversity of GM, and its influence on both the composition and functionality of GM is subject to a multitude of influencing factors. These alterations, occurring across the lifespan, are intricately shaped by a complex interplay of variables, including but not limited to diet, genetics, gender, SES, disease state, pregnancy, geography, and environmental exposures [82]. In a population-based study concentrating on school-aged children in Latin America, age exhibited a positive correlation with heightened diversity in children of normal weight, while displaying a negative correlation in those who were overweight and obese [81]. Likewise, investigations on six-year-old children underscored the impact of early life determinants, such as exclusive breastfeeding, antibiotic utilization between ages 1–2 years, and the mode of delivery, on the alpha diversity of GM [83]. Acknowledging the growing importance of early-life GM in human health, a study emphasized their pivotal role in diverse aspects, encompassing the digestion of food substances for nutrient release, regulation of pathogens, modulation of the immunity, and impact on organs such as the liver, brain, and endocrine system [84]. Findings from a murine investigation reaffirmed the influence of age on GM composition, with gender also exerting a discernible effect, while geographical location demonstrated a relatively modest impact on diversity. Remarkably, shifts in serum cytokine expression, specifically IL-10 and IL-6, were noted, revealing distinct regulatory patterns in the context of both gender and aging [85]. Fundamentally, age exerts a substantive influence on modulating the diversity of GM, as the configuration of GM undergoes alterations driven by a multifaceted interplay of diverse factors, spanning individual behavioral patterns to environmental exposures.

5.3.6 Gender

Gender emerges as a noteworthy factor capable of influencing GM diversity, as highlighted by various studies. An exhaustive review article investigating the correlation between diet and GM diversity identified gender as one of the individual parameters manifesting robust associations with GM diversity [86]. Contrasting findings emerged from a study conducted on Chinese forest musk deer, where gender did not significantly impact bacterial diversity within GM [87]. In contrast, an examination centered on hyperlipidemic outbred mice unveiled noteworthy gender-related distinctions. Female mice exhibited elevated microbial alpha diversity in comparison to their male counterparts, and beta diversity analysis revealed substantial dissimilarities in the overarching microbial community structure between the two genders [88]. Investigating irritable bowel syndrome (IBS), another study found that females were more susceptible to IBS and exhibited more severe gut dysbiosis compared to males [89]. These findings collectively underscore the potential influence of gender on GM diversity. However, it is crucial to note that the extent of this influence may vary based on factors such as the animal species under investigation and other contextual considerations.

5.4 INTERACTIONS BETWEEN PHTHALATES AND THE GUT MICROBIOME

5.4.1 Impact of Phthalates on Gut Microbial Composition

PAEs have attracted interest because of their widespread presence and possible negative impacts on human health. Recent investigations have begun to disclose the complex relationship between PAE exposure and GM, showing how they can affect microbial composition and host health. Research using various experimental models, from mice to in vitro systems, has consistently shown that exposure to PAEs leads to significant changes in gut microbial composition [90]. A study by Li et al. (2014) using mouse models showed significant changes in the levels of important microbial species after being exposed to DEHP, a frequently encountered PAE compound. There were significant decreases in friendly bacteria like Lactobacillus and Bifidobacterium, leading to changes in the overall variety and abundance of microbes. In exploring the intricate reasons behind changes in gut microbial composition caused by PAEs, it was found that when PAEs are ingested or absorbed, they go through metabolic changes in the body, producing active byproducts that can influence the growth and metabolic processes of microbes. In addition, PAEs could potentially affect microbial cellular integrity or metabolic processes, disrupting the balance of GM [91]. New findings indicate that exposure to PAEs may trigger gut inflammation and weaken the protective barrier of the gut, creating conditions that promote an imbalance in gut bacteria. Findings from human epidemiological studies continue to support the association between PAE exposure and alterations in GM composition. Another study by Wang et al. (2020) in adult cohorts found important links between urinary levels of PAEs metabolites and changes in gut microbial diversity and abundance. The results highlight the importance of PAEs-induced dysbiosis in actual human populations and emphasize the possible implications for public health interventions. The effects of PAEs-induced dysbiosis go beyond disruptions in the gut microbial ecosystem, impacting the overall health of the host and their susceptibility to diseases. The GM is crucial in various physiological processes, such as maintaining metabolic balance, regulating the immune system, and preserving the integrity of the gut barrier. Changes in GM caused by PAEs could potentially increase the risk of various health conditions such as metabolic syndrome, GIT issues, immune system imbalances, and neurodevelopmental disorders [92].

5.4.2 Mechanisms and Influence of Phthalates on the Gut Microbiome

Exploring the complex relationship between PAEs and GM reveals a web of intricate molecular pathways that impact microbial composition and function. PAEs are synthetic chemicals often used as plasticizers and can be present in a variety of consumer products. They have the potential to enter the human body through ingestion, inhalation, or dermal exposure. After entering the body, PAEs go through metabolism, resulting in the creation of bioactive metabolites that have the potential to impact the GM either directly or indirectly. One way of interaction includes the direct antimicrobial effects of PAE metabolites. For example, DEHP, a commonly used PAE compound, can be converted to MEHP in the body through metabolism. Research has demonstrated that MEHP can hinder the growth of specific gut microbial species, showcasing antimicrobial activity [93, 94]. Direct inhibition may result in changes in microbial diversity and abundance, ultimately affecting the overall composition of GM. Additionally, PAEs have the potential to impact GM by changing the availability of necessary nutrients and substrates for microbial growth and metabolism. PAEs might also interfere with the absorption and metabolism of dietary components in the gut, causing alterations in nutrient availability for gut microbes. Moreover, PAEs can influence host-microbe interactions by impacting the expression of genes related to nutrient transport and metabolism in the gut epithelium. Changes in nutrient availability and interactions between the host and microbes can influence the structure and activity of GM. Another way interactions occur is through the adjustment of host immune responses by PAEs, which can have an indirect impact on GM [92, 95]. Research has linked PAE exposure to immune dysregulation, such as inflammation and changes in immune cell function. Such effects can influence the gut environment and change the GM composition. In addition, inflammation in the gut can foster the growth of harmful microbes and hinder the growth of helpful bacteria. Furthermore, PAEs are linked to disrupting the gut epithelial barrier's integrity, potentially causing heightened permeability and leakage of microbial products into the bloodstream. This occurrence, referred to as "leaky gut," has the potential to initiate widespread immune reactions and inflammation, which can, in turn, affect the makeup and operation of GM [96].

Trillions of microorganisms residing in the GIT have attracted significant interest for their impact on different aspects of human health and disease. The GM plays a crucial role in metabolizing environmental chemicals, such as PAEs. PAEs are a group of synthetic chemicals commonly utilized in the manufacturing of plastics, personal care items, and various consumer goods. Due to the potential health risks associated with their widespread presence in the environment, there is a growing interest in studying how GM influences their metabolism. PAEs are not broken down directly by human enzymes. Instead, they are biotransformed mainly in the liver, where they are transformed into metabolites that can be eliminated [97]. Recent studies have brought attention to GM involvement in PAE metabolism, showing that gut bacteria can break down PAEs using different enzymatic reactions. One important way in which gut bacteria impact PAE metabolism is through hydrolysis. Some gut bacteria have esterase enzymes that can break down the ester bonds in PAE molecules, which are esters. Research has demonstrated that bacteria belonging to the *Bacteroides* and *Clostridium* genera are capable of breaking down PAEs like DEP and di-n-butyl phthalate (DBP), resulting in the release of alcohol and phthalic acid components. Metabolites can be further metabolized by gut bacteria or absorbed into the bloodstream for processing by the host. In addition, gut bacteria play a role in reducing PAEs by transforming them into secondary metabolites. Specific bacteria have enzymes that can transform DEHP into MEHP, a metabolite with higher toxicity. On the other hand, certain gut bacteria might help in the breakdown of PAE metabolites, resulting in the creation of less harmful or easily excretable substances. GM composition performs an essential part in shaping the scope and characteristics of PAE metabolism. Research has shown variations in gut microbial communities among individuals, which are linked to differences in PAE metabolism and toxicity [98]. When GM is disrupted, for example, by antibiotic treatment or dysbiosis, it can impact how PAEs are metabolized and eliminated, potentially affecting their overall exposure and harmful effects [99]. Table 5.1 enlists the research done on GM and PAEs.

5.4.3 Bidirectional Relationship between Phthalates and the Gut Microbiome

There is a bidirectional relationship between PAEs and GMs. PAEs can impact the composition and activity of GM, and vice versa, GM can affect the metabolism and effects of PAEs. This intricate relationship has significant consequences for both human health and environmental well-being. PAEs have the potential to impact GM through different pathways. PAEs can have direct antimicrobial effects on gut bacteria. Research has indicated that specific PAEs, like DEP and DBP, may hinder the growth of helpful gut bacteria and encourage the spread of harmful species. When the microbial balance in the gut is disrupted, it can result in dysbiosis, which involves changes in the composition and diversity of GM. This has been linked to different health issues such as inflammatory bowel disease (IBD) and obesity. Moreover, PAEs have the potential to impact the GM by altering host physiology and immune response [90]. Exposure to PAEs has been associated with changes in gut barrier function, intestinal permeability, and immune responses, potentially leading to an environment that promotes dysbiosis. PAEs can interfere with tight junction proteins in the intestinal epithelium, causing higher intestinal permeability and allowing bacteria and microbial products to cross the gut barrier. These factors can lead to immune responses and inflammation, exacerbating dysbiosis and gut dysfunction. On the other hand, GM can impact the metabolism and effects of PAEs. Researchers have discovered that gut bacteria have a wide range of enzymes that can break down PAEs using various processes like hydrolysis, reduction, and oxidation [105]. Furthermore, GM composition and activity can differ significantly among individuals, resulting in variations in PAE metabolism and vulnerability to their negative impacts. Research suggests that GM could potentially help in detoxifying PAEs and reducing their negative impact. Research has indicated that gut bacteria can convert PAEs into metabolites that are less harmful or easier to eliminate from the body, potentially decreasing the risk of negative health effects. On the other hand, specific gut bacteria might transform PAEs into more harmful metabolites, worsening their negative impact on the host's health. In addition, GM can influence how the body responds to PAE exposure, affecting the likelihood of developing PAE-related diseases. For instance, gut bacteria have the potential to engage with the host immune system and impact inflammatory reactions to PAEs. Research suggests that dysbiosis-induced inflammation could worsen the negative impact of PAEs on gut health and functions in the onset of chronic inflammatory conditions like IBD [106].

TABLE 5.1
Phthalates and the gut microbiome.

Chemical	Model	Effect on Gut Microbiota	Conclusions	Refs
DEHP	Infants	↓ *Rothia* spp. and *Bifidobacterium longum*	Exposure to DEHP at dosages commonly used in medical settings changed gut microbiota and reduced diversity, potentially impacting immunological responses in the future.	[100]
DEP, methylparaben, triclosan	Rats	↑ Bacteroidetes (*Prevotella*), Elusimicrobia in PND 62 ↓ Firmicutes (*Bacilli*) in PND 62	Exposure to low concentrations of chemicals throughout the postnatal period in rats altered the gut microbiota by the time they reached adolescence (PND 62). Microbiota variations decrease with age (PND 181).	[101]
MEHP	Mice	↑Firmicutes; *Intestinimonas*, *Holdemanella*, *Coprobacter*, and *Parasutterella* ↓Verrucomicrobia; *Akkermansia*, *Tannerella*, and *Alloprevotella* genus	Postnatal treatment to 0.05 mg/kg MEHP resulted in changes in gray matter and caused adipocyte hypertrophy, as well as cholesterol accumulation, accumulation, and transit.	[102, 103]
DEHP	Mice	↑ *Lachnoclostridium* (*in vivo*) ↓ *Akkermansia*, *Odoribacter*, and *Clostridium sensu stricto* (*in vivo*) ↑ *Alistipes*, *Paenibacillus*, and *Lachnoclostridium* (*in vitro* cultured cecal microbiota)	The control group showed lower fecal α-diversity compared to the 7-day-DEHP group. Higher levels of p-cresol were observed in the 7-day-DEHP therapy groups (10 and 100 μM) compared to the control group. Reduced production of butyrate after 7 days of exposure to 100 μM DEHP.	[102]
DMP, DEP, and DBP combined	Asian carp species silver and bighead	*Achromobacter aegrifaciens*, *Pseudomonas japonica*, *Bacillus subtilis*, and *Pseudomonas putida* isolated and enriched	Asian carp isolated bacteria successfully removed phthalates from experimental settings.	[103, 104]

DBP, dibutyl phthalate; DEHP, di-2-ethylhexyl phthalate; DEP, diethyl phthalate; DMP, dimethyl phthalate; MEHP, mono-(2-ethylhexyl) phthalate; GM, gut microbiome

5.5 EFFECTS OF PHTHALATES ON GUT HEALTH

5.5.1 Disruption of Gut Barrier Function

Understanding the gut barrier is essential as it controls the flow of nutrients and prevents harmful substances from entering the bloodstream. When the gut barrier function is disrupted, it can result in heightened intestinal permeability, commonly referred to as "leaky gut," which is associated with the development of different GIT and systemic conditions. These synthetic chemicals, known as PAEs, are frequently utilized in plastics, personal care items, and food packaging. Research has demonstrated their ability to interfere with gut barrier function through various pathways. PAEs can disturb gut barrier function by affecting tight junction proteins, which play a crucial role in preserving the integrity of the intestinal epithelium. Seals are created by tight junctions between neighboring epithelial cells to control the movement of molecules and ions in the paracellular pathway. PAEs have the potential to disrupt the expression and formation of tight junction proteins, resulting in the weakening of tight junctions and heightened permeability of the gut barrier [93]. Several studies have shown that being exposed to PAEs like DEP and DBP can reduce the levels of certain proteins that maintain barrier function, leading to a weakened barrier. Moreover, PAEs have the potential to trigger oxidative stress (OS) and inflammation in the intestinal epithelium, leading to the impairment of gut barrier function. When there is an imbalance between the production of reactive oxygen species (ROS) and the antioxidant defense mechanisms of cells, OS occurs. Research has demonstrated that PAEs can elevate ROS production and reduce antioxidant enzyme activity in the gut, resulting in oxidative harm to lipids, proteins, and DNA within intestinal epithelial cells. OS may compromise tight junction integrity and worsen barrier dysfunction. In addition, PAEs have the potential to trigger inflammation in the gut mucosa, leading to additional challenges in maintaining barrier function. PAEs-induced inflammation involves the activation of immune cells like macrophages and dendritic cells, along with the release of pro-inflammatory cytokines and chemokines. Studying chronic inflammation in the gut reveals its impact on epithelial cell turnover, mucosal healing, and barrier dysfunction [92, 107]. For instance, animal research has demonstrated that PAE exposure can boost the generation of pro-inflammatory cytokines like interleukin-6 (IL-6), tumor necrosis factor-alpha (TNF-α), and IL-1β in the gut mucosa, resulting in inflammation and barrier disruption. Furthermore, PAEs have the potential to change the composition and function of GM, leading to potential impacts on gut barrier function. Changes in GM composition and diversity, known as dysbiosis, have been linked to higher intestinal permeability and barrier dysfunction. PAE exposure can lead to an imbalance in gut bacteria, affecting the production of SCFAs that are essential for gut health and immune function. Furthermore, specific gut bacteria might convert PAEs into harmful or inflammatory substances that worsen gut barrier dysfunction [108, 109].

5.5.2 Inflammatory Effects of Phthalates

PAEs have been linked to inflammatory effects on the gut. Research has shown a connection between prolonged PAE exposure and the onset and worsening of inflammatory issues in the digestive system via different pathways. PAEs can trigger gut inflammation by activating immune cells and causing the release of pro-inflammatory cytokines and chemokines. PAEs can activate various immune cells like macrophages, dendritic cells, and lymphocytes, resulting in the release of cytokines such as IL-6, TNF-α, IL-1β, and IL-8. These pro-inflammatory substances can attract more immune cells to the inflammation site and enhance tissue damage and inflammation in the gut lining. In addition, PAEs can worsen inflammation in the gut by disturbing the equilibrium between pro-inflammatory and anti-inflammatory signals [97]. Additionally, PAEs have the potential to trigger OS in the gut mucosa, leading to increased inflammation and tissue damage. When there is an imbalance between the production of ROS and the antioxidant defense mechanisms of cells, OS occurs. Research has demonstrated that PAEs can elevate ROS production and reduce antioxidant enzyme activity in the gut epithelium, resulting in oxidative harm to lipids, proteins, and DNA. When OS occurs, it can trigger inflammatory signaling pathways and lead to the release of pro-inflammatory cytokines, which intensify the inflammatory response in the gut [110, 111]. Studies on animals have shown that PAEs can cause inflammation in the gut. For instance, in a study with mice exposed to DEP and DEHP, an uptick in the expression of pro-inflammatory cytokines like TNF-α, IL-6, and IL-1β was noted in the colonic mucosa. It appears that PAEs may trigger an inflammatory reaction in the gut, leading to tissue damage and inflammation. PAEs have the potential to impact immune cells, OS, and gut inflammation by changing GM composition and function. Changes in GM microbial composition and diversity, known as dysbiosis, have been linked to higher vulnerability to inflammatory conditions like IBD and IBS. PAEs have the potential to disturb GM equilibrium, resulting in dysbiosis and immune response dysregulation in the gut lining. Exposure to PAEs has been demonstrated to decrease the levels of helpful bacteria like Bifidobacterium and Lactobacillus while encouraging the proliferation of harmful species [111].

5.5.3 Implications for Gut-Related Diseases

PAEs can impact gut health by interfering with the gut barrier, causing inflammation, and changing GM. The impact of PAEs on the GIT may serve in the advancement and worsening of different gut-related conditions such as IBD, IBS, and colorectal cancer (CRC). One of the key findings regarding PAEs and gut-related diseases is their impact on gut barrier function. PAEs have the potential to disrupt the expression and formation of tight junction proteins, crucial for preserving the integrity of the intestinal epithelium. This disturbance in tight junctions may result in heightened intestinal permeability, commonly referred to as "leaky gut," enabling detrimental substances to breach the gut barrier and access the bloodstream [112]. Research suggests that heightened intestinal permeability plays a role in the onset of IBD like Crohn's disease and ulcerative colitis, along with contributing to systemic inflammation and autoimmune conditions. Research indicates that exposure to PAEs like DEP and DBP may impact gut barrier integrity and raise intestinal permeability in animal models. These effects could potentially lead to the onset of gut-related diseases as they enable luminal antigens, microbial toxins, and inflammatory mediators to enter the bloodstream and initiate immune responses in the gut mucosa. Additionally, PAEs have the potential to trigger inflammation in the GIT, impacting the development and advancement of gut-related conditions. PAEs can activate immune cells and induce the production of pro-inflammatory cytokines and chemokines, resulting in damage and inflammation in the gut mucosa [113]. Research has linked chronic inflammation in the gut to the onset of IBDs like Crohn's disease and ulcerative colitis, along with a higher likelihood of CRC. Research on animals has shown that being exposed to PAEs like DEHP can trigger inflammation and OS in the colon. This can result in symptoms similar to colitis and a higher risk of CRC. It appears that inflammation in the gut caused by PAEs could play a role in the development and advancement of gut-related diseases through damaging tissues, disrupting immune responses, and creating a pro-carcinogenic setting. Moreover, PAEs have the potential to change the structure and operation of GM, impacting gut-related conditions. Changes in the microbial composition and diversity of GM, known as dysbiosis, have been linked to the onset and worsening of different gut-related conditions such as IBD, IBS, and CRC. Consider this: PAEs exposure can disturb the gut bacteria balance, causing dysbiosis and changes in the levels of helpful and harmful species. Such alterations in GM have the potential to impact host-microbe interactions, immune function, and metabolic pathways, which play a role in the development of gut-related diseases [95].

5.6 MODULATION OF GUT MICROBIOME FOR DETOXIFICATION OF PHTHALATES

5.6.1 Role of Gut Microbiome in Phthalate Metabolism and Detoxification Processes

GM is crucial in the metabolism and detoxification processes of PAEs, a group of synthetic chemicals commonly found in plastics, personal care products, and food packaging. Recent investigations have revealed that GM play a significant role in PAE metabolism and detoxification, revealing a complex relationship between GM and exposure to environmental chemicals. PAEs are esters that can be transformed through processes like hydrolysis, reduction, and oxidation. When PAEs are in the gut, they can be broken down by enzymes produced by different types of bacteria living in the GIT. PAE molecules undergo hydrolysis by GM, where ester bonds are broken to release alcohol and phthalic acid moieties [114]. Numerous gut bacterial species have been recognized for their ability to break down PAEs. For instance, bacteria belonging to the Bacteroides and Clostridium genera have esterase enzymes capable of breaking down PAE esters like DEP and DBP. Enzymes facilitate the decomposition of ester bonds in PAE molecules, releasing the alcohol and phthalic acid parts. After being hydrolyzed, PAE metabolites may undergo additional metabolism by gut bacteria or be absorbed into the bloodstream for further processing by the host. Moreover, gut bacteria can help break down PAEs into secondary metabolites. More specifically, specific bacteria have enzymes that can convert DEHP into MEHP, a metabolite that is more toxic than the original compound. On the other hand, certain gut bacteria might help in converting PAE metabolites into less harmful or easily excretable compounds [93]. Moreover, GM plays a role in affecting the bioavailability and toxicity of PAEs by interacting with host physiology and immune function. Modifications in GM diversity and functioning have been linked to alterations in PAE metabolism and increased vulnerability to their negative impacts. Disruption of GM, such as through antibiotic treatment or dysbiosis-related conditions, could potentially change the metabolism and disposition of PAEs, resulting in alterations in their systemic exposure and toxicity. Furthermore, specific gut bacteria have metabolic pathways capable of neutralizing PAEs or reducing their negative impacts. For example, certain bacterial species can break down PAEs into compounds that are less harmful or easier to eliminate using ester hydrolysis, reduction, and oxidation. Exploring detoxification pathways may include the involvement of particular enzymes produced by gut bacteria, leading to the conversion of PAEs into less harmful or more easily excreted metabolites [98]. Moreover, GM might be involved in regulating how the body reacts to PAE exposure, impacting the likelihood of developing PAE-related illnesses. An imbalance in GM can worsen the impact of PAEs on gut health, leading to the onset of chronic inflammatory conditions like IBD and IBS. Aside from being involved in PAE metabolism and detoxification, GM can influence the availability of PAEs and their byproducts by impacting intestinal permeability and absorption. When the gut barrier function is disrupted, such as through dysbiosis or changes in tight junction integrity, it can lead to higher absorption of PAEs and their metabolites into the bloodstream, raising their systemic exposure and potential toxicity [115].

5.6.2 Strategies for Enhancing Gut Microbiome-Mediated Detoxification of Phthalates

To improve the detoxification of PAEs through GM, it is important to focus on adjusting GM composition and activity to enhance its ability to metabolize and detoxify PAEs efficiently. Implementing dietary changes, adding probiotics, using prebiotics, and adjusting lifestyle habits can help improve GM [115, 116].

5.6.2.1 Dietary Interventions

Dietary fiber supports the growth and activity of beneficial gut bacteria by serving as a substrate. Eating a diet full of fruits, vegetables, whole grains, and legumes can help promote the growth of helpful bacteria that can break down PAEs. Various foods like berries, nuts, seeds, tea, and dark chocolate contain polyphenols that have antioxidant and anti-inflammatory properties and could support the growth of beneficial gut bacteria. By incorporating foods rich in polyphenols into your diet, you can help promote a healthy GM and boost its ability to detoxify. These fermented foods are rich in beneficial probiotic bacteria that support gut health, including yogurt, kefir, sauerkraut, kimchi, and kombucha. By incorporating fermented foods into your diet, you can introduce helpful bacteria to your gut, which can improve its capacity to process PAEs and other toxins from the environment [117].

5.6.2.2 Probiotic Supplementation

Probiotics are living microorganisms that provide health advantages when ingested in sufficient quantities. Some probiotic strains, such as Lactobacillus and Bifidobacterium species, have demonstrated enzymatic activities that can metabolize PAEs. Using probiotics with these particular strains could potentially improve GM's ability to detoxify PAEs. It is crucial to carefully select probiotic supplements, taking into account strain specificity, dosage, and viability. Seeking guidance from a health-care professional or registered dietitian can assist in choosing the most suitable probiotic supplement according to individual requirements and health conditions [117, 118].

5.6.2.3 Prebiotic Administration

Non-digestible dietary fibers nourish beneficial gut bacteria, supporting their growth and function. Supplementing with prebiotics can specifically boost the growth of helpful bacteria that can break down PAEs. Inulin, oligofructose, galactooligosaccharides (GOS), and resistant starch are all types of prebiotic compounds. By including prebiotic-rich foods or dietary supplements in your diet, you can help maintain a healthy GM and improve its ability to detoxify. Prebiotic supplements come in different forms such as powders, capsules, and functional foods [117].

5.6.2.4 Lifestyle Modifications

Engaging in physical activity has been linked to a broader and more robust GM. Regular exercise can enhance gut health by boosting microbial diversity and metabolic activity. By including both aerobic and resistance exercises in your regimen, you can potentially improve the health of your GM and boost its capacity to metabolize PAEs. Long-term stress may have a detrimental effect on GM, resulting in an imbalance and compromised immune response. Engaging in stress management techniques like mindfulness meditation, yoga, deep breathing exercises, and spending time in nature can help enhance a healthy GM and aid in its detoxification processes [116].

5.6.3 Potential Therapeutic Interventions Targeting the Gut Microbiome in Phthalate Exposure

Approaching GM as a potential therapeutic option for PAE exposure requires implementing methods to adjust the types and functions of gut bacteria to lessen the negative impacts of PAE exposure on overall health. Such interventions may involve probiotic supplementation, prebiotic administration, fecal microbiota transplantation (FMT), dietary adjustments, and lifestyle changes [116].

5.6.3.1 Probiotic Supplementation

Probiotics are living microorganisms that provide health advantages when ingested in sufficient quantities. Some probiotic strains have demonstrated enzymatic activities that can metabolize PAEs and decrease their toxicity. Consider this: *Lactobacillus* and *Bifidobacterium* species have been extensively researched for their capacity to break down PAEs via hydrolysis and various enzymatic pathways. Using probiotics with these particular strains could potentially rebalance GM and improve the detoxification of PAEs. It is crucial to choose probiotic strains that have demonstrated effectiveness in metabolizing PAEs and to seek guidance from health-care professionals regarding the correct dosage and duration [118, 119].

5.6.3.2 Prebiotic Administration

Non-digestible dietary fibers selectively promote the growth and activity of beneficial gut bacteria. Prebiotics can boost the growth of bacteria that break down PAEs, leading to improved detoxification of these compounds by GM. Inulin, oligofructose, GOS, and resistant starch are all types of prebiotic compounds. By including prebiotic-rich foods or dietary supplements in your diet, you can help promote a healthy GM and improve its ability to metabolize and detoxify PAEs efficiently [118, 119].

5.6.3.3 Fecal Microbiota Transplantation (FMT)

Transferring fecal matter from a healthy donor to a recipient can help restore the balance of GM and treat a range of GI disorders. Exploring the potential of FMT beyond treating specific conditions like *Clostridioides difficile* infection and IBD could also be beneficial in addressing the negative impacts of PAE exposure on gut health. Investigating the therapeutic possibilities of FMT in PAE exposure is currently limited, yet shows promise. Utilizing FMT from healthy donors could potentially introduce bacteria that metabolize PAEs into the recipient's gut. This may help improve detoxification processes and reduce the negative impacts of PAE exposure [118].

5.6.3.4 Dietary Modifications

Implementing dietary strategies to support a balanced GM can enhance the effectiveness of probiotic and prebiotic approaches in reducing the impact of PAE exposure. Eating a diet high in fiber, polyphenols, and fermented foods can help improve GM diversity and metabolic activity, boosting its ability to metabolize and detoxify PAEs. Furthermore, steering clear of processed foods, artificial additives, and PAE-containing products can assist in reducing PAE exposure and lessening their effects on gut health [119].

5.6.3.5 Lifestyle Interventions

Factors such as stress, sleep, and physical activity play a role in shaping GM composition and function. By implementing stress management techniques, making quality sleep a priority, and participating in regular exercise, you can boost the health of your GM and improve its ability to withstand PAE exposure. By making conscious choices to reduce exposure to environmental toxins and pollutants, such as PAEs, you can promote gut health and enhance your overall well-being. Opt for natural and organic products, switch to PAEs-free personal care items, and cut down on plastic usage [119].

5.7 FUTURE DIRECTIONS AND CHALLENGES

Recent advancements in research concerning GM and PAEs have shed light on the intricate relationship between environmental exposures and the composition and function of gut microbes. Nonetheless, several notable research gaps persist, necessitating further exploration. Longitudinal studies examining the prolonged effects of PAE exposure on GM composition and subsequent implications for human health are lacking. Such studies are essential for comprehending the temporal dynamics of PAEs-induced alterations in GM and their potential contribution to disease progression [120]. Although some studies have identified links between PAE exposure and changes in gut microbial diversity and functionality, the underlying mechanisms remain poorly understood [121]. Further research is warranted to elucidate the mechanistic pathways through which PAEs interact with GM, including their effects on microbial metabolism, host immune response, and gut barrier integrity.

Despite recent investigations into the metabolic capabilities of gut microbes in PAE degradation and biotransformation, our knowledge of the specific microbial taxa and enzymatic pathways involved remains limited. Additional research is needed to characterize the microbial metabolism of various PAE compounds and their metabolites within the gut environment. PAE exposure may disrupt the delicate balance between the host and GM, potentially leading to dysbiosis and related health consequences [122]. However, more research is necessary to explore the bidirectional interactions between PAEs, GM, and host physiology, including the influence of host genetic factors on these interactions. Addressing these research gaps will not only enhance our awareness of the intricate relation between PAEs and GM but also inform strategies for mitigating the adverse health effects associated with PAE exposure.

5.8 CONCLUSION

The multifaceted relationship between PAEs and GM underscores the intricate interplay between environmental exposures and host health. PAEs, commonly found in various consumer products, exert significant influence on gut microbial composition and function through diverse mechanisms, with profound implications for human health. These chemicals undergo metabolic transformations upon ingestion or absorption, generating bioactive metabolites that directly impact gut microbial diversity and abundance. Concurrently, dysbiosis induced by PAE exposure or other factors can exacerbate the negative impacts on gut health, potentially leading to chronic inflammatory conditions like IBD. Understanding the bidirectional relationship between PAEs and GM is crucial for elucidating their collective impact on human health and devising strategies to mitigate adverse effects. Future research endeavors should focus on unraveling the intricate molecular pathways underlying PAEs-GM interactions, elucidating their implications for host physiology, and exploring targeted interventions to modulate GM-mediated detoxification of PAEs. The impact of PAEs on gut health extends beyond disruptions in gut barrier function and inflammatory effects to encompass implications for gut-related diseases. Chronic inflammation in the gut, triggered by PAE exposure, serves a significant function in the commencement and progression of IBD, IBS, and CRC. Additionally, PAEs-induced alterations in GM composition and diversity further exacerbate gut barrier dysfunction and inflammation, creating an environment conducive to the development of gut-related diseases. Addressing these research gaps through longitudinal studies and mechanistic investigations will not only enhance our understanding of the relationship between PAEs and GM but also inform strategies for mitigating the adverse health effects associated with PAE exposure. Overall, a comprehensive understanding of the interactions between PAEs and the GM is essential for safeguarding gut health and mitigating the risks posed by environmental exposures.

REFERENCES

1. A. Giuliani, M. Zuccarini, A. Cichelli, H. Khan, M. Reale, Critical review on the presence of phthalates in food and evidence of their biological impact, *International Journal of Environmental Research and Public Health*, 17 (2020) 5655.
2. K.S. Alharbi, Y. Singh, W. Hassan Almalki, S. Rawat, O. Afzal, A.S. Alfawaz Altamimi, I. Kazmi, F.A. Al-Abbasi, S.I. Alzarea, S.K. Singh, S. Bhatt, D.K. Chellappan, K. Dua, G. Gupta, Gut microbiota disruption in COVID-19 or post-COVID illness association with severity biomarkers: a possible role of pre / pro-biotics in manipulating microflora, *Chemico–Biological Interactions,* 358 (2022) 109898.
3. M.S. Hussain, G. Gupta, R. Mishra, N. Patel, S. Gupta, S.I. Alzarea, I. Kazmi, P. Kumbhar, J. Disouza, H. Dureja, N. Kukreti, S.K. Singh, K. Dua, Unlocking the secrets: volatile organic compounds (VOCs) and their devastating effects on lung cancer, *Pathology, Research and Practice*, 255 (2024) 155157.
4. Y. Wang, H. Qian, Phthalates and their impacts on human health, *Healthcare*, 9 (2021) 603.
5. F. Arrigo, F. Impellitteri, G. Piccione, C. Faggio, Phthalates and their effects on human health: focus on erythrocytes and the reproductive system, Comparative biochemistry and physiology. *Toxicology & Pharmacology: CBP*, 270 (2023) 109645.

6. M.S. Hussain, P. Sharma, D.S. Dhanjal, N. Khurana, M. Vyas, N. Sharma, M. Mehta, M.M. Tambuwala, S. Satija, S.S. Sohal, B.G.G. Oliver, H.S. Sharma, Nanotechnology based advanced therapeutic strategies for targeting interleukins in chronic respiratory diseases, *Chemico–Biological Interactions*, 348 (2021) 109637.
7. T. Ulanova, T. Karnazhitskaya, A. Zorina, Techniques for determining phthalates in food products aimed for infants' nutrition: an overview, *АНАЛИЗ РИСКА ЗДОРОВЬЮ*, 1 (2019) 161.
8. S. Ashique, G. De Rubis, E. Sirohi, N. Mishra, M. Rihan, A. Garg, R.J. Reyes, B. Manandhar, S. Bhatt, N.K. Jha, T.G. Singh, G. Gupta, S.K. Singh, D.K. Chellappan, K.R. Paudel, P.M. Hansbro, B.G. Oliver, K. Dua, Short chain fatty acids: fundamental mediators of the gut-lung axis and their involvement in pulmonary diseases, *Chemico–Biological Interactions*, 368 (2022) 110231.
9. B. Bashir, S. Alam, N. Khandale, D. Birla, S. Vishwas, N.K. Pandey, G. Gupta, K.R. Paudel, H. Dureja, P. Kumar, T.G. Singh, G. Kuppusamy, F.C. Zacconi, T.J.A. Pinto, M. Dhanasekaran, M. Gulati, K. Dua, S.K. Singh, Opening avenues for treatment of neurodegenerative disease using post-biotics: breakthroughs and bottlenecks in clinical translation, *Ageing Research Reviews*, 95 (2024) 102236.
10. R. Sender, S. Fuchs, R. Milo, Revised estimates for the number of human and bacteria cells in the body, *PLoS Biology,* 14 (2016) e1002533.
11. S. Subramanian, S. Huq, T. Yatsunenko, R. Haque, M. Mahfuz, M.A. Alam, A. Benezra, J. DeStefano, M.F. Meier, B.D. Muegge, M.J. Barratt, L.G. VanArendonk, Q. Zhang, M.A. Province, W.A. Petri, Jr., T. Ahmed, J.I. Gordon, Persistent gut microbiota immaturity in malnourished Bangladeshi children, *Nature*, 510 (2014) 417–421.
12. M.S. Hussain, A. Sharma, R. Kumar, Prebiotics and probiotics: a focused review of applications in respiratory disorders, *Carpathian Journal of Food Science and Technology*, 15 (2023) 183–207.
13. B.D. Anderson, J.E. Bisanz, Challenges and opportunities of strain diversity in gut microbiome research, *Frontiers in Microbiology,* 14 (2023) 1117122.
14. S. Bhatt, J. Kanoujia, S. Mohana Lakshmi, C.R. Patil, G. Gupta, D.K. Chellappan, K. Dua, Role of brain-gut-microbiota axis in depression: emerging therapeutic avenues, *CNS & Neurological Disorders Drug Targets*, 22 (2023) 276–288.
15. D.K. Chellappan, Q.L. Sze Ning, S.K. Su Min, S.Y. Bin, P.J. Chern, T.P. Shi, S.W. Ee Mei, T.H. Yee, O.J. Qi, L. Thangavelu, S. Rajeshkumar, P. Negi, J. Chellian, R. Wadhwa, G. Gupta, T. Collet, P.M. Hansbro, K. Dua, Interactions between microbiome and lungs: paving new paths for microbiome based bio-engineered drug delivery systems in chronic respiratory diseases, *Chemico–Biological Interactions*, 310 (2019) 108732.
16. S. Hussain, M.A. Khan, R. Rajan, J. Jyoti, S. Sharma, S.K. Sahu, Nanorobots: the future of healthcare, *AIP Conference Proceedings*, AIP Publishing, 2023.
17. J.W. Arnold, J. Roach, M.A. Azcarate-Peril, Emerging technologies for gut microbiome research, *Trends in Microbiology*, 24 (2016) 887–901.
18. D. Graf, R. Di Cagno, F. Fåk, H.J. Flint, M. Nyman, M. Saarela, B. Watzl, Contribution of diet to the composition of the human gut microbiota, *Microbial Ecology in Health and Disease*, 26 (2015) 26164.
19. A.W. Walker, J. Ince, S.H. Duncan, L.M. Webster, G. Holtrop, X. Ze, D. Brown, M.D. Stares, P. Scott, A. Bergerat, P. Louis, F. McIntosh, A.M. Johnstone, G.E. Lobley, J. Parkhill, H.J. Flint, Dominant and diet-responsive groups of bacteria within the human colonic microbiota, *The ISME Journal*, 5 (2011) 220–230.
20. F. Bäckhed, R.E. Ley, J.L. Sonnenburg, D.A. Peterson, J.I. Gordon, Host-bacterial mutualism in the human intestine, *Science* (New York, N.Y.), 307 (2005) 1915–1920.
21. D.L. Topping, P.M. Clifton, Short-chain fatty acids and human colonic function: roles of resistant starch and nonstarch polysaccharides, *Physiological Reviews*, 81 (2001) 1031–1064.
22. L. Corrie, A. Awasthi, J. Kaur, S. Vishwas, M. Gulati, I.P. Kaur, G. Gupta, N. Kommineni, K. Dua, S.K. Singh, Interplay of gut microbiota in polycystic ovarian syndrome: role of gut microbiota, *Mechanistic Pathways and Potential Treatment Strategies, Pharmaceuticals* (Basel, Switzerland), 16 (2023) 197.
23. L. Corrie, H. Singh, M. Gulati, S. Vishwas, D.K. Chellappan, G. Gupta, A.C. Paiva-Santos, F. Veiga, F. Alotaibi, A. Alam, R.D. Eri, P. Prasher, J. Adams, K.R. Paudel, K. Dua, S.K. Singh, Polysaccharide-fecal microbiota-based colon-targeted self-nanoemulsifying drug delivery system of curcumin for treating polycystic ovarian syndrome, *Naunyn-Schmiedeberg's Archives of Pharmacology*, 397 (2024) 6721–6743.
24. T. Schettler, Human exposure to phthalates via consumer products, *International Journal of Andrology*, 29 (2006) 134–139.
25. K. Dua, G. Gupta, R. Awasthi, D.K. Chellappan, Why is there an emerging need to look for a suitable drug delivery platform in targeting and regulating microbiota?, *Panminerva Medica*, 60 (2018) 136–137.

26. M.T. Das, S.S. Kumar, P. Ghosh, G. Shah, S.K. Malyan, S. Bajar, I.S. Thakur, L. Singh, Remediation strategies for mitigation of phthalate pollution: challenges and future perspectives, *Journal of Hazardous Materials*, 409 (2021) 124496.
27. K.V. Naveen, K. Saravanakumar, X. Zhang, A. Sathiyaseelan, M.H. Wang, Impact of environmental phthalate on human health and their bioremediation strategies using fungal cell factory-a review, *Environmental Research*, 214 (2022) 113781.
28. G. Singhvi, V. Girdhar, S. Patil, G. Gupta, P.M. Hansbro, K. Dua, Microbiome as therapeutics in vesicular delivery, *Biomedicine & Pharmacotherapy = Biomedecine & Pharmacotherapie*, 104 (2018) 738–741.
29. S.P. Den Braver-Sewradj, A. Piersma, E.V.S. Hessel, An update on the hazard of and exposure to diethyl hexyl phthalate (DEHP) alternatives used in medical devices, *Critical Reviews in Toxicology*, 50 (2020) 650–672.
30. D.K. Chellappan, R.R. Bhandare, A.B. Shaik, K. Prasad, N.A.A. Suhaimi, W.S. Yap, A. Das, P. Banerjee, N. Ghosh, T. Guith, A. Das, S. Balakrishnan, M. Candasamy, J. Mayuren, K. Palaniveloo, G. Gupta, S.K. Singh, K. Dua, Vaccine for diabetes-where do we stand?, *International Journal of Molecular Sciences*, 23 (2022).
31. G. Malarvannan, M. Onghena, S. Verstraete, E. van Puffelen, A. Jacobs, I. Vanhorebeek, S. Verbruggen, K.F.M. Joosten, G. Van den Berghe, P.G. Jorens, A. Covaci, Phthalate and alternative plasticizers in indwelling medical devices in pediatric intensive care units, *Journal of Hazardous Materials*, 363 (2019) 64–72.
32. A. Broe, A. Pottegård, J. Hallas, T.P. Ahern, R.F. Lamont, P. Damkier, Phthalate exposure from drugs during pregnancy and possible risk of preterm birth and small for gestational age, *European Journal of Obstetrics, Gynecology, and Reproductive Biology*, 240 (2019) 293–299.
33. D.K. Chellappan, N.S. Sivam, K.X. Teoh, W.P. Leong, T.Z. Fui, K. Chooi, N. Khoo, F.J. Yi, J. Chellian, L.L. Cheng, R. Dahiya, G. Gupta, G. Singhvi, S. Nammi, P.M. Hansbro, K. Dua, Gene therapy and type 1 diabetes mellitus, *Biomedicine & Pharmacotherapy = Biomedecine & Pharmacotherapie*, 108 (2018) 1188–1200.
34. G. Singh, R. Gollapalli, A. Blinder, F. Gallo, M. Patel, A case study demonstrating the migration of diethyl phthalate from an ancillary component to the drug product, *Journal of Pharmaceutical and Biomedical Analysis*, 164 (2019) 574–580.
35. L. Casas, M.F. Fernández, S. Llop, M. Guxens, F. Ballester, N. Olea, M.B. Irurzun, L.S. Rodríguez, I. Riaño, A. Tardón, M. Vrijheid, A.M. Calafat, J. Sunyer, Urinary concentrations of phthalates and phenols in a population of Spanish pregnant women and children, *Environ Int*, 37 (2011) 858–866.
36. D.K. Chellappan, W.S. Yap, N.A. Bt Ahmad Suhaimi, G. Gupta, K. Dua, Current therapies and targets for type 2 diabetes mellitus, *Panminerva Medica*, 60 (2018) 117–131.
37. G. Gupta, T. de Jesus Andreoli Pinto, D.K. Chellappan, A. Mishra, H. Malipeddi, K. Dua, A clinical update on metformin and lung cancer in diabetic patients, *Panminerva Medica*, 60 (2018) 70–75.
38. S.M. Kim, J.A. Yoo, J.M. Baek, K.H. Cho, Diethyl phthalate exposure is associated with embryonic toxicity, fatty liver changes, and hypolipidemia via impairment of lipoprotein functions, *Toxicology In Vitro: An International Journal Published in Association with BIBRA*, 30 (2015) 383–393.
39. S.M. Praveena, C. Siok Fong, A.F. Amaruddin, Phthalates in children toys available in Malaysian market: quantification and potential human health risk, *The Journal of Steroid Biochemistry and Molecular Biology*, 213 (2021) 105955.
40. A. Pagoni, O.S. Arvaniti, O.I. Kalantzi, Exposure to phthalates from personal care products: urinary levels and predictors of exposure, *Environmental Research*, 212 (2022) 113194.
41. G. Gupta, G. Singhvi, D.K. Chellappan, S. Sharma, A. Mishra, R. Dahiya, T. de Jesus Andreoli Pinto, K. Dua, Peroxisome proliferator-activated receptor gamma: promising target in glioblastoma, *Panminerva Medica*, 60 (2018) 109–116.
42. C. Philippat, D. Bennett, A.M. Calafat, I.H. Picciotto, Exposure to select phthalates and phenols through use of personal care products among Californian adults and their children, *Environmental Research*, 140 (2015) 369–376.
43. A. Saini, C. Thaysen, L. Jantunen, R.H. McQueen, M.L. Diamond, From clothing to laundry water: investigating the fate of phthalates, brominated flame retardants, and organophosphate Esters, *Environmental Science & Technology*, 50 (2016) 9289–9297.
44. F. Zeng, Y. Lin, K. Cui, J. Wen, Y. Ma, H. Chen, F. Zhu, Z. Ma, Z. Zeng, Atmospheric deposition of phthalate esters in a subtropical city, *Atmospheric Environment*, 44 (2010) 834–840.
45. M.A. Javed Shaikh, R. S, H. Singh, S. Rawat, S. Pathak, A. Mishra, G. Gupta, Role of various gene expressions in etiopathogenesis of type 2 diabetes mellitus, *Advances in Mind-Body Medicine* 35 (2021) 31–39.

46. S.K. Sah, V.P. Samuel, S. Dahiya, Y. Singh, R.M. Gilhotra, G. Gupta, A. Mishra, R.K. Sharma, G.S. Kumar, N. SreeHarsha, D.K. Chellappan, K. Dua, A contemporary biological pathway of islet amyloid polypeptide for the management of diabetic dementia, *Chemico–Biological Interactions*, 306 (2019) 117–122.
47. Y.M. Lee, J.E. Lee, W. Choe, T. Kim, J.Y. Lee, Y. Kho, K. Choi, K.D. Zoh, Distribution of phthalate esters in air, water, sediments, and fish in the Asan Lake of Korea, *Environ Int*, 126 (2019) 635–643.
48. J. Liu, W. Wang, J. Zhu, Y. Li, L. Luo, Y. Huang, W. Zhang, Di(2-ethylhexyl) phthalate (DEHP) influences follicular development in mice between the weaning period and maturity by interfering with ovarian development factors and microRNAs, *Environmental Toxicology*, 33 (2018) 535–544.
49. C. Casals-Casas, B. Desvergne, Endocrine disruptors: from endocrine to metabolic disruption, *Annual Review of Physiology*, 73 (2011) 135–162.
50. E.E. Hatch, J.W. Nelson, M.M. Qureshi, J. Weinberg, L.L. Moore, M. Singer, T.F. Webster, Association of urinary phthalate metabolite concentrations with body mass index and waist circumference: a cross-sectional study of NHANES data, 1999–2002, *Environmental Health: A Global Access Science Source*, 7 (2008) 27.
51. G. Muscogiuri, A. Colao, Phtalates: new cardiovascular health disruptors?, *Archives of Toxicology*, 91 (2017) 1513–1517.
52. P. Ventrice, D. Ventrice, E. Russo, G. De Sarro, Phthalates: European regulation, chemistry, pharmacokinetic and related toxicity, *Environmental Toxicology and Pharmacology*, 36 (2013) 88–96.
53. J.J. Jaakkola, T.L. Knight, The role of exposure to phthalates from polyvinyl chloride products in the development of asthma and allergies: a systematic review and meta-analysis, *Environmental Health Perspectives*, 116 (2008) 845–853.
54. V.P. Samuel, R. Dahiya, Y. Singh, G. Gupta, S.K. Sah, S.K. Gubbiyappa, D.K. Chellappan, K. Dua, Metformin: a salutary candidate for colorectal cancer treatment in patients with diabetes, *Journal of Environmental Pathology, Toxicology and Oncology: Official Organ of the International Society for Environmental Toxicology and Cancer*, 38 (2019) 133–141.
55. S.J. Genuis, S. Beesoon, R.A. Lobo, D. Birkholz, Human elimination of phthalate compounds: blood, urine, and sweat (BUS) study, *The Scientific World Journal,* 2012 (2012) 615068.
56. Y. Wang, H. Zhu, K. Kannan, A review of biomonitoring of phthalate exposures, *Toxics*, 7 (2019) 21.
57. J.M. Weiss, Å. Gustafsson, P. Gerde, Å. Bergman, C.H. Lindh, A.M. Krais, Daily intake of phthalates, MEHP, and DINCH by ingestion and inhalation, *Chemosphere*, 208 (2018) 40–49.
58. M. Wittassek, J. Angerer, Phthalates: metabolism and exposure, *International Journal of Andrology*, 31 (2008) 131–138.
59. V.P. Samuel, G. Gupta, R. Dahiya, D.A. Jain, A. Mishra, K. Dua, Current update on preclinical and clinical studies of resveratrol, a naturally occurring phenolic compound, *Critical Reviews in Eukaryotic Gene Expression*, 29 (2019) 529–537.
60. S.H. Shahcheraghi, A.A.A. Aljabali, M.S. Al Zoubi, V. Mishra, N.B. Charbe, Y.A. Haggag, G. Shrivastava, A.G. Almutary, A.M. Alnuqaydan, D. Barh, K. Dua, D.K. Chellappan, G. Gupta, M. Lotfi, Á. Serrano-Aroca, B. Bahar, Y.K. Mishra, K. Takayama, P.K. Panda, H.A. Bakshi, M.M. Tambuwala, Overview of key molecular and pharmacological targets for diabetes and associated diseases, *Life Sciences*, 278 (2021) 119632.
61. L.K. Ursell, J.L. Metcalf, L.W. Parfrey, R. Knight, Defining the human microbiome, *Nutrition Reviews*, 70 Suppl 1 (2012) S38–S44.
62. M. Kumar, M.S. Hussain, S. Raj, R. Verma, S. Sharma, S.K. Sahu, An overview of treatment modalities and management aspects for obesity, *Current Nutrition and Food Science*, 19 (2023) 105–113.
63. F. Mobeen, V. Sharma, P. Tulika, Enterotype variations of the healthy human gut microbiome in different geographical regions, *Bioinformation*, 14 (2018) 560–573.
64. J. Qin, R. Li, J. Raes, M. Arumugam, K.S. Burgdorf, C. Manichanh, T. Nielsen, N. Pons, F. Levenez, T. Yamada, D.R. Mende, J. Li, J. Xu, S. Li, D. Li, J. Cao, B. Wang, H. Liang, H. Zheng, Y. Xie, J. Tap, P. Lepage, M. Bertalan, J.M. Batto, T. Hansen, D. Le Paslier, A. Linneberg, H.B. Nielsen, E. Pelletier, P. Renault, T. Sicheritz-Ponten, K. Turner, H. Zhu, C. Yu, S. Li, M. Jian, Y. Zhou, Y. Li, X. Zhang, S. Li, N. Qin, H. Yang, J. Wang, S. Brunak, J. Doré, F. Guarner, K. Kristiansen, O. Pedersen, J. Parkhill, J. Weissenbach, P. Bork, S.D. Ehrlich, J. Wang, A human gut microbial gene catalogue established by metagenomic sequencing, *Nature*, 464 (2010) 59–65.
65. R. Sender, S. Fuchs, R. Milo, Are we really vastly outnumbered? revisiting the ratio of bacterial to host cells in humans, *Cell*, 164 (2016) 337–340.

66. M.A.J. Shaikh, G. Gupta, O. Afzal, M.M. Gupta, A. Goyal, A.S.A. Altamimi, S.I. Alzarea, W.H. Almalki, I. Kazmi, P. Negi, S.K. Singh, K. Dua, Sodium alginate-based drug delivery for diabetes management: a review, *International Journal of Biological Macromolecules*, 236 (2023) 123986.
67. B. Pan, X. Han, K. Yu, H. Sun, R. Mu, C.A. Lian, Geographical distance, host evolutionary history and diet drive gut microbiome diversity of fish across the Yellow River, *Molecular Ecology*, 32 (2023) 1183–1196.
68. G. Lilli, C. Sirot, H. Campbell, D. Brophy, C. Graham, I. George, Geographic origin and host's phylogeny are predictors of the gut mucosal microbiota diversity and composition in Mediterranean scorpionfishes (Scorpaena sp.), *Frontiers in Marine Science*, 10 (2023) 1286706.
69. R. Thapa, O. Afzal, A.S. Alfawaz Altamimi, A. Goyal, W.H. Almalki, S.I. Alzarea, I. Kazmi, V. Jakhmola, S.K. Singh, K. Dua, R. Gilhotra, G. Gupta, Galangin as an inflammatory response modulator: an updated overview and therapeutic potential, *Chemico–Biological Interactions*, 378 (2023) 110482.
70. N.D. Youngblut, G.H. Reischer, W. Walters, N. Schuster, C. Walzer, G. Stalder, R.E. Ley, A.H. Farnleitner, Host diet and evolutionary history explain different aspects of gut microbiome diversity among vertebrate clades, *Nature Communications*, 10 (2019) 2200.
71. J. Tiwari, G. Gupta, T. de Jesus Andreoli Pinto, R. Sharma, K. Pabreja, Y. Matta, N. Arora, A. Mishra, R. Sharma, K. Dua, Role of microRNAs (miRNAs) in the pathophysiology of diabetes mellitus, *Panminerva Medica*, 60 (2018) 25–28.
72. A.M. Shankregowda, P. Siriyappagouder, M. Kuizenga, T.M.P. Bal, Y. Abdelhafiz, C. Eizaguirre, J.M.O. Fernandes, V. Kiron, J.A.M. Raeymaekers, Host habitat rather than evolutionary history explains gut microbiome diversity in sympatric stickleback species, *Frontiers in Microbiology*, 14 (2023) 1232358.
73. G.L.L. Xin, Y.P. Khee, T.Y. Ying, J. Chellian, G. Gupta, A.P. Kunnath, S. Nammi, T. Collet, P.M. Hansbro, K. Dua, D.K. Chellappan, Current status on immunological therapies for type 1 diabetes mellitus, *Current Diabetes Report*, 19 (2019) 22.
74. M. Tufail, C. Wu, M.S. Hussain, Dietary, addictive and habitual factors, and risk of colorectal cancer, *Nutrition*, 120 (2024) 112334.
75. Y. Zhang, H. Chen, M. Lu, J. Cai, B. Lu, C. Luo, M. Dai, Habitual diet pattern associations with gut microbiome diversity and composition: results from a Chinese adult cohort, *Nutrients*, 14 (2022) 2639.
76. Z. Miao, W. Du, C. Xiao, C. Su, W. Gou, L. Shen, J. Zhang, Y. Fu, Z. Jiang, Z. Wang, X. Jia, J.S. Zheng, H. Wang, Gut microbiota signatures of long-term and short-term plant-based dietary pattern and cardiometabolic health: a prospective cohort study, *BMC Medicine*, 20 (2022) 204.
77. Z. Chen, D. Radjabzadeh, A. Ikram, A. Uitterlinden, R. Kraaij, T. Voortman, Dietary patterns and gut microbiome composition in a large population-based cohort, *Current Developments in Nutrition*, 4 (2020) 514–514.
78. D. Yu, S.M. Nguyen, Y. Yang, W. Xu, H. Cai, J. Wu, Q. Cai, J. Long, W. Zheng, X.O. Shu, Long-term diet quality is associated with gut microbiome diversity and composition among urban Chinese adults, *The American Journal of Clinical Nutrition*, 113 (2021) 684–694.
79. I. Zuniga-Chaves, S. Eggers, A.E. Kates, N. Safdar, G. Suen, K.M.C. Malecki, Neighborhood socioeconomic status is associated with low diversity gut microbiomes and multi-drug resistant microorganism colonization, *NPJ Biofilms and Microbiomes*, 9 (2023) 61.
80. C.R. Lewis, K.S. Bonham, S.H. McCann, A.R. Volpe, V. D'Sa, M. Naymik, M.D. De Both, M.J. Huentelman, K. Lemery-Chalfant, S.K. Highlander, S.C.L. Deoni, V. Klepac-Ceraj, Family SES is associated with the gut microbiome in infants and children, *Microorganisms*, 9 (2021) 1608.
81. S. Moran-Ramos, B.E. Lopez-Contreras, R. Villarruel-Vazquez, E. Ocampo-Medina, L. Macias-Kauffer, J.N. Martinez-Medina, H. Villamil-Ramirez, P. León-Mimila, B.E. Del Rio-Navarro, I. Ibarra-Gonzalez, M. Vela-Amieva, F.J. Gomez-Perez, R. Velazquez-Cruz, J. Salmeron, Z. Reyes-Castillo, C. Aguilar-Salinas, S. Canizales-Quinteros, Environmental and intrinsic factors shaping gut microbiota composition and diversity and its relation to metabolic health in children and early adolescents: a population-based study, *Gut Microbes*, 11 (2020) 900–917.
82. E.D. Lephart, F. Naftolin, Estrogen action and gut microbiome metabolism in dermal health, *Dermatology and Therapy*, 12 (2022) 1535–1550.
83. S. Gschwendtner, H. Kang, E. Thiering, S. Kublik, B. Fösel, H. Schulz, S. Krauss-Etschmann, J. Heinrich, A. Schöler, M. Schloter, M. Standl, Early life determinants induce sustainable changes in the gut microbiome of six-year-old children, *Scientific Reports*, 9 (2019) 12675.
84. A. Sarkar, J.Y. Yoo, S. Valeria Ozorio Dutra, K.H. Morgan, M. Groer, The association between early-life gut microbiota and long-term health and diseases, *Journal of Clinical Medicine*, 10 (2021) 459.

85. S.E. Webster, D. Vos, T.L. Rothstein, N.E. Holodick, Modulation of microbiome diversity and cytokine expression is influenced in a sex-dependent manner during aging, *Frontiers in Microbiomes*, 1 (2022) 994464.
86. H. Daniel, Diet and gut microbiome and the "Chicken or Egg" problem, *Frontiers in Nutrition*, 8 (2021) 828630.
87. G. Zhao, T. Ma, W. Tang, D. Li, S.K. Mishra, Z. Xu, Q. Wang, H. Jie, Gut microbiome of chinese forest musk deer examined across gender and age, *BioMed Research International*, 2019 (2019) 9291216.
88. M. Kim, M.N. Huda, E. Que, E. Gertz, B. Bennett, Sexual dimorphism of atherosclerosis by gut microbiome in a hyperlipidemic diversity outbred F1 mouse population, *Current Developments in Nutrition*, 4 (2020) 1569–1569.
89. Q. Su, H.M. Tun, Q. Liu, Y.K. Yeoh, J.W.Y. Mak, F.K. Chan, S.C. Ng, Gut microbiome signatures reflect different subtypes of irritable bowel syndrome, *Gut Microbes*, 15 (2023) 2157697.
90. R. Almamoun, P. Pierozan, L. Manoharan, O. Karlsson. Altered gut microbiota community structure and correlated immune system changes in dibutyl phthalate exposed mice, *Ecotoxicology Environmental Safety* 262 (2023) 115321.
91. L. Li, T. Zhang, X.-S. Qin, W. Ge, H.-G. Ma, L.-L. Sun, Z.-M. Hou, H. Chen, P. Chen, G.-Q. Qin, W. Sin, X.-F. Zhang. Exposure to diethylhexyl phthalate (DEHP) results in a heritable modification of imprint genes DNA methylation in mouse oocytes, *Molecular Biology Reports* 41 (2014) 1227–1235.
92. G. Wang, Q. Chen, P. Tian, L. Wang, X. Li, Y.-k. Lee, J. Zhao, H. Zhang, W. Chen, Gut microbiota dysbiosis might be responsible to different toxicity caused by Di-(2-ethylhexyl) phthalate exposure in murine rodents, *Environmental Pollution* 261 (2020) 114164.
93. T.-N. Yang, X.-N. Li, Y.-X. Wang, X.-Y. Ma, J.-L. Li, Disrupted microbiota-barrier-immune interaction in phthalates-mediated barrier defect in the duodenum, *Chemosphere* 308 (2022) 136275.
94. Y. Deng, Z. Yan, R. Shen, M. Wang, Y. Huang, H. Ren, Y. Zhang, B. Lemos, Microplastics release phthalate esters and cause aggravated adverse effects in the mouse gut, *Environment International* 143 (2020) 105916.
95. S.P. Goyal, C. Saravanan, An insight into the critical role of gut microbiota in triggering the phthalate-induced toxicity and its mitigation using probiotics, *The Science of the Total Environment* 904 (2023) 166889.
96. Q. Chen, Q. Kong, P. Tian, Y. He, J. Zhao, H. Zhang, G. Wang, W. Chen, Lactic acid bacteria alleviate di-(2-ethylhexyl) phthalate-induced liver and testis toxicity via their bio-binding capacity, antioxidant capacity and regulation of the gut microbiota, *Ecotoxicology and Environmental Safety* 305 (2022) 119197.
97. Z. Xiong, Y. Zeng, J. Zhou, R. Shu, X. Xie, Z. Fu, Exposure to dibutyl phthalate impairs lipid metabolism and causes inflammation via disturbing microbiota-related gut–liver axis, *Acta Biochimica et Biophysica Sinica* 52 (2020) 1382–1393.
98. D. Xiong, Y. Chen, S. Zhu, L. Liu, L. Zhao, C. Zeng, Y. Li, H. Wang, L. Tu, K. Zou, Exploring the relationship between urinary phthalate metabolites and Crohn's disease via oxidative stress, and the potential moderating role of gut microbiota: a conditional mediation model, *Free Radical Biology and Medicine* 208 (2023) 468–477.
99. Y. Zhang, Z. Yang, X. Li, P. Song, J. Wang, Effects of diisononyl phthalate exposure on the oxidative stress and gut microorganisms in earthworms (*Eisenia fetida*), *Science of the Total Environment* 822 (2022) 153563.
100. Y.-N. Yang, Y.-C.S. Yang, I.-H. Lin, Y.-Y. Chen, H.-Y. Lin, C.-Y. Wu, Y.-T. Su, Y.-J. Yang, S.-N. Yang, J.-L.J.F. Suen, Phthalate exposure alters gut microbiota composition and IgM vaccine response in human newborns, *Food and Chemical Toxicology*, 132 (2019) 110700.
101. J. Hu, V. Raikhel, K. Gopalakrishnan, H. Fernandez-Hernandez, L. Lambertini, F. Manservisi, L. Falcioni, L. Bua, F. Belpoggi, S. Teitelbaum, Effect of postnatal low-dose exposure to environmental chemicals on the gut microbiome in a rodent model, *Microbiome* 4 (2016) 1–11.
102. C. Wang, S. Yue, Z. Hao, G. Ren, D. Lu, Q. Zhang, M. Zhao, Pubertal exposure to the endocrine disruptor mono-2-ethylhexyl ester at body burden level caused cholesterol imbalance in mice, *Environmental Pollution* 244 (2019) 657–666.
103. K. Chiu, G. Warner, R.A. Nowak, J.A. Flaws, W. Mei, The impact of environmental chemicals on the gut microbiome, *Toxicological Sciences* 176 (2020) 253–284.
104. S.A. Kolb, E.J. O'Loughlin, T.C. Gsell, Characterization of phthalate-degrading bacteria from Asian carp microbiomes and riverine sediments, *International Biodeterioration & Biodegradation* 143 (2019) 104727.

105. F. Giambò, C. Costa, M. Teodoro, C. Fenga, Role-playing between environmental pollutants and human gut microbiota: a complex bidirectional interaction, *Frontiers in Medicine* 9 (2022) 810397.
106. Y. Zhou, W. Guo, L. Lei, Y. Sun, R. Li, Y. Guo, L. Yang, X. Shi, J. Han, B. Zhou, Bis (2-ethylhexyl)-tetrabromophthalate induces zebrafish obesity by altering the brain-gut axis and intestinal microbial composition, *Environmental Pollution* 290 (2021) 118127.
107. T.-F. Shum, L. Wang, J. Chiou, Impact of plasticizer on the intestinal epithelial integrity and tissue-repairing ability within cells in the proximity of the human gut microbiome, *International Journal of Environmental Research and Public Health* 20 (2023) 2152.
108. S.T. Bashir, K. Chiu, E. Zheng, A. Martinez, J. Chiu, K. Raj, S. Stasiak, N.Z.E. Lai, R.B. Arcanjo, J.A. Flaws, Subchronic exposure to environmentally relevant concentrations of di-(2-ethylhexyl) phthalate differentially affects the colon and ileum in adult female mice, *Chemosphere* 309 (2022) 136680.
109. J. Wang, F. Qiu, Z. Zhang, Y. Liu, Q. Zhou, S. Dai, S. Xiang, C. Wei, Clostridium butyricum alleviates DEHP plasticizer-induced learning and memory impairment in mice via gut–brain axis, *Journal of Food and Agricultural Chemistry* 71 (2023) 18524–18537.
110. K. Chiu, S.T. Bashir, R.A. Nowak, W. Mei, J.A. Flaws, Subacute exposure to di-isononyl phthalate alters the morphology, endocrine function, and immune system in the colon of adult female mice, *Scientific Reports* 10 (2020) 18788.
111. H. Su, P. Yuan, H. Lei, L. Zhang, D. Deng, L. Zhang, X. Chen, Long-term chronic exposure to di-(2-ethylhexyl)-phthalate induces obesity via disruption of host lipid metabolism and gut microbiota in mice, *Chemosphere* 287 (2022) 132414.
112. K.J. Groh, B. Geueke, J. Muncke,Food contact materials and gut health: implications for toxicity assessment and relevance of high molecular weight migrants, *Food and Chemical Toxicology*, 109 (2017) 1–18.
113. D. Nalage, T. Sontakke, A. Biradar, V. Jogdand, R. Kale, S. Harke, R. Kale, P. Dixit, The impact of environmental toxins on the animal gut microbiome and their potential to contribute to disease, *Food and Chemistry Advances* 3 (2023) 100497.
114. X. Cheng, J. Chen, X. Guo, H. Cao, C. Zhang, G. Hu, Y. Zhuang, Disrupting the gut microbiota/metabolites axis by Di-(2-ethylhexyl) phthalate drives intestinal inflammation via AhR/NF-κB pathway in mice, *Environmental Pollution* 343 (2024) 123232.
115. X. Yang, J. Xu, X. Chen, M. Yao, M. Pei, Y. Yang, P. Gao, C. Zhang, Z. Wang, Co-exposure of butyl benzyl phthalate and TiO_2 nanomaterials (anatase) in Metaphire guillelmi: gut health implications by transcriptomics, *Journal of Environmental Management* 354 (2024) 120429.
116. M.C. Arroyo, T.G. Barrera, B.C. Leblic, A.A. Borrego, M.C. Collado, A review of the impact of xenobiotics from dietary sources on infant health: early life exposures and the role of the microbiota, *Journal of Environmental Pollution* 269 (2021) 115994.
117. S.P. Claus, J.R. Swann, technology, Nutrimetabonomics: applications for nutritional sciences, with specific reference to gut microbial interactions, *Annual Review of Food and Science and Technology* 4 (2013) 381–399.
118. A. Cox, Z. Bomstein, A. Jayaraman, C. Allred, The intestinal microbiota as mediators between dietary contaminants and host health, *Experimental Biology and Medicine* 248 (2023) 2131–2150.
119. V. Ramírez, P. González-Palacios, M.A. Baca, P.J. González-Domenech, M. Fernández-Cabezas, M.J. Álvarez-Cubero, L. Rodrigo, A. Rivas, Effect of exposure to endocrine disrupting chemicals in obesity and neurodevelopment: the genetic and microbiota link, *The Science of the Total Environment* 852 (2022) 158219.
120. M. Vacca, F.M. Calabrese, F. Loperfido, B. Maccarini, R.M. Cerbo, E. Sommella, E. Salviati, L. Voto, M. De Angelis, G. Ceccarelli, I. Di Napoli, B. Raspini, D. Porri, E. Civardi, F. Garofoli, P. Campiglia, H. Cena, R. De Giuseppe, Maternal exposure to endocrine-disrupting chemicals: analysis of their impact on infant gut microbiota composition, *Biomedicines*, 12 (2024) 234.
121. A.N. Buerger, D.T. Dillon, J. Schmidt, T. Yang, J. Zubcevic, C.J. Martyniuk, J.H. Bisesi, Jr., Gastrointestinal dysbiosis following diethylhexyl phthalate exposure in zebrafish (Danio rerio): altered microbial diversity, functionality, and network connectivity, *Environmental Pollution*, 265 (2020) 114496.
122. R. Almamoun, P. Pierozan, L. Manoharan, O. Karlsson, Altered gut microbiota community structure and correlated immune system changes in dibutyl phthalate exposed mice, *Ecotoxicology and Environmental Safety*, 262 (2023) 115321.

6 Pesticides and the Gut Microbiome

*Muhammad Mustafa[1], Bibi Nazia Murtaza[2], Sheza Javaid[1], Fatima Mohsin[1], Imran Kazmi[3], and Muhammad Shahid Nadeem[3]**

[1]Forman Chrisitan College (A Chartered University) Lahore, Pakistan
[2]Department of Zoology, Abbottabad University of Science and Technology (AUST), Abbottabad, Pakistan
[3]Department of Biochemistry, faculty of Science, King Abdulaziz University Jeddah, Saudi Arabia
*Corresponding Author: Muhammad Shahid Nadeem.
Email id: shahid.nadeem.hu@gmail.com

6.1 INTRODUCTION

The spatial diversity and distribution of species within an ecosystem define their behavior and relationships. Within a biological community, the species present in close proximity are more likely to induce bilateral influences on each other's systems and functions [1]. In line with macro ecosystems, the gastrointestinal tract hosts various regions and microenvironments that provide optimal conditions for the growth of microbial communities. The gradient of physical and biochemical conditions of the gut such as oxygen concentration, pH, nutrients, mucus density, peristaltic mixing, immune factors, and antimicrobial peptides (AMPs) is associated with the presence of specific microbial elements in that region [2–5]. In humans, a normal gut microbiota harbors a dynamic population of microbes (around 100 trillion) including bacteria, yeast, and viruses [6, 7]. The dominant bacterial phyla in the human gut are Bacteroidetes and Firmicutes, accounting for 90% of total gut microbiota, while other phyla include Proteobacteria, Verrumicrobia, Actinobacteria, Fusobacteria, and Cyanobacteria (Figure 6.1) [8, 9]. The metagenomic analysis shows that about 40% of microbial genes in an individual are shared among half of the population, evidencing the existence of a core microbiome [10]. In trans-kingdom symbiosis, gut bacteria are the key regulators of metabolic and immune homeostasis in the host, and in turn, the host provides niche and nutrients for microbial survival. In addition, the eubiotic (healthy) microbiome relays specific functions in nutrient, xenobiotic, and drug metabolism, immunomodulation, circadian rhythmicity, antimicrobial protection, and maintenance of gut mucosal integrity and structure in the host [11].

In humans, gut colonization occurs in pre-natal stages and is influenced by placental microbiota profile [12]. The colonization of gut microbiota and the growth of certain bacterial genera in infants indicate the maturation of the microbiome [13, 14]. However, with host development and aging, the composition and spatial differences in gut bacteria turn out to be quite heterogenous. The exposome entails the non-genomic influence of environmental exposures from the pre-natal period and throughout the life-course on health and disease [15]. Although the composition of the indigenous gut microbial population in adults is relatively steady, it is subject to perturbation by changing lifestyle and environmental elements [16]. The altered composition of gut microbiota, and dysbiosis, leads to the pathogenesis of various metabolic, neurological, inflammatory, cardiovascular, and immune disorders [17, 18]. Microbial dysbiosis can result from exposure to several environmental factors including drugs, pathogens, diet, and toxins [19]. Humans are continuously

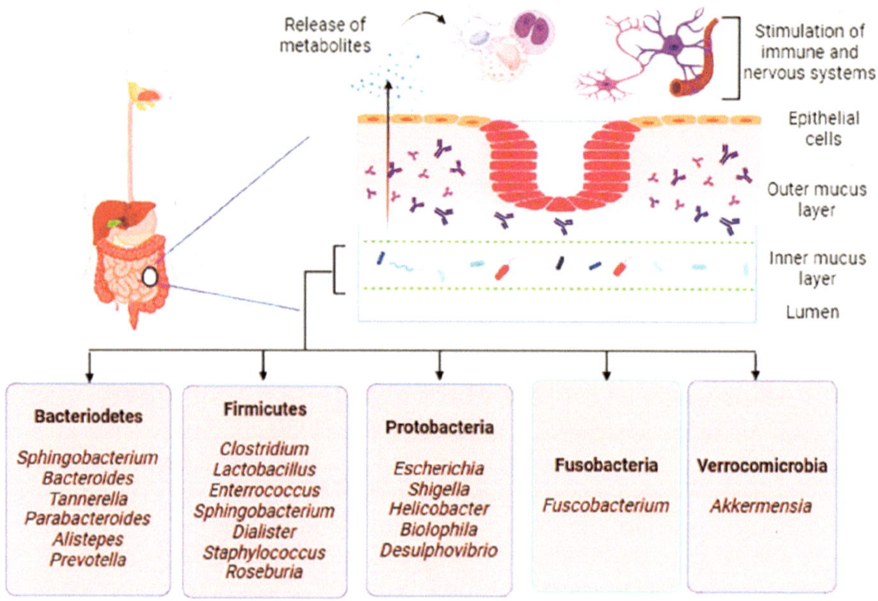

FIGURE 6.1 Human gut physiology and associated microbiome.

exposed to environmental chemicals and their metabolites, with 300 of them being reported in biological samples. Even though diet significantly alters the gut microbial landscape, environmental chemicals are the chief contaminants of our food and are often disregarded [20]. With the increase in world population, the demand for higher crop yield and better agricultural products has also intensified. In parallel, the use of pesticides has also amplified to comply with the global demand for food as well as yield economic benefits. However, the surge in uncontrolled and unreasonable use of pesticides poses a potential toxic threat not only to agricultural products but also to the whole ecosystem. Moreover, the pesticide residues, owing to their high soil persistence, also make their way to the non-target species [21, 22]. In addition, pesticide residues are also present in other environmental media, including the atmosphere and aquatic bodies. Consequently, the exposure to pesticide toxicity has increased in non-targeted organisms, mainly animals and humans via the ingestion of infected food commodities and water. A substantial number of studies have reported the toxicological significance of pesticide exposure to various diseases such as cancer, asthma, hormonal imbalance, and allergies [23]. Accordingly, exposure to multiple pesticides also triggers the dysbiosis of gut microflora, altering the microbial composition, functioning, and host's metabolic profile [24, 25].

6.2 ROLE OF GUT MICROBIOTA IN MAINTAINING HOST HEALTH

Gut microbiota and host share a symbiotic relationship, officiating various benefits to host health. The gut microbiome profile, initially assimilated at the time of birth, continues to develop with age [26, 27]. The diversity of the gut microbiome is influenced by a number of factors, including delivery mode during birth, the mother's microbiota, and the host's lifestyle as well as environmental factors during later age [28, 29]. The gut microbiome is involved in providing health-endorsing advantages to its host, some of which include regulating the immune system, protection against pathogens, production of vitamins, contaminants' detoxification, modulation of the nervous system, maintaining the structural integrity of gut mucosa, and, most importantly, food metabolism, all of which is done by the production of various metabolites [9, 30]. The latest evidence has elucidated the evolution of the host's gut microbiome, resulting in the colonization of beneficial microorganisms, known as commensals, performing protective roles against disease development, in a major way by optimizing immune functions by influencing the release of certain chemokines, cytokines, and interferons [31, 32]. The gut microbiome is also known to be a reason for altered medication responses varying from one individual to another by controlling drug metabolism and activation [33]. Moreover, it participates in the digestion and uptake of nutrients present in food by directly aiding the functions of gastrointestinal mucosal enzymes and the liver [34]. All this current evidence strongly dictates the health-promoting roles of commensal gut microbiome, not only to strengthen gut health but also to provide immunomodulatory effects as well as protection against diseases. Impaired diversity of gut microbiome can lead to significant metabolic disorders, cardiac diseases, inflammatory issues, and gut dysfunctioning as well as neurological impairments [35].

Trillions of different types of microorganisms residing in the gut produce various metabolites crucial for the maintenance of optimized host health. With recent advancements in metabolomic techniques, including mass spectrometry-coupled capillary electrophoresis, tandem mass spectrometry, nuclear magnetic resonance, and ultra-pressure liquid chromatography as well as high-performance liquid chromatography, the identification and profiling of metabolites, for instance, bile acids, choline metabolites, and short chain fatty acids (SCFAs), has become easier and has applications in assessing the host health and progression of diseases [36–39]. Therefore, it is important to discover the host microbiome's metabolite signaling to understand the complete pathophysiology of diseases and to target metabolic pathways being affected.

6.3 CLASSIFICATION OF METABOLITES

Microbes located in the gut constitute almost more than 22 million genes, which exceeds even the number of total genes present in humans [40]. These genes are involved in the production of such enzymes, which are capable of fermenting and digesting a number of compounds or nutrients that stay undigested either because of the incapability of human enzymes to digest them or because they escaped from digestive juices released from the gastrointestinal tract [41]. During the process, a variety of microbial metabolites having diverse functions are being produced, which are classified into three main groups [42].

- Metabolites that are initially produced by the host and are modified by gut microbes like taurine and secondary bile acids.
- Metabolites that are a byproduct of food and are synthesized by the activity of gut microbiota, for instance, indole derivatives, polyamines, and short-chain fatty acids (SCFAs). The metabolites released from gut microbiome and their effects are depicted in Figure 6.2.
- De novo synthesized metabolites by gut microbiota (originally produced by gut microbiome), including polysaccharide A.

The chemical structures of different metabolites released by gut microbiome are similar to each other, hence they possess similar properties and perform similar kinds of functions in the host body. Therefore, a single metabolite can perform diverse roles, have more than one targeted region, and hence can influence the efficiency of several tissues and organs. For example, butyrate, an SFCA metabolite produced by the fermentation of fibers present in the diet, has various roles, including modulation of the immune response and maintenance of an intestinal barrier as well as acting as an energy and nutrient source for epithelial cells of the colon [43]. Following are some common groups of metabolites released by the gut microbiome.

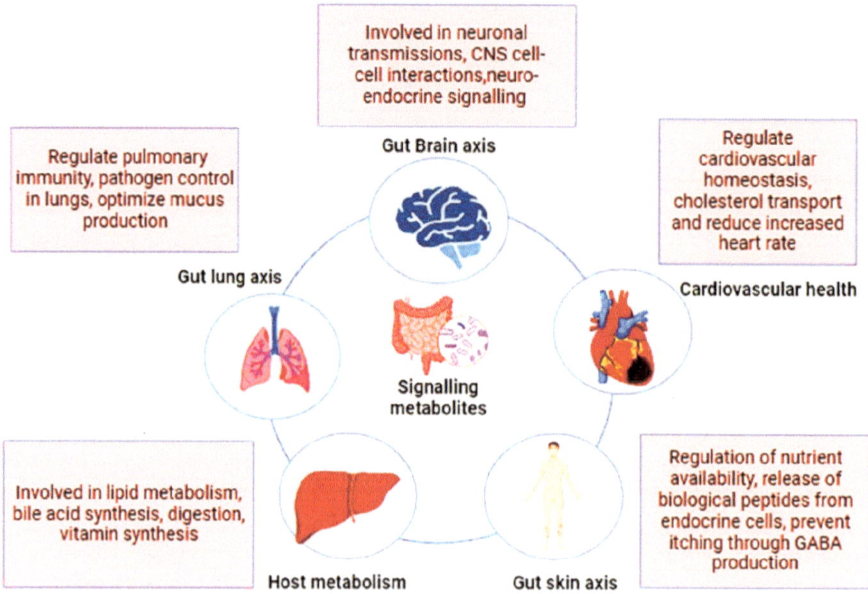

FIGURE 6.2 Metabolites released from gut microbiome and their effects on host health.

6.3.1 Short-Chain Fatty Acids (SCFAS)

SCFAs are volatile carboxylic fatty acids, having less than six carbons in a linear arrangement, and are organic in nature [44]. Some of the common metabolites that fall under the category of SCFAs include acetate, valerate, isovalerate, butyrate, iso-butyrate, propionate, hexanoate, and 2-methyl-propionate [45, 46]. They are involved in maintaining gut and overall host health by directly aiming two signaling pathways, Histone Deacetylases (HDACs) and G-coupled protein receptors (GPCRs) [44]. Their common targeted genes include HDAC3, HDAC1, GPR109A, GPR91, GPR81, GPR43, and GPR41 [47–49]. SCFAs regulate important bodily functions such as maintaining the integrity of the gut barrier, production of hormones in the gut, energy metabolism, regulation of circadian cycles, maintenance of water and sodium balance, inhibition of proinflammatory cytokine release, modulation of the nervous system, and immune responses [50–52].

6.3.2 Choline Metabolites

Choline is an essential nutrient required in the body. Choline metabolites include both water-soluble and lipid-soluble compounds. Choline is naturally present in the diet and it is further processed by gut microbiota to produce trimethylamine (TMA), dimethylglycine, dimethylamine, and methylamine [53]. TMA is further transported to the liver, where it is processed by hepatic enzymes and is converted to trimethylamine N-oxide (TMAO) [54, 55]. Choline metabolites are involved in the activation of NLRP3 inflammasome, protein kinase C, and nuclear factor kappa B signaling pathway and are also responsible for maintaining the structural and functional integrity of membranes and optimization of brain functions [56, 57]. Moreover, abnormal amounts of these metabolites play a role in promoting inflammatory responses, dysfunctioning of mitochondria, and inhibition of bile acid synthesis [58].

6.3.3 Bile Acids

Bile acids are originally produced in the liver by the synthesis of its precursor cholesterol and is further processed by gut microbiota to produce certain metabolites, for instance cholate, glycocholate, hypocholate, lithocholate, hyodeoxycholate, taurocholate, taurohyocholate, taurodeoxycholate, tauro-α muricholate, and ursodeoxycholate [59–61]. Bile acid metabolites target constitutive androstane receptor (CAR), muscarinic acetylcholine receptor (mAChR), formyl-peptide receptor (FPR), and sphingosine-1 phosphate receptor 2 (S1PR2) [62]. Some important bodily functions that involve bile acid metabolites are the facilitation of vitamin and lipid absorption, glucose, and maintenance of amino acid and lipid homeostasis as well as fluid and electrolyte balance control in the intestines [63]. Moreover, these metabolites also influence gut microbial composition as well as neurotransmission and regulation of genes responsible for circadian rhythms [64–66].

6.3.4 Tryptophan and Indole Derivatives

Tryptophan is one of the 20 essential amino acids, which produces two types of derivatives upon its metabolism; the first ones contain an intact indole ring and the second type breaks down the ring to form other compounds and signaling molecules [67]. Some of the common tryptophan-derived indole metabolites produced by gut microbiota include indole acetic acid, indole pyruvic acid, indole-3 lactic acid, indole-3-acetamide, indoxyl sulfuric acid, and serotonin [68]. These metabolites influence functions of the intestinal barrier, formation of biofilm, secretion of hormones from the gut, drug resistance, and immunomodulatory responses as well as the motility of the gastrointestinal track [69, 70] by targeting downstream effector genes, including aryl hydrocarbon receptors (AhR), retinoid-related orphan receptor gamma-t (RORγt), and pregnane X receptor (PXR) [71–73].

6.3.5 Neurotransmitters

The gut microbiome is also associated with the production of neurotransmitter metabolites, including dopamine, gamma-aminobutyric acid (GABA), and catecholamines [74]. They act on adrenergic, GABA, and 5-HTT receptors and are involved in the regulation of stress and memory-associated responses and nervous-related immune functions, and they regulate the motility of the gut [75].

6.3.6 Lipids

Triglycerides, cholesterol, conjugated fatty acids, phosphatidylcholines, and lipopolysaccharides (LPS) are common lipid metabolites produced by gut microbiota [76]. LPS directly targets Toll-like receptor 4 (TLR4), which is involved in defense against foreign bodies by recognizing pathogen-associated molecular patterns [77]. Moreover, these metabolites regulate inflammation, profiles of lipoproteins, and hyperinsulinemia and also act as precursor molecules for the synthesis of bile acids in the liver [78, 79].

6.4 PESTICIDE TOXICOLOGY

Being omnipresent in our environment, the residues of pesticides inevitably emerge as toxicological agents in our ecosystem. The chronic exposure to pesticides, even at low dosage, produces a cumulative and a cocktail effect. By interacting with various other environmental pollutants, pesticides hold the potential to produce a pronounced effect within its host, especially the intestinal flora [80, 81]. Pesticides can be classified based on their occurrence, whether they exist naturally or are chemically synthesized, based on their chemical structures (organophosphates, organochlorides, etc.) and, lastly, based on their targets (insecticides, fungicides, herbicides). Here, we discuss some major classes of pesticides and their impact on gut microbiota of the non-target species. Table 6.1 treats the impacts of pesticides on gut microbiota in non-target hosts.

TABLE 6.1
Effects of pesticides on gut microbiota in non-target hosts.

Types	Pesticides	Chemical Structure	Gut Microbial Dysbiosis	Outcomes	References
Insecticides	Chlorpyrifos	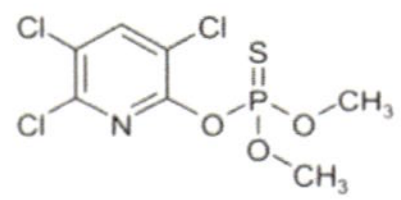	↑ *Bacteroides* ↑ *Enterococcus* ↓ *Lactobacillus* ↓ *Bifidobacterium*	Insulin resistance, Pro-obesity phenotype, Inflammation, Abnormal intestinal permeability	[82–84]
	Diazinon	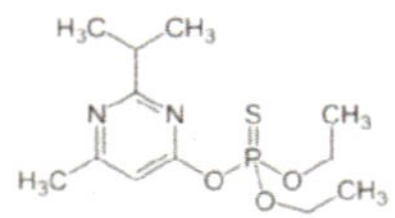	↑ *Fusobacteria* ↑ *Firmicutes* ↑ *Bacteroidetes*	Activation of stress response pathway, metabolic disorders	[85, 86]
	Trichlorfon	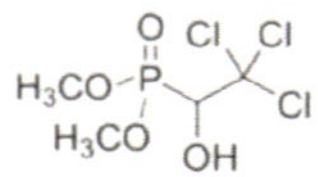	↑ *Proteobacteria* ↓ *Lactobacillus* ↓ *Akkermansia* ↓ *Bifidobacterium*	Inflammation, Damage to intestinal barrier, Oxidative stress	[87, 88]
	Neonicotinoids (Nitenpyram)	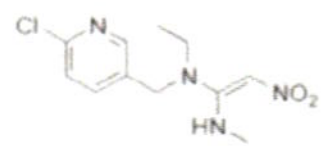	↓ *Proteobacteria*	Abnormal lipid metabolism, Immune disorders	[89, 90]
	Pyrethroids (Cis-bifenthrin)	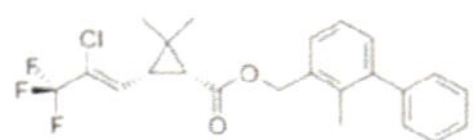	↑ *Plesiomonas* ↑ *Aeromonas* ↓ *Firmicutes*	Abnormal cholesterol metabolism, Hepatic metabolic syndrome	[91]
	Dichlorodiphenyldichloroethylene	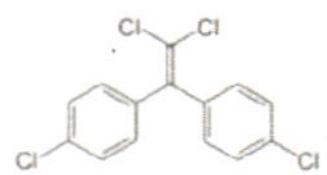	↓ *Bacteriodetes:Firmicutes*	Lipid metabolism disorder, Hepatic dysfunction	[92, 93]
Fungicides	Carbendazim	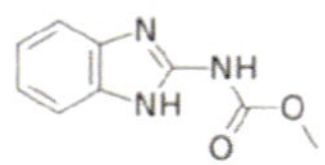	↑ *Proteobacteria* ↑ *Firmicutes* ↑ *Actinobacteria* ↓ *Bacteroidetes*	Cellular apoptosis, Hyperlipidemia, Obesity, Hepatic metabolic disorder	[94, 95]

(Continued)

TABLE 6.1 (CONTINUED)
Effects of pesticides on gut microbiota in non-target hosts.

	Imazalil	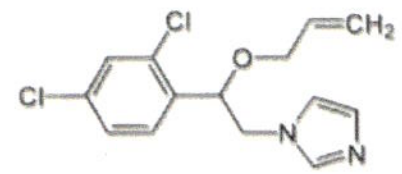	↑ *Deltaproteobacteria* ↑ *Desulfovibrio* ↓ *Actinobacteria* ↓ *Firmicutes* ↓ *Bacteroidetes* ↓ *Lactobacillus* ↓ *Bifidobacterium*	Colonic inflammation, Bacterial infection, Metabolic disorders	[96, 97]
	Epoxiconazole	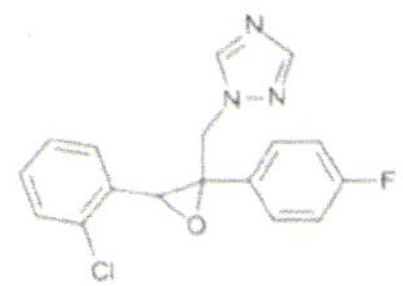	↑ *Bacteroidetes* ↑ *Proteobacteria* ↓ *Firmicutes*	Mucus secretion disorder, Intestinal barrier dysfunction, glycolipid metabolism disorders	[98, 99]
	Propamocarb	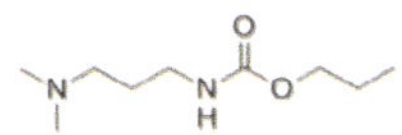	↑ *Chloroflexi* ↑ *Bacteroidetes* ↑ *Actinobacteria*	Hepatic bile acid disorders, Enterohepatic metabolic disorders, Disrupted NO/NOS pathway, Cardiovascular disorders	[100, 101]
Herbicides	Glyphosate	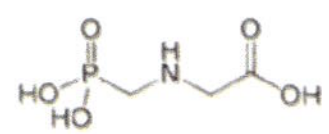	↓ *Lactobacillus* ↓ *Corynebacterium* ↓ *Bacteroidetes*	Neurological disorders, Inflammation, Potential cardiovascular disorders	[102–104]
	Triazine (Atrazine)	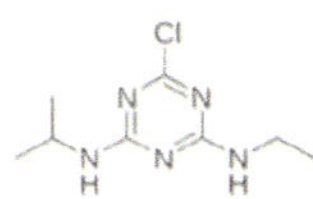	↓ *Bifidobacterium* ↓ *Lactobacillus*	Disrupted Autophagy & Apoptosis, Hepatic metabolic disorders,	[105, 106]
	2,4 Dichlorophenoxyacetic Acid	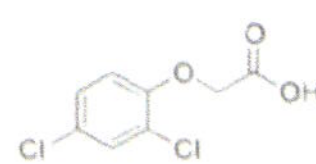	↓ *Chloroflexi* ↓ *Bacteroidetes* ↓ *Spirochaetes* ↓ *Thermotogea*	Metabolic disorders	[107]

6.4.1 Organophosphate Pesticides (OPPs)

Organophosphate pesticides are organic degradable compounds containing carbon-phosphorus (C-P) bonds, usually amide, thiol, or ester derivatives of phosphonic, phosphoric, phosphinic, or thiophosphoric acids with two organic and additional side chains [108]. Due to their biodegradable nature, organophosphate pesticides are an attractive alternative to persistent pesticides and are the most widely applied insecticides around the globe. Although organophosphate pesticides are relatively less present in soil, water, and food, they are highly soluble in water and are acutely toxic [109]. Acute or chronic exposure to organophosphate pesticides induces varying levels of toxicity in insects as well as in animals and humans [110, 111]. According to the toxicokinetics, organophosphate pesticides are absorbed instantaneously due to their lipophilic structures via ocular, dermal, respiratory, and gastrointestinal routes with varying degrees of toxicity symptoms [112]. Mechanistically, OPPs are potent covalent inhibitors of carbolic ester hydrolases, including acetylcholinesterase (AChE), which hydrolyzes a neurotransmitter, acetylcholine (ACh). Subsequently, acetylcholine accumulates in cholinergic synapses and neuromuscular junctions. The continuous overstimulation of acetylcholine receptors, i.e., muscarinic and nicotinic receptors, drives the manifestation of symptoms related to cholinergic toxidrome (dizziness, convulsions, diarrhea, hypersalivation, sweating, etc.) [113, 114]. Therefore, the immediate effect of acute OPP intoxication leads to autonomic and neuromuscular dysfunction and various CNS effects. OPP exposure also engenders significant risk to metabolic endocrine, neurological, and hepato-renal disorders in the body. Studies investigating the association between OPP exposure and gut microbiota have revealed that intestinal bacteria degrade OPPs. Chronic exposure to the toxicant induces gut microbial dysbiosis and intestinal dysfunctions [115]. For instance, the gut microbial OPP degradation produces short-chain fatty acids (acetic acid), which stimulates gluconeogenesis and induces glucose intolerance. A study shows that high levels of OPP residues in plasma correlated with fecal esterase activity and enhanced acetate levels in humans who were likely to develop diabetes [116].

6.4.2 Organochlorine Pesticides (OCPs)

Organochlorine pesticides belong to a class of chlorinated chemical compounds, commonly used for the protection of crops against insects and pests. They are non-polar in nature and contain at least one chlorine covalently bonded in their chemical structure, making them attain higher molecular weight and getting denser than H_2O. The OCPs have the ability to get absorbed in the body through various routes, including the nasal, skin, and specifically through the gut [117]. Endosulfan, cyclodienes, lindane, toxaphene, hexachlorocyclohexane, dicofol, methoxychlor, and mirex are common OCPs [118, 119]. OCPs have been known to be associated with adverse health effects such as metabolic issues, neuro-behavioral disorders, immunological diseases, and even cancer [120, 121]. Prolonged chronic exposure can increase the concentration of OCPs in tissues leading to enhanced production of hepatic enzymes involved in the metabolism of drugs [122]. OCPs alter the relative abundance and composition of the gut microbiome, specifically enhancing the activity of bile salt hydrolase enzyme in lactobacillus species. Furthermore, OCP exposure was linked to changed composition of bile acids, decreased expression of genes associated with reabsorption of bile acids, increased hydrophobicity, and increased synthesis of bile acids. Prolonged OCP exposure led to gut microbiome impairment, resulting in altered metabolite profiling and causing metabolic disorders that pose serious disease risks [93].

6.4.3 Insecticides

Insecticides, the major contributor of agriproducts, are broadly employed in the agricultural domain in order to kill or mitigate pest insects. The widespread application of insecticides not only targets the desired pests but also certain non-target species (such as honeybees) experience chronic or acute exposure to these insecticides. This non-specific action of the pesticides makes them a persistent environmental toxicant with adverse health implications. Among the several endpoints of insecticide-induced toxicity, gut microbiome is one of the most overlooked subjects.

6.4.4 Chlorpyrifos

Chlorpyrifos (O, O-diethyl-O-(3,5,6-trichloro-2-pyridinyl) phosphorothionate) is one of the routinely used OPPs and targets soil-borne pests. The high dietary exposure of chlorpyrifos exceeds the acute reference (ARfD) dose in food products [123]. The chronic, low-dose intoxication of CPF on an in vitro gut microbiota simulator (SHIME) indicated a decline in the growth of beneficial microbes (*Lactobacillus* and *Bifidobacterium*) and an increase in *Bacteroides* and *Enterococcus* [82]. Chlorpyrifos (CPF) has been shown to possess the potential to induce obesity and insulin resistance in humans via disrupting mucus-microbial interaction. Long-term exposure to CPF disrupts the microbial balance and induces low-grade inflammation via the entry of lipopolysaccharides (LPS), present in the cell wall of gram-negative bacteria, in the body. The LPS mediates the production of pro-inflammatory cytokines by binding to Toll-like receptor 4 (TLR4) on the innate immune cells. The elevated levels of LPS in plasma trigger obesity and insulin resistance in the body [83, 124]. Furthermore, CPF exposure in rats significantly alters the composition of gut microbiota, with the abundance of bacteria related to diabetic and obese phenotypes. With the reduction in C-peptide, amylin, and serum insulin prior to CPF exposure, rats exhibit pro-obesity phenotypes with a normal fat diet. In addition, the amplification of gut bacteria associated with islet injury and neurotoxicity has also been observed in rats [125]. Chlorpyrifos-induced disruptions in the gut microbial community are evident in rats fed on a high-fat diet, stimulating the growth of bacterial species that regulate the endocrine system via the hypothalamic-pituitary-adrenal (HPA) axis as well as immune response. In newly weaned rats exposed to chlorpyrifos, the concentrations of appetite-regulating hormones – peptide tyrosine-tyrosine (PYY) and glucagon-like peptide 1 (GLP-1) – that are linked with gut microbiota in shaping appetite are relatively higher [126]. The increase in intestinal permeability and gut dysbiosis in rodent models has been extensively investigated. Chronic CPF exposure also induces alterations in urine metabolites linked to short-chain fatty acids, energy, bile acid, and phenyl derivates in mice. Such variations subsequently correlate to perturbed gut microbiome, which, in turn, leads to intestinal permeability and inflammation [84]. The toxic effects of CPF ingestion are also evident during pre- and post-natal periods as it is permeable across the placental barrier. In rats subjected to CPF *in utero* and after birth, an apparent variation was observed in the gene expression of tight junctions (TJ) and their transmembrane proteins, occluding and claudins, in ileum and colon and thus altering gut absorbency and disrupting gut barrier function. The exposed mice exhibited an increment in Proteobacteria with a decrease in Bacteroidetes phylum. The translocation of gut bacteria also falls under the influence of CPF [127]. Compelling evidence suggests that peri-gestational exposure to CPF causes gut microbial disruption, accompanied by metabolic disturbances and imbalance in lipid and glucose markers [128]. Likewise, pre-natal exposure holds the potential to delay the maturation of the digestive tract, alter the mucosal barrier (mucin-2), and enhance bacterial translocation in the liver and spleen. Physiologically, the rat pups treated with CPF were apparently smaller than non-CPF-treated pups. Internally, microbial dysbiosis in the gut is associated with changes in gene expression of microbial pattern recognition receptors (TLR-2 and 4) and stimulation of innate immune response [129]. The detrimental effects of CPF are not limited to terrestrial life, but aquatic organisms also bear the brunt of the unregulated usage of the toxicants. By way of an example, a study on CPF-exposed zebrafish showed an abundance in the proliferation of α*-Proteobacteria*, β*-Proteobacteria*, and γ *-Proteobacteria* and gut oxidative stress [130].

6.4.5 Diazinon

Another extensively used OPP in agricultural pest control has been detected in groundwater and drinking water, making its way to the food chain. In the environment, diazinon degradation is primarily affected by the pH and microbial degradation in soil [131]. Nonetheless, diazinon and its related metabolites have been effectively identified in various environmental media and in human urine. Subsequently, the ingested diazinon residues have been linked to altered functional metagenome and microbial flora. Correspondingly, diazinon exposure in mice leads to the modification of the gut microbial composition along with its associated metabolic profiles in a sex-specific manner, with greater microbial perturbations in males [132]. Additionally, the diazinon exposure-induced gut microbial profile modulates quorum sensing, which regulates bacterial composition, proliferation, and microbial functional genes. In response to diazinon, the activation of stress response signaling pathways and blighted carbohydrate metabolism of gut bacteria have been detected in exposed mice [85]. Furthermore, in a study demonstrating the effect of diazinon toxicity on intestinal microbiome in a crucian carp (*Carassius auratus gibelio*), it was revealed that its exposure concentration (300μg/L) significantly affected the microbial diversity. Accordingly, the toxicant exposure multiplied the count of Fusobacteria, Firmicutes, and Bacteroidetes in the observed organism [86].

6.4.6 Trichlorfon

The detriment of another widely administered OPP, trichlorfon, has been characterized to affect the heterogeneity and decrease microbial growth in common carp. According to multiple studies, the intestinal probiotic ratio of certain bacteria declined, especially *Lactobacillus*, *Akkermansia*, and *Bifidobacterium*, whereas the Proteobacteria population increased in avian (quail) gut and carp gut microbiota upon trichlorfon exposure [87, 88].

6.4.7 Neonicotinoids

A principal class of synthetic insecticides, neonicotinoids, targets the sucking insects on agricultural plants and currently accounts for approximately 30% of insecticide sales worldwide. Being systemic insecticides, the translocation of neonicotinoids to plant tissues directly renders them lethal to herbivore insects and indirectly from viruses transmitted via insects [133]. Structurally, neonicotinoid stem from its prototype, nithiazine, which is composed of a nitro methylene functional group and has now been replaced by various substituents. The insecticidal potency of neonicotinoids lies in its capacity to disrupt cholinergic neurotransmission in the central nervous system (CNS). Neonicotinoids mimic the action of a neurotransmitter, namely acetylcholine, by binding to nicotinic acetylcholine receptors (nAChRs) in invertebrates [134]. In this process, the neurons are continuously overstimulated, which leads to the death of the target pest. In general, vertebrates (including humans) have relatively smaller number of nicotinic receptors with higher binding affinity to neonicotinoids than invertebrates, and the fundamental differences in receptors confers selective toxicity to invertebrates [135]. From 2010 onwards, due to their undesirable effects on pollinator insects, honeybees, and wild bees, they are considered an environmental risk [136]. In honeybees (*Apis mellifera* L.), exposure to nitenpyram (a neonicotinoid) decreases the overall survival rate due to the induction of gut dysbiosis. The alterations in the proliferation of key microbiotas and differential gene expression of genes involved in the maintenance of metabolic homeostasis and immunity increase the insects' mortality rate [89]. Likewise, the dose-dependent exposure of honeybees (at different developmental stages) to another neonicotinoid, thiacloprid, significantly reduced the absolute abundance of core gut bacteria and exposed the bees to be vulnerable to the pathogenic bacteria [137]. Furthermore, imidacloprid (IMI) exposure to *Drosophila melanogaster* led to microbial dysregulation, with the increment in the indigenous population of *Acetobacter* and *Lactobacillus* species [138]. In pregnant mice, ubiquitous exposure to nitenpyram caused abnormal lipid metabolism, and the growth of *Proteobacteria* declined after prenatal exposure to the neonicotinoid [90]. Studies have also reported that the sub-chronic exposure of clothianidin (CLO) to mice stimulated intestinal gut microbial perturbation, accompanied by the reduction of short-chain fatty acid-producing bacterial species. The CLO exposure also caused the suppression of α-defensin (microbiota regulator) secretion from Paneth cells in the jejunum and ileum, as part of the mechanism involved in dysbiosis [139]. In pediatric type 1 diabetes (T1D), the gut microbial variations lead to a low count of butyrate-producing microbes in children exposed to neonicotinoids, which further contributes to the increased risk of TID onset [140].

6.4.8 Pyrethroid/ Permethrins (PEM)

Pyrethroids are the synthetic analogs of naturally occurring compounds derived from plants and permethrins. The insecticidal properties of pyrethroid are determined by its potential to bind and interrupt the functioning of voltage-gated sodium channels in the nerves of insects [141]. By interacting with the sodium channels on the neuronal membrane, the pyrethroid creates a condition of hyperexcitability due to action potential which causes tremors in insects. Alternatively, pyrethroids also block chloride channels of the GABA-A receptor and induce a seizure in the target insect [142]. Mounting evidence suggests that the synthetic pyrethroids impart neurotoxicity, developmental, and reproductive toxicity as well as gut dysbiosis in non-target species [143–145]. As an example, upon exposure of environmentally relevant concentrations of cis-bifenthrin to an African clawed frog (*Xenopus laevis*), the gut microbial ecosystem was perturbed as several genera belonging to Firmicutes phylum declined whereas the prevalence of certain pathogenic bacteria increased (*Plesiomonas* and *Aeromonas*). As total bile acid is dehydroxylated by intestinal bacteria, the intestinal metabolites involved in cholesterol metabolism, bile acid biosynthesis, and bile secretion were also considerably altered, thus stimulating liver metabolic syndrome [91]. According to a study, low cypermethrin exposure induces multiple organ toxicity in pubertal male rats [146]. The low-dose exposure of permethrin showed a reduced abundance of *Prevotella*, *Porphyromonas*, and *Bacteroides* and an increased *Lactobacillus* and *Enterobacteriaceae* population in fecal microbiota of postnatal rat post four months of the follow-up period [147].

6.4.9 Dichlorodiphenyldichloroethylene

Dichlorodiphenyldichloroethylene (DDE) belongs to organochlorine pesticides, with a chemical formula of $C14H8Cl4$. Its use as an insecticide has been discontinued due to its toxic effects as an environmental pollutant as well as the potential harm it causes to the health of crop consumers. It works by inhibiting the transport of $Ca+$ by blocking the activity of ATPase in the insect shell glands, resulting in its thinning and breakage [148]. Several studies have shown how DDE is involved in posing harmful health effects and has become a cause of many diseases, including neurodevelopmental, metabolic, digestive as well as respiratory disorders [148–151]. Its toxic effects have also been associated with risk of different types of cancer, for instance, adenocarcinoma, breast, lung, prostate, and hepatocellular carcinoma. DDE also targets microbial species of the gut, leading to GM dysbiosis. In an experimental study to explore the opposing effects of pectin against DDE, eight-week exposure of this insecticide led to a decline in gut *Cyanobacteria*, *Defferribacteres*, and *Proteobacteri*, but an increased number of Bacteroidetes was observed, coupled with insulin resistance, hyperglycemia, and increased body fat [152]. DDE-induced gut dysbiosis led to a decline in Bacteriodetes to Firmicutes ratio, along with altered concentrations of lipid metabolites, including phosphatidylcholine, phosphatidylethanolamine, phosphatidylserine, and triacylglycerol. Researchers have stated that these altered metabolite concentrations can possibly be a consequence of GM dysbiosis [92]. Recent studies have indicated that these metabolite concentration shifts can lead to changes in the composition of bile acids, causing hepatic dysfunction [93].

6.4.10 Other Insecticides

Dieldrin is another common organochlorine insecticide, having a chemical formula C12H8Cl6O. Dieldrin works as a neurotoxin; it blocks the GABA receptors and disrupts synaptic signaling, targeting the central nervous system [153]. According to a study in 2020, Dieldrin exposure led to a drastic decline in Betaprotobacteria, Clostridia, Verrucomicrobiae, and Firmicutes abundance, later involved in the resorption of energy, thus regulating metabolism [154]. Betaprotobacteria are associated with regulated immune functions and reduced risk of kidney stone formation in their host [155]. Another insecticide, Endosulfan, acts as an insecticide by inducing endocrine dysfunction leading to developmental disorders [156]. Endosulfan toxicity in the body is also linked with unfavorable GM modifications, leading to metabolic health issues. Endosulfan-exposed animal models fed with high-fat diets showed hindered adipogenesis and impaired glucose metabolism as well as affected expression of genes associated with hepatic functioning, laying the foundation for the development of metabolic diseases [25]. Furthermore, an experimental study conducted on mouse models to evaluate the toxic effects of endosulfan on GM dysbiosis showed various pathways being affected, including amino acid, energy, and lipid metabolisms. Moreover, metabolic profiling revealed altered threonine, serine, and glycine metabolisms, along with hepatic inflammation and necrosis [157]. Indoxocarb, having a chemical formula C22H17ClF3N3O7, is an oxodiazine insecticide used to control larval pests. It also directly targets the nervous system by inhibiting the flow of Na+ ions to the nerve cells [158]. It possesses multiple toxic effects on its consumer's health as a persistent compound taken through diet [159]. However, the exact mechanisms involving altered gut microflora associated with harmful health effects still need to be evaluated.

6.4.11 Fungicides

Fungicides are biocidal compounds extensively used to selectively kill or inhibit fungal growth or their spores. Fungicides are directly applied to the plants either to elicit systemic toxicity in the developing fungi or to prevent fungal infestation to crops or seeds [160]. The intrinsic activity of fungicides depends upon the chemical structure, affinity to bind to specific sites, and then inhibition of enzyme activity or signal transduction pathway in the host [161]. The ubiquitous exposure of residues of fungicides to mammals through the environment (air, water, and soil) inevitably influences the composition of endosymbiotic bacteria in the gut. In humans, the traces of fungicide enter the gut primarily through food consumption as a result of post-harvest treatments [162]. A substantial number of studies have emphasized the modulatory effect of fungicide toxicity on the gut microbiota.

6.4.12 Carbendazim (CBZ)

Carbendazim (methyl 2-benzimidazole carbamate) is a systemic broad-spectrum benzimidazole fungicide that is widely used in controlling fungal diseases and as a preservative in both agricultural and industrial domains [163]. In fungi, carbendazim (CBZ) binds to tubulin and inhibits the microtubule assembly and nuclear division [164]. In a non-specific host, CBZ is an endocrine disruptor that triggers cell apoptosis and immunotoxicity, even in the embryonic state [165]. CBZ is also associated with inducing mammalian toxicity in various endpoints, especially hepatic metabolic disorder. According to a study, the oral administration of CBZ for 28 days caused hepatic lipid accumulation and increased triglyceride levels with the occurrence of inflammation in mice. The bioaccumulation of CBZ in the gastrointestinal tract and its interaction with the gut remodeled the microbial landscape as it decreased the growth of Bacteroidetes and increased the relative abundance of Proteobacteria, Firmicutes, and Actinobacteria [94]. Moreover, chronic CBZ exposure (14 weeks) to mice led to gut dysbiosis as well as hyperlipidemia and metabolic disorder-induced obesity, thus ascribing CBZ as an obesogenic compound [95]. Researchers have also revealed that sub-chronic CBZ exposure affects not only lipid metabolism but also glycolysis in adult zebrafish (*Danio rerio*). In addition, it was also revealed that the growth of Actinobacteria, Bacteroidetes, Firmicutes, Verrucomicrobia, and α-Proteobacteria, γ-Proteobacteria was observed to decrease significantly in the host organism.

6.4.13 Imazalil (IMZ)

Imazalil (IMZ), belonging to the class of imidazole, is a systemic fungicide that is toxic to certain plant pathogens, particularly the *Penicillium* species. The primary antifungal activity of IMZ is associated with its potential to inhibit demethylation in the biosynthesis of ergosterol [166]. It is commonly used in vegetables and fruit plantations and as a post-harvest treatment to avoid rot. Lately, numerous researchers have revealed that the fungitoxicity of IMZ via ingestion disrupts the balance of gut microbiome in mammals. For instance, IMZ intake reduces the abundance of Actinobacteria, Firmicutes, and Bacteroidetes at the phylum level. At the genus level, IMZ promoted the growth of the *Deltaproteobacteria* and *Desulfovibrio* population, whereas a decline in *Lactobacillus* and *Bifidobacterium* was noted in the cecal contents and feces of mice. Apart from gut microbiota dysbiosis, IMZ exposure further elevates lipocalin-2 levels and colonic inflammation in the affected organism [96]. According to another study, the changes in the microecosystem in the gut upon chronic IMZ exposure, coupled with cystic fibrosis transmembrane conductance regulator (CFTR) dysregulation, allow abnormal bacterial colonization, which ultimately leads to bacterial infection in mice. Subsequently, IMZ-induced gut dysbiosis also perturbed the intestinal barrier function and mucus secretion in both the ileum and the colon [167]. Another research reported that the IMZ treatment contributes to developmental abnormalities and disrupted locomotive activity in the early stages of development in zebrafish [168]. Interestingly, the metabolomic analysis revealed that the administration of a high concentration of IMZ for 21 days changed the metabolite profile related to glycolysis, lipid, and amino acid metabolism. It could be established that these modifications were associated with IMZ-induced dysbiosis due to changes in microbial composition at the phylum level, including an increment in a number of Firmicutes and Fusobacteria and a decline in Bacteroidetes and Proteobacteria abundance [97].

6.4.14 Epoxiconazole

Epoxiconazole, a triazole-based fungicide, is a globally used broad-spectrum fungicide that inhibits the development of fungal cell walls by interfering with ergosterol synthesis. As with other triazole derivatives, its antifungal potential is mediated through the inhibition of lanosterol-C14 demethylase (CYP51), which, in turn, leads to the accretion of methylated sterols in the cell wall, thus impairing its function [169]. In the initial reports linking liver toxicity with epoxiconazole, changes in the gut microbial composition were considered as an early sign of potential liver damage. Consequently, it was determined that Lachnospiraceae and Enterobacteriaceae were selectively enriched in the population at the family level with the epoxiconazole exposure in rats. Likewise, gut *Bacteroidetes* and *Proteobacteria* multiplied in abundance while Firmicutes decreased in number [98]. Later, it was analyzed that the chronic exposure to epoxiconazole altered the microbiome composition, coupled with a decrease in mucus secretion, imbalance in intestinal barrier function, and glycolipid metabolism in mice. Just like certain other pesticides, epoxiconazole increased the proliferation of pathogenic bacteria, *Heliobacter* and *Alistipes*, whereas the Firmicutes to Bacteroidetes ratio was increased [99]. In a study investigating the synergistic effect of epoxiconazole and microplastic toxicity, it was found that epoxiconazole caused intestinal barrier dysfunction by targeting gut dysbiosis in mice. It added to the accumulation of polystyrene in the liver, which, conversely disrupted the metabolic clearance of epoxiconazole [170]. Other triazole derivatives have also been demonstrated to affect the composition of endosymbiotic bacteria in host species. As an example, difenoconazole ingestion amplified the number of certain gut microbes such as *Aeromonas*, *Bacteroides*,

Enterobacteriaceae, and Firmicutes in zebrafish. This gut dysbiosis was also found to participate in the hepatotoxicity and lipid metabolism in the toxicant-exposed aquatic life [171].

6.4.15 Propamocarb (PM)

Propamocarb (PM) is a systemic fungicide with protective action against *Phytophthora* diseases of nursery crops [172]. Mechanistically, it is speculated that it interferes with the biosynthesis of oomycetes membranes. It modulates membrane permeability, causing an efflux of intracellular compounds from mycelium [173]. It was observed that the propamocarb exposure affects metabolism by altering gut microbiota-related pathways and metabolite profile. For example, hepatic lipid metabolism in zebrafish was determined to be altered, accompanied by gut dysbiosis following PM exposure [174]. In addition, high doses of PM exposure elicited a major influence on the fecal metabolite, including short-chain fatty acids, succinate, trimethylamine, and bile acids, with a marked shift in the overall microbial structure in mice [175]. A separate study also confirmed that chronic PM exposure perturbed the microbial community in mice as Chloroflexi, Bacteroidetes, and Actinobacteria phyla and *Oscillospira*, *Prevotella*, *Parabacteroides*, and *Dorea* genera multiplied. The study also showed that hepatic bile acid, glycolysis, β-oxidation, triglycerol synthesis, and fatty acid transportation were greatly disturbed in mice administered with PM-treated drinking water. Interestingly, there was a high prevalence of trimethylamine (atherosclerosis-promoting molecule) in feces, which disturbs the cardiac NO/NOS pathway and increases Nf-κb transcriptional levels, thus ultimately posing a risk of cardiovascular disease [100]. Furthermore, the gut microbiota analysis in mice subjected to PM-infused drinking water for 24 weeks revealed an enrichment of *Paeniclostridium*, *Clostridioides*, and *Allobaculum*, which are closely related to atherosclerosis [101]. Overall, several studies strongly hint that PM-induced gut dysbiosis may promote cardiovascular and enterohepatic metabolic disorders by interfering with lipid metabolism and stimulating inflammatory response. All this evidence clearly suggests that the fungicide-induced gut microbial changes could be an early indicator for monitoring the health risks and identifying the prognosis of associated diseases in the non-target species.

6.4.16 Herbicides

With respect to commercially used pesticides, herbicides serve as a distinct class of agrochemicals. Apart from being anti-weed compounds, the biological activities of herbicides extend beyond their effect on the target organism. The (eco)toxicological relevant amounts of herbicides are translocated to the unintended host (mainly humans and animal) through multiple environmental channels [176]. The xenobiotic exposure inadvertently disrupts and modifies the gut microbial viability and function. A number of studies have pointed out the impact of herbicide residues on the diversity of microorganism consortia within the gut of the host, affecting the host metabolism and intestinal homeostasis as well as the gut-brain axis [177].

6.4.17 Glyphosate

The most common herbicide used to eradicate unsolicited vegetation is Glyphosate ($C_3H_8NO_5P$). Being organophosphorus in nature, it works by blocking the 5-enolpyruvylshikimate-3-phosphate enzyme (EPSP), resulting in the interruption of the shikimate pathway. It inhibits the production of amino acids essential for plant growth and synthesis of proteins [178]. Continued use of this pesticide has been linked to various health issues, including alterations in gut microbiome. Recent research has indicated that Glyphosate exposure can lead to decreased amounts of beneficial GM phyla, for instance, Firmicutes and Proteobacteria [179]. It can potentially result in increased vulnerability to harmful pathogens present in GIT [180]. Experimental studies have observed that

GLY's presence in an organism's digestive tract is associated with a change in the apparent structure of the duodenum and jejunum villi. Moreover, the mRNA expression profiles revealed increased amounts of various inflammatory cytokines, including TNF-α, Caspase-3, NF- κB, MAPK3, IL-1, and IL-6, demonstrating the ability of GLY to disturb the signaling pathways in the body. All these effects were a result of a significant decline in Firmicutes abundance and enrichment of pathogenic microbes [181]. Moreover, GLY-associated GM alterations can result in increased intestinal inflammation, diagnosed by enhanced expression of proinflammatory factors, including Lipocalin-2 and CD4+ T cells [102]. GLY exposure has also been linked to several neurological disorders as a result of gut microbiome disruption. Experimental models exposed to chronic glyphosate levels resulted in a drastic decrease in beneficial gut biota, including *Lactobacillus*, *Corynebacterium*, and Bacteroidetes, causing the development of depression and anxiety-like behaviors [103]. Furthermore, it has been hypothesized that abnormal GM balance and altered SCFAs' composition occurring as a result of maternal exposure to GLY during pregnancy can lead to increased chances of a child having autism. Researchers added that this GLY-associated autism-like behavior might be because of enhanced epoxide hydrolase activity in the brain of exposed individuals [182]. Intake of chronic glyphosate concentrations can even lead to disrupted or failed embryonic development along with metabolic dysfunctioning as a result of GIT microbiome's dysbiosis, demonstrating the potential toxic effects GLY poses on host health [183]. Another experimental study revealed that glyphosate-associated GM abnormalities resulted in an increased abundance of metabolite homocysteine, which acts as a risk factor for inflammatory and cardiovascular diseases [104].

6.4.18 Triazine

Triazine, having a chemical formula H_3N_3, is another inorganic compound actively found in many herbicides commercially available. Recently, the toxic impressions of triazene herbicides on non-targeted species have raised concerns. Several studies have indicated the role of triazene herbicides in influencing the diversity of gut microbiota in exposed species. Acquaintance of increased atrazine (ATR) levels can result in altered size and composition of the GM community, with an apparent reduction in *Bifidobacterium* and *Lactobacillus* species. This change in GM diversity led to the weakening of immune responses and greater pathogens' susceptibility [105]. An experimental study evaluated the role of atrazine in GM dysbiosis, causing hepatotoxic effects. GM alterations associated with ATR exposure resulted in abnormal amounts of different metabolites in liver tissues, causing metabolic issues and disruption of autophagy and apoptosis-related genes. This highlights the importance of GM diversity and associated metabolites in the regulation of normal physiological functions through the gut-liver axis and maintains normal hepatic functioning [106]. Research conducted to evaluate the potential harmful effects of prometryn, a Triazene constituting herbicide, stated that prometryn accumulation in intestines can lead to a visible decline in microbial diversity, resulting in chronic pathological damages to GIT. Moreover, increased malonaldehyde and hydrogen peroxide concentrations were reported in the experimental group treated with prometryn [184].

6.4.19 2,4 Dichlorophenoxyacetic Acid

2,4 Dichlorophenoxyacetic acid (2,4 D), an organic compound belonging to the phenoxy group of herbicides, has a chemical formula $C_8H_6Cl_2O_3$. Recent research has been focused on evaluating the toxic impacts of 2,4 D on gut microbiome dysbiosis and the potential effects it has on host health. Experimental models treated with a low dose of 2,4 D showed altered metabolite profiles from the control group. It badly affected the abundance of Chloroflexi, Bacteroidetes, Spirochaetes, Chlorobi, and Thermotogea and changed acylcarnitine concentrations in the plasma of mice exposed to 2,4 D. Furthermore, observations revealed its toxic effects on some metabolic pathways, for instance, metabolism of amino acids and carbohydrates as well as degradation of urea [107]. It showed that

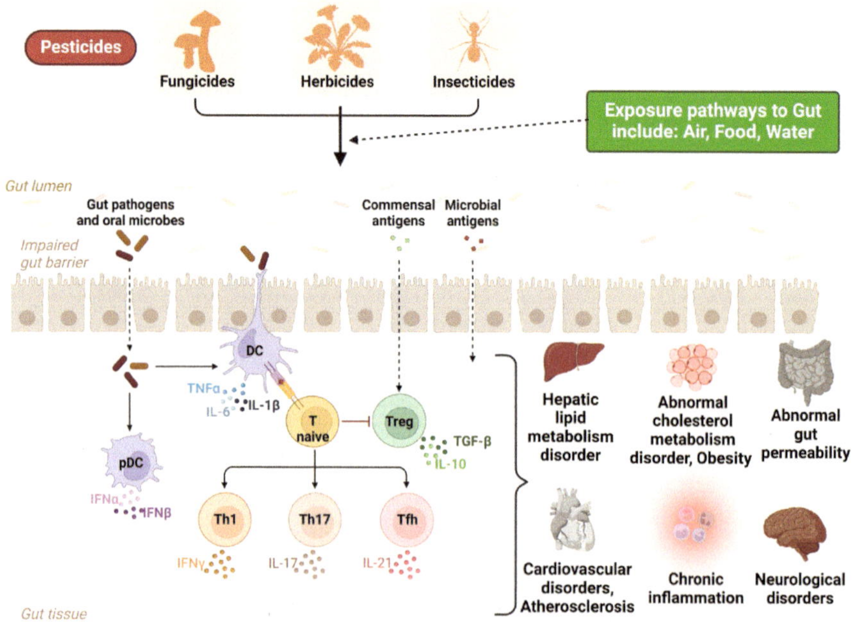

FIGURE 6.3 Effects of various types of pesticides on the gut microbiota in non-target hosts.

pesticide toxins play a significant role in gut microbial diversity, which, in turn, impacts various physiological processes in host organisms.

Depending upon the intended target, pesticides are further broadly classified in several other groups, including bactericides, larvicides, rodenticides, etc. However, their explication demands a discrete space and is, therefore, beyond the scope of this chapter. All in all, exposure to pesticides as food contaminants and environmental pollutants affects the gut microbial composition, which is a key factor in spurring diseases in off-target organisms. Apart from their health implications, the excessive and unwarranted use of insecticides can be greatly concerning as they can induce insecticide resistance in the targeted host organism. For instance, researchers have pointed out that mosquito microbiota contributes to insecticide detoxification when exposed to pyrethroids (alphacypermethrin, permethrins, and deltamethrin) and, thus, augments insecticide resistance in the host [185]. Figure 6.3 shows the effects of pesticides on the gut microbiota in non-target hosts. Therefore, serious measures to ensure regulated usage of pesticides need to be in place for public health safety and the ecosystem.

6.5 GUT MICROBIOTA DYSBIOSIS AS A BIOMARKER FOR PESTICIDE EXPOSURE

Gut microbiome diversity is significant for the maintenance of normal physiological processes in the host body. Continuous increments in the use of pesticides to increase yield and to meet productivity requirements have potentially aggravated the toxic effects it holds for its consumer's well-being [186]. With recent advancements highlighting the impacts of pesticide exposure on gut microflora, GM dysbiosis can be used as a biomarker to determine pesticide toxicity. Toxic endpoints for each pesticide's exposure need to be evaluated and established in a dose-dependent manner in order to report their toxicity levels. Recent methodologies for the assessment of a chemical's toxicity on gut microflora involve the integration of insect and animal models [187]. Meta-omics-based gene and metabolite profiling done through sequencing and mass-spectrometry analysis, respectively, are common methods used to determine altered GM diversity. The precise assessment of GM dysbiosis and gut toxicity in response to specific pesticides can help us understand how gut microbes react to exposure to a pollutant, providing mechanistic insights into pesticide-induced GM distress and potential biomarkers for diagnosis of GM-associated illnesses. For instance, in the case of diazinon toxicity, the signaling pathways in GM species change, causing a metabolite profile shift, resulting in the release of metabolites and causing neurotoxicity in a gender-dependent manner. So these GM-derived neurotoxic metabolites can serve as biomarkers to indicate gut toxicity caused by diazinon exposure [132].

The risk assessment for pesticide toxicology depends on robust and definite endpoints, which need to be defined. To achieve that purpose, it is important to define what a healthy gut microbiome means, what is GM toxicity and related alterations, and how it impacts host health. Moreover, there is a need to identify suitable biomarkers and define proper conditions for the use of that biomarker, define values for diagnosis and abnormality, establish associated mechanisms, and state the conditions under which biomarker applicability is most suitable. A report has been published by EFSA, which notes the significance of GM diversity and its implications in the assessment of environmental pollutants' toxicity along with its adverse health impacts [188]. However, the need continues to define limitations, study gaps and inclusion and exclusion criteria as well as standardization of methods related to data processing and interpretation. Gut microflora's influence on host health and how it is being impacted by pesticide exposure-related toxicity is a diverse mechanism, including many other environmental as well as physiological factors that include diet patterns, age, host genetics, gastric motility, geographical location, and antibiotic intake, etc. Considerable effort is required to completely understand this diversity of mechanisms involved in GM toxicity and host health [188]. Literature research performed to assess the potential adverse effects of pesticide exposure on GM and host well-being share a common gap. Most of the studies lack standardization as well as optimization of in-vivo and in-vitro procedures and analytical approaches. Standardization is a critical step as it enables the reproducibility of experimentations and comparability of results. Furthermore, results obtained from in-vitro studies should be validated in in-vivo models. In addition, experiments determining pesticide toxicology must be designed in a way that measures the appropriate amounts of chemical residues and its relevant response, leading to the establishment of dose-response statistics. It will ultimately lead to the determination of chronic toxin exposure levels.

6.6 CHEMICAL RISK ASSESSMENTS, ISSUES, AND CHALLENGES IN PESTICIDE TOXICOLOGY MANAGEMENT

With ongoing controversies related to the risks carried by the use of pesticides, the world has become more aware of their opposing impacts on consumer health. Despite this fact, it has become almost inevitable to lessen the risk associated with the use of pesticides because producers believe that lowering their use would result in decreased productivity or increased costs in terms of pesticide substitution with other pest control methodologies [189]. Therefore, strategies developed to reduce the risk associated with pesticide use will eventually lead to continuous increases in the costs of agricultural commodities, putting a burden on farmers and agricultural producers. Similar facts have been stated by a study [189] which says that requirements to reduce the environmental and health-related effects posed by the use of pesticides would cause agricultural prices to go higher and higher. This inclined shift in prices is directly linked with increased needs for cogent pesticides, which have gone through various screening steps for their application approval and their safety being tested through animal models and trials first.

Agricultural research scientists, amid the wave of pesticide toxicology, have developed safety measures for the management and reduction of negative side effects of these environmental pollutants used for protection and increased crop yields. Integrated crop management (ICM) has issued guidelines related to safe farming practices and to ensure the production of safe crop products, keeping in view the environmental stability and consumer's health [190, 191]. ICM has promoted the use of complementary measures for the management of pests in crops, for instance, the use of crop breeds that are resistant to fungi and insects and the use of biological procedures as well as less toxic physical methods. ICM allows the use of a specific pesticide only through a program called integrated pest management (IPM). Each pesticide goes through a particular selection criterion, followed by firm instructions for their application on crops, and then their residual levels are analyzed for a certain time period [192]. The pesticides that get approved are biologically active and effective against a wide range of pests, have lower risks for developing resistance, are user and environment friendly (lesser toxicity, no residues on agricultural products, convenient application procedures, increased storage span, reduced mobility in soil) as well as are economically profitable (cost effective) [193]. Moreover, the pesticide application practices can be optimized to ensure environmental and consumer safety, including not exceeding the recommended dose of active ingredients as well as modification of the cultivation system to control pests' exposure to crops, leading to reduced pesticide requirements [194]. The trend of crop protection from chemicals, including pesticides, has changed for the past several years through the development and formulation of recent chemicals and through analyzing their relationship with other factors, including environmental as well as human health. Advanced agricultural research has ensured the synthesis of new pesticides that have a novel mechanism of action and are relatively more effective and safer to be used for crop protection [195]. Thus, the use of new pesticides coupled with the right application techniques and appropriate doses can lead to better toxicology management and improved efficiency of active ingredients, ensuring better environmental protection from pollutant exposure and promoting reduced risk to consumer's health.

6.7 CONCLUSION

Apart from the molecular pathways elucidated here, several other cellular processes are expected to be influenced by pesticide toxicity. Moreover, the correlation between the microbial composition in the gut and pesticide toxicity needs to be further investigated in humans. In recent studies, there is evidence of pesticide contamination in utero and its trans-generational impacts [80]. What's more? Populations that are exposed to pesticides at multigenerational levels are prone to confer microbiome shifts. In the agroecosystem, high residual and occupational exposure of pesticides to farmers and front-line workers raises serious concerns for their health. In addition, the percolating contamination of pesticides in our food chain also needs to be closely monitored. Therefore, the long-term consequences of pesticide pollution-induced gut dysbiosis on public health should be addressed and mitigative measures need to be implemented effectively. Moreover, the proper assessment of newly designed or currently used pesticides needs transparent and consistent evaluation of their use, and associated effects must be studied. Furthermore, standards and guidelines must be set for data interpretation generated from the experiments that are focused on studying the relationship between the potential use of various pesticides and their impacts on population and environment.

REFERENCES

1. G. McCallum, C. Tropini, The gut microbiota and its biogeography, *Nat Rev Microbiol*, 22 (2024) 105–118.
2. C. Tropini, How the physical environment shapes the microbiota, *mSystems*, 6 (2021) e0067521.
3. G.A. O'May, N. Reynolds, A.R. Smith, A. Kennedy, G.T. Macfarlane, Effect of pH and antibiotics on microbial overgrowth in the stomachs and duodena of patients undergoing percutaneous endoscopic gastrostomy feeding, *J Clin Microbiol*, 43 (2005) 3059–3065.
4. J. Cremer, I. Segota, C.Y. Yang, M. Arnoldini, J.T. Sauls, Z. Zhang, E. Gutierrez, A. Groisman, T. Hwa, Effect of flow and peristaltic mixing on bacterial growth in a gut-like channel, *Proc Natl Acad Sci USA*, 113 (2016) 11414–11419.
5. M. Ţufail, C. Wu, M.S. Hussain, Dietary, addictive and habitual factors, and risk of colorectal cancer, *Nutrition*, 120 (2024) 112334.
6. E. Thursby, N. Juge, Introduction to the human gut microbiota, *Biochem J*, 474 (2017) 1823–1836.

7. O. Cho, M. Matsukura, T. Sugita, Molecular evidence that the opportunistic fungal pathogen *Trichosporon asahii* is part of the normal fungal microbiota of the human gut based on rRNA genotyping, *Int J Infect Dis*, 39 (2015) 87–88.
8. R.E. Ley, P.J. Turnbaugh, S. Klein, J.I. Gordon, Microbial ecology: human gut microbes associated with obesity, *Nature*, 444 (2006) 1022–1023.
9. E. Rinninella, P. Raoul, M. Cintoni, F. Franceschi, G.A.D. Miggiano, A. Gasbarrini, M.C. Mele, What is the healthy gut microbiota composition? A changing ecosystem across age, environment, diet, and diseases, *Microorganisms*, 7 (2019) 14.
10. S. Carding, K. Verbeke, D.T. Vipond, B.M. Corfe, L.J. Owen, Dysbiosis of the gut microbiota in disease, *Microb Ecol Health Dis*, 26 (2015) 26191.
11. S.M. Jandhyala, R. Talukdar, C. Subramanyam, H. Vuyyuru, M. Sasikala, D. Nageshwar Reddy, Role of the normal gut microbiota, *World J Gastroenterol*, 21 (2015) 8787–8803.
12. H.E. Al-Rashidi, Gut microbiota and immunity relevance in eubiosis and dysbiosis, *Saudi J Biol Sci*, 29 (2022) 1628–1643.
13. M. Depner, D.H. Taft, P.V. Kirjavainen, K.M. Kalanetra, A.M. Karvonen, S. Peschel, E. Schmausser-Hechfellner, C. Roduit, R. Frei, R. Lauener, A. Divaret-Chauveau, J.C. Dalphin, J. Riedler, M. Roponen, M. Kabesch, H. Renz, J. Pekkanen, F.M. Farquharson, P. Louis, D.A. Mills, E. von Mutius, P.s. group, M.J. Ege, Maturation of the gut microbiome during the first year of life contributes to the protective farm effect on childhood asthma, *Nat Med*, 26 (2020) 1766–1775.
14. M.S. Hussain, N. Sharma, N. Khurana, An update on eating disorders, *Curr Nutri Food Sci*, 20 (2024) 167–174.
15. C.P. Wild, Complementing the genome with an "exposome": the outstanding challenge of environmental exposure measurement in molecular epidemiology, *Cancer Epidemiol Biomarkers Prev*, 14 (2005) 1847–1850.
16. L. Dethlefsen, D.A. Relman, Incomplete recovery and individualized responses of the human distal gut microbiota to repeated antibiotic perturbation, *Proc Natl Acad Sci USA*, 108 Suppl 1 (2011) 4554–4561.
17. V. Tremaroli, F. Backhed, Functional interactions between the gut microbiota and host metabolism, *Nature*, 489 (2012) 242–249.
18. M.S. Hussain, G. Sharma, The burden of cardiovascular diseases due to COVID-19 pandemic, *Thorac Cardiovasc Surge*, 72 (2024) 40–50.
19. M.S. Hussain, G. Gupta, R. Mishra, N. Patel, S. Gupta, S.I. Alzarea, I. Kazmi, P. Kumbhar, J. Disouza, H. Dureja, N. Kukreti, S.K. Singh, K. Dua, Unlocking the secrets: volatile organic compounds (VOCs) and their devastating effects on lung cancer, *Pathol Res Pract*, 255 (2024) 155157.
20. K. Chiu, G. Warner, R.A. Nowak, J.A. Flaws, W. Mei, The impact of environmental chemicals on the gut microbiome, *Toxicol Sci*, 176 (2020) 253–284.
21. V. Silva, H.G.J. Mol, P. Zomer, M. Tienstra, C.J. Ritsema, V. Geissen, Pesticide residues in European agricultural soils — a hidden reality unfolded, *Sci Total Environ*, 653 (2019) 1532–1545.
22. D.B. Barr, R. Allen, A.O. Olsson, R. Bravo, L.M. Caltabiano, A. Montesano, J. Nguyen, S. Udunka, D. Walden, R.D. Walker, G. Weerasekera, R.D. Whitehead, Jr., S.E. Schober, L.L. Needham, Concentrations of selective metabolites of organophosphorus pesticides in the United States population, *Environ Res*, 99 (2005) 314–326.
23. M.C. Turner, D.T. Wigle, D. Krewski, Residential pesticides and childhood leukemia: a systematic review and meta-analysis, *Environ Health Perspect*, 118 (2010) 33–41.
24. R. Liu, J. Hong, X. Xu, Q. Feng, D. Zhang, Y. Gu, J. Shi, S. Zhao, W. Liu, X. Wang, H. Xia, Z. Liu, B. Cui, P. Liang, L. Xi, J. Jin, X. Ying, X. Wang, X. Zhao, W. Li, H. Jia, Z. Lan, F. Li, R. Wang, Y. Sun, M. Yang, Y. Shen, Z. Jie, J. Li, X. Chen, H. Zhong, H. Xie, Y. Zhang, W. Gu, X. Deng, B. Shen, X. Xu, H. Yang, G. Xu, Y. Bi, S. Lai, J. Wang, L. Qi, L. Madsen, J. Wang, G. Ning, K. Kristiansen, W. Wang, Gut microbiome and serum metabolome alterations in obesity and after weight-loss intervention, *Nat Med*, 23 (2017) 859–868.
25. J. Yan, D. Wang, Z. Meng, S. Yan, M. Teng, M. Jia, R. Li, S. Tian, C. Weiss, Z. Zhou, W. Zhu, Effects of incremental endosulfan sulfate exposure and high fat diet on lipid metabolism, glucose homeostasis and gut microbiota in mice, *Environ Pollut*, 268 (2021) 115697.
26. C. Milani, S. Duranti, F. Bottacini, E. Casey, F. Turroni, J. Mahony, C. Belzer, S. Delgado Palacio, S. Arboleya Montes, L. Mancabelli, G.A. Lugli, J.M. Rodriguez, L. Bode, W. de Vos, M. Gueimonde, A. Margolles, D. van Sinderen, M. Ventura, The first microbial colonizers of the human gut: composition, activities, and health implications of the infant gut microbiota, *Microbiol Mol Biol Rev*, 81 (2017) 10–128.

27. M. Kumar, M.S. Hussain, S. Raj, R. Verma, S. Sharma, S.K. Sahu, An overview of treatment modalities and management aspects for obesity, *Curr Nutri Food Sci*, 19 (2023) 105–113.
28. V. Lazar, L.M. Ditu, G.G. Pircalabioru, A. Picu, L. Petcu, N. Cucu, M.C. Chifiriuc, Gut Microbiota, host organism, and diet trialogue in diabetes and obesity, *Front Nutr*, 6 (2019) 21.
29. A. Gagliardi, V. Totino, F. Cacciotti, V. Iebba, B. Neroni, G. Bonfiglio, M. Trancassini, C. Passariello, F. Pantanella, S. Schippa, Rebuilding the gut microbiota ecosystem, *Int J Environ Res Public Health*, 15 (2018) 1679.
30. J. Mazuryk, K. Klepacka, W. Kutner, P.S. Sharma, Glyphosate: impact on the microbiota-gut-brain axis and the immune-nervous system, and clinical cases of multiorgan toxicity, *Ecotoxicol Environ Saf*, 271 (2024) 115965.
31. J. Chow, S.M. Lee, Y. Shen, A. Khosravi, S.K. Mazmanian, Host-bacterial symbiosis in health and disease, *Adv Immunol*, 107 (2010) 243–274.
32. M.S. Hussain, P. Sharma, D.S. Dhanjal, N. Khurana, M. Vyas, N. Sharma, M. Mehta, M.M. Tambuwala, S. Satija, S.S. Sohal, B.G.G. Oliver, H.S. Sharma, Nanotechnology based advanced therapeutic strategies for targeting interleukins in chronic respiratory diseases, *Chem-Biol Interact*, 348 (2021) 109637.
33. R.K. Aziz, S.M. Hegazy, R. Yasser, M.R. Rizkallah, M.T. ElRakaiby, Drug pharmacomicrobiomics and toxicomicrobiomics: from scattered reports to systematic studies of drug-microbiome interactions, *Expert Opin Drug Metab Toxicol*, 14 (2018) 1043–1055.
34. K. Oliphant, E. Allen-Vercoe, Macronutrient metabolism by the human gut microbiome: major fermentation by-products and their impact on host health, *Microbiome*, 7 (2019) 91.
35. L. Lin, J. Zhang, Role of intestinal microbiota and metabolites on gut homeostasis and human diseases, *BMC Immunol*, 18 (2017) 2.
36. A. Milshteyn, D.A. Colosimo, S.F. Brady, Accessing bioactive natural products from the human microbiome, *Cell Host Microbe*, 23 (2018) 725–736.
37. C.N. Brocker, T. Velenosi, H.K. Flaten, G. McWilliams, K. McDaniel, S.K. Shelton, J. Saben, K.W. Krausz, F.J. Gonzalez, A.A. Monte, Metabolomic profiling of metoprolol hypertension treatment reveals altered gut microbiota-derived urinary metabolites, *Hum Genomics*, 14 (2020) 10.
38. S. Hussain, M.A. Khan, R. Rajan, J. Jyoti, S. Sharma, S.K. Sahu, Nanorobots: The future of healthcare, *AIP Conference Proceedings*, AIP Publishing, 2023.
39. M.R. Islam, A. Rauf, S. Akash, A. Kumer, M.S. Hussain, S. Akter, J.K. Gupta, L.T. Ansari, M.M.M.I. Raj, T.B. Emran, Recent perspective on the potential role of phytocompounds in the prevention of gastric cancer, *Process Biochem*, 135 (2023) 83–101.
40. J. Qin, R. Li, J. Raes, M. Arumugam, K.S. Burgdorf, C. Manichanh, T. Nielsen, N. Pons, F. Levenez, T. Yamada, D.R. Mende, J. Li, J. Xu, S. Li, D. Li, J. Cao, B. Wang, H. Liang, H. Zheng, Y. Xie, J. Tap, P. Lepage, M. Bertalan, J.M. Batto, T. Hansen, D. Le Paslier, A. Linneberg, H.B. Nielsen, E. Pelletier, P. Renault, T. Sicheritz-Ponten, K. Turner, H. Zhu, C. Yu, S. Li, M. Jian, Y. Zhou, Y. Li, X. Zhang, S. Li, N. Qin, H. Yang, J. Wang, S. Brunak, J. Dore, F. Guarner, K. Kristiansen, O. Pedersen, J. Parkhill, J. Weissenbach, H.I.T.C. Meta, P. Bork, S.D. Ehrlich, J. Wang, A human gut microbial gene catalogue established by metagenomic sequencing, *Nature*, 464 (2010) 59–65.
41. J. Xu, H.B. Chen, S.L. Li, Understanding the molecular mechanisms of the interplay between herbal medicines and gut microbiota, *Med Res Rev*, 37 (2017) 1140–1185.
42. T.S. Postler, S. Ghosh, Understanding the holobiont: how microbial metabolites affect human health and shape the immune system, *Cell Metab*, 26 (2017) 110–130.
43. W. Feng, H. Ao, C. Peng, Gut microbiota, short-chain fatty acids, and herbal medicines, *Front Pharmacol*, 9 (2018) 1354.
44. J. He, P. Zhang, L. Shen, L. Niu, Y. Tan, L. Chen, Y. Zhao, L. Bai, X. Hao, X. Li, S. Zhang, L. Zhu, Short-chain fatty acids and their association with signalling pathways in inflammation, glucose and lipid metabolism, *Int J Mol Sci*, 21 (2020) 6356.
45. E.S. Chambers, T. Preston, G. Frost, D.J. Morrison, Role of gut microbiota-generated short-chain fatty acids in metabolic and cardiovascular health, *Curr Nutr Rep*, 7 (2018) 198–206.
46. J.M.R. Alamil, K.R. Paudel, Y. Chan, D. Xenaki, J. Panneerselvam, S.K. Singh, M. Gulati, N.K. Jha, D. Kumar, P. Prasher, G. Gupta, R. Malik, B.G. Oliver, P.M. Hansbro, K. Dua, D.K. Chellappan, Rediscovering the therapeutic potential of agarwood in the management of chronic inflammatory diseases, *Molecules*, 27 (2022) 3038.
47. S. Karaki, R. Mitsui, H. Hayashi, I. Kato, H. Sugiya, T. Iwanaga, J.B. Furness, A. Kuwahara, Short-chain fatty acid receptor, GPR43, is expressed by enteroendocrine cells and mucosal mast cells in rat intestine, *Cell Tissue Res*, 324 (2006) 353–360.

48. D. Wanders, E.C. Graff, R.L. Judd, Effects of high fat diet on GPR109A and GPR81 gene expression, *Biochem Biophys Res Commun*, 425 (2012) 278–283.
49. I. Kimura, D. Inoue, T. Maeda, T. Hara, A. Ichimura, S. Miyauchi, M. Kobayashi, A. Hirasawa, G. Tsujimoto, Short-chain fatty acids and ketones directly regulate sympathetic nervous system via G protein-coupled receptor 41 (GPR41), *Proc Natl Acad Sci USA*, 108 (2011) 8030–8035.
50. Y. Tahara, M. Yamazaki, H. Sukigara, H. Motohashi, H. Sasaki, H. Miyakawa, A. Haraguchi, Y. Ikeda, S. Fukuda, S. Shibata, Gut microbiota-derived short chain fatty acids induce circadian clock entrainment in mouse peripheral tissue, *Sci Rep*, 8 (2018) 1395.
51. B. Dalile, L. Van Oudenhove, B. Vervliet, K. Verbeke, The role of short-chain fatty acids in microbiota-gut-brain communication, *Nat Rev Gastroenterol Hepatol*, 16 (2019) 461–478.
52. R.J.F. Felizardo, I.K.M. Watanabe, P. Dardi, L.V. Rossoni, N.O.S. Camara, The interplay among gut microbiota, hypertension and kidney diseases: the role of short-chain fatty acids, *Pharmacol Res*, 141 (2019) 366–377.
53. J. Liu, Y. Tan, H. Cheng, D. Zhang, W. Feng, C. Peng, Functions of gut microbiota metabolites, current status and future perspectives, *Aging Dis*, 13 (2022) 1106–1126.
54. W. Zhu, J.A. Buffa, Z. Wang, M. Warrier, R. Schugar, D.M. Shih, N. Gupta, J.C. Gregory, E. Org, X. Fu, L. Li, J.A. DiDonato, A.J. Lusis, J.M. Brown, S.L. Hazen, Flavin monooxygenase 3, the host hepatic enzyme in the metaorganismal trimethylamine N-oxide-generating pathway, modulates platelet responsiveness and thrombosis risk, *J Thromb Haemost,* 16 (2018) 1857–1872.
55. K.S. Alharbi, T. Ali, Y. Singh, A.S. Ali Al-Ghamdi, I. Kazmi, F.A. Al-Abbasi, S.I. Alzarea, O. Afzal, A.S. Alfawaz Altamimi, S.K. Singh, D.K. Chellappan, K. Dua, G. Gupta, Biochemical interaction of pyrvinium in gentamicin-induced acute kidney injury by modulating calcium dyshomeostasis and mitochondrial dysfunction, *Chem-Biol Interact*, 363 (2022) 110020.
56. E. Sanchez-Lopez, Z. Zhong, A. Stubelius, S.R. Sweeney, L.M. Booshehri, L. Antonucci, R. Liu-Bryan, A. Lodi, R. Terkeltaub, J.C. Lacal, A.N. Murphy, H.M. Hoffman, S. Tiziani, M. Guma, M. Karin, Choline uptake and metabolism modulate macrophage IL-1beta and IL-18 production, *Cell Metab*, 29 (2019) 1350–1362 e1357.
57. S.H. Zeisel, M.D. Niculescu, Perinatal choline influences brain structure and function, *Nutr Rev*, 64 (2006) 197–203.
58. S. Yang, X. Li, F. Yang, R. Zhao, X. Pan, J. Liang, L. Tian, X. Li, L. Liu, Y. Xing, M. Wu, Gut microbiota-dependent marker TMAO in promoting cardiovascular disease: inflammation mechanism, clinical prognostic, and potential as a therapeutic target, *Front Pharmacol*, 10 (2019) 1360.
59. O. Ramirez-Perez, V. Cruz-Ramon, P. Chinchilla-Lopez, N. Mendez-Sanchez, The role of the gut microbiota in bile acid metabolism, *Ann Hepatol*, 16 Suppl 1 (2017) S21–S26.
60. A. Wahlstrom, S.I. Sayin, H.U. Marschall, F. Backhed, Intestinal crosstalk between bile acids and microbiota and its impact on host metabolism, *Cell Metab*, 24 (2016) 41–50.
61. A.A. Bhat, R. Gilhotra, Y. Singh, S. Sharma, T. Jesus Andreoli Pinto, H.G. Ferraz, S.K. Singh, K. Dua, G. Gupta, Advanced drug-delivery approaches in managing P53-mediated lung diseases remodeling, *Nanomedicine (London, England)*, 18 (2023) 583–587.
62. S.J. Keely, A. Urso, A.V. Ilyaskin, C. Korbmacher, N.W. Bunnett, D.P. Poole, S.E. Carbone, Contributions of bile acids to gastrointestinal physiology as receptor agonists and modifiers of ion channels, *Am J Physiol Gastrointest Liver Physiol*, 322 (2022) G201–G222.
63. J.C. Poland, C.R. Flynn, Bile acids, their receptors, and the gut microbiota, *Physiology (Bethesda)*, 36 (2021) 235–245.
64. K. Govindarajan, J. MacSharry, P.G. Casey, F. Shanahan, S.A. Joyce, C.G. Gahan, Unconjugated bile acids influence expression of circadian genes: a potential mechanism for microbe-host crosstalk, *PLoS One*, 11 (2016) e0167319.
65. M. McMillin, S. DeMorrow, Effects of bile acids on neurological function and disease, *FASEB J*, 30 (2016) 3658–3668.
66. K. Dua, G. Gupta, D.K. Chellapan, M. Bebawy, T. Collet, Nanoparticle-based therapies as a modality in treating wounds and preventing biofilm, *Panmin Med*, 60 (2018) 237–238.
67. A. Fiore, P.J. Murray, Tryptophan and indole metabolism in immune regulation, *Curr Opin Immunol*, 70 (2021) 7–14.
68. A. Agus, J. Planchais, H. Sokol, Gut microbiota regulation of tryptophan metabolism in health and disease, *Cell Host Microbe*, 23 (2018) 716–724.
69. J. Zhang, S. Zhu, N. Ma, L.J. Johnston, C. Wu, X. Ma, Metabolites of microbiota response to tryptophan and intestinal mucosal immunity: a therapeutic target to control intestinal inflammation, *Med Res Rev*, 41 (2021) 1061–1088.

70. W. Roth, K. Zadeh, R. Vekariya, Y. Ge, M. Mohamadzadeh, tryptophan metabolism and gut-brain homeostasis, *Int J Mol Sci,* 22 (2021) 2973.
71. J. Gao, K. Xu, H. Liu, G. Liu, M. Bai, C. Peng, T. Li, Y. Yin, Impact of the gut microbiota on intestinal immunity mediated by tryptophan metabolism, *Front Cell Infect Microbiol*, 8 (2018) 13.
72. G. Vazquez-Gomez, J. Petras, Z. Dvorak, J. Vondracek, Aryl hydrocarbon receptor (AhR) and pregnane X receptor (PXR) play both distinct and common roles in the regulation of colon homeostasis and intestinal carcinogenesis, *Biochem Pharmacol*, 216 (2023) 115797.
73. G. Gupta, R. Dahiya, Y. Singh, A. Mishra, A. Verma, S.K. Gothwal, A.A.A. Aljabali, H. Dureja, P. Prasher, P. Negi, D.N. Kapoor, R. Goyal, M.M. Tambuwala, D.K. Chellappan, K. Dua, Monotherapy of RAAS blockers and mobilization of aldosterone: A mechanistic perspective study in kidney disease, *Chem-Biol Interact*, 317 (2020) 108975.
74. Y. Wang, N. Li, J.J. Yang, D.M. Zhao, B. Chen, G.Q. Zhang, S. Chen, R.F. Cao, H. Yu, C.Y. Zhao, L. Zhao, Y.S. Ge, Y. Liu, L.H. Zhang, W. Hu, L. Zhang, Z.T. Gai, Probiotics and fructo-oligosaccharide intervention modulate the microbiota-gut brain axis to improve autism spectrum reducing also the hyper-serotonergic state and the dopamine metabolism disorder, *Pharmacol Res*, 157 (2020) 104784.
75. L.H. Morais, H.L.t. Schreiber, S.K. Mazmanian, The gut microbiota-brain axis in behaviour and brain disorders, *Nat Rev Microbiol*, 19 (2021) 241–255.
76. M. Schoeler, R. Caesar, Dietary lipids, gut microbiota and lipid metabolism, *Rev Endocr Metab Disord*, 20 (2019) 461–472.
77. M. Molteni, S. Gemma, C. Rossetti, The role of toll-like receptor 4 in infectious and noninfectious inflammation, *Mediat Inflamm*, 2016 (2016) 6978936.
78. F. Di Lorenzo, C. De Castro, A. Silipo, A. Molinaro, Lipopolysaccharide structures of Gram-negative populations in the gut microbiota and effects on host interactions, *FEMS Microbiol Rev*, 43 (2019) 257–272.
79. S.S. Imam, F.A. Al-Abbasi, S. Hosawi, M. Afzal, M.S. Nadeem, M.M. Ghoneim, S. Alshehri, S.I. Alzarea, A. Alquraini, G. Gupta, I. Kazmi, Role of platelet rich plasma mediated repair and regeneration of cell in early stage of cardiac injury, *Regen Ther*, 19 (2022) 144–153.
80. N. Djekkoun, J.D. Lalau, V. Bach, F. Depeint, H. Khorsi-Cauet, Chronic oral exposure to pesticides and their consequences on metabolic regulation: role of the microbiota, *Eur J Nutr*, 60 (2021) 4131–4149.
81. M. Jeyaraman, S. Muthu, A. Bapat, R. Jain, E.S. Sushmitha, A. Gulati, T. Channaiah Anudeep, S.J. Dilip, N.K. Jha, D. Kumar, K.K. Kesari, S. Ojha, S. Dholpuria, G. Gupta, H. Dureja, D.K. Chellappan, S.K. Singh, K. Dua, S.K. Jha, Bracing NK cell based therapy to relegate pulmonary inflammation in COVID-19, *Heliyon*, 7 (2021) e07635.
82. C. Joly, J. Gay-Queheillard, A. Leke, K. Chardon, S. Delanaud, V. Bach, H. Khorsi-Cauet, Impact of chronic exposure to low doses of chlorpyrifos on the intestinal microbiota in the Simulator of the human intestinal microbial ecosystem (SHIME) and in the rat, *Environ Sci Pollut Res Int*, 20 (2013) 2726–2734.
83. Y. Liang, J. Zhan, D. Liu, M. Luo, J. Han, X. Liu, C. Liu, Z. Cheng, Z. Zhou, P. Wang, Organophosphorus pesticide chlorpyrifos intake promotes obesity and insulin resistance through impacting gut and gut microbiota, *Microbiome*, 7 (2019) 19.
84. Y. Zhao, Y. Zhang, G. Wang, R. Han, X. Xie, Effects of chlorpyrifos on the gut microbiome and urine metabolome in mouse (*Mus musculus*), *Chemosphere*, 153 (2016) 287–293.
85. B. Gao, X. Bian, L. Chi, P. Tu, H. Ru, K. Lu, Editor's Highlight: Organophosphate diazinon altered quorum sensing, cell motility, stress response, and carbohydrate metabolism of gut microbiome, *Toxicol Sci*, 157 (2017) 354–364.
86. J. Tang, W. Wang, Y. Jiang, W. Chu, Diazinon exposure produces histological damage, oxidative stress, immune disorders and gut microbiota dysbiosis in crucian carp (*Carassius auratus gibelio*), *Environ Pollut*, 269 (2021) 116129.
87. X. Chang, X. Wang, J. Feng, X. Su, J. Liang, H. Li, J. Zhang, Impact of chronic exposure to trichlorfon on intestinal barrier, oxidative stress, inflammatory response and intestinal microbiome in common carp (*Cyprinus carpio* L.), *Environ Pollut*, 259 (2020) 113846.
88. E. Crisol-Martinez, L.T. Moreno-Moyano, N. Wilkinson, T. Prasai, P.H. Brown, R.J. Moore, D. Stanley, A low dose of an organophosphate insecticide causes dysbiosis and sex-dependent responses in the intestinal microbiota of the Japanese quail (*Coturnix japonica*), *PeerJ*, 4 (2016) e2002.
89. L. Zhu, S. Qi, X. Xue, X. Niu, L. Wu, Nitenpyram disturbs gut microbiota and influences metabolic homeostasis and immunity in honey bee (*Apis mellifera* L.), *Environ Pollut*, 258 (2020) 113671.
90. S. Yan, S. Tian, Z. Meng, J. Yan, M. Jia, R. Li, Z. Zhou, W. Zhu, Imbalance of gut microbiota and fecal metabolites in offspring female mice induced by nitenpyram exposure during pregnancy, *Chemosphere*, 260 (2020) 127506.

91. M. Li, T. Liu, T. Yang, J. Zhu, Y. Zhou, M. Wang, Q. Wang, Gut microbiota dysbiosis involves in host non-alcoholic fatty liver disease upon pyrethroid pesticide exposure, *Environ Sci Ecotechnol*, 11 (2022) 100185.
92. Y. Liang, D. Liu, J. Zhan, M. Luo, J. Han, P. Wang, Z. Zhou, New insight into the mechanism of POP-induced obesity: evidence from DDE-altered microbiota, *Chemosphere*, 244 (2020) 125123.
93. Q. Liu, W. Shao, C. Zhang, C. Xu, Q. Wang, H. Liu, H. Sun, Z. Jiang, A. Gu, Organochloride pesticides modulated gut microbiota and influenced bile acid metabolism in mice, *Environ Pollut*, 226 (2017) 268–276.
94. Y. Jin, Z. Zeng, Y. Wu, S. Zhang, Z. Fu, Oral exposure of mice to carbendazim induces hepatic lipid metabolism disorder and gut microbiota dysbiosis, *Toxicol Sci*, 147 (2015) 116–126.
95. C. Jin, Z. Zeng, C. Wang, T. Luo, S. Wang, J. Zhou, Y. Ni, Z. Fu, Y. Jin, Insights into a possible mechanism underlying the connection of carbendazim-induced lipid metabolism disorder and gut microbiota dysbiosis in mice, *Toxicol Sci*, 166 (2018) 382–393.
96. C. Jin, Z. Zeng, Z. Fu, Y. Jin, Oral imazalil exposure induces gut microbiota dysbiosis and colonic inflammation in mice, *Chemosphere*, 160 (2016) 349–358.
97. C. Jin, T. Luo, Z. Zhu, Z. Pan, J. Yang, W. Wang, Z. Fu, Y. Jin, Imazalil exposure induces gut microbiota dysbiosis and hepatic metabolism disorder in zebrafish, *Comp Biochem Physiol C Toxicol Pharmacol*, 202 (2017) 85–93.
98. C. Xu, Q. Liu, F. Huan, J. Qu, W. Liu, A. Gu, Y. Wang, Z. Jiang, Changes in gut microbiota may be early signs of liver toxicity induced by epoxiconazole in rats, *Chemotherapy*, 60 (2014) 135–142.
99. Y. Weng, T. Xu, C. Wang, Y. Jin, Oral exposure to epoxiconazole disturbed the gut micro-environment and metabolic profiling in male mice, *Metabolites*, 13 (2023) 522.
100. S. Wu, T. Luo, S. Wang, J. Zhou, Y. Ni, Z. Fu, Y. Jin, Chronic exposure to fungicide propamocarb induces bile acid metabolic disorder and increases trimethylamine in C57BL/6J mice, *Sci Total Environ*, 642 (2018) 341–348.
101. C. Jin, Y. Weng, Y. Zhang, Z. Bao, G. Yang, Z. Fu, Y. Jin, Propamocarb exposure has the potential to accelerate the formation of atherosclerosis in both WT and ApoE(-/-) mice accompanied by gut microbiota dysbiosis, *Sci Total Environ*, 800 (2021) 149602.
102. P.C. Lehman, N. Cady, S. Ghimire, S.K. Shahi, R.L. Shrode, H.J. Lehmler, A.K. Mangalam, Low-dose glyphosate exposure alters gut microbiota composition and modulates gut homeostasis, *Environ Toxicol Pharmacol*, 100 (2023) 104149.
103. Y. Aitbali, S. Ba-M'hamed, N. Elhidar, A. Nafis, N. Soraa, M. Bennis, Glyphosate based-herbicide exposure affects gut microbiota, anxiety and depression-like behaviors in mice, *Neurotoxicol Teratol*, 67 (2018) 44–49.
104. J. Hu, C. Lesseur, Y. Miao, F. Manservisi, S. Panzacchi, D. Mandrioli, F. Belpoggi, J. Chen, L. Petrick, Low-dose exposure of glyphosate-based herbicides disrupt the urine metabolome and its interaction with gut microbiota, *Sci Rep*, 11 (2021) 3265.
105. K. Wang, M. Cai, J. Sun, H. Chen, Z. Lin, Z. Wang, Q. Niu, T. Ji, Atrazine exposure can dysregulate the immune system and increase the susceptibility against pathogens in honeybees in a dose-dependent manner, *J Hazard Mater*, 452 (2023) 131179.
106. B. Liu, Q. Zeng, H. Chen, J. Liao, Y. Bai, Q. Han, N. Qiao, S. Wang, K. Mehmood, R. Hussain, B.Z. Ahmed, Z. Tang, H. Zhang, Y. Li, The hepatotoxicity of altrazine exposure in mice involves the intestinal microbiota, *Chemosphere*, 272 (2021) 129572.
107. P. Tu, B. Gao, L. Chi, Y. Lai, X. Bian, H. Ru, K. Lu, Subchronic low-dose 2,4-D exposure changed plasma acylcarnitine levels and induced gut microbiome perturbations in mice, *Sci Rep*, 9 (2019) 4363.
108. G.K. Sidhu, S. Singh, V. Kumar, D.S. Dhanjal, S. Datta, J. Singh, Toxicity, monitoring and biodegradation of organophosphate pesticides: a review, *Crit Rev Environ Sci Technol*, 49 (2019) 1135–1187.
109. S. Kumar, G. Kaushik, J.F. Villarreal-Chiu, Scenario of organophosphate pollution and toxicity in India: a review, *Environ Sci Pollut Res Int*, 23 (2016) 9480–9491.
110. H. Mali, C. Shah, B.H. Raghunandan, A.S. Prajapati, D.H. Patel, U. Trivedi, R.B. Subramanian, Organophosphate pesticides an emerging environmental contaminant: Pollution, toxicity, bioremediation progress, and remaining challenges, *J Environ Sci*, 127 (2023) 234–250.
111. N.K. Jha, S. Ojha, S.K. Jha, H. Dureja, S.K. Singh, S.D. Shukla, D.K. Chellappan, G. Gupta, S. Bhardwaj, N. Kumar, M. Jeyaraman, R. Jain, S. Muthu, R. Kar, D. Kumar, V.K. Goswami, J. Ruokolainen, K.K. Kesari, S.K. Singh, K. Dua, Evidence of Coronavirus (CoV) pathogenesis and emerging pathogen SARS-CoV-2 in the Nervous system: a review on neurological impairments and manifestations, *J Mole Neurosci: MN*, 71 (2021) 2192–2209.

112. T.C. Kwong, Organophosphate pesticides: biochemistry and clinical toxicology, *Ther Drug Monit*, 24 (2002) 144–149.
113. P. Roman, D. Cardona, L. Sempere, F. Carvajal, Microbiota and organophosphates, *Neurotoxicology*, 75 (2019) 200–208.
114. M. Balali-Mood, H. Saber, Recent advances in the treatment of organophosphorous poisonings, *Iran J Med Sci*, 37 (2012) 74–91.
115. M.K. Harishankar, C. Sasikala, M. Ramya, Efficiency of the intestinal bacteria in the degradation of the toxic pesticide, chlorpyrifos, *3 Biotech*, 3 (2013) 137–142.
116. G. Velmurugan, T. Ramprasath, K. Swaminathan, G. Mithieux, J. Rajendhran, M. Dhivakar, A. Parthasarathy, D.D. Babu, L.J. Thumburaj, A.J. Freddy, V. Dinakaran, S.S. Puhari, B. Rekha, Y.J. Christy, S. Anusha, G. Divya, K. Suganya, B. Meganathan, N. Kalyanaraman, V. Vasudevan, R. Kamaraj, M. Karthik, B. Jeyakumar, A. Abhishek, E. Paul, M. Pushpanathan, R.K. Rajmohan, K. Velayutham, A.R. Lyon, S. Ramasamy, Gut microbial degradation of organophosphate insecticides-induces glucose intolerance via gluconeogenesis, *Genome Biol*, 18 (2017) 8.
117. R. Jayaraj, P. Megha, P. Sreedev, Organochlorine pesticides, their toxic effects on living organisms and their fate in the environment, *Interdiscip Toxicol*, 9 (2016) 90–100.
118. J. Bi, Q. Liu, G. Fan, Q. Fang, X. Zhang, X. Qin, M. Wu, Z. Wan, Y. Lv, Y. Wang, L. Song, Exposure to organochlorine pesticides and polychlorinated biphenyls, adherence to an ideal cardiovascular health, and arterial stiffness among Chinese adults, *Environ Geochem Health*, 46 (2023) 10.
119. I. Kazmi, F.A. Al-Abbasi, M. Afzal, M.S. Nadeem, H.N. Altayb, G. Gupta, Phosphodiesterase-4 inhibitor roflumilast-mediated protective effect in sepsis-induced late-phase event of acute kidney injury: a narrative review, *Pharmaceuticals (Basel, Switzerland)*, 15 (2022).
120. M. Maisano, T. Cappello, S. Oliva, A. Natalotto, A. Giannetto, V. Parrino, P. Battaglia, T. Romeo, A. Salvo, N. Spano, A. Mauceri, PCB and OCP accumulation and evidence of hepatic alteration in the Atlantic bluefin tuna, *T. thynnus*, from the Mediterranean Sea, *Mar Environ Res*, 121 (2016) 40–48.
121. R.C.P. Go, M.J. Corley, G.W. Ross, H. Petrovitch, K.H. Masaki, A.K. Maunakea, Q. He, M.I. Tiirikainen, Genome-wide epigenetic analyses in Japanese immigrant plantation workers with Parkinson's disease and exposure to organochlorines reveal possible involvement of glial genes and pathways involved in neurotoxicity, *BMC Neurosci*, 21 (2020) 31.
122. E.J. Mrema, F.M. Rubino, G. Brambilla, A. Moretto, A.M. Tsatsakis, C. Colosio, Persistent organochlorinated pesticides and mechanisms of their toxicity, *Toxicology*, 307 (2013) 74–88.
123. A. European Food Safety, The 2016 European Union report on pesticide residues in food, *EFSA J*, 16 (2018) e05348.
124. Y. Singh, G. Gupta, S.K. Singh, K. Dua, Biochemical study of monkeypox zoonotic disease associated human skin dysbiosis and soft tissue injury — Correspondence, *Int J Surg*, 107 (2022) 106977.
125. B. Fang, J.W. Li, M. Zhang, F.Z. Ren, G.F. Pang, Chronic chlorpyrifos exposure elicits diet-specific effects on metabolism and the gut microbiome in rats, *Food Chem Toxicol*, 111 (2018) 144–152.
126. J.W. Li, B. Fang, G.F. Pang, M. Zhang, F.Z. Ren, Age- and diet-specific effects of chronic exposure to chlorpyrifos on hormones, inflammation and gut microbiota in rats, *Pestic Biochem Physiol*, 159 (2019) 68–79.
127. C. Joly Condette, H. Khorsi-Cauet, P. Morliere, L. Zabijak, J. Reygner, V. Bach, J. Gay-Queheillard, Increased gut permeability and bacterial translocation after chronic chlorpyrifos exposure in rats, *PLoS One*, 9 (2014) e102217.
128. N. Djekkoun, F. Depeint, M. Guibourdenche, H. El Khayat El Sabbouri, A. Corona, L. Rhazi, J. Gay-Queheillard, L. Rouabah, F. Hamdad, V. Bach, M. Benkhalifa, H. Khorsi-Cauet, Chronic perigestational exposure to chlorpyrifos induces perturbations in gut bacteria and glucose and lipid markers in female rats and their offspring, *Toxics*, 10 (2022) 138.
129. C. Joly Condette, V. Bach, C. Mayeur, J. Gay-Queheillard, H. Khorsi-Cauet, Chlorpyrifos exposure during perinatal period affects intestinal microbiota associated with delay of maturation of digestive tract in rats, *J Pediatr Gastroenterol Nutr*, 61 (2015) 30–40.
130. X. Wang, M. Shen, J. Zhou, Y. Jin, Chlorpyrifos disturbs hepatic metabolism associated with oxidative stress and gut microbiota dysbiosis in adult zebrafish, *Comp Biochem Physiol C Toxicol Pharmacol*, 216 (2019) 19–28.
131. V. Aggarwal, X. Deng, A. Tuli, K.S. Goh, Diazinon-chemistry and environmental fate: a California perspective, *Rev Environ Contam Toxicol*, 223 (2013) 107–140.
132. B. Gao, X. Bian, R. Mahbub, K. Lu, Sex-specific effects of organophosphate diazinon on the gut microbiome and its metabolic functions, *Environ Health Perspect*, 125 (2017) 198–206.

133. N. Simon-Delso, V. Amaral-Rogers, L.P. Belzunces, J.M. Bonmatin, M. Chagnon, C. Downs, L. Furlan, D.W. Gibbons, C. Giorio, V. Girolami, D. Goulson, D.P. Kreutzweiser, C.H. Krupke, M. Liess, E. Long, M. McField, P. Mineau, E.A. Mitchell, C.A. Morrissey, D.A. Noome, L. Pisa, J. Settele, J.D. Stark, A. Tapparo, H. Van Dyck, J. Van Praagh, J.P. Van der Sluijs, P.R. Whitehorn, M. Wiemers, Systemic insecticides (neonicotinoids and fipronil): trends, uses, mode of action and metabolites, *Environ Sci Pollut Res Int*, 22 (2015) 5–34.
134. E. Taillebois, A. Cartereau, A.K. Jones, S.H. Thany, Neonicotinoid insecticides mode of action on insect nicotinic acetylcholine receptors using binding studies, *Pestic Biochem Physiol*, 151 (2018) 59–66.
135. M. Tomizawa, J.E. Casida, Selective toxicity of neonicotinoids attributable to specificity of insect and mammalian nicotinic receptors, *Annu Rev Entomol*, 48 (2003) 339–364.
136. M. Ihara, K. Matsuda, Neonicotinoids: molecular mechanisms of action, insights into resistance and impact on pollinators, *Curr Opin Insect Sci*, 30 (2018) 86–92.
137. Y.J. Liu, N.H. Qiao, Q.Y. Diao, Z. Jing, R. Vukanti, P.L. Dai, Y. Ge, Thiacloprid exposure perturbs the gut microbiota and reduces the survival status in honeybees, *J Hazard Mater*, 389 (2020) 121818.
138. B.A. Daisley, M. Trinder, T.W. McDowell, H. Welle, J.S. Dube, S.N. Ali, H.S. Leong, M.W. Sumarah, G. Reid, Neonicotinoid-induced pathogen susceptibility is mitigated by Lactobacillus plantarum immune stimulation in a Drosophila melanogaster model, *Sci Rep*, 7 (2017) 2703.
139. S. Yonoichi, Y. Hara, Y. Ishida, A. Shoda, M. Kimura, M. Murata, S. Nunobiki, M. Ito, A. Yoshimoto, Y. Mantani, T. Yokoyama, T. Hirano, Y. Ikenaka, Y. Yokoi, T. Ayabe, K. Nakamura, N. Hoshi, Effects of exposure to the neonicotinoid pesticide clothianidin on alpha-defensin secretion and gut microbiota in mice, *J Vet Med Sci*, 86 (2024) 277–284.
140. Z.R. Xu, X.X. Yuan, R.M. Chen, H.Y. Wei, L.Q. Chen, H.W. Du, G.M. Li, Y. Yang, X.J. Chen, X. Fang, F.H. Luo, Association between new onset type 1 diabetes and real-world antibiotics and neonicotinoids' exposure-related gut microbiota perturbation, *World J Pediatr*, 18 (2022) 671–679.
141. D.M. Soderlund, Molecular mechanisms of pyrethroid insecticide neurotoxicity: recent advances, *Arch Toxicol*, 86 (2012) 165–181.
142. L.G. Costa, The neurotoxicity of organochlorine and pyrethroid pesticides, *Handb Clin Neurol*, 131 (2015) 135–148.
143. Y. Hu, Y. Zhang, A. Vinturache, Y. Wang, R. Shi, L. Chen, K. Qin, Y. Tian, Y. Gao, Effects of environmental pyrethroids exposure on semen quality in reproductive-age men in Shanghai, China, *Chemosphere*, 245 (2020) 125580.
144. D. Xiang, L. Zhong, S. Shen, Z. Song, G. Zhu, M. Wang, Q. Wang, B. Zhou, Chronic exposure to environmental levels of cis-bifenthrin: enantioselectivity and reproductive effects on zebrafish (*Danio rerio*), *Environ Pollut*, 251 (2019) 175–184.
145. J.T. Magnuson, Z. Cryder, N.E. Andrzejczyk, G. Harraka, D.C. Wolf, J. Gan, D. Schlenk, Metabolomic profiles in the brains of juvenile steelhead (*Oncorhynchus mykiss*) following bifenthrin treatment, *Environ Sci Technol*, 54 (2020) 12245–12253.
146. S. Yue, Q. Yuan, Q. Shen, Y. Xu, P. Wang, M. Si, M. Zhao, Multiomics implicate gut microbiota in low cypermethrin (CP) exposure induced multiorgan toxicological effects in pubertal male rats, *J Hazard Mater*, 458 (2023) 131721.
147. C. Nasuti, M.M. Coman, R.A. Olek, D. Fiorini, M.C. Verdenelli, C. Cecchini, S. Silvi, D. Fedeli, R. Gabbianelli, Changes on fecal microbiota in rats exposed to permethrin during postnatal development, *Environ Sci Pollut Res Int*, 23 (2016) 10930–10937.
148. B.S. Zhorov, K. Dong, Elucidation of pyrethroid and DDT receptor sites in the voltage-gated sodium channel, *Neurotoxicology*, 60 (2017) 171–177.
149. G.E. Howell, 3rd, C. Mulligan, E. Meek, J.E. Chambers, Effect of chronic p,p'-dichlorodiphenyldichloroethylene (DDE) exposure on high fat diet-induced alterations in glucose and lipid metabolism in male C57BL/6H mice, *Toxicology*, 328 (2015) 112–122.
150. J. Sunyer, M. Torrent, L. Munoz-Ortiz, N. Ribas-Fito, D. Carrizo, J. Grimalt, J.M. Anto, P. Cullinan, Prenatal dichlorodiphenyldichloroethylene (DDE) and asthma in children, *Environ Health Perspect*, 113 (2005) 1787–1790.
151. B. Eskenazi, A.R. Marks, A. Bradman, L. Fenster, C. Johnson, D.B. Barr, N.P. Jewell, In utero exposure to dichlorodiphenyltrichloroethane (DDT) and dichlorodiphenyldichloroethylene (DDE) and neurodevelopment among young Mexican American children, *Pediatrics*, 118 (2006) 233–241.
152. J. Zhan, Y. Liang, D. Liu, X. Ma, P. Li, W. Zhai, Z. Zhou, P. Wang, Pectin reduces environmental pollutant-induced obesity in mice through regulating gut microbiota: a case study of p,p′-DDE, *Environ Int*, 130 (2019) 104861.

153. T. Raisch, S. Raunser, The modes of action of ion-channel-targeting neurotoxic insecticides: lessons from structural biology, *Nat Struct Mol Biol*, 30 (2023) 1411–1427.
154. Q. Hua, O. Adamovsky, H. Vespalcova, J. Boyda, J.T. Schmidt, M. Kozuch, S.L.M. Craft, P.E. Ginn, S. Smatana, E. Budinska, M. Persico, J.H. Bisesi, Jr., C.J. Martyniuk, Microbiome analysis and predicted relative metabolomic turnover suggest bacterial heme and selenium metabolism are altered in the gastrointestinal system of zebrafish (*Danio rerio*) exposed to the organochlorine dieldrin, *Environ Pollut*, 268 (2021) 115715.
155. J. Jalanka-Tuovinen, A. Salonen, J. Nikkila, O. Immonen, R. Kekkonen, L. Lahti, A. Palva, W.M. de Vos, Intestinal microbiota in healthy adults: temporal analysis reveals individual and common core and relation to intestinal symptoms, *PLoS One*, 6 (2011) e23035.
156. R.G. Menezes, T.F. Qadir, A. Moin, H. Fatima, S.A. Hussain, M. Madadin, S.B. Pasha, F.A. Al Rubaish, S. Senthilkumaran, Endosulfan poisoning: an overview, *J Forensic Legal Med*, 51 (2017) 27–33.
157. P. Zhang, W. Zhu, D. Wang, J. Yan, Y. Wang, Z. Zhou, L. He, A combined NMR- and HPLC-MS/MS-based metabolomics to evaluate the metabolic perturbations and subacute toxic effects of endosulfan on mice, *Environ Sci Pollut Res*, 24 (2017) 18870–18880.
158. K.D. Wing, M. Sacher, Y. Kagaya, Y. Tsurubuchi, L. Mulderig, M. Connair, M. Schnee, Bioactivation and mode of action of the oxadiazine indoxacarb in insects, *Crop Protect*, 19 (2000) 537–545.
159. F. Giambò, M. Teodoro, C. Costa, C. Fenga, Toxicology and microbiota: How do pesticides influence gut microbiota? A review, *Int J Environ Res Public Health,* 18 (2021) 5510.
160. P.K. Gupta, Toxicity of fungicides, *Veterinary Toxicology*, Ramesh C. Gupta (ed.), Elsevier, Academic Press, 2018, pp. 569–580.
161. I. Yamaguchi, M. Fujimura, Recent topics on action mechanisms of fungicides, *J Pesticide Sci*, 30 (2005) 67–74.
162. Y. Pico, M.A. El-Sheikh, A.H. Alfarhan, D. Barcelo, Target vs non-target analysis to determine pesticide residues in fruits from Saudi Arabia and influence in potential risk associated with exposure, *Food Chem Toxicol*, 111 (2018) 53–63.
163. M. Veerappan, I. Hwang, M. Pandurangan, Effect of cypermethrin, carbendazim and their combination on male albino rat serum, *Int J Exp Pathol*, 93 (2012) 361–369.
164. L.C. Davidse, Benzimidazole fungicides: mechanism of action and biological impact, *Annu Rev Phytopathol*, 24 (1986) 43–65.
165. J. Jiang, S. Wu, C. Wu, X. An, L. Cai, X. Zhao, Embryonic exposure to carbendazim induces the transcription of genes related to apoptosis, immunotoxicity and endocrine disruption in zebrafish (*Danio rerio*), *Fish Shellfish Immunol*, 41 (2014) 493–500.
166. M.R. Siegel, N. Ragsdale, Antifungal mode of action of imazalil, *Pesticide Biochem Physiol*, 9 (1978) 48–56.
167. C. Jin, J. Xia, S. Wu, W. Tu, Z. Pan, Z. Fu, Y. Wang, Y. Jin, Insights into a possible influence on gut microbiota and intestinal barrier function during chronic exposure of mice to Imazalil, *Toxicol Sci*, 162 (2018) 113–123.
168. Y. Jin, Z. Zhu, Y. Wang, E. Yang, X. Feng, Z. Fu, The fungicide imazalil induces developmental abnormalities and alters locomotor activity during early developmental stages in zebrafish, *Chemosphere*, 153 (2016) 455–461.
169. S. Schneider, T. Hofmann, S. Stinchcombe, M.C. Moreno, I. Fegert, V. Strauss, S. Groters, E. Fabian, J. Thiaener, K.C. Fussell, B. van Ravenzwaay, Species differences in developmental toxicity of epoxiconazole and its relevance to humans, *Birth Defects Res B Dev Reprod Toxicol*, 98 (2013) 230–246.
170. W. Sun, S. Yan, Z. Meng, S. Tian, M. Jia, S. Huang, Y. Wang, Z. Zhou, J. Diao, W. Zhu, Combined ingestion of polystyrene microplastics and epoxiconazole increases health risk to mice: based on their synergistic bioaccumulation in vivo, *Environ Int*, 166 (2022) 107391.
171. J. Jiang, L. Chen, S. Wu, L. Lv, X. Liu, Q. Wang, X. Zhao, Effects of difenoconazole on hepatotoxicity, lipid metabolism and gut microbiota in zebrafish (*Danio rerio*), *Environ Pollut*, 265 (2020) 114844.
172. J. Hu, C. Hong, E.L. Stromberg, G.W. Moorman, Effects of propamocarb hydrochloride on mycelial growth, sporulation, and infection by phytophthora nicotianae isolates from virginia nurseries, *Plant Dis*, 91 (2007) 414–420.
173. G. Papavizas, N. O'Neill, J. Lewis, Fungistatic activity of propyl-N-(adimethylaminopropyl) carbamate on Pythium spp. and its reversal by sterols, *Phytopathology*, 68 (1978) 1667–1671.
174. R. Zhang, Z. Pan, X. Wang, M. Shen, J. Zhou, Z. Fu, Y. Jin, Short-term propamocarb exposure induces hepatic metabolism disorder associated with gut microbiota dysbiosis in adult male zebrafish, *Acta Biochim Biophys Sin,* 51 (2019) 88–96.

175. S. Wu, C. Jin, Y. Wang, Z. Fu, Y. Jin, Exposure to the fungicide propamocarb causes gut microbiota dysbiosis and metabolic disorder in mice, *Environ Pollut*, 237 (2018) 775–783.
176. A. Kortekamp, Herbicides and environment, BoD–Books on Demand, 2011.
177. R. Matsuzaki, E. Gunnigle, V. Geissen, G. Clarke, J. Nagpal, J.F. Cryan, Pesticide exposure and the microbiota-gut-brain axis, *ISME J*, 17 (2023) 1153–1166.
178. R. Mesnage, N. Defarge, J. Spiroux de Vendomois, G.E. Seralini, Potential toxic effects of glyphosate and its commercial formulations below regulatory limits, *Food Chem Toxicol*, 84 (2015) 133–153.
179. N. Blot, L. Veillat, R. Rouze, H. Delatte, Glyphosate, but not its metabolite AMPA, alters the honeybee gut microbiota, *PLoS One*, 14 (2019) e0215466.
180. E.V.S. Motta, M. Mak, T.K. De Jong, J.E. Powell, A. O'Donnell, K.J. Suhr, I.M. Riddington, N.A. Moran, Oral or topical exposure to glyphosate in herbicide formulation impacts the gut microbiota and survival rates of honey bees, *Appl Environ Microbiol*, 86 (2020) e01150-20.
181. Q. Tang, J. Tang, X. Ren, C. Li, Glyphosate exposure induces inflammatory responses in the small intestine and alters gut microbial composition in rats, *Environ Pollut*, 261 (2020) 114129.
182. Correction for Pu et al., Maternal glyphosate exposure causes autism-like behaviors in offspring through increased expression of soluble epoxide hydrolase, *Proc Natl Acad Sci USA*, 118 (2021).
183. A. Suppa, J. Kvist, X. Li, V. Dhandapani, H. Almulla, A.Y. Tian, S. Kissane, J. Zhou, A. Perotti, H. Mangelson, K. Langford, V. Rossi, J.B. Brown, L. Orsini, Roundup causes embryonic development failure and alters metabolic pathways and gut microbiota functionality in non-target species, *Microbiome*, 8 (2020) 170.
184. F. Zeng, L. Wu, X. Ren, B. Xu, S. Cui, M. Li, W. Chen, Y. Han, T. Ren, Effects of chronic prometryn exposure on antioxidative status, intestinal morphology, and microbiota in sea cucumber (*Apostichopus japonicus*), *Comp Biochem Physiol C Toxicol Pharmacol*, 250 (2021) 109187.
185. N. Dada, J.C. Lol, A.C. Benedict, F. Lopez, M. Sheth, N. Dzuris, N. Padilla, A. Lenhart, Pyrethroid exposure alters internal and cuticle surface bacterial communities in *Anopheles albimanus*, *ISME J*, 13 (2019) 2447–2464.
186. M.S. Hussain, A. Sharma, R. Kumar, Prebiotics and probiotics: a focused review of applications in respiratory disorders, *Carpa J Food Sci Technol*, 15 (2023) 183–207.
187. T.S.B. Schmidt, J. Raes, P. Bork, The human gut microbiome: from association to modulation, *Cell*, 172 (2018) 1198–1215.
188. C. Merten, R. Schoonjans, D. Di Gioia, C. Peláez, Y. Sanz, D. Maurici, T. Robinson, Exploring the need to include microbiomes into EFSA's scientific assessments, Wiley Online Library, 2020, pp. e18061.
189. C.J. Morrison Paul, V.E. Ball, R.G. Felthoven, A. Grube, R.F. Nehring, Effective costs and chemical use in United States agricultural production: using the environment as a "free" input, *Am J Agric Econ*, 84 (2002) 902–915.
190. A. Frangenberg, Integrated Crop Management as fundamental basis for sustainable production, *Pflanzenschutz-Nachrichten Bayer*, 53 (2000) 131–153.
191. I.N. Tsakiris, T.G. Danis, I.A. Stratis, D. Nikitovic, I.A. Dialyna, A.K. Alegakis, A.M. Tsatsakis, Monitoring of pesticide residues in fresh peaches produced under conventional and integrated crop management cultivation, *Food Addit Contam*, 21 (2004) 670–677.
192. F.E. Nwilene, K. Nwanze, A. Youdeowei, Impact of integrated pest management on food and horticultural crops in Africa, *Entomologia Experimentalis et Applicata*, 128 (2008) 355–363.
193. C. Palacios Xutuc, *Manual to Train Trainers on Safe and Correct Use of Plant Protection Products and Integrated Pest Management (IPM)*, CropLife Latin America: Guatemala City, Guatemala, (2010).
194. J. Bürger, F. de Mol, B. Gerowitt, The "necessary extent" of pesticide use—thoughts about a key term in German pesticide policy, *Crop Protection*, 27 (2008) 343–351.
195. A. Bhattacharyya, S.R. Barik, P. Ganguly, New pesticide molecules, formulation technology and uses: present status and future challenges, *J Plant Prot Sci*, 1 (2009) 9–15.

7 Organic Pollutants and the Gut Microbiome

Rajiv Dahiya[1], Sunita Dahiya[2], Mudasir Maqbool[3], Zulfkar Qadrie[4], Md Sadique Hussain[5], Neelam Singla[6], Sarita Rawat[7], Mahaveer Singh[8], Ayeh Bani Saeid[9,10], and Gaurav Gupta[11,12]

[1]School of Pharmacy, Faculty of Medical Sciences, The University of the West Indies, St. Augustine, Trinidad and Tobago, West Indies
[2]Department of Pharmaceutical Sciences, School of Pharmacy, University of Puerto Rico, Medical Sciences Campus, San Juan, Puerto Rico, USA
[3]Department of Pharmaceutical Sciences, University of Kashmir, Srinagar, Jammu and Kashmir, India
[4]Department of Pharmacology, Government Medical College, Baramulla, Jammu and Kashmir, India
[5]Uttaranchal Institute of Pharmaceutical Sciences, Uttaranchal University, Dehradun, Uttarakhand, India
[6]School of Pharmacy, Suresh Gyan Vihar University, Jaipur, Rajasthan, India
[7]Faculty of Pharmacy and Sciences, Amrapali University, Nainital, Uttarakhand, India
[8]School of Pharmacy and Technology Management, SVKMs, NMIMS University, Shirpur Campus, Maharastra, India
[9]Faculty of Health, Australian Research Centre in Complementary and Integrative Medicine, University of Technology Sydney, Ultimo, Australia
[10]Discipline of Pharmacy, Graduate School of Health, University of Technology Sydney, NSW, Australia
[11]Centre for Research Impact & Outcome, Chitkara College of Pharmacy, Chitkara University, Rajpura, Punjab, India
[12]Centre of Medical and Bio-allied Health Sciences Research, Ajman University, Ajman, United Arab Emirates
*Corresponding Author: Gaurav Gupta.
Email: gauravpharma25@gmail.com

7.1 INTRODUCTION: OVERVIEW OF ORGANIC POLLUTANTS

In recent times, the rapid expansion of various industries, such as petrochemicals, plastics, synthetic fibers, coking, dyeing, and printing, has led to an increased prevalence of highly concentrated organic pollutants (OPs). The discharge of these contaminants has aggravated environmental pollution [1, 2]. The intensification of pollution has underscored the critical importance of pollution control within the broader framework of environmental preservation. The treatment of toxic OPs, often referred to as "three-hazardous" pollutants due to their carcinogenic, teratogenic, and mutagenic properties, has become a matter of paramount significance [3, 4].

A broad spectrum of substances falls under the category of OPs, posing potential risks to both the environment and human health. These pollutants stem from diverse origins and encompass compounds such as residues from agrochemicals and pharmaceuticals, personal care products, illicit drugs, industrial contaminants, flame retardants, plasticizers, polycyclic aromatic hydrocarbons

 DOI: 10.1201/9781003489221-7

(PAHs), halogenated PAHs, dioxins, and other persistent OPs (POPs) [5–8]. They infiltrate the environment through human activities, including discharges from wastewater treatment facilities, uncontrolled discharge of residues, degradation of precursor substances, accumulation of plastic debris along beaches, industrial operations, and thermal processes within industrial settings. For instance, plastic debris found on beaches can act as reservoirs of OPs like flame retardants and plasticizers [5, 6, 8]. The presence of OPs poses substantial risks to ecosystems and human health due to their toxic and potentially carcinogenic nature. OPs can contaminate soil and aquatic environments, resulting in adverse impacts on various forms of life. A comprehensive understanding of the effects of OPs is imperative for devising strategies to mitigate their adverse effects on the environment [7, 9].

Pollutants regulatory frameworks governing the management and mitigation of OPs vary across different regions and are primarily aimed at curbing their release into the environment. Regulatory measures typically entail monitoring pollutant levels, establishing emission standards for industrial activities, implementing effective waste management protocols, and promoting environmentally friendly remediation techniques, such as microbial degradation [7, 10]. OPs encompass a wide array of substances originating from diverse sources, including industrial operations and the accumulation of plastic waste. These pollutants can exert detrimental effects on ecosystems and human health, underscoring the importance of regulatory interventions to minimize their environmental impact.

7.2 SIGNIFICANCE OF THE GUT MICROBIOME

The human gastrointestinal tract (GIT) contains an immense number of bacterial cells, estimated to be in a ratio of around 1:1 with our own cells [11, 12]. Consequently, they form a dynamic living population exhibiting metabolic activity akin to the liver. There is a growing inclination to designate the microbiota as a novel human organ. Given their close association with the highly biologically active intestinal mucosa, characterized by a high cellular turnover rate, their impact on human health, and their role in disease prevention or precipitation can be significant. The gut microbiota (GM) establishes a symbiotic relationship with the host, contributing to the maintenance of health and metabolic homeostasis through the generation of various metabolites. The collective genetic material of these microorganisms inhabiting the gut is referred to as the metagenome. The establishment of GM commences at birth and is largely completed within three years in healthy individuals; however, it is susceptible to modification by environmental factors, particularly the composition and volume of the diet, and antibiotic therapy [13–15]. GM is integral to functions like food breakdown, fermentation, metabolic pathways (MPs), and synthesis of vitamins. Additionally, it aids in the development and regulation of the immune system and ensures a beneficial equilibrium of intestinal microorganisms [16, 17].

Research on the microbiome has witnessed rapid global expansion, offering new perspectives on both human and animal diseases and interventions designed to address them [18, 19]. Bacterial populations colonize the entire length of the human gut, from the oral cavity to the rectum. However, the density and composition of resident microbial communities vary significantly across anatomical sites and are influenced by transit rates, host secretions, environmental conditions, substrate availability, and the integrity of the gut wall [20]. Gram-positive Firmicutes and gram-negative Bacteroidetes are the predominant bacterial groups in the overall microbiome [21]. Bacteria residing in the distal regions of the gut contribute to host health by synthesizing vitamins and essential amino acids and producing vital metabolites from indigestible food components [22]. Metabolites, particularly short-chain fatty acids (SCFAs), result from bacterial fermentation of indigestible carbohydrates, predominantly fiber. These SCFAs act as the primary energy source for colonic epithelial cells, thereby fortifying the mucosal barrier and defense mechanisms [23].

7.3 RATIONALE FOR STUDYING THE INTERACTION BETWEEN ORGANIC POLLUTANTS AND THE GM

Investigating the interaction between OPs and the GM is crucial given their potential ramifications for human well-being and the ecosystem. OPs are widespread in our surroundings and can amass in diverse environments, including the human body. The GM plays a pivotal role in maintaining host well-being by contributing to metabolism, immune function, and nutrient absorption. Recent studies have indicated that OPs can disturb the composition and function of the GM, leading to dysbiosis and adverse health consequences, such as inflammation, metabolic disorders, and compromised immune responses (Figure 7.1). Comprehending the intricate interplay between OPs and the GM is essential for unraveling the mechanisms behind OPs-induced toxicity and devising strategies to mitigate their detrimental effects on human health and environmental ecosystems. This chapter delves into the intricate relationship between OPs and the GM, shedding light on their reciprocal impact on human health. Employing an interdisciplinary approach, this chapter aims to explore the effects of OPs on the composition, diversity, and functionality of the GM community, assessing the repercussions of dysbiosis triggered by OPs on host physiology, immune function, and susceptibility to related ailments, reviewing recent advancements in therapeutic interventions aimed at modulating the GM to alleviate the adverse consequences of OPs, and identifying areas of knowledge deficiency and propose future research avenues to gain a deeper understanding of the intricate interplay between OPs and the GM. These insights can inform strategies for environmental remediation and the development of personalized healthcare approaches.

7.4 COMMON ORGANIC POLLUTANTS

OPs encompass a wide array of substances categorized based on their origins and chemical compositions (Figure 7.2). The mentioned categories are typical examples of OPs:

Pesticides: These are substances utilized for pest control in agricultural settings, households, and public health initiatives. They comprise insecticides, herbicides, fungicides, and rodenticides [24].

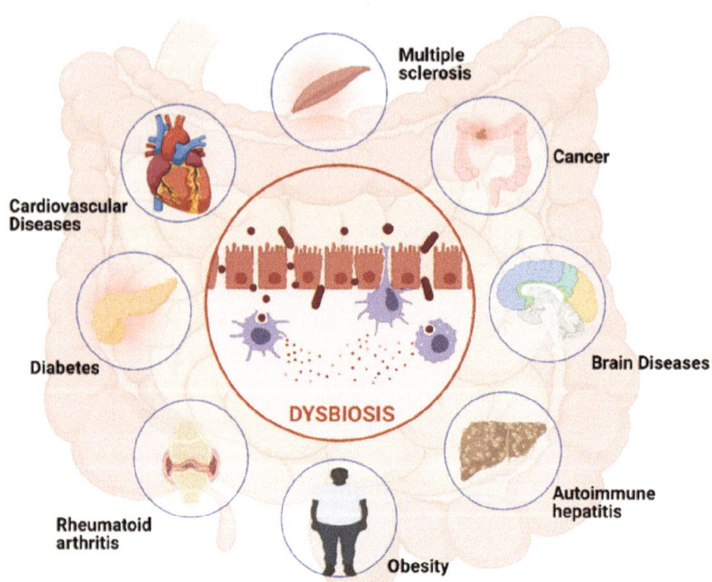

FIGURE 7.1 Dysbiosis leads to various adverse health outcomes.

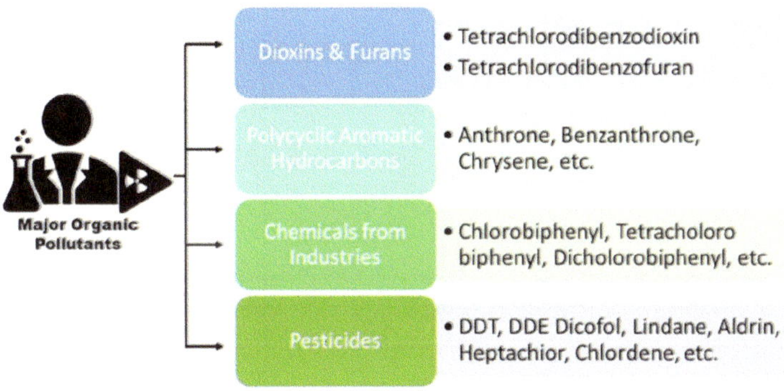

FIGURE 7.2 Common classes of organic pollutants along with examples.

Solvents: Organic solvents find applications across various industries for purposes such as cleaning, degreasing, and participating in chemical reactions. Examples include benzene, toluene, xylene, and chlorinated solvents like trichloroethylene (TCE) [25].

Pharmaceuticals: Pharmaceutical compounds enter the environment through diverse pathways like human and animal excretion or improper disposal. They encompass antibiotics, hormones, painkillers, and other prescription medications [26].

Polychlorinated Biphenyls (PCBs): PCBs constitute a group of synthetic organic chemicals extensively employed in electrical equipment, hydraulic fluids, and industrial processes. They are POPs with the capacity for environmental bioaccumulation [27].

Dioxins: Dioxins are highly toxic organic compounds generated as byproducts of chlorine-involved industrial processes like waste incineration and chemical production. They are recognized for their environmental persistence and potential for bioaccumulation [28].

These categories exemplify some of the prominent OPs with the potential to exert detrimental effects on both human health and the environment.

7.5 FACTORS INFLUENCING GUT MICROBIOME COMPOSITION

GM is predominantly composed of species from the prokaryotic domain, encompassing bacteria and other unicellular microbes devoid of specialized organelles. Additionally, fungi, parasites, archaea, and viruses contribute to the complexity of this environment [29]. From a broader perspective, humans can be categorized based on their enterotypes, representing distinct GM compositions observed in specific populations [30]. Within a single individual, the collective GM species have been demonstrated to possess approximately 3.3 million genes. This vast gene repertoire, when contrasted with the human genome's roughly 23,000 genes, underscores the extensive magnitude and potential impact of these species on human health [31]. Furthermore, the human body harbors nearly an equivalent number of bacterial cells as human cells [32].

The diversity of the GM is intricately molded by various determinants, encompassing geographical distance, host evolutionary lineage, dietary patterns, SES, age, and potentially gender (Figure 7.3). These factors collectively influence the configuration of the GM, thereby holding potential implications for the understanding of human wellness.

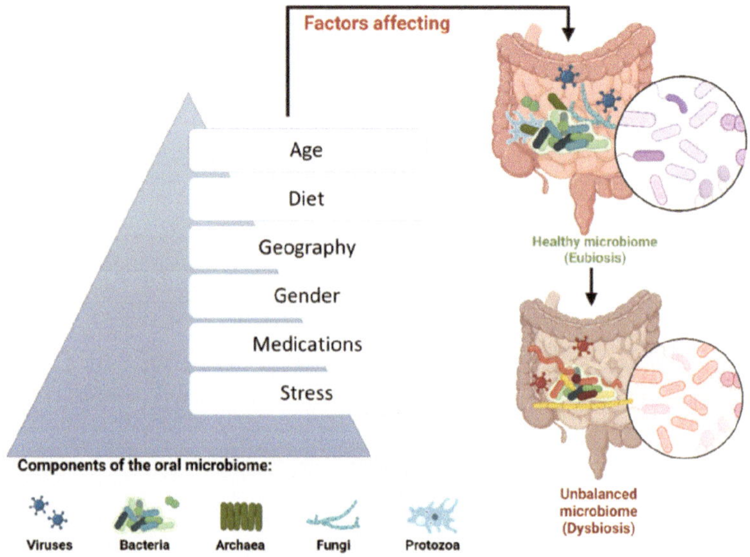

FIGURE 7.3 Factors influencing the composition of gut microbiota.

7.5.1 Geographical Distance

Geographical distance emerges as a notable determinant influencing the diversity of GM, as demonstrated by studies across diverse animal species. For instance, an inquiry into fish species originating from the Yellow River underscored geographical distance as the primary factor delineating the composition of the GM. The GM community structure demonstrated an augmented dissimilarity concomitant with the widening geographical span [33, 34]. Likewise, an investigation involving Mediterranean scorpionfishes underscored the significance of geographical provenance in influencing the diversity and constitution of GM communities. This influence surpassed the impact of host phylogenetic relatedness within the studied fish group [35]. These cumulative observations emphasize the influence of geographical distance in molding the composition of GM across diverse animal species.

7.5.2 Host Evolutionary History

Numerous investigations have probed the impact of host evolutionary lineage on the diversity of the GM, elucidating distinctive insights within diverse vertebrate clades. In a study concentrating on fish species from the Yellow River in China, host phylogeny emerged as a notable determinant influencing the diversity of the GM community. While geographical distance predominantly governed the GM, resulting in an escalated dissimilarity in GM community structure with greater geographical separation, the GM communities demonstrated heightened uniformity at higher taxonomic resolutions of host classification, indicative of the predominant influence of homogeneous selection in facilitating community convergence. Although the host diet exerted a substantial impact on GM diversity, it elucidated only a constrained variation across the assessed metrics [33]. In a thorough examination of an extensive dataset encompassing the distal gut 16S rRNA microbiomes of a diverse array of animals, including 80% wild specimens across Mammalia, Aves, Amphibia, Reptilia, and Actinopterygii, the study methodically unraveled the influences of host evolutionary lineage and diet on GM diversity. The results emphasized that each factor distinctly shaped facets of diversity, with evolutionary history emerging as the primary determinant predicting GM differentiation among the investigated vertebrate clades [36]. Additional revelations were gleaned from an examination of two phylogenetically linked fish, namely the three-spined stickleback and the nine-spined stickleback. In this context, habitat emerged as a potent influencer of GM diversity in both host species, surpassing the explanatory efficacy of host species in the variability of GM composition. The investigation underscored that robust, habitat-specific effects took precedence over considerations such as geographical distance and historical freshwater colonization, emphasizing the principal association of GM with localized environmental conditions [37]. These collective inquiries collectively attest that host evolutionary lineage, as manifested in the host's phylogenetic relationships, profoundly molds the diversity of the GM across a spectrum of vertebrate clades, encompassing fish and other wildlife. The amalgamated results imply that the configuration of the GM in diverse animal species is intricately modulated by the interplay of both host evolutionary history and localized environmental conditions.

7.5.3 Diet

Diet emerges as a central determinant influencing the diversity of the GM, and numerous studies have scrutinized the intricate relationship between dietary patterns and the composition of the GM. For instance, an investigation involving a substantial cohort of asymptomatic Chinese adults unveiled that a general dietary pattern characterized by diminished refined grain consumption exhibited a correlation with heightened abundances of the genus Anaerostipes, along with a specific species within it. The study further delineated 17 associations elucidating the linkages between the intake of various food groups and specific genera and species. This was exemplified by adverse associations between red meat consumption and the relative abundances of the genus Weissella, including an uncultured species within it [38]. Comparably, an investigation conducted within an extensive population-based cohort in the Netherlands established a correlation between elevated diet quality and enhanced alpha diversity, elucidating a portion of the variability in beta diversity. The overarching quality of the diet demonstrated associations with the relative abundance of four families, specifically Ruminococcaceae, Erysipelotrichaceae, Lachnospiraceae, and Christensenellaceae [39, 40]. A separate investigation concentrating on urban Chinese adults showcased that sustained diet quality, spanning eight food groups, exhibited a positive correlation with microbiome alpha-diversity and the prevalence of Firmicutes, Tenericutes, and Actinobacteria, along with specific genera/species within these phyla. Notably, this included *Coprococcus*, *Bifidobacterium* (specifically *Bifidobacterium adolescentis*), *Faecalibacterium* (specifically *Faecalibacterium prausnitzii*), and the order RF39 [36, 41, 42]. These collective findings underscore the influence of dietary patterns in shaping the composition of the GM, suggesting potential implications for human health and disease.

7.5.4 Socioeconomic Status (SES)

Several investigations have explored the correlation between SES and the diversity of the GM, yielding noteworthy insights. One study detected a link between the SES of neighborhoods and the diversity of the GM. Notably, it revealed that lower diversity in GM and the presence of multidrug resistant microorganisms were associated with neighborhoods characterized by lower SES. This research underscores the conceivable influence of the socioeconomic environment as a determinant affecting the diversity of the GM [43]. An additional investigation explored the relationship between family SES and host genetics in conjunction with the composition of the GM in infants and children. The findings suggested that family SES played a substantial role in predicting a latent variable reflecting the abundance of six genera identified through whole-genome shotgun sequencing. While SES did not predict alpha diversity, it did exert an influence on beta diversity, underscoring the impact of malleable environmental factors, including family SES, on the initial composition of the GM [44]. In a study conducted on a population of school-aged children in Latin

America, associations between 14 clinical and environmental covariates and inter-individual variability in GM composition were discerned. Significantly, indicators of SES were identified as influential factors affecting the prevalence of the most abundant taxa and the distribution of enterotypes. Furthermore, age exhibited a positive correlation with heightened diversity in children of normal weight but manifested a negative correlation in overweight and obese children, thereby emphasizing the role of SES in molding the composition of GM in children [45]. In aggregate, these results imply an association between SES, encompassing both neighborhood and family levels, and the diversity of the GM. This highlights the impact of social and environmental determinants on the intricate configuration of the GM.

7.5.5 Age

Age emerges as a crucial determinant significantly shaping the diversity of the GM, and its influence on both the composition and the functionality of the GM is subject to a multitude of influencing factors. These alterations, occurring across the lifespan, are intricately shaped by a complex interplay of variables, including but not limited to diet, genetics, gender, SES, disease state, pregnancy, geography, and environmental exposures [46]. In a population-based study concentrating on school-aged children in Latin America, age exhibited a positive correlation with heightened diversity in children of normal weight, while displaying a negative correlation in those who were overweight and obese [45]. Likewise, investigations on six-year-old children underscored the impact of early life determinants, such as exclusive breastfeeding, antibiotic utilization between ages 1 and 2 years, and the mode of delivery, on the alpha diversity of the GM [47]. Acknowledging the growing importance of early-life GM in human health, a study emphasized their pivotal role in diverse aspects, encompassing the digestion of food substances for nutrient release, regulation of pathogens, modulation of the immune system, and impact on systems such as the liver, brain, and endocrine system [48]. Findings from a murine investigation reaffirmed the influence of age on GM composition, with gender also exerting a discernible effect, while geographical location demonstrated a relatively modest impact on diversity. Remarkably, shifts in serum cytokine expression, specifically IL-10 and IL-6, were noted, revealing distinct regulatory patterns in the context of both gender and aging [49]. Fundamentally, age exerts a substantive influence on modulating the diversity of the GM, as the configuration of the GM undergoes alterations driven by a multifaceted interplay of diverse factors, spanning individual behavioral patterns to environmental exposures.

7.5.6 Gender

Gender emerges as a noteworthy factor capable of influencing GM diversity, as highlighted by various studies. An exhaustive review article investigating the correlation between diet and GM diversity identified gender as one of the individual parameters manifesting robust associations with GM diversity [50]. Contrasting findings emerged from a study conducted on Chinese forest musk deer, where gender did not significantly impact bacterial diversity within the GM [51]. In contrast, an examination centered on hyperlipidemic outbred mice unveiled noteworthy gender-related distinctions. Female mice exhibited elevated GM alpha diversity in comparison to their male counterparts, and beta diversity analysis revealed substantial dissimilarities in the overarching GM community structure between the two genders [52]. Investigating irritable bowel syndrome (IBS), another study found that females were more susceptible to IBS and exhibited more severe gut dysbiosis compared to males [53]. These findings collectively underscore the potential influence of gender on GM diversity. However, it is crucial to note that the extent of this influence may vary based on factors such as the animal species under investigation and other contextual considerations.

7.6 INTERACTIONS BETWEEN ORGANIC POLLUTANTS AND THE GUT MICROBIOME

The GM of humans is a vital component of both gut and overall health. A healthy gut maintains a delicate balance of bacteria, and the kind, amount, and proportion of GM in a normal condition are determined by genetic and environmental factors [54]. The equilibrium of the GM can be upset by several variables, mentioned above, and can lead to a variety of disorders [55]. Consequently, OP and GM interact in both directions and aid in the development of weariness.

Characterizing how interacting with pollutants and toxins contributes to the onset and advancement of many illnesses, including diabetes, cancer, and obesity, as well as to the disturbance of immunological and reproductive homeostasis, is a topic of increasing interest. The creation of a dynamic, bidirectional link between GM and OPs is thought to be a crucial factor in mitigating their harmful effects. This relationship is significant from a chemical perspective. Indeed, it has been shown that several environmental chemicals influence GM composition. These factors may be crucial in determining an individual's unique microbiotype.

7.6.1 Impact of OPs on GM Composition

OPs are widespread in the environment because of industrial activities, agricultural practices, and other human actions. Although the negative consequences on human health have been extensively studied, new findings are shedding light on how they affect GM, which is essential for maintaining host health and balance [56, 57]. Recent findings indicate that OPs may disturb the balance and operation of GM, resulting in dysbiosis, inflammation, and heightened vulnerability to different illnesses. One instance of OPs affecting GM composition involves the prevalent herbicide glyphosate (GLP), frequently detected in agricultural products. Studies have demonstrated that GLP decreases GM diversity in both animal models and human populations [57, 58]. According to research conducted by Lozupone et al. (2012), exposure to GLP led to a reduction in beneficial bacteria like *Bifidobacteria* and *Lactobacilli*, and an increase in potentially harmful bacteria such as *Clostridia*. These industrial chemicals, PCBs and PBDEs, have been linked to changes in GM. Research indicates that PCBs and PBDEs exposure can alter the balance of certain bacterial species in GM, possibly leading to dysbiosis and inflammation [59]. An investigation conducted by Cai et al. (2020) revealed a connection between PBDE exposure and changes in GM composition. This resulted in a rise in Proteobacteria levels and a decline in Bacteroidetes levels. In addition, OPs may affect the gut barrier's integrity, which is essential for blocking the passage of harmful substances from the gut into the bloodstream. Research has demonstrated that Bisphenol A (BPA) and phthalates (PAEs), which are often present in plastics and various consumer goods, can interfere with the regulation of tight junction proteins in the intestinal epithelium [60, 61]. This disruption can result in heightened permeability of the gut barrier.

By disrupting the barrier, it enables microbial products and toxins to enter the bloodstream, leading to inflammation and possibly playing a role in the development of conditions like inflammatory bowel disease (IBD) and metabolic syndrome (MS). Furthermore, OPs may disrupt the balance between beneficial and pathogenic microbes in the gut by interfering with host-microbiota interactions. PAHs, present in both air pollution and grilled or smoked foods, have been demonstrated to disrupt the metabolism of bile acids by gut bacteria, resulting in dysbiosis and inflammation [62, 63].

Organophosphate pesticides have been linked to alterations in the production of SCFAs by GM, crucial for gut health and immune function. Long-term exposure to OPs may lead to mild inflammation in the gut, resulting in higher levels of pro-inflammatory cytokines and immune cell activity. The inflammatory response may continue to disturb GM composition and function, leading to a cycle of dysbiosis and inflammation. Several studies have demonstrated that being exposed to polychlorinated dibenzo-p-dioxins (PCDDs) and polychlorinated dibenzofurans (PCDFs), which are environmental pollutants produced during industrial activities, may trigger inflammation in the intestines and change the population of certain microorganisms in the gut [63–65].

7.6.2 Mechanisms of Interaction between OPs and the GM

Delving into the complex ways in which OPs interact with GM is essential for grasping their effects on human health. Various OPs such as pesticides, industrial chemicals, and POPs have been demonstrated to disturb the balance and operation of GM, resulting in dysbiosis and related health consequences. OPs may disrupt the variety and quantity of GM communities, resulting in dysbiosis [57, 63]. For example, research conducted by Popli, Shivani, et al. (2022) discovered that GLP exposure led to a decrease in GM diversity and alterations in the distribution of certain bacterial groups [57]. In another research conducted by Zhang et al. (2022), exposure to PCBs was linked to changes in GM composition in rats, resulting in dysbiosis and compromised metabolic health. OPs may affect the gut barrier, leading to increased intestinal permeability and the movement of harmful substances into the bloodstream [66]. For instance, BPA has been demonstrated to interfere with tight junction proteins in the intestinal epithelium, resulting in heightened gut permeability [56]. Liu et al. (2022) found that exposure to BPA led to gut barrier dysfunction and changes in GM composition in mice, emphasizing the connection between gut barrier health and GM balance. OPs have the potential to disrupt the delicate balance between beneficial and pathogenic GM by interfering with host-microbiota interactions. Exposure to PAHs has been demonstrated to change the metabolism of bile acids by gut bacteria, resulting in dysbiosis and inflammation [67].

According to a study conducted by Jin et al. (2018), exposure to PAHs was found to be linked to alterations in bile acid metabolism and GM composition in mice, indicating a potential connection between environmental pollutants, host metabolism, and microbial dysbiosis. Long-term exposure to OPs may lead to mild inflammation in the gut, which can disturb the balance and operation of GM [67, 68]. Exposure to PCDDs has been demonstrated to trigger intestinal inflammation and change the population of certain microbial species in the gut. According to research conducted by Defois et al. (2018), exposure to PCDDs was linked to gut inflammation and dysbiosis in rats, emphasizing the impact of environmental pollutants on GM composition and inflammation. Studying the impact of OPs on GM composition and function is crucial for understanding its effects on human health. Research has connected dysbiosis to the onset of several conditions, such as obesity, IBD, MS, and mental health disorders [69, 70]. In research conducted by Feng et al. (2020), they found that exposure to organophosphate pesticides led to changes in GM composition and a higher likelihood of developing MS in individuals. This indicates a possible connection between environmental toxins, gut imbalance, and metabolic well-being [71].

By utilizing different enzymatic processes, GM can convert OPs into metabolites that may have different properties, such as increased or decreased toxicity, when compared to the original compounds. Research has indicated that gut bacteria have enzymes that can break down PAHs into forms that are less harmful or easier to eliminate [72, 73]. For example, specific strains of gut bacteria from the genus Clostridium have been discovered to convert benzo[a]pyrene (BaP), a cancer-causing PAH, into less harmful hydroxylated metabolites. In addition, enzymes from GM, like cytochrome P450 monooxygenases, have the ability to oxidize PAHs, resulting in the production of hydroxylated metabolites that could have different biological effects. PCBs are well known for their toxicity and long-lasting presence in the environment. Research has shown that GM can impact the metabolism of PCBs by undergoing reductive dechlorination, a method that involves removing chlorine atoms from the PCB molecule one by one [62, 73].

Anaerobic bacteria like *Dehalococcoides spp.* have demonstrated the ability to facilitate the dechlorination of PCBs, transforming heavily chlorinated congeners into metabolites that are less harmful or non-toxic. Microbial processes can influence the availability and harmful effects of PCBs in the organism. Studies have shown that gut bacteria can break down OPs through processes like hydrolysis and detoxification. Specific bacterial types from the Pseudomonas and

Bacillus groups create organophosphorus hydrolases, which are enzymes that can break down the phosphate ester bonds found in OP molecules [74]. Microbial enzymes help in detoxifying and eliminating OP residues from the GIT, thus minimizing their negative impact on host health. Industrial chemicals like TCE and perchloroethylene (PCE), are commonly found in contaminated groundwater and soil. Research indicates that GM can metabolize chlorinated solvents via reductive dechlorination and other enzymatic reactions. Consider anaerobic bacteria like Dehalococcoides spp. which have reductive dehalogenases that can remove chlorine atoms from TCE and PCE molecules, resulting in the production of ethene and ethane, which are less toxic [75].

Microbial transformations are crucial for naturally reducing chlorinated solvent contamination in the environment. These heavy metals (HMs), such as mercury, lead, and cadmium, are commonly found in the environment and are notorious for their harmful effects on health. Research has shown that GM plays a role in detoxifying HMs through different methods such as metal sequestration, enzymatic reduction, and biotransformation. Specific types of bacteria like Lactobacillus spp. and Bacillus spp. generate metal-binding proteins and enzymes that can transform toxic metal ions into less harmful forms. Microbial processes play a crucial role in reducing the harmful impact of HMs exposure in the GIT [76, 77].

7.7 EFFECTS OF ORGANIC POLLUTANTS ON GUT HEALTH

Understanding the importance of the gut barrier, which consists of a single layer of epithelial cells that line the intestinal mucosa, is crucial for preserving intestinal balance and blocking the passage of harmful substances into the bloodstream. Concerns arise from the potential implications for human health due to the disruption of gut barrier function by OPs. Numerous research studies have explored how different OPs interfere with gut barrier function, revealing insights into their negative impact on gastrointestinal (GI) health. Studies have indicated that BPA exposure has the potential to interfere with gut barrier function by changing the expression and distribution of tight junction proteins, which play a crucial role in preserving the integrity of the intestinal epithelium [56, 77]. Research conducted by Di Ciaula et al. (2020) revealed that exposure to BPA reduced the levels of occludin and zonula occludens-1 (ZO-1), which are crucial elements of tight junction complexes, in the intestinal epithelium of mice. This resulted in heightened intestinal permeability. Research has shown that PCB exposure can interfere with gut barrier function by triggering oxidative stress (OS) and inflammation in the intestinal lining [78, 79]. In another study conducted by Ruan et al. (2020), it was demonstrated that exposure to a combination of PCB congeners led to an increase in intestinal permeability in mice. This was associated with an increase in pro-inflammatory cytokines and activation of the NF-κB signaling pathway in the intestinal epithelium [63, 80]. Additionally, in a study by Wei et al. (2020) it was shown that GLP exposure caused disruption to the intestinal barrier in zebrafish embryos. This led to higher permeability and the leakage of fluorescein isothiocyanate (FITC)-dextran into circulation. The disturbance led to changes in GM, with a reduction in beneficial bacteria like Lactobacillus and an increase in pathogenic bacteria such as Proteobacteria. These compounds, found in a range of industrial and consumer goods, have been demonstrated to interfere with gut barrier function [81]. In a study conducted by Bonato et al. (2020), it was discovered that exposure to perfluorooctane sulfonate (PFOS) led to an increase in intestinal permeability in mice. This was observed alongside a decrease in tight junction proteins and an increase in inflammatory cytokines in the intestinal epithelium [82]. Cheng et al. (2024) showed that exposure to di(2-ethylhexyl) phthalate (DEHP) led to an increase in intestinal permeability in rats. This was linked to a decrease in occludin and ZO-1 expression, along with elevated levels of reactive oxygen species (ROS) in the intestinal mucosa. Disruption of gut barrier function by OPs involves changes in tight junction integrity, OS, inflammation, and dysbiosis of GM. This research emphasizes the significance of comprehending the mechanisms behind gut barrier disruption caused by OPs and their impact on GI and systemic health. Additional investigation is required to clarify the precise molecular pathways at play and to create approaches to reduce the negative impacts of OPs on gut barrier function and general well-being [83, 84].

7.7.1 Inflammatory Effects of OPs on the Gut

Research on the impact of OPs on gut inflammation has been receiving more attention lately. Various OPs such as BPA, PCBs, GLP, perfluorinated compounds (PFCs), and PAEs have been linked to triggering and worsening inflammatory reactions in the GIT. Research indicates that BPA exposure can trigger the production of pro-inflammatory cytokines and chemokines in the intestinal mucosa, resulting in immune cell infiltration and tissue damage [85, 86]. Research conducted by Wu et al. (2024) showed that BPA exposure stimulated the expression of interleukin-8 (IL-8) and tumor necrosis factor-alpha (TNF-α) in human intestinal epithelial cells, suggesting an inflammatory reaction. PCBs, environmental contaminants recognized for their long-lasting nature and harmful effects, have also been linked to gut inflammation. In a study by Wang et al. (2019) it was demonstrated that exposure to a combination of PCB congeners led to an upregulation of IL-6 and IL-1β, in the intestinal mucosa of mice. The inflammatory response was linked to the activation of the nuclear factor-kappa B (NF-κB) cascade [87, 88]. Studies indicate that exposure to GLP can change GM composition, resulting in dysbiosis and the growth of pro-inflammatory bacteria [89, 90]. Another research conducted by Maddalon et al. (2021) showed that exposure to GLP led to an increase in pathogenic bacteria like Escherichia coli and Clostridium spp. in the poultry gut, leading to intestinal inflammation and epithelial damage [91]. In a study by Maxwell, Joseph R., et al. (2015), it was discovered that exposure to PFOS, a common PFC, triggered the production of IL-17 and IL-23 in the intestinal mucosa of mice. The inflammatory response led to the activation of the NF-κB signaling pathway and the infiltration of immune cells into the gut tissue [92].

7.7.2 Implications for Gut-Related Diseases

OPs play a crucial role in gut-related diseases by interfering with different physiological processes in the GIT, which can result in the onset or worsening of GI disorders (GIDs). Numerous studies have revealed the impact of OPs on the development of IBD, IBS, and colorectal cancer (CRC) [93–95]. Studies have shown that OPs play a role in the onset and advancement of IBD by impacting gut barrier function, immune system regulation, and inflammation. Research has indicated that exposure to BPA can interfere with gut barrier function and trigger inflammation in the intestinal mucosa, playing a role in the development of IBD. In animal models, PCBs have been linked to heightened intestinal permeability and immune cell infiltration, which can worsen gut inflammation and possibly lead to the onset of IBD [96]. OPs have been linked to the development of IBS due to their impact on gut motility, visceral hypersensitivity, and immune response. For example, research has demonstrated that being exposed to GLP can disturb GM and trigger visceral hypersensitivity in animal models, replicating symptoms seen in individuals with IBS. Moreover, research has linked PFCs to changes in gut motility and visceral pain in animal experiments, indicating a possible connection to the onset of IBS [97].

CRC ranks as the third most prevalent cancer globally and is impacted by a range of environmental factors, such as OPs. Studies indicate that being exposed to OPs like PAEs and PCBs could potentially raise the likelihood of developing CRC by impacting inflammation, OS, and DNA damage in the GIT. Moreover, there is a connection between OPs and changes in GM composition, potentially influencing the development of gut-related diseases [98, 99]. An imbalance in GM, known as dysbiosis, has been linked to the onset of IBD, IBS, and CRC. Research indicates that GLP exposure can disturb the balance of gut bacteria, potentially fostering the development of harmful bacteria, which may result in inflammation and gut dysbiosis. PAEs have been linked to changes in GM composition and a higher risk of intestinal inflammation and CRC [100–102].

7.8 MODULATION OF GUT MICROBIOME FOR DETOXIFICATION OF ORGANIC POLLUTANTS

GM is crucial for detoxification processes, aiding in the metabolism and elimination of drugs, environmental pollutants, and dietary compounds. The complex relationship between GM and xenobiotics includes a wide range of microbial enzymes and MPs, which play a crucial role in maintaining the host's health and overall well-being. Delving into GM detoxification function, it is crucial to consider how it processes dietary polyphenols. Antioxidant properties are found in polyphenols, which are linked to various health advantages. However, their absorption in the small intestine is frequently restricted, affecting their bioavailability [103, 104]. Research indicates that GM can process polyphenols by breaking down sugars, removing hydroxyl groups, and breaking the ring structure, which can improve how these compounds are absorbed and utilized in the body. Specific bacterial species, such as those in the *Clostridium* genus, produce metabolites with enhanced antioxidant and anti-inflammatory effects. Moreover, GM is essential for metabolizing dietary fibers, which are complex carbohydrates that resist digestion by host enzymes. When GM ferment dietary fibers, they produce SCFAs. These SCFAs act as a fuel for intestinal epithelial cells and have diverse physiological impacts [105, 106]. Research has demonstrated that SCFAs play a role in regulating host metabolism, immune function, and inflammation, which can impact overall health. In addition, SCFAs can impact drug metabolism through the control of drug-metabolizing enzymes in the liver and other tissues. For example, butyrate has been discovered to boost the expression of cytochrome P450 in the liver, which helps in the metabolism of xenobiotics and detoxification. In addition, GM plays a crucial role in transforming environmental pollutants like HMs, pesticides, and industrial chemicals into less harmful or easier to eliminate forms. Microorganisms in the gut have different enzymes and MPs that can break down harmful substances, which helps lower their impact on the body and potential harm to the host. Additionally specific gut bacteria such as Lactobacillus and Bifidobacterium species have demonstrated the ability to detoxify HMs by either binding to them or facilitating their conversion into less harmful forms [107].

Certain gut bacteria can break down pesticides and industrial chemicals using various processes like hydrolysis, oxidation, and reduction. This helps in deactivating and removing these substances from the body. In addition, GM plays a role in shaping the metabolism and effectiveness of therapeutic drugs, which is referred to as pharmacomicrobiomics. When studying GM, it is important to consider how they can break down drugs, which can affect how the drugs work in the body [108]. Specific gut bacteria can convert prodrugs like sulfasalazine into active metabolites that produce therapeutic benefits. In addition, GM can impact drug metabolism through their interactions with host enzymes and transporters, which can change how drugs are absorbed, distributed, and eliminated. Studying GM composition reveals how individuals differ in drug metabolism and response, emphasizing the need to include microbial factors in personalized medicine strategies [64, 109].

7.8.1 Strategies for Enhancing GM-Mediated Detoxification

Exploring the potential of improving GM-mediated detoxification shows great promise in enhancing human health and reducing the negative impacts of xenobiotic exposure. Several approaches have been suggested to adjust GM to enhance its detoxification abilities, including dietary changes and the use of probiotics and prebiotics. One approach includes dietary adjustments focused on enhancing the development of helpful GM that play a role in detoxification processes. Research has demonstrated that dietary elements like fiber, polyphenols, and fermented foods play a crucial role in promoting a balanced GM and improving detoxification processes [104]. Wilson et al. (2020) showed that adopting a high-fiber, low-fat, plant-based diet led to positive alterations in GM composition. This included a rise in beneficial butyrate-producing bacteria and a decrease in harmful species. Through the dietary pattern, a connection was found between decreased inflammation markers and better metabolic health, indicating a possible contribution to improving detoxification.

In addition, foods high in polyphenols have been proven to influence GM and improve the body's ability to detoxify [110]. Dueñas et al. (2015) revealed that administering a polyphenol-rich grape extract resulted in positive alterations in GM composition, leading to higher levels of beneficial bacteria like Bifidobacterium and Lactobacillus. These changes led to improvements in metabolic parameters and markers of OS, suggesting improved detoxification abilities. Aside from dietary interventions, using probiotics and prebiotics is being considered as a way to improve GM-mediated detoxification [111].

Probiotics are living microorganisms that provide health benefits when consumed in sufficient quantities, whereas prebiotics are indigestible dietary fibers that specifically promote the growth of beneficial gut bacteria. Various research studies have explored the impact of probiotic supplementation on xenobiotic metabolism and detoxification [42, 112]. Research conducted by Kumar et al. (2015) showed that giving rats the probiotic strain Lactobacillus plantarum helped improve the removal of HMs like cadmium and lead by increasing their excretion through feces. In the same way, research has demonstrated that adding prebiotics can alter the composition of GM and enhance detoxification capabilities [113–115]. Another study discovered that adding inulin, a prebiotic fiber, resulted in positive alterations in GM composition. This included a rise in beneficial butyrate-producing bacteria and a decrease in levels of potentially harmful species. These changes led to improvements in metabolic parameters and markers of inflammation, suggesting improved detoxification abilities.

Furthermore, fecal microbiota transplantation (FMT) is an advanced method for influencing GM and improving detoxification capabilities. Transferring fecal matter from a healthy donor to a recipient aims to restore a healthy GM composition through FMT [116, 117]. Various research studies have explored the potential of FMT in boosting detoxification processes and enhancing health results. Ruud et al. (2017) showed that FMT from lean donors enhanced insulin sensitivity and metabolic parameters in obese recipients, indicating a possible contribution to improving detoxification and metabolic health. In the same way, FMT has demonstrated the ability to alter GM makeup and alleviate symptoms in individuals with IBD, suggesting its promise as a treatment to boost detoxification in diseased conditions. Moreover, utilizing enzymes and metabolites derived from microbes as potential therapeutic agents shows potential in improving GM-mediated detoxification [118, 119]. Enzymes produced by microorganisms, such as glucuronidases and sulfatases, are essential for breaking down conjugated metabolites and aiding in the elimination of xenobiotics from the body. Numerous studies have explored the therapeutic potential of microbial enzymes in improving detoxification processes. Purdel et al. (2023) showed that adding microbial glucuronidase to the diet enhanced the effectiveness of chemotherapy drugs in cancer patients by boosting their metabolism and activation. Studies have demonstrated that microbial metabolites like SCFAs and secondary bile acids can impact host metabolism and immune function, affecting detoxification processes. For example, SCFAs were discovered to control the expression and function of drug-metabolizing enzymes in the liver, resulting in improved detoxification and drug clearance [120–122].

7.8.2 Potential Therapeutic Interventions Targeting the GM

In recent years, there has been considerable focus on therapeutic interventions aimed at GM because of their ability to influence host health and potentially treat different diseases. These strategies focus on altering GM composition and function using dietary changes, probiotics, prebiotics, FMT, microbial-based treatments, and drugs targeting specific microbes. Adjusting what we eat can be a simple and effective way to influence GM. Several dietary patterns, like the Mediterranean diet, high-fiber diet, and plant-based diet, have shown links to positive alterations in GM composition and better health results [123]. Wu et al. (2020) examined how following a Mediterranean diet impacts the composition of GM. The study revealed that sticking to this dietary regimen resulted in higher levels of beneficial bacteria like *Bifidobacterium* and *Lactobacillus*, while reducing the presence of potentially harmful species. These changes led to improvements in metabolic parameters and a

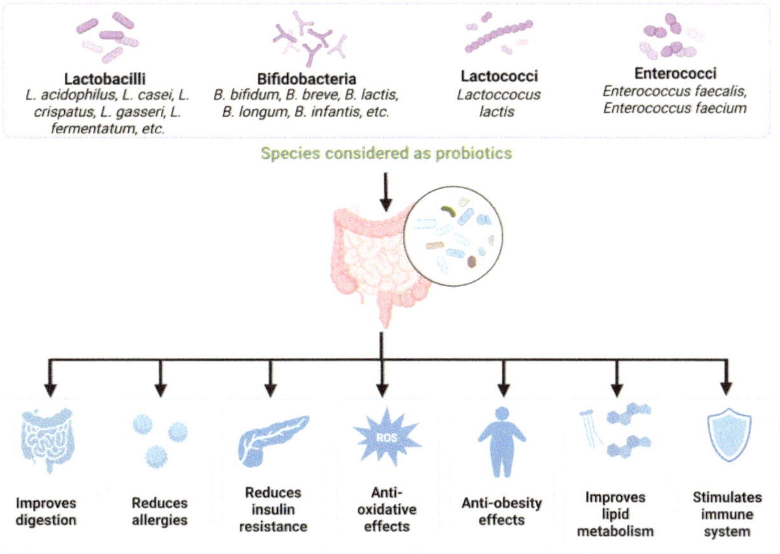

FIGURE 7.4 Probiotics-based interventions show positive effects targeting gut microbiome diversity.

decreased risk of cardiovascular disease and MS. Probiotics, when consumed in sufficient quantities, has been a focus of research for their possible therapeutic uses in influencing GM (Figure 7.4) [124–127]. Several studies have examined the effectiveness of probiotic supplementation in different health conditions, such as GIDs, MS, and immune-related diseases. Szajewska et al. (2015) examined the impact of probiotics on preventing antibiotic-associated diarrhea in children. The results showed that probiotic supplementation notably decreased the likelihood of developing diarrhea when compared to a placebo. In the same vein, probiotics have demonstrated effectiveness in alleviating symptoms in individuals with IBS and IBD, indicating a possible impact on GM and relief of GI symptoms [128]. Research has shown that adding prebiotics to the diet can change the balance of bacteria in the gut and enhance digestive health [117]. Vandeputte et al. (2017) examined the impact of inulin supplementation on GM composition in overweight and obese adults. The study revealed that inulin supplementation resulted in higher levels of butyrate-producing bacteria and lower levels of potentially harmful species. These modifications correlated with enhancements in metabolic parameters and markers of inflammation, suggesting a possible contribution of prebiotics in enhancing gut health [129].

In clinical trials, FMT has proven to be a very effective treatment for recurrent Clostridioides difficile infection (CDI), with cure rates surpassing 90%. In addition, FMT has demonstrated potential in addressing various GIDs like IBD, IBS, and constipation [130]. A randomized controlled trial conducted by Costello et al. (2017) explored the impacts of FMT on individuals with ulcerative colitis. The study revealed that FMT resulted in notable enhancements in disease activity and mucosal healing when compared to a placebo. This indicates a possible function for FMT in influencing GM and easing symptoms of IBD. Novel approaches involving microbial enzymes, metabolites, and bacteriophages are being explored for targeting GM and treating a range of diseases [131–134].

7.9 CHALLENGES AND LIMITATIONS

Exploring the relationship between OPs and GM comes with various challenges and constraints because of the intricate microbial community, the range of pollutants, and the ever-changing host-microbe interactions. Understanding the mechanisms behind pollutant-microbiome interactions

(PMIs) and assessing their impact on human health is hindered by these challenges. Dealing with the intricate and diverse GM composition among different individuals and populations poses a significant obstacle. Understanding GM can be complex due to various factors such as genetics, diet, age, and environmental influences, which can complicate determining the direct impact of OPs exposure on microbial composition [64]. Khoo et al. (2024) delved into the impacts of exposure to PAHs on GM composition in mice. The study revealed that variations in microbial reactions to PAH exposure were affected by initial GM composition and host factors like age and sex. These results highlight the importance of conducting extensive, long-term research to consider differences between individuals and pinpoint reliable microbial patterns linked to exposure to OPs [135–137].

One of the challenges lies in the wide variety and intricate nature of OPs, which include a broad spectrum of chemical classes and structures. Various OPs exhibit differences in their physical and chemical characteristics, levels of toxicity, and mechanisms of action, posing challenges in drawing broad conclusions that apply to all compounds. Lim et al. (2021) examined how exposure to PCBs and PBDEs impacted GM composition in mice. The study revealed that PCBs and PBDEs triggered different microbial reactions, indicating unique effects on GM. This study emphasizes the significance of taking into account the wide range of OPs and their specific interactions with GM in research on PMIs [138]. Moreover, the ever-changing relationship between hosts and microbes poses difficulties in understanding the timing of changes in GM caused by OPs. OPs may cause immediate or long-term impacts on GM, resulting in temporary or lasting changes in microbial makeup and activity. The significance of conducting longitudinal studies to understand the temporal dynamics of PMIs and their impact on host health is underscored by these results. Furthermore, the absence of consistent methodologies in researching PMIs creates difficulties in comparing outcomes between studies and confirming replicable results. When experimental protocols, sample processing techniques, and analytical methods vary, it can lead to bias and complicate the interpretation of results [72]. Johnson et al. (2020) examined various DNA extraction techniques for analyzing GM in mice exposed to PAHs. The study revealed that the extraction method had a notable impact on the microbial community composition and diversity metrics, underscoring the need for standardized methodologies to guarantee reliable and consistent outcomes [139]. Moreover, ethical concerns and practical limitations restrict the possibility of conducting human studies on PMIs. When conducting research on humans, it is common to use observational data or analyze existing cohorts retrospectively. However, these methods can be influenced by various confounding factors and biases. In addition, conducting controlled experimental studies with human subjects may lead to ethical considerations regarding exposure to OPs and possible health risks for participants [140–142].

7.10 FUTURE DIRECTIONS AND CHALLENGES

Exploring new frontiers in OPs and GM is broadening our knowledge of the complex relationships between environmental toxins and microbial populations in the digestive system. Various aspects are explored such as the effects of OPs on GM composition and function, the involvement of microbial metabolism in pollutant detoxification, and the potential implications for human health. Exploring the impact of OPs on GM composition and diversity is a growing area of study. Research has shown that being exposed to environmental pollutants like PAHs, pesticides, and HMs can disturb GM equilibrium, resulting in dysbiosis and changes in the microbial community structure [143–147]. Zhang et al. (2021) explored how exposure to the pesticide chlorpyrifos impacted GM composition in mice. The study revealed that chlorpyrifos exposure led to dysbiosis, marked by a reduction in beneficial bacteria like *Bifidobacterium* and *Lactobacillus* and an increase in pathogenic species. Exposure to HMs like cadmium and lead can change GM composition and diversity in both humans and animals, indicating a possible connection between pollutant exposure and GM imbalance [148]. Delving into a new research field involves investigating how GM influences the processing of OPs.

Studying the microbial communities in the GIT reveals a variety of MPs that can break down various xenobiotics, such as OPs. Giri et al. (2024) delved into the metabolism of the flame retardant PBDEs by GM. The study revealed that specific bacterial species, including *Bacteroides* and *Clostridium*, could transform PBDEs into metabolites with lower toxicity. Research indicates that the metabolism of GM can impact the availability and harmful effects of PCBs and PAHs, suggesting a possible detoxification process for these pollutants. Moreover, recent studies are revealing the effects of OPs and GM interactions on human health. Research has shown a connection between dysbiosis caused by exposure to OPs and a range of GIDs such as IBD, IBS, and colorectal cancer [149–151]. Balaguer-Trias et al. (2022) delved into the link between exposure to PCBs and GM dysbiosis in humans. The study revealed that PCB exposure led to changes in GM makeup, such as a reduction in butyrate-producing bacteria and a rise in pro-inflammatory species. These changes were associated with an elevated risk of IBD and other GIDs, indicating a possible connection between exposure to OPs, imbalance in GM, and development of diseases [152]. Furthermore, new studies are investigating how the gut-liver axis influences the impact of OPs on host metabolism and health. Studying the liver's function in detoxifying xenobiotics and metabolizing environmental contaminants is essential. Recent findings indicate a potential link between GM and hepatic metabolism and detoxification pathways [73]. Another research conducted by Li, Meng, et al. (2022) explored the impact of pesticide exposure on the gut–liver axis in mice. The study revealed that exposure to the pesticide altered GM and liver gene expression, resulting in disrupted lipid metabolism and liver steatosis. This study emphasizes the complex relationship among environmental pollutants, GM, and host metabolism, and how it can impact liver health and disease. Moreover, recent studies are investigating the possibility of using therapeutic treatments that focus on GM to reduce the negative impact of OPs on human health. Exploring methods like probiotics, prebiotics, dietary interventions, and microbial-derived therapeutics shows potential in adjusting GM to improve detoxification and lower OPs-related toxicity [153–155].

There are various promising opportunities for further research and practical use in the realm of OPs and the GM. These avenues could enhance our knowledge of how pollutants interact with the GM and impact both human health and environmental sustainability. Different areas of research can be explored such as creating new treatments, identifying biomarkers, and developing strategies for bioremediation, along with studying the connections between the gut and the lungs and brain. Exploring new treatments that focus on GM could help reduce the negative impact of OPs on human health. Exploring probiotics, prebiotics, dietary interventions, and microbial-derived therapeutics shows potential in adjusting GM to improve detoxification and lower OPs-related toxicity [140, 156]. Bhatt et al. (2021) explored the impact of probiotic supplementation on GM and host metabolism in mice exposed to the flame retardant hexabromocyclododecane (HBCD). The study revealed that probiotic treatment restored GM diversity and enhanced metabolic parameters, indicating a possible therapeutic strategy for addressing OPs-induced dysbiosis and MS. One promising direction for future investigation involves studying the connection between the gut and the lungs, particularly in relation to exposure to OPs and its impact on respiratory health. Recent findings indicate that GM could impact pulmonary immunity and inflammation, with an association between microbial imbalance and respiratory conditions like asthma and chronic obstructive pulmonary disease (COPD) [157–160]. Anand et al. (2018) explored how depleting GM affected pulmonary immune responses in mice exposed to air pollution. The study revealed that microbial imbalance worsened airway inflammation and OS, emphasizing the importance of gut-lung axis interactions in OPs-related respiratory conditions. In addition, potential investigations could delve into how the gut-brain axis influences the impact of OPs on neurodevelopment and cognitive function [161]. There is increasing evidence indicating that GM can impact brain function and behavior via various pathways, with an imbalance in GM linked to conditions like depression, anxiety, and autism spectrum disorders. Zhang et al. (2021) explored the impact of GM transplantation on depressive behavior in mice exposed to PCBs. The study revealed that microbial transplantation reduced PCB-induced depressive symptoms, indicating a possible treatment strategy for addressing OPs-induced neurobehavioral disorders. In addition, upcoming studies could investigate how GM could be utilized in bioremediation and environmental sustainability initiatives. The diverse MPs found in GM communities within the GIT can metabolize and detoxify OPs. Bioremediation strategies facilitated by microbes show potential for cleaning up polluted environments [162]. In the future, there could be a focus on creating new biomarkers to evaluate exposure to pollutants and imbalances in the microbial community in people. Identifying individuals at risk of OPs-related health effects and monitoring the efficacy of therapeutic interventions can be achieved through biomarkers like microbial signatures, metabolites, and inflammatory markers [162–165].

7.11 CONCLUSION

OPs and GM have a complex relationship that highlights the importance of environmental toxicology and human health. Given the prevalence of OPs in our environment because of industrial activities and urbanization, they present considerable obstacles to both human health and ecosystem stability. Exploring how pollutants interact with GM provides valuable insights into the mechanisms behind toxicity and the potential for detoxification and health protection mediated by microbes. Studying this field has revealed different ways in which OPs disrupt GM and impact host physiology. These mechanisms encompass changes in microbial composition, disturbance of gut barrier function, triggering of inflammatory reactions, and adjustment of detoxification processes. OPs can worsen intestinal inflammation, disrupt metabolic balance, and play a role in the development of chronic diseases by impacting GM.

Despite advancements in comprehending PMIs, there are still numerous knowledge gaps and obstacles to address. Understanding the intricate GM, range of OPs, and ever-changing host-microbe interactions can be challenging when trying to uncover the mechanisms and assess their impact on human health. To address these challenges and enhance our knowledge of PMIs, it is crucial to establish methodological standardization, conduct longitudinal studies, and adopt interdisciplinary approaches. In addition, the impacts of interactions between pollutants and the microbiome go beyond human health to also affect environmental sustainability and ecosystem health. GM is essential in transforming and detoxifying pollutants, impacting how contaminants behave and are accessible in the environment. Exploring bioremediation methods using microbes shows great potential in addressing pollution and improving the health of ecosystems.

Considering these discoveries, upcoming research paths in the realm of OPs and GM involve different areas, such as creating new treatments, investigating the gut-lung and gut-brain connections, utilizing microbial-based bioremediation, and developing biomarkers. Exploring these avenues presents thrilling chances to tackle the intricate issues caused by OPs and enhance environmental and human well-being. Studying OPs and GM is a dynamic field that combines environmental toxicology, microbiology, and human health. Through understanding the mechanisms behind PMIs and their impact on health and environmental sustainability, researchers can help create new approaches to reduce risks from OPs and safeguard public health. Further exploration in this field is crucial to tackle the intricate issues presented by OPs and enhance our comprehension of their effects on human health and the environment.

REFERENCES

1. S. Zahmatkesh, M. Hajiaghaei-Keshteli, A. Bokhari, S. Sundaramurthy, B. Panneerselvam, Y. Rezakhani, Wastewater treatment with nanomaterials for the future: A state-of-the-art review, *Environmental Research*, 216 (2023) 114652.
2. H.R. Safford, K. Shapiro, H.N. Bischel, Wastewater analysis can be a powerful public health tool—if it's done sensibly, *Proceedings of the National Academy of Sciences*, 119 (2022) e2119600119.
3. J. Cai, B. Niu, Q. Xie, N. Lu, S. Huang, G. Zhao, J. Zhao, Accurate removal of toxic organic pollutants from complex water matrices, *Environmental Science & Technology*, 56 (2022) 2917–2935.

4. H. Guo, W. Hu, Z. Xu, S. Guo, D. Qiao, X. Wang, H. Xu, W. Yan, How to improve lead dioxide anodes performance in organic wastewater treatment: Review and prospect, *Process Safety and Environmental Protection*, 164 (2022) 189–207.
5. R. Álvarez-Ruiz, Y. Picó, D. Sadutto, J. Campo, Development of multi-residue extraction procedures using QuEChERS and liquid chromatography tandem mass spectrometry for the determination of different types of organic pollutants in mussel, *Analytical and Bioanalytical Chemistry*, 413 (2021) 4063–4076.
6. X. Cao, L. Wang, Y. Zhang, Y. Li, C. Zhu, X. Zheng, Y. Sun, D. Li, L. Mo, B. Mai, Occurrence of organic pollutants in plastics on beach: Stranded foams can be sources of pollutants in islands, *The Science of the Total Environment*, 707 (2020) 136119.
7. S. Kumari, S. Das, Bacterial enzymatic degradation of recalcitrant organic pollutants: Catabolic pathways and genetic regulations, *Environmental Science and Pollution Research International*, 30 (2023) 79676–79705.
8. L. Yang, S. Wang, X. Peng, M. Zheng, Y. Yang, K. Xiao, G. Liu, Gas chromatography-Orbitrap mass spectrometry screening of organic chemicals in fly ash samples from industrial sources and implications for understanding the formation mechanisms of unintentional persistent organic pollutants, *The Science of the Total Environment*, 664 (2019) 107–115.
9. S.K. Samanta, O.V. Singh, R.K. Jain, Polycyclic aromatic hydrocarbons: Environmental pollution and bioremediation, *Trends in Biotechnology*, 20 (2002) 243–248.
10. M.S. Hussain, G. Gupta, R. Mishra, N. Patel, S. Gupta, S.I. Alzarea, I. Kazmi, P. Kumbhar, J. Disouza, H. Dureja, N. Kukreti, S.K. Singh, K. Dua, Unlocking the secrets: Volatile organic compounds (VOCs) and their devastating effects on lung cancer, *Pathology, Research and Practice*, 255 (2024) 155157.
11. R. Sender, S. Fuchs, R. Milo, Revised estimates for the number of human and bacteria cells in the body, *PLoS Biology*, 14 (2016) e1002533.
12. M. Kumar, M.S. Hussain, S. Raj, R. Verma, S. Sharma, S.K. Sahu, An overview of treatment modalities and management aspects for obesity, *Current Nutrition and Food Science*, 19 (2023) 105–113.
13. S. Subramanian, S. Huq, T. Yatsunenko, R. Haque, M. Mahfuz, M.A. Alam, A. Benezra, J. DeStefano, M.F. Meier, B.D. Muegge, M.J. Barratt, L.G. VanArendonk, Q. Zhang, M.A. Province, W.A. Petri, Jr., T. Ahmed, J.I. Gordon, Persistent gut microbiota immaturity in malnourished Bangladeshi children, *Nature*, 510 (2014) 417–421.
14. S. Hussain, M.A. Khan, R. Rajan, J. Jyoti, S. Sharma, S.K. Sahu, Nanorobots: The future of healthcare, *AIP Conference Proceedings*, AIP Publishing, 2023.
15. M.R. Islam, A. Rauf, S. Akash, A. Kumer, M.S. Hussain, S. Akter, J.K. Gupta, L.T. Ansari, M.M.M.I. Raj, T.B. Emran, Recent perspective on the potential role of phytocompounds in the prevention of gastric cancer, *Process Biochemistry*, 135 (2023) 83–101.
16. B.D. Anderson, J.E. Bisanz, Challenges and opportunities of strain diversity in gut microbiome research, *Frontiers in Microbiology,* 14 (2023) 1117122.
17. S. Zhu, Y. Jiang, K. Xu, M. Cui, W. Ye, G. Zhao, L. Jin, X. Chen, The progress of gut microbiome research related to brain disorders, *Journal of Neuroinflammation*, 17 (2020) 25.
18. J.W. Arnold, J. Roach, M.A. Azcarate-Peril, Emerging technologies for gut microbiome research, *Trends in Microbiology*, 24 (2016) 887–901.
19. G. Reid, N. Nduti, W. Sybesma, R. Kort, T.R. Kollmann, R. Adam, H. Boga, E.M. Brown, A. Einerhand, H. El-Nezami, G.B. Gloor, Kavere, II, J. Lindahl, A. Manges, W. Mamo, R. Martin, A. McMillan, J. Obiero, P.A. Ochieng, A. Onyango, S. Rulisa, E. Salminen, S. Salminen, A. Sije, J.R. Swann, W. van Treuren, D. Waweru, S.J. Kemp, Harnessing microbiome and probiotic research in sub-Saharan Africa: Recommendations from an African workshop, *Microbiome*, 2 (2014) 12.
20. D. Graf, R. Di Cagno, F. Fåk, H.J. Flint, M. Nyman, M. Saarela, B. Watzl, Contribution of diet to the composition of the human gut microbiota, *Microbial Ecology in Health and Disease*, 26 (2015) 26164.
21. A.W. Walker, J. Ince, S.H. Duncan, L.M. Webster, G. Holtrop, X. Ze, D. Brown, M.D. Stares, P. Scott, A. Bergerat, P. Louis, F. McIntosh, A.M. Johnstone, G.E. Lobley, J. Parkhill, H.J. Flint, Dominant and diet-responsive groups of bacteria within the human colonic microbiota, *The ISME Journal*, 5 (2011) 220–230.
22. F. Bäckhed, R.E. Ley, J.L. Sonnenburg, D.A. Peterson, J.I. Gordon, Host-bacterial mutualism in the human intestine, *Science*, 307 (2005) 1915–1920.
23. D.L. Topping, P.M. Clifton, Short-chain fatty acids and human colonic function: Roles of resistant starch and nonstarch polysaccharides, *Physiological Reviews*, 81 (2001) 1031–1064.
24. F. Quarcoo, C. Bonsi, N. Tackie, Pesticides, the environment, and human health, In *Tech Europe*, Rijeka, 2014.

25. C.J. Clarke, W.C. Tu, O. Levers, A. Bröhl, J.P. Hallett, Green and sustainable solvents in chemical processes, *Chemical Reviews*, 118 (2018) 747–800.
26. K. Kümmerer, The presence of pharmaceuticals in the environment due to human use—present knowledge and future challenges, *Journal of Environmental Management*, 90 (2009) 2354–2366.
27. N. Othman, Z. Ismail, M.I. Selamat, S.H. Sheikh Abdul Kadir, N.A. Shibraumalisi, A review of polychlorinated biphenyls (PCBs) pollution in the air: Where and how much are we exposed to?, *International Journal of Environmental Research and Public Health*, 19 (2022) 13923.
28. L. Rodenburg, J. Guo, R. Christie, Polychlorinated biphenyls in pigments: Inadvertent production and environmental significance, *Coloration Technology*, 131 (2015) 353–369.
29. L.K. Ursell, J.L. Metcalf, L.W. Parfrey, R. Knight, Defining the human microbiome, *Nutrition Reviews*, 70 Suppl 1 (2012) S38–S44.
30. F. Mobeen, V. Sharma, P. Tulika, Enterotype variations of the healthy human gut microbiome in different geographical regions, *Bioinformation*, 14 (2018) 560–573.
31. J. Qin, R. Li, J. Raes, M. Arumugam, K.S. Burgdorf, C. Manichanh, T. Nielsen, N. Pons, F. Levenez, T. Yamada, D.R. Mende, J. Li, J. Xu, S. Li, D. Li, J. Cao, B. Wang, H. Liang, H. Zheng, Y. Xie, J. Tap, P. Lepage, M. Bertalan, J.M. Batto, T. Hansen, D. Le Paslier, A. Linneberg, H.B. Nielsen, E. Pelletier, P. Renault, T. Sicheritz-Ponten, K. Turner, H. Zhu, C. Yu, S. Li, M. Jian, Y. Zhou, Y. Li, X. Zhang, S. Li, N. Qin, H. Yang, J. Wang, S. Brunak, J. Doré, F. Guarner, K. Kristiansen, O. Pedersen, J. Parkhill, J. Weissenbach, P. Bork, S.D. Ehrlich, J. Wang, A human gut microbial gene catalogue established by metagenomic sequencing, *Nature*, 464 (2010) 59–65.
32. R. Sender, S. Fuchs, R. Milo, Are we really vastly outnumbered? Revisiting the ratio of bacterial to host cells in humans, *Cell*, 164 (2016) 337–340.
33. B. Pan, X. Han, K. Yu, H. Sun, R. Mu, C.A. Lian, Geographical distance, host evolutionary history and diet drive gut microbiome diversity of fish across the Yellow River, *Molecular Ecology*, 32 (2023) 1183–1196.
34. Y. Ge, Z. Jing, Q. Diao, J.Z. He, Y.J. Liu, Host species and geography differentiate honeybee gut bacterial communities by changing the relative contribution of community assembly processes, *mBio*, 12 (2021) e0075121.
35. G. Lilli, C. Sirot, H. Campbell, D. Brophy, C. Graham, I. George, Geographic origin and host's phylogeny are predictors of the gut mucosal microbiota diversity and composition in Mediterranean scorpionfishes (Scorpaena sp.), *Frontiers in Marine Science*, 10 (2023) 1286706.
36. N.D. Youngblut, G.H. Reischer, W. Walters, N. Schuster, C. Walzer, G. Stalder, R.E. Ley, A.H. Farnleitner, Host diet and evolutionary history explain different aspects of gut microbiome diversity among vertebrate clades, *Nature Communications,* 10 (2019) 2200.
37. A.M. Shankregowda, P. Siriyappagouder, M. Kuizenga, T.M.P. Bal, Y. Abdelhafiz, C. Eizaguirre, J.M.O. Fernandes, V. Kiron, J.A.M. Raeymaekers, Host habitat rather than evolutionary history explains gut microbiome diversity in sympatric stickleback species, *Frontiers in Microbiology*, 14 (2023) 1232358.
38. Y. Zhang, H. Chen, M. Lu, J. Cai, B. Lu, C. Luo, M. Dai, Habitual diet pattern associations with gut microbiome diversity and composition: Results from a Chinese adult cohort, *Nutrients*, 14 (2022) 2639.
39. Z. Miao, W. Du, C. Xiao, C. Su, W. Gou, L. Shen, J. Zhang, Y. Fu, Z. Jiang, Z. Wang, X. Jia, J.S. Zheng, H. Wang, Gut microbiota signatures of long-term and short-term plant-based dietary pattern and cardiometabolic health: A prospective cohort study, *BMC Medicine*, 20 (2022) 204.
40. Z. Chen, D. Radjabzadeh, A. Ikram, A. Uitterlinden, R. Kraaij, T. Voortman, Dietary patterns and gut microbiome composition in a large population-based cohort, *Current Developments in Nutrition*, 4 (2020) 514–514.
41. D. Yu, S.M. Nguyen, Y. Yang, W. Xu, H. Cai, J. Wu, Q. Cai, J. Long, W. Zheng, X.O. Shu, Long-term diet quality is associated with gut microbiome diversity and composition among urban Chinese adults, *The American Journal of Clinical Nutrition*, 113 (2021) 684–694.
42. M.S. Hussain, A. Sharma, R. Kumar, Prebiotics and probiotics: A focused review of applications in respiratory disorders, *Carpathian Journal of Food Science and Technology,* 15 (2023) 183–207.
43. I. Zuniga-Chaves, S. Eggers, A.E. Kates, N. Safdar, G. Suen, K.M.C. Malecki, Neighborhood socioeconomic status is associated with low diversity gut microbiomes and multi-drug resistant microorganism colonization, *NPJ Biofilms and Microbiomes,* 9 (2023) 61.
44. C.R. Lewis, K.S. Bonham, S.H. McCann, A.R. Volpe, V. D'Sa, M. Naymik, M.D. De Both, M.J. Huentelman, K. Lemery-Chalfant, S.K. Highlander, S.C.L. Deoni, V. Klepac-Ceraj, Family SES is associated with the gut microbiome in infants and children, *Microorganisms*, 9 (2021) 1608.

45. S. Moran-Ramos, B.E. Lopez-Contreras, R. Villarruel-Vazquez, E. Ocampo-Medina, L. Macias-Kauffer, J.N. Martinez-Medina, H. Villamil-Ramirez, P. León-Mimila, B.E. Del Rio-Navarro, I. Ibarra-Gonzalez, M. Vela-Amieva, F.J. Gomez-Perez, R. Velazquez-Cruz, J. Salmeron, Z. Reyes-Castillo, C. Aguilar-Salinas, S. Canizales-Quinteros, Environmental and intrinsic factors shaping gut microbiota composition and diversity and its relation to metabolic health in children and early adolescents: A population-based study, *Gut Microbes*, 11 (2020) 900–917.
46. E.D. Lephart, F. Naftolin, Estrogen action and gut microbiome metabolism in dermal health, *Dermatology and Therapy*, 12 (2022) 1535–1550.
47. S. Gschwendtner, H. Kang, E. Thiering, S. Kublik, B. Fösel, H. Schulz, S. Krauss-Etschmann, J. Heinrich, A. Schöler, M. Schloter, M. Standl, Early life determinants induce sustainable changes in the gut microbiome of six-year-old children, *Scientific Reports*, 9 (2019) 12675.
48. A. Sarkar, J.Y. Yoo, S. Valeria Ozorio Dutra, K.H. Morgan, M. Groer, The association between early-life gut microbiota and long-term health and diseases, *Journal of Clinical Medicine*, 10 (2021) 459.
49. S.E. Webster, D. Vos, T.L. Rothstein, N.E. Holodick, Modulation of microbiome diversity and cytokine expression is influenced in a sex-dependent manner during aging, *Frontiers in Microbiomes*, 1 (2022) 994464.
50. H. Daniel, Diet and gut microbiome and the "chicken or egg" problem, *Frontiers in Nutrition*, 8 (2021) 828630.
51. G. Zhao, T. Ma, W. Tang, D. Li, S.K. Mishra, Z. Xu, Q. Wang, H. Jie, Gut microbiome of chinese forest musk deer examined across gender and age, *BioMed Research International,* 2019 (2019) 9291216.
52. C.S. Fuziwara, K.C. Saito, E.T. Kimura, Thyroid follicular cell loss of differentiation induced by microRNA miR-17-92 cluster is attenuated by CRISPR/Cas9n gene silencing in anaplastic thyroid cancer, *Thyroid*, 30 (2020) 81–94.
53. Q. Su, H.M. Tun, Q. Liu, Y.K. Yeoh, J.W.Y. Mak, F.K. Chan, S.C. Ng, Gut microbiome signatures reflect different subtypes of irritable bowel syndrome, *Gut Microbes*, 15 (2023) 2157697.
54. N. Kamada, G. Núñez, Role of the gut microbiota in the development and function of lymphoid cells, *Journal of Immunology*, 190 (2013) 1389–1395.
55. S. Krishnan, N. Alden, K. Lee, Pathways and functions of gut microbiota metabolism impacting host physiology, *Current Opinion in Biotechnology*, 36 (2015) 137–145.
56. Y. Tian, W. Gui, B. Rimal, I. Koo, P.B. Smith, R.G. Nichols, J. Cai, Q. Liu, A.D. Patterson, Metabolic impact of persistent organic pollutants on gut microbiota, *Gut Microbes*, 12 (2020) 1848209.
57. S. Popli, P.C. Badgujar, T. Agarwal, B. Bhushan, V. Mishra, Persistent organic pollutants in foods, their interplay with gut microbiota and resultant toxicity, *Science of The Total Environment*, 832 (2022) 155084.
58. Y. Jin, S. Wu, Z. Zeng, Z. Fu, Effects of environmental pollutants on gut microbiota, *Environmental Pollution*, 222 (2017) 1–9.
59. C.A. Lozupone, J.I. Stombaugh, J.I. Gordon, J.K. Jansson, R. Knight, Diversity, stability and resilience of the human gut microbiota, *Nature*, 489 (2012) 220–230.
60. K. Cai, Q. Song, W. Yuan, J. Ruan, H. Duan, Y. Li, J. Li, Human exposure to PBDEs in e-waste areas: A review, *Environmental Pollution*, 267 (2020) 115634.
61. C. Lozupone, K. Faust, J. Raes, J.J. Faith, D.N. Frank, J. Zaneveld, J.I. Gordon, R. Knight, Identifying genomic and metabolic features that can underlie early successional and opportunistic lifestyles of human gut symbionts, *Genome Research*, 22 (2012) 1974–1984.
62. D. Wang, J. Ren, Z. Tan, J. You, Gut microbial profiles in nereis succinea and their contribution to the degradation of organic pollutants, *Environmental Science & Technology*, 54 (2020) 6235–6243.
63. L. Zhang, R.G. Nichols, J. Correll, I.A. Murray, N. Tanaka, P.B. Smith, T.D. Hubbard, A. Sebastian, I. Albert, Hatzakis, Persistent organic pollutants modify gut microbiota–host metabolic homeostasis in mice through aryl hydrocarbon receptor activation, *Environmental Health Perspectives*, 123 (2015) 679–688.
64. A.E. Lindell, M. Zimmermann-Kogadeeva, K.R. Patil, Multimodal interactions of drugs, natural compounds and pollutants with the gut microbiota, *Nature Reviews Microbiology*, 20 (2022) 431–443.
65. X.-D. Li, L. Xin, W.-T. Rong, X.-Y. Liu, W.-A. Deng, Y.-C. Qin, X.-L. Li, Effect of heavy metals pollution on the composition and diversity of the intestinal microbial community of a pygmy grasshopper (Eucriotettix oculatus), *Ecotoxicology and Environmental Safety*, 223 (2021) 112582.
66. Y. Zhang, R. Cui, G. Shi, Y. Dai, J. Dong, Q. Wu, H. Zhang, J. Dai, Dioxin-like polychlorinated biphenyl 126 (PCB126) disrupts gut microbiota-host metabolic dysfunction in mice via aryl hydrocarbon receptor activation, *Ecotoxicology and Environmental Safety*, 236 (2022) 113448.

67. R. Liu, D. Cai, X. Li, B. Liu, J. Chen, X. Jiang, H. Li, Z. Li, K. Teerds, J. Sun, Effects of Bisphenol A on reproductive toxicity and gut microbiota dysbiosis in male rats, *Ecotoxicology and Environmental Safety*, 239 (2022) 113623.
68. Y. Jin, J. Xia, Z. Pan, J. Yang, W. Wang, Z. Fu, Polystyrene microplastics induce microbiota dysbiosis and inflammation in the gut of adult zebrafish, *Environmental Pollution*, 235 (2018) 322–329.
69. C. Defois, J. Ratel, G. Garrait, S. Denis, O. Le Goff, J. Talvas, P. Mosoni, E. Engel, P. Peyret, Food chemicals disrupt human gut microbiota activity and impact intestinal homeostasis as revealed by in vitro systems, *Scientific Reports*, 8 (2018) 11006.
70. N. Djekkoun, J.-D. Lalau, V. Bach, F. Depeint, H. Khorsi-Cauet, Chronic oral exposure to pesticides and their consequences on metabolic regulation: Role of the microbiota, *European Journal of Nutrition*, 60 (2021) 4131–4149.
71. P. Feng, X. Xiao, T. Zhou, X.J Li, , Effects of the bio-accumulative environmental pollutants on the gut microbiota, *Gut Remediation of Environmental Pollutants* (2020) 109–143.
72. K. Chiu, G. Warner, R.A. Nowak, J.A. Flaws, W. Mei, The impact of environmental chemicals on the gut microbiome, *Toxicological Sciences*, 176 (2020) 253–284.
73. R. Cruz, J.D. Palmeira, Z.E. Martins, M.A. Faria, H. Ferreira, A. Marques, S. Casal, S.C. Cunha, Multidisciplinary approach to determine the effect of polybrominated diphenyl ethers on gut microbiota, *Environmental Pollution*, 260 (2020) 113920.
74. M.M. El-Dalatony, X. Li, Environmental pollutants that can be metabolized by the host (gut microbiota), *Gut Remediation of Environmental Pollutants* (2020) 145–168.
75. C. Wang, Y. Zhao, Y. Jin, The emerging PFOS alternative OBS exposure induced gut microbiota dysbiosis and hepatic metabolism disorder in adult zebrafish, *Comparative Biochemistry and Physiology Part C: Toxicology & Pharmacology*, 230 (2020) 108703.
76. H. Su, J. Liu, G. Wu, Z. Long, J. Fan, Z. Xu, J. Liu, Z. Yu, M. Cao, N. Liao, Homeostasis of gut microbiota protects against polychlorinated biphenyl 126-induced metabolic dysfunction in liver of mice, *Science of the Total Environment*, 720 (2020) 137597.
77. G. Wang, S. Sun, X. Wu, S. Yang, Y. Wu, J. Zhao, H. Zhang, W. Chen, Intestinal environmental disorders associate with the tissue damages induced by perfluorooctane sulfonate exposure, *Ecotoxicology and Environmental Safety*, 197 (2020) 110590.
78. A. Di Ciaula, J. Baj, G. Garruti, G. Celano, M. De Angelis, H.H. Wang, D.M. Di Palo, L. Bonfrate, D.Q. Wang, P. Portincasa, Liver steatosis, gut-liver axis, microbiome and environmental factors. A never-ending bidirectional cross-talk, *Journal of Clinical Medicine*, 9 (2020) 2648.
79. A. Gaylord, L. Trasande, K. Kannan, K.M. Thomas, S. Lee, M. Liu, J. Levine, Persistent organic pollutant exposure and celiac disease: A pilot study, *Environmental Research*, 186 (2020) 109439.
80. J. Ruan, J. Guo, Y. Huang, Y. Mao, Z. Yang, Z. Zuo, Adolescent exposure to environmental level of PCBs (Aroclor 1254) induces non-alcoholic fatty liver disease in male mice, *Environmental Research*, 181 (2020) 108909.
81. H. Wei, S. Wang, E.G. Xu, J. Liu, X. Li, Z. Wang, Synergistic toxicity of microcystin-LR and Cu to zebrafish, *Science of the Total Environment*, 713 (2020) 136393.
82. M. Bonato, F. Corrà, M. Bellio, L. Guidolin, L. Tallandini, P. Irato, G. Santovito, PFAS environmental pollution and antioxidant responses: An overview of the impact on human field, *International Journal of Environmental Research and Public Health*, 17 (2020) 8020.
83. X. Cheng, J. Chen, X. Guo, H. Cao, C. Zhang, G. Hu, Y. Zhuang, Disrupting the gut microbiota/metabolites axis by Di-(2-ethylhexyl) phthalate drives intestinal inflammation via AhR/NF-κB pathway in mice, *Environmental Pollution*, 343 (2024) 123232.
84. C.G. Sree, V. Buddolla, B.A. Lakshmi, Y.-J. Kim, Phthalate toxicity mechanisms: An update, *Comparative Biochemistry and Physiology Part C: Toxicology & Pharmacology*, 263 (2023) 109498.
85. M.C. Petriello, J.B. Hoffman, O. Vsevolozhskaya, A.J. Morris, B. Hennig, Dioxin-like PCB 126 increases intestinal inflammation and disrupts gut microbiota and metabolic homeostasis, *Environmental Pollution*, 242 (2018) 1022–1032.
86. O. Sandys, A. Te Velde, Sciences, Raising the alarm: Environmental factors in the onset and maintenance of chronic (low-grade) inflammation in the gastrointestinal tract, *Digestive Diseases and Sciences*, 67 (2022) 4355–4368.
87. J. Wu, H. Chen, J. Xu, M.S.U. Rahman, S. Li, J. Wang, S. Huang, C.C. Han, S. Xu, Y. Liu, The lull before microplastics pollution outbreaks: Some implications for human health and control strategies, *Nonotoday*, 54 (2024) 102062.

88. M.S. Hussain, O. Afzal, G. Gupta, A.S.A. Altamimi, W.H. Almalki, S.I. Alzarea, I. Kazmi, N. Kukreti, S. Gupta, K. Sulakhiya, S.K. Singh, K. Dua, Probing the links: Long non-coding RNAs and NF-κB signalling in atherosclerosis, *Pathology, Research and Practice*, 249 (2023) 154773.
89. C. Wang, M.C. Petriello, B. Zhu, B. Hennig, PCB 126 induces monocyte/macrophage polarization and inflammation through AhR and NF-κB pathways, *Toxicology and Applied Pharmacology*, 367 (2019) 71–81.
90. D. Liu, J.T. Perkins, M.C. Petriello, B. Hennig, Exposure to coplanar PCBs induces endothelial cell inflammation through epigenetic regulation of NF-κB subunit p65, *Toxicology and Applied Pharmacology*, 289 (2015) 457–465.
91. A. Maddalon, V. Galbiati, C. Colosio, S. Mandić-Rajčević, E. Corsini, Glyphosate-based herbicides: Evidence of immune-endocrine alteration, *Toxicology*, 459 (2021) 152851.
92. J.R. Maxwell, Y. Zhang, W.A. Brown, C.L. Smith, F.R. Byrne, M. Fiorino, E. Stevens, J. Bigler, J.A. Davis, J.B. Rottman, Differential roles for interleukin-23 and interleukin-17 in intestinal immunoregulation, *Immunity*, 43 (2015) 739–750.
93. K.J. Groh, B. Geueke, J. Muncke, , Food contact materials and gut health: Implications for toxicity assessment and relevance of high molecular weight migrants, *Food and Chemical Toxicology*, 109 (2017) 1–18.
94. N. Zhang, Z. Ju, T. Zuo, Time for food: The impact of diet on gut microbiota and human health, *Nutrition*, 51 (2018) 80–85.
95. M. Tufail, C. Wu, M.S. Hussain, Dietary, addictive and habitual factors, and risk of colorectal cancer, *Nutrition*, 120 (2024) 112334.
96. D. Nalage, T. Sontakke, A. Biradar, V. Jogdand, R. Kale, S. Harke, R. Kale, P. Dixit, The impact of environmental toxins on the animal gut microbiome and their potential to contribute to disease, *The Journal of Food Composition and Analysis*, 3 (2023) 100497.
97. M. Sendra, E. Sparaventi, B. Novoa, A.J. Figueras, An overview of the internalization and effects of microplastics and nanoplastics as pollutants of emerging concern in bivalves, *Science of the Total Environment*, 753 (2021) 142024.
98. K. Sarkar, P.C. Sil, Effect of diet, pharmaceuticals, and environmental toxicants on gut microbiota imbalance and increased intestinal membrane permeability, *Toxicological Risk Assessment and Multi-System Health Impacts from Exposure*, Elsevier, 2021, pp. 403–413.
99. M.S. Hussain, G. Gupta, M. Afzal, S.M. Alqahtani, V.P. Samuel, W. Hassan Almalki, I. Kazmi, S.I. Alzarea, S. Saleem, H. Dureja, S.K. Singh, K. Dua, L. Thangavelu, Exploring the role of lncrna neat1 knockdown in regulating apoptosis across multiple cancer types: A review, *Pathology, Research and Practice*, 252 (2023) 154908.
100. C. Stolfi, C. Maresca, G. Monteleone, F.Laudisi, Implication of intestinal barrier dysfunction in gut dysbiosis and diseases, *Biomedicines*, 10 (2022) 289.
101. R. Thapa, O. Afzal, A.A. Bhat, A. Goyal, A.S. Alfawaz Altamimi, W.H. Almalki, S.I. Alzarea, I. Kazmi, S.K. Singh, K. Dua, L. Thangavelu, G. Gupta, New horizons in lung cancer management through ATR/CHK1 pathway modulation, *Future Medicinal Chemistry,* 15 (2023) 1807–1818.
102. R. Thapa, G. Gupta, A.A. Bhat, W.H. Almalki, S.I. Alzarea, I. Kazmi, S. Saleem, R. Khan, N. Altwaijry, H. Dureja, S.K. Singh, K. Dua, A review of Glycogen Synthase Kinase-3 (GSK3) inhibitors for cancers therapies, *International Journal of Biological Macromolecules,* 253 (2023) 127375.
103. T.-Z. Jing, F.-H. Qi, Z.-Y. Wang, Most dominant roles of insect gut bacteria: Digestion, detoxification, or essential nutrient provision?, *Microbiome*, 8 (2020) 1–20.
104. S.L. Collins, A.D. Patterson, The gut microbiome: An orchestrator of xenobiotic metabolism, *Acta Pharmaceutica Sinica B*, 10 (2020) 19–32.
105. I.D. Wilson, J.K. Nicholson, Gut microbiome interactions with drug metabolism, efficacy, and toxicity, *Translational Research*, 179 (2017) 204–222.
106. N. Koppel, V. Maini Rekdal, E.P. Balskus, Chemical transformation of xenobiotics by the human gut microbiota, *Science*, 356 (2017) eaag2770.
107. A. Pant, T.K. Maiti, D. Mahajan, B. Das, Human gut microbiota and drug metabolism, *Microbial Ecology*, 86 (2023) 97–111.
108. S. Hertli, P. Zimmermann, Molecular interactions between the intestinal microbiota and the host, *Molecular Microbiology*, 117 (2022) 1297–1307.
109. R. Thapa, O. Afzal, M. Afzal, G. Gupta, A.A. Bhat, W. Hassan Almalki, I. Kazmi, S.I. Alzarea, S. Saleem, P. Arora, S.K. Singh, K. Dua, From LncRNA to metastasis: The MALAT1-EMT axis in cancer progression, *Pathology, Research and Practice*, 253 (2024) 154959.

110. A.S. Wilson, K.R. Koller, M.C. Ramaboli, L.T. Nesengani, S. Ocvirk, C. Chen, C.A. Flanagan, F.R. Sapp, Z.T. Merritt, F. Bhatti, Diet and the human gut microbiome: An international review, *Digestive Diseases and Sciences*, 65 (2020) 723–740.
111. M. Dueñas, I. Muñoz-González, C. Cueva, A. Jiménez-Girón, F. Sánchez-Patán, C. Santos-Buelga, M. Moreno-Arribas, B. Bartolomé, A survey of modulation of gut microbiota by dietary polyphenols, *BioMed Research International*, 2015 (2015) 850902.
112. U. Etxeberria, A. Fernández-Quintela, F.I. Milagro, L. Aguirre, J.A. Martínez, M.P. Portillo, Impact of polyphenols and polyphenol-rich dietary sources on gut microbiota composition, *Journal of Agricultural and Food Chemistry*, 61 (2013) 9517–9533.
113. A. Kumar, D. Kumar, Characterization of Lactobacillus isolated from dairy samples for probiotic properties, *Anaerobe*, 33 (2015) 117–123.
114. P. Sharma, M. Mehta, D.S. Dhanjal, S. Kaur, G. Gupta, H. Singh, L. Thangavelu, S. Rajeshkumar, M. Tambuwala, H.A. Bakshi, D.K. Chellappan, K. Dua, S. Satija, Emerging trends in the novel drug delivery approaches for the treatment of lung cancer, *Chemico-Biological Interactions*, 309 (2019) 108720.
115. V. Subramaniyan, S. Fuloria, G. Gupta, D.H. Kumar, M. Sekar, K.V. Sathasivam, K. Sudhakar, K.S. Alharbi, W.H. Al-Malki, O. Afzal, I. Kazmi, F.A. Al-Abbasi, A.S.A. Altamimi, N.K. Fuloria, A review on epidermal growth factor receptor's role in breast and non-small cell lung cancer, *Chemico-Biological Interactions*, 351 (2022) 109735.
116. A.E. Mohr, E. Gumpricht, D.D. Sears, K.L. Sweazea, Recent advances and health implications of dietary fasting regimens on the gut microbiome, *American Journal of Physiology-Gastrointestinal and Liver Physiology*, 320 (2021) G847–G863.
117. D.A. Schupack, R.A. Mars, D.H. Voelker, J.P. Abeykoon, P.C. Kashyap, The promise of the gut microbiome as part of individualized treatment strategies, *Nature Reviews Gastroenterology & Hepatology*, 19 (2022) 7–25.
118. R.S. Kootte, E. Levin, J. Salojärvi, L.P. Smits, A.V. Hartstra, S.D. Udayappan, G. Hermes, K.E. Bouter, A.M. Koopen, J. Holst, Improvement of insulin sensitivity after lean donor feces in metabolic syndrome is driven by baseline intestinal microbiota composition, *Cell Metabolism*, 26 (2017) 611–619. e616.
119. A.M. Koopen, E.L. Almeida, I. Attaye, J.J. Witjes, E. Rampanelli, S. Majait, M. Kemper, J.H. Levels, A.W. Schimmel, H. Herrema, Effect of fecal microbiota transplantation combined with Mediterranean diet on insulin sensitivity in subjects with metabolic syndrome, *Frontiers in Microbiology*, 12 (2021) 662159.
120. C. Purdel, A. Ungurianu, I. Adam-Dima, D. Margină, Exploring the potential impact of probiotic use on drug metabolism and efficacy, *Biomedicine & Pharmacotherapy*, 161 (2023) 114468.
121. S. Rohilla, M. Singh, S. Priya, W.H. Almalki, S.M. Haniffa, V. Subramaniyan, S. Fuloria, N.K. Fuloria, M. Sekar, S.K. Singh, N.K. Jha, D.K. Chellappan, P. Negi, K. Dua, G. Gupta, Exploring the mechanical perspective of a new anti-tumor agent: Melatonin, *Journal of Environmental Pathology, Toxicology and Oncology*, 42 (2023) 1–16.
122. A. Sharma, D. Shambhwani, S. Pandey, J. Singh, H. Lalhlenmawia, M. Kumarasamy, S.K. Singh, D.K. Chellappan, G. Gupta, P. Prasher, K. Dua, D. Kumar, Advances in lung cancer treatment using nanomedicines, *ACS Omega*, 8 (2023) 10–41.
123. P.D. Cani, N.M. Delzenne, The gut microbiome as therapeutic target, *Pharmacology & Therapeutics*, 130 (2011) 202–212.
124. W. Wu, Q. Kong, P. Tian, Q. Zhai, G. Wang, X. Liu, J. Zhao, H. Zhang, Y.K. Lee, W. Chen, Targeting gut microbiota dysbiosis: Potential intervention strategies for neurological disorders, *Engineering*, 6 (2020) 415–423.
125. L. Schröder, S. Kaiser, B. Flemer, J. Hamm, F. Hinrichsen, D. Bordoni, P. Rosenstiel, F. Sommer, Nutritional targeting of the microbiome as potential therapy for malnutrition and chronic inflammation, *Nutrients*, 12 (2020) 3032.
126. M. Purohit, G. Gupta, O. Afzal, A.S.A. Altamimi, S.I. Alzarea, I. Kazmi, W.H. Almalki, M. Gulati, I.P. Kaur, S.K. Singh, K. Dua, Janus kinase/signal transducers and activator of transcription (JAK/STAT) and its role in Lung inflammatory disease, *Chemico-Biological Interactions*, 371 (2023) 110334.
127. S. Rohilla, M. Singh, S.I. Alzarea, W.H. Almalki, F.A. Al-Abbasi, I. Kazmi, O. Afzal, A.S.A. Altamimi, S.K. Singh, D.K. Chellappan, K. Dua, G. Gupta, Recent developments and challenges in molecular-targeted therapy of non-small-cell lung cancer, *Journal of Environmental Pathology, Toxicology and Oncology*, 42 (2023) 27–50.
128. H. Szajewska, M. Kołodziej, Systematic review with meta-analysis: Lactobacillus rhamnosus GG in the prevention of antibiotic-associated diarrhoea in children and adults, *Alimentary Pharmacology & Therapeutics*, 42 (2015) 1149–1157.

129. D. Vandeputte, G. Falony, S. Vieira-Silva, J. Wang, M. Sailer, S. Theis, K. Verbeke, J. Raes, Prebiotic inulin-type fructans induce specific changes in the human gut microbiota, *Gut*, 66 (2017) 1968–1974.
130. E. Reigadas, J. van Prehn, M. Falcone, F. Fitzpatrick, M.J. Vehreschild, E.J. Kuijper, E. Bouza, How to: Prophylactic interventions for prevention of Clostridioides difficile infection, *Clinical Microbiology and Infection*, 27 (2021) 1777–1783.
131. S. Costello, W. Soo, R. Bryant, V. Jairath, A. Hart, J. Andrews, Systematic review with meta-analysis: Faecal microbiota transplantation for the induction of remission for active ulcerative colitis, *Alimentary Pharmacology & Therapeutics*, 46 (2017) 213–224.
132. S.P. Costello, P.A. Hughes, O. Waters, R.V. Bryant, A.D. Vincent, P. Blatchford, R. Katsikeros, J. Makanyanga, M.A. Campaniello, C. Mavrangelos, Effect of fecal microbiota transplantation on 8-week remission in patients with ulcerative colitis: A randomized clinical trial, *JAMA*, 321 (2019) 156–164.
133. J. Kaur, M. Gulati, L. Corrie, A. Awasthi, N.K. Jha, D.K. Chellappan, G. Gupta, R. MacLoughlin, B.G. Oliver, K. Dua, S.K. Singh, Role of nucleic acid-based polymeric micelles in treating lung diseases, *Nanomedicine*, 17 (2022) 1951–1960.
134. R. Pradhan, G. Singhvi, S.K. Dubey, G. Gupta, K. Dua, MAPK pathway: A potential target for the treatment of non-small-cell lung carcinoma, *Future Medicinal Chemistry*, 11 (2019) 793–795.
135. S.C. Khoo, N. Zhang, L.-I. Vijitra, M.S. Goh, C. Sonne, N.L. Ma, Exploring environmental exposomes and the gut-brain nexus: Unveiling the impact of pesticide exposure, *Environmental Research* (2024) 118441.
136. M.S. Hussain, O. Afzal, G. Gupta, A. Goyal, W.H. Almalki, I. Kazmi, S.I. Alzarea, A.S. Alfawaz Altamimi, N. Kukreti, A. Chakraborty, S.K. Singh, K. Dua, Unraveling NEAT1's complex role in lung cancer biology: A comprehensive review, *EXCLI Journal*, 23 (2024) 34–52.
137. M.S. Hussain, A.S.A. Altamimi, M. Afzal, W.H. Almalki, I. Kazmi, S.I. Alzarea, S. Saleem, P. Prasher, B. Oliver, S.K. Singh, R. MacLoughlin, K. Dua, G. Gupta, From carcinogenesis to therapeutic avenues: LncRNAs and mTOR crosstalk in lung cancer, *Pathology, Research and Practice*, 253 (2024) 155015.
138. J.J. Lim, M. Dutta, J.L. Dempsey, H.-J. Lehmler, J. MacDonald, T. Bammler, C. Walker, T.J. Kavanagh, H. Gu, S. Mani, Neonatal exposure to BPA, BDE-99, and PCB produces persistent changes in hepatic transcriptome associated with gut dysbiosis in adult mouse livers, *Toxicological Sciences*, 184 (2021) 83–103.
139. A. DeBofsky, Y. Xie, C. Grimard, A.J. Alcaraz, M. Brinkmann, M. Hecker, J.P. Giesy, Differential responses of gut microbiota of male and female fathead minnow (*Pimephales promelas*) to a short-term environmentally-relevant, aqueous exposure to benzo [a] pyrene, *Chemosphere*, 252 (2020) 126461.
140. Z. Qu, L. Zhang, R. Hou, X. Ma, J. Yu, W. Zhang, C. Zhuang, Exposure to a mixture of cigarette smoke carcinogens disturbs gut microbiota and influences metabolic homeostasis in A/J mice, *Chemico-Biological Interactions*, 344 (2021) 109496.
141. M.A. Hakami, A. Hazazi, F.R. Khan, O. Abdulaziz, K. Alshaghdali, A. Abalkhail, S.A. Nassar, B.I.A. Omar, F. Almarshadi, G. Gupta, A.S. Binshaya, PVT1 lncRNA in lung cancer: A key player in tumorigenesis and therapeutic opportunities, *Pathology, Research and Practice*, 253 (2024) 155019.
142. M.S. Hussain, O. Afzal, G. Gupta, A.S.A. Altamimi, W.H. Almalki, S.I. Alzarea, I. Kazmi, N.K. Fuloria, M. Sekar, D.U. Meenakshi, L. Thangavelu, A. Sharma, Long non-coding RNAs in lung cancer: Unraveling the molecular modulators of MAPK signaling, *Pathology, Research and Practice*, 249 (2023) 154738.
143. W. Jia, H. Li, L. Zhao, J.K. Nicholson, Gut microbiota: A potential new territory for drug targeting, *Nature Reviews Drug Discovery*, 7 (2008) 123–129.
144. C.M. Cullen, K.K. Aneja, S. Beyhan, C.E. Cho, S. Woloszynek, M. Convertino, S.J. McCoy, Y. Zhang, M.Z. Anderson, D. Alvarez-Ponce, Emerging priorities for microbiome research, *Frontiers in Microbiology*, 11 (2020) 136.
145. M. Shamoon, N.M. Martin, C.L. O'Brien, Recent advances in gut microbiota mediated therapeutic targets in inflammatory bowel diseases: Emerging modalities for future pharmacological implications, *Pharmacological Research*, 148 (2019) 104344.
146. G. Gupta, D.K. Chellappan, T. de Jesus Andreoli Pinto, P.M. Hansbro, M. Bebawy, K. Dua, Tumor suppressor role of miR-503, *Panminerva Medica*, 60 (2018) 17–24.
147. G. Gupta, T. de Jesus Andreoli Pinto, D.K. Chellappan, A. Mishra, H. Malipeddi, K. Dua, A clinical update on metformin and lung cancer in diabetic patients, *Panminerva Medica*, 60 (2018) 70–75.
148. Y. Zhang, Q. Jia, C. Hu, M. Han, Q. Guo, S. Li, C. Bo, Y. Zhang, X. Qi, L. Sai, Effects of chlorpyrifos exposure on liver inflammation and intestinal flora structure in mice, *Toxicology Research*, 10 (2021) 141–149.

149. S. Giri, G. Lamichhane, D. Khadka, H.P. Devkota, Microplastics contamination in food products: Occurrence, analytical techniques and potential impacts on human health, *Current Research in Biotechnology*, 7 (2024) 100190.
150. A.A. Bhat, R. Thapa, A. Goyal, V. Subramaniyan, D. Kumar, S. Gupta, S.K. Singh, K. Dua, G. Gupta, Curcumin-based nanoformulations as an emerging therapeutic strategy for inflammatory lung diseases, *Future Medicinal Chemistry*, 15 (2023) 583–586.
151. Y. Chan, V.S.R. Raju Allam, K.R. Paudel, S.K. Singh, M. Gulati, M. Dhanasekaran, P.K. Gupta, N.K. Jha, H.P. Devkota, G. Gupta, P.M. Hansbro, B.G.G. Oliver, D.K. Chellappan, K. Dua, Nutraceuticals: Unlocking newer paradigms in the mitigation of inflammatory lung diseases, *Critical Reviews in Food Science and Nutrition*, 63 (2023) 3302–3332.
152. J. Balaguer-Trias, D. Deepika, M. Schuhmacher, V. Kumar, Impact of contaminants on microbiota: Linking the gut–brain axis with neurotoxicity, *International Journal of Environmental Research and Public Health*, 19 (2022) 1368.
153. M. Li, T. Liu, T. Yang, J. Zhu, Y. Zhou, M. Wang, Q. Wang, Gut microbiota dysbiosis involves in host non-alcoholic fatty liver disease upon pyrethroid pesticide exposure, *Environmental Science and Ecotechnology*, 11 (2022) 100185.
154. A.A. Bhat, G. Gupta, K.S. Alharbi, O. Afzal, A.S.A. Altamimi, W.H. Almalki, I. Kazmi, F.A. Al-Abbasi, S.I. Alzarea, D.K. Chellappan, S.K. Singh, R. MacLoughlin, B.G. Oliver, K. Dua, Polysaccharide-based nanomedicines targeting lung cancer, *Pharmaceutics*, 14 (2022) 2788.
155. A.A. Bhat, G. Gupta, A. Goyal, R. Thapa, W.H. Almalki, I. Kazmi, S.I. Alzarea, N. Kukreti, M. Sekar, D.U. Meenakshi, S.K. Singh, R. MacLoughlin, K. Dua, Unwinding circular RNA's role in inflammatory pulmonary diseases, *Naunyn-Schmiedeberg's Archives of Pharmacology*, 397 (2023) 2567–2588.
156. F. Giambò, C. Costa, M. Teodoro, C. Fenga, Role-playing between environmental pollutants and human gut microbiota: A complex bidirectional interaction, *Frontiers in Medicine*, 9 (2022) 810397.
157. P. Bhatt, S. Gangola, G. Bhandari, W. Zhang, D. Maithani, S. Mishra, S. Chen, New insights into the degradation of synthetic pollutants in contaminated environments, *Chemosphere*, 268 (2021) 128827.
158. W. Feng, H. Ao, C. Peng, D. Yan, Gut microbiota, a new frontier to understand traditional Chinese medicines, *Pharmacological Research*, 142 (2019) 176–191.
159. A.A. Bhat, O. Afzal, N. Agrawal, R. Thapa, W.H. Almalki, I. Kazmi, S.I. Alzarea, A.S.A. Altamimi, N. Kukreti, A. Chakraborty, S.K. Singh, K. Dua, G. Gupta, A comprehensive review on the emerging role of long non-coding RNAs in the regulation of NF-κB signaling in inflammatory lung diseases, *International Journal of Biological Macromolecules*, 253 (2023) 126951.
160. A.A. Bhat, R. Gilhotra, Y. Singh, S. Sharma, T. Jesus Andreoli Pinto, H.G. Ferraz, S.K. Singh, K. Dua, G. Gupta, Advanced drug-delivery approaches in managing P53-mediated lung diseases remodeling, *Nanomedicine*, 18 (2023) 583–587.
161. S. Anand, S.S. Mande, Diet, microbiota and gut-lung connection, *Frontiers in Microbiology*, 9 (2018) 2147.
162. W. Zhang, Z. Sun, Q. Zhang, Z. Sun, Y. Su, J. Song, B. Wang, R. Gao, Preliminary evidence for an influence of exposure to polycyclic aromatic hydrocarbons on the composition of the gut microbiota and neurodevelopment in three-year-old healthy children, *BMC Pediatrics*, 21 (2021) 1–13.
163. Q. Maqsood, A. Sumrin, R. Waseem, M. Hussain, M. Imtiaz, N. Hussain, Bioengineered microbial strains for detoxification of toxic environmental pollutants, *Environmental Research*, 227 (2023) 115665.
164. K.S. Alharbi, O. Afzal, W.H. Almalki, I. Kazmi, M.A. Javed Shaikh, L. Thangavelu, M. Gulati, S.K. Singh, N.K. Jha, P.K. Gupta, D.K. Chellappan, B.G. Oliver, K. Dua, G. Gupta, Nuclear factor-kappa B (NF-κB) inhibition as a therapeutic target for plant nutraceuticals in mitigating inflammatory lung diseases, *Chemico-Biological Interactions*, 354 (2022) 109842.
165. A.A. Bhat, O. Afzal, M. Afzal, G. Gupta, R. Thapa, H. Ali, W. Hassan Almalki, I. Kazmi, S.I. Alzarea, S. Saleem, V.P. Samuel, S.K. Gubbiyappa, V. Subramaniyan, MALAT1: A key regulator in lung cancer pathogenesis and therapeutic targeting, *Pathology, Research and Practice*, 253 (2024) 154991.

8 Bisphenols and the Gut Microbiome

Abida Khan[1,2] *and Mohd Imran*[1,2]*

[1]Department of Pharmaceutical Chemistry, College of Pharmacy, Northern Border University, Rafha, Saudi Arabia

[2]Center for Health Research, Northern Border University, Arar, Saudi Arabia

*Corresponding Author: Mohd Imran.
Email id: imran.pchem@gmail.com

8.1 INTRODUCTION

A bridge atom connects the two phenol rings in chemical bisphenols. The compounds BPA, bisphenol F, BPS, bisphenol B, and dimethyl-BPA are the members of the bisphenol family. BPA is one among them that is widely used in the production of epoxy resins and polycarbonate plastics, and it is mass-produced [1]. Food packaging is one area where polycarbonate polymers are widely used. Epoxy resins coat the inside of a variety of metallic objects, such as water supply pipes, bottle caps, and food cans. Plastic that contains BPA is utilized in the production of infant care products, including feeding bottles and formula food containers [2]. Additionally, BPA is used in the production of sales receipts and dental implants. Other bisphenol group members function similarly to BPA in the plastics industry and are mimics of BPA. Plastics that come into touch with food are widely used in daily life [3]. These plastic containers are affordable, durable, and simple to use, making them ideal for food and drink storage, serving, and transportation. These containers constantly leak BPA into the food, especially when they are subjected to high temperatures, non-neutral pH, and recycling [4]. When food is contaminated with BPA, the first organ to be exposed to the substance is the gut. In addition to food, BPA may enter the body through the placenta, skin, and breathed air. Since BPA is ubiquitous in the environment, it is inevitable that it will come into contact with biological systems [5]. Numerous investigations have revealed that the majority of human bodily fluid samples from Asian and American populations contain BPA. BPA is an estrogenic compound that is a member of the endocrine-disrupting chemical (EDC) family [6]. It has a long history of being linked to immune system malfunction, metabolic problems, neurological abnormalities, developmental anomalies, and breast and prostate malignancies. Research showing the harmful effects of BPA on child development pushed several international government agencies to ban the use of BPA in plastic products intended for baby care [7]. The plastics industry swiftly created BPA-free plastic for food product packaging and storage to abide with these regulations. These polymers are produced using BPA's structural and chemical equivalents. BPS is the most widely used BPA alternative in plastic items [8]. Compared to BPA, BPS is less photosensitive and more heat resistant. It was therefore expected to leak from plastic products less frequently. As a safer substitute for BPA, manufacturers created BPS. However, as BPS use has increased, 81% of people in Asia and the United States have tested positive for BPS in their urine. Similar hormone action is seen in BP and BPS [9]. Only a small amount of research has examined the effects of bisphenols on gut function, even though the gut is the first organ to be exposed to them and that oral intake is the most common mode of exposure [10]. This article summarizes research showing how bisphenols, especially BPA, affect gut flora. The effects of bisphenols, particularly BPA, on gut motility and other gut processes might be the subject of separate debate [11].

DOI: 10.1201/9781003489221-8

8.2 BISPHENOLS: TYPES AND PROPERTIES

Bisphenols encompass a diverse group of synthetic compounds characterized by the presence of two hydroxyphenyl functionalities. Various types of bisphenols exist, with BPA being one of the most widely recognized. BPA is extensively utilized in the production of polycarbonate plastics and epoxy resins, contributing to the manufacturing of plastic bottles, food containers, and dental sealants [12, 13]. Another prominent member of this class is BPS, often considered a BPA alternative. It shares similarities with BPA and is frequently found in thermal paper products, as well as certain plastics. Additionally, BPF and BPAF are noteworthy derivatives, each exhibiting unique properties that find applications in various industrial processes [14].

Chemically, bisphenols are characterized by their aromatic structure and the presence of two hydroxyphenyl groups. The core structure typically consists of two benzene rings connected by a carbon bridge, creating a distinct molecular configuration [15]. This structural arrangement imparts specific chemical properties to bisphenols, such as stability and resilience to environmental factors. The hydroxyphenyl groups contribute to their reactivity and ability to interact with other molecules, a feature that influences their biological effects [16]. In industrial settings, bisphenols play a crucial role in the production of polycarbonate plastics, which are renowned for their transparency, durability, and heat resistance. Polycarbonates find application in a myriad of consumer products, ranging from eyewear lenses to electronic devices [17]. Epoxy resins, another major application of bisphenols, are utilized as coatings for food and beverage cans, ensuring the preservation and safety of packaged goods. Moreover, the incorporation of bisphenols in the synthesis of flame retardants and thermal paper adds to their versatility in industrial processes [18]. On the consumer front, bisphenols are prevalent in everyday items. Plastic containers, often used for food storage and packaging, frequently contain BPA, posing concerns about its potential leaching into food and beverages [19]. The dental industry relies on bisphenol-based materials for dental sealants and composites, impacting patient exposure. Thermal paper, commonly used in receipts and tickets, contains bisphenols like BPS, raising questions about the extent of human exposure through skin contact [20]. As the demand for plastic-based products continues to rise, the industrial and consumer applications of bisphenols persist. While these compounds contribute significantly to the functionality and convenience of various materials, the potential health implications associated with their widespread use warrant a closer examination. Understanding the types and chemical properties of bisphenols is crucial for evaluating their impact on human health, particularly concerning the intricate relationship between bisphenols and the gut microbiome [21].

8.3 EXPLORING THE COMPLEXITIES OF THE GUT MICROBIOME: INSIGHTS INTO COMPOSITION, FUNCTIONS, AND INFLUENTIAL FACTORS

The intestinal microbiota, an intricate ecosystem located in the digestive tract, constitutes a dynamic assembly of trillions of microorganisms, encompassing bacteria, viruses, fungi, and archaea [22]. The constitution of the intestinal microbiota is remarkably varied, featuring myriad species, and it is profoundly personalized, influenced by a blend of genetic and environmental elements. Firmicutes and Bacteroidetes are two dominant bacterial phyla, with numerous genera and species within each contributing to the intricate balance of this microbial community [23]. These microorganisms, collectively referred to as gut microbes, engage in a symbiotic relationship with the host, exerting profound influences on human physiology. The functions of gut microbes span a wide range of essential activities, including the fermentation of undigested carbohydrates, the production of SCFAs that serve as an energy source for colonocytes, and the synthesis of vitamins like B and K [24]. Gut microbes also play a pivotal role in the development and regulation of the immune system, aiding in defense against pathogens while maintaining immune tolerance to beneficial microbes and dietary antigens. Furthermore, the gut microbiome contributes to the metabolism of dietary compounds, influences nutrient absorption, and participates in the detoxification of certain substances. The health of the gut microbiome is influenced by a myriad of factors, encompassing both internal and external elements [25]. Antibiotic use, dietary habits, age, genetics, and environmental exposures all shape the composition and functionality of the gut microbiome. A balanced and diverse microbiome is generally associated with good health, while dysbiosis, an imbalance in microbial composition, has been linked to various health conditions, including obesity, inflammatory bowel diseases, and metabolic disorders [26]. Nutritional selections, specifically the intake of foods abundant in dietary fiber, fermented items, and prebiotics, play a role in sustaining a robust gastrointestinal microbiome by furnishing substrates for microbial fermentation and fostering the proliferation of advantageous bacteria [27]. Conversely, diets abundant in processed foods and deficient in fiber could adversely influence the gastrointestinal microbiome, potentially resulting in dysbiosis and related health complications. Additionally, lifestyle factors such as stress, sleep patterns, and physical activity levels have been implicated in influencing the gut microbiome [28]. Recognizing the multifaceted interplay between the gut microbiome's composition, its diverse functions, and the array of factors impacting its health is imperative for comprehending the intricate relationship between gut microbes and overall human well-being. Efforts to maintain a balanced and resilient gut microbiome through lifestyle and dietary interventions hold promise for promoting optimal health and preventing a spectrum of diseases linked to microbial dysregulation [29].

8.4 BISPHENOLS AND GUT MICROBIOME INTERACTION

8.4.1 Effect of Bisphenol A on Rat Gut Muscle Contractility In Vitro

A small number of research discuss how BPA affects GI motility. These are all rat-based experimental investigations. Rats were given BPA orally or continuously fed to evaluate its effect on gastrointestinal motility [30]. Following the treatment period, the contractility of various gut segments from rats was measured in an organ bath preparation or in other studies where the animals were sacrificed at the start of the investigation and the prepared gut segments were acutely exposed in vitro to BPA in the organ bath preparation before the gut segments' contractility was measured [31]. A study showing the detrimental effects of BPA on the rat duodenum's in vitro function was released in 2013. In this study, BPA (50 mg/kg/day) was given orally to several groups of adult rats for 20 and 30 days. Following the conclusion of the therapy, the animals were slaughtered, and duodenal contractions were recorded in a prepared organ bath [32]. Treatment for 20 and 30 days resulted in a statistically significant decrease in the frequency and amplitude of duodenal contractions, respectively. Acetylcholinesterase (AChE) activity in the duodenal tissue homogenate increased significantly in both treatment groups [33]. Thus, it was postulated that BPA reduces the frequency and amplitude of duodenal contractions by increasing the activity of AChE in smooth muscle endplates and the efferents of the surrounding Myenteric Auerbach's plexus [34]. Furthermore, researchers discovered that duodenal tissue slices from BPA-exposed groups had higher calcium (Ca) salt accumulation. These findings were made using von Kossa staining. They proposed an additional mechanism by which BPA reduces duodenal contractility: a reduction in the availability of free Ca2+ in smooth muscle cells and local myenteric terminals [35]. Later, in 2014, it was discovered that the activity of antioxidant enzymes was found to be considerably reduced when rats were exposed to BPA for identical amounts of time and days. This suggested that the rat duodenum was experiencing higher oxidative stress. According to its toxicological profile, BPA may already be an oxidative stressor [36]. For example, there are several instances of oxidative stress caused by BPA in rats' liver, kidney, and epididymal cells. BPA's oxidative stress on the smooth muscle of the duodenum may be another way in which this chemical inhibits the duodenum's ability to contract [37]. Moreover, there was an increase in the expression of the nitric oxide synthase (NOS) enzyme in the muscularis layer, indicating that BPA may promote the production of nitric oxide (NO) in myenteric neurons, hence reducing duodenal contractility. It is noteworthy that BPA reduced rat atrial contractility via a NO signaling pathway that was dependent on guanylyl cyclase [38]. More research was done on NO's function in reducing intestinal contractility brought on by BPA. L-NAME, an inhibitor of NO synthase, and methylene blue, a soluble guanylyl cyclase (sGC) blocker, substantially undid the BPA-induced decrease in duodenal contractions [39]. The NO donor sodium nitroprusside also markedly increased the contractions. Therefore, the primary cause of the BPA-induced reduction in movement may be the stimulation of myenteric efferents, which produce the muscle relaxant NO [40]. By triggering NO synthase to produce NO in non-adrenergic non-cholinergic neurons and sGC in smooth muscle cells to initiate NO signaling cascades, BPA may reduce duodenal contractions [41]. Furthermore, the α-adrenergic receptor blocker phentolamine significantly reversed the BPA-induced inhibition of movement in the previously mentioned investigation, indicating that norepinephrine-secreting motor neurons may be involved in the BPA-induced inhibition of movement [42]. Rats treated to a relatively modest oral dosage of BPA (50 μg/kg/day for 28 days) in a later study had reduced spontaneous and agonist-induced in-vitro intestinal contractility. In this study, rats given BPA showed a significantly reduced ACh-induced contractile response when compared to control rats. Treatment with atropine (Atrp) reduced the ACh-evoked response in rats fed a normal diet by as much as 60%, but it inhibited the ACh-induced response decrease in rats fed BPA by 100% [43]. Atrp is known to inhibit just muscarinic receptors. Thus, the researchers postulated that the nicotinic-type neuro plexus may be the site of the ACh receptor alteration. Therefore, BPA protected the muscarinic receptor in smooth muscle while inhibiting the activity of the ACh-mediated receptor (nicotinic type alone) in the neural plexus. This allowed Atrp to produce a 100% suppression of ACh-induced contraction in rats that were given BPA [44]. Furthermore, the scientists found that although serotonin, a crucial regulator of gut contractility in a healthy gut, increased the gut tissue's contractile tension in control rats but had no effect on animals given BPA [45]. It was proposed that changes in serotonin receptor activation brought on by BPA were the cause of seratonin's inefficiency [46]. The impact of BPA on the colon and ileum contractility in adult and neonatal rats was investigated by Nirja et al. Because it has not reached functional maturity, the newborn gut may be more vulnerable to BPA's harmful effects. Few studies have been conducted on the impact of BPA on the intestinal motility of newborns [47]. Using a force transducer and a computerized data collection system, isometric contractions from dissected colon and ileum sections were recorded in organ bath preparation for each of these experiments [48]. The BPA bath dosages were applied cumulatively (1–100 μm) to the gastrointestinal segments. Following the administration of each dose of BPA, there was a concentration-dependent reduction in the maximum height of contractions in the ileal and colon tissue [49]. There was a significant decrease in contractile tension at concentrations of 30 and 100 μm bath. An analogous reduction in contractile frequency was seen upon increasing the BPA content in the bath to 30 and 100 μm. The frequency decreased to 0 at a bath concentration of 100 μm, indicating that there was no spontaneous contraction [50]. Furthermore, compared to the adult gut, the effects of BPA were more severe in the newborn gut [51]. This study also looked at the processes underlying the inhibition of intestinal smooth muscle contractions caused by BPA. Given that estrogen has been reported to impact the contractile activity of gut muscle, the author hypothesized that the observed reduction in gut contractile activity could be related to the estrogen-like effect of BPA [52]. The two types of ERs that estrogen acts via are estrogen receptor (ERα) and estrogen receptor (ERβ). In the gut, ERβ is expressed. It has been suggested that BPA might possibly act on the enteric nervous system's neuronal components [53]. The researchers also tried to look into the function of NO in light of earlier findings that BPA slows duodenal motility via NO-mediated mechanisms and decreases atrial activity through NO pathways. Therefore, a second series of studies was designed in which gut tissue was exposed to one of the following four antagonists in order to explore additional possible mechanisms of action of BPA [54, 55]. Tests using distinct antagonists demonstrated that the depressive impacts of BPA might not be facilitated by ERs, NO-dependent pathways, cholinergic pathways, or the enteric nervous system. Next, a theory was put out suggesting that BPA's ER-independent effect on intestinal smooth muscle might be the cause of its detrimental effects on gut contractility [56]. In the past, it was hypothesized that estrogen had a comparable ER-independent impact via blocking or opening potassium channels. However, in this experiment, it was unclear if the same mechanisms were in charge of BPA's inhibitory action [57].

It has also been documented that BPA inhibits the function of cardiac and uterine muscles in addition to the muscles of the stomach. In vitro, BPA reduced rat atrial contractility through processes mediated by NOcGMP [58]. There was a dose-dependent reduction in spontaneous uterine contractions due to BPA (0.1–10 μm) and its negative effect on uterine contractility. The spontaneous and acetylcholine-evoked uterine contractions were considerably reduced after a 28-day chronic oral exposure to BPA (2 μg/kg/day) [59]. The persistent disclosure of detrimental health impacts caused by BPA has raised concerns about whether goods that come into touch with food should include it [60]. The use of BPA in infant feeding bottles is prohibited by law in several countries [61]. Owing to regulations on its marketing and growing knowledge of the detrimental effects of BPA on health, producers have introduced BPS, an analogue of BPA, as a safer alternative. BPA-free thermal paper could really contain BPS. BPS is harmful to human health and is being studied for its toxicity [62]. In contrast to BPA, little research has been done on how BPS affects gastrointestinal motility. Like BPA, BPS is referred to as EDC. Additionally, it possesses estrogenic properties [63]. BPS was formerly thought to be a safe alternative to BPA because of its exceptional stability at high temperatures, tolerance to sunlight, and noticeably decreased estrogenic activity, as reported by certain older research [64]. Recent research has shown that BPS and BPA have comparable hormonal potencies. Moreover, it has been demonstrated that BPS and BPA have many of the same in vitro and in vivo toxicities [65]. In certain investigations, BPS has been reported to exhibit broader toxicological manifestations and may induce additional alterations that might result in disruptions not observed with BPA. In contrast to BPA, BPS increased levels of 17α—OH progesterone, which is known to impair gastrointestinal motility [66]. To investigate the distinct effects of BPS and BPA on the contractility of gastrointestinal smooth muscle, Sharma and Mandal measured the bath concentration (0.1–100 μm) response of BPS and BPA on in vitro contractility of the stomach and small intestine [67]. Similar to BPA, BPS reduced the spontaneous contractile activity in both gastrointestinal tissues [68]. The results of comparing the effects of BPA and BPS showed that there is a comparable chance that both substances will reduce the small intestine's and stomach's natural ability to contract [69]. It is also possible to investigate if the effects of both bisphenols in the newborn gut are comparable to those in the adult gut. Investigations into the mechanism underlying BPS's inhibitory effect on gut muscle tissue are also necessary [70]. Figure 8.1 illustrates the detrimental effects of BPA on gut microbiota.

To further understand the differences between BPA and BPS's effects on the contractile function of gut smooth muscle, these researchers also looked at the effects of a 28-day oral exposure to 50 mg/kg/day of BPA and BPS on the contractility of the small intestine and the stomach in vitro [71]. The results showed that long-term BPA consumption reduced the small intestine's and the stomach's spontaneous contractile activity and ACh-induced contractions. But only in the colon and not in the stomach did long-term BPS intake suppress spontaneous contractile activity. Nonetheless, the contractions brought on by Ach were lessened in the stomach and intestine. According to a recent report, BPS prevented the duodenum from contracting after acute exposure by reducing the activity of excitatory cholinergic myenteric efferents. It also exposed the rats to varying doses of plastic toxins and prolonged feeding times [72]. Gut dysmotility might be the outcome of decreased gut contractility. Therefore, thorough consideration of BPS as a safe BPA alternative has to be given again [73]. Table 8.1 summarizes experimental investigations on BPA's impact on gastrointestinal (GI) motility.

8.4.2 Effect of Bisphenol A on Gut Functions Other Than Motility

It is widely known that estrogens play a critical role in establishing and maintaining the intestinal barrier. Gender differences exist in intestinal permeability and inflammation, with women experiencing higher rates of irritable bowel syndrome [74]. Stomach discomfort related to irritable bowel syndrome in women might change over their menstrual cycle [75]. In the human colonic mucosa, both in fetal and in adult forms, the ERs are abundantly expressed. Intestinal epithelial cells' viscerosensitivity and inflammation are regulated by ERs. The central nervous system receives peripheral information from the sense neurons [76]. Rats' visceral nociceptive response is increased and

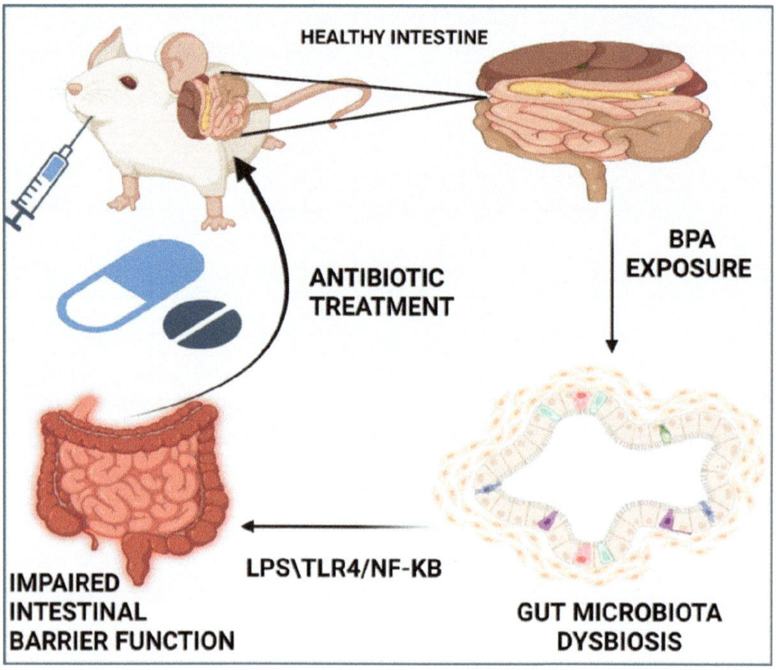

FIGURE 8.1 This image illustrates the detrimental effects of BPA on gut microbiota, intestinal barrier disruption, and inflammation, emphasizing the importance of understanding and controlling BPA exposure for disease prevention.

epithelial barrier function is altered by BPA binding to ERs. Moreover, BPA exposure during pregnancy may increase the likelihood that a female child may experience severe colonic inflammation as an adult [77, 78].

In a research study, BPA treatment (20 mg/kg/day) during pregnancy reduced pre-delivery mice's blood Ca levels and elevated the duodenum and ileum's mRNA expressions of the proteins occludin and junction adhesion molecular A (JAM-A) [79]. These proteins are in charge of closing gaps between cells at tightly connected apical junctions [80]. This data indicates that 20 mg of BPA reduced the quantity of calcium absorbed by paracellular gaps and, consequently, the calcium levels in serum. Notably, estrogen elevates JAM-A and occluding expression. Calcium is necessary for a number of physiological functions, such as the development of bones, neurotransmission, and muscle contraction [81].

When BPA was given to the pig stomach wall at relatively low doses (0.05 and 0.5 mg/kg body weight/day for 28 days), the neurochemical coding of the neural structures changed. Every region of the stomach wall had less cholinergic neurons after BPA was administered [82]. In the mucosal, muscular, and enteric myenteric ganglia, there were more vasoactive intestinal polypeptide-positive neurological structures; in the submucous ganglia, there were less of these structures. Each of these many substances has a part in regulating the digestive system's motility [83, 84]. Furthermore, it may be said that SP, GAL, and CART have anti-apoptotic, neurotropic, and anti-inflammatory properties, respectively [85]. The mechanism that generated the observed modifications is likely linked to the neuroprotective and adaptive processes that comprise the response to the neurotoxic and pro-inflammatory action of BPA. When BPA was administered at relatively low quantities, the neurochemical coding of the nerve structures in the stomach wall of pigs was altered [86–88]. For 28 days, the doses were 0.05 and 0.5 mg/kg body weight/day. Treatment with BPA reduced the number of cholinergic neurons in the stomach wall and increased the number

TABLE 8.1
Summary of experimental investigations on BPA's impact on gastrointestinal (GI) motility in rats, revealing mechanisms, gender-specific aspects, serotonin interactions, neonatal vulnerability, ER-independent actions, and a comparison with BPS effects.

Study Focus	Experimental Methods	Findings	Gender-Specific Impact
Oral BPA Exposure	Organ Bath Preparation	Reduced Duodenal Contractions, AChE Activity Increase	Not Addressed
Neonatal Vulnerability	Serotonin Interaction	Altered Response to Serotonin, Severe Effects in Newborn Rats	Limited Studies on Neonatal Gut
ER-Independent Action	BPS Comparison	Potential ER-Independent Impact on Smooth Muscle	Investigated Similar Inhibitory Effects
Oxidative Stress Effect	Antioxidant Enzyme Activity Assessment	Reduced Enzyme Activity, Higher Oxidative Stress	Not Investigated
Contractile Mechanism	NO Pathway Inhibition	NO-Dependent Pathways Not Evident, ER-Independent Impact	Phentolamine Reversed Inhibition
Short-Term BPA Exposure	Muscarinic Receptor Response Assessment	Reduced ACh-Induced Response, Protected Muscarinic Receptor	Not Investigated in Females
BPS and BPA Comparison	Long-Term Exposure Contractility Study	Both Reduced Spontaneous and ACh-Induced Contractions	More Severe Effects in Newborn Gut

of neuronal structures containing SP, GAL, and CART peptide [89]. The amount of nerve cells in the muscular and mucosal layers, as well as the enteric myenteric ganglia, that tested positive for vasoactive intestinal polypeptide increased. However, in the submucous ganglia, there was a decrease in the quantity of these cells [90]. Each of these substances controls the digestive system's motility. Anti-apoptotic, neurotropic, and anti-inflammatory qualities are possessed by SP, GAL, and CART. The changes that were seen most likely included neuroprotective and adaptive mechanisms that functioned as a response to the pro-inflammatory and neurotoxic effects of BPA [91]. In a similar vein, it was shown that BPA had an impact on the enteric neurons immunoreactive to neuregulin 1, a well-known component involved in maintaining proper nervous system development [92, 93]. This impact was observed in the big intestine of the pig's enteric nervous system. Additionally, several research findings have reported on the impact of BPA on the gut microbiota. Research found that the gut microbiota of mice administered BPA showed a substantial reduction in species diversity [94–96]. In this work, oral BPA consumption in conjunction with food influenced the composition of the gut microbial community in mice that was comparable to that of a high-fat and high-sucrose diet. In another study, BPA was linked to weight increase and glucose intolerance, but it had no discernible effect on the gut flora following bariatric or metabolic surgery [97]. Table 8.2 summarizes the BPA's diverse impacts in gut microbiota diversity

TABLE 8.2
An overview of BPA's diverse impacts, from pregnancy-related changes to neurochemical coding in pigs' stomach walls and alterations in gut microbiota diversity.

BPA Impact on Pregnancy	Mice Blood Ca Levels	Neurochemical Coding in Pigs' Stomach Wall	Gut Microbiota Diversity Reduction
Reduced Blood Ca Levels in Mice	Decreased Ca Absorption in Duodenum and Ileum	Cholinergic Neuron Reduction, Altered Peptide Levels	Species Diversity Reduction in Mice
Elevated mRNA Expression of Occludin and JAM-A	Estrogen-Mediated Protein Expression Increase	Neurotoxic and Adaptive Responses to BPA Action	Similar Impact to High-Fat and High-Sucrose Diet
Possible Impact on Severe Colonic Inflammation	Calcium Essential for Physiological Functions	Impact on Neuregulin 1-Related Enteric Neurons	Linked to Weight Increase, Glucose Intolerance
Connection to ERs and Visceral Nociceptive Response	Estrogen Role in Barrier Function and Inflammation	Involvement of SP, GAL, CART in Neurotropic Processes	No Significant Effect Following Bariatric or Metabolic Surgery

8.5 HEALTH IMPLICATIONS

8.5.1 Potential Health Risks Associated with Bisphenol Exposure

Exposure to bisphenols, particularly the widely recognized Bisphenol A (BPA), has raised concerns due to its potential health risks. BPA is commonly used in the production of polycarbonate plastics and epoxy resins, leading to its presence in various everyday items such as plastic bottles, food containers, and dental materials [98]. The concern stems from the ability of bisphenols to leach into food and beverages, thereby entering the human body. Research has indicated that BPA can mimic the structure and function of estrogen, a hormone that plays a crucial role in various physiological processes [99]. This estrogenic activity has been linked to several health risks, including endocrine disruption, reproductive issues, and an increased risk of certain cancers [100]. One primary concern associated with bisphenol exposure is its potential impact on endocrine function [101]. BPA, in particular, is known as an EDC, meaning it can interfere with the normal hormonal balance in the body. Studies have suggested that BPA may affect the endocrine system by binding to estrogen receptors, leading to hormonal imbalances. This disruption has been associated with adverse reproductive outcomes, such as infertility, and developmental issues, including abnormalities in the male and female reproductive systems [102–104]. Moreover, the potential for bisphenol exposure to contribute to cardiovascular diseases has been a subject of investigation. Some studies have suggested a link between BPA exposure and an increased risk of hypertension, coronary artery disease, and other cardiovascular issues [105]. The mechanisms underlying this association are not fully understood, but it is postulated that the estrogenic effects of BPA may play a role in influencing cardiovascular health. The impact of bisphenols on the development of metabolic disorders has also been a focus of research [106]. There is evidence suggesting that exposure to BPA may contribute to obesity, insulin resistance, and diabetes. Animal studies have shown that BPA can disrupt adipocyte function, leading to changes in fat metabolism and storage. Additionally, BPA may interfere with insulin signaling pathways, potentially contributing to insulin resistance, a key factor in the development of type 2 diabetes. Furthermore, emerging research has explored the potential link between bisphenol exposure and neurodevelopmental disorders. Some studies suggest that prenatal and early-life exposure to BPA may be associated with behavioral and cognitive deficits in children [107]. The neurotoxic effects of bisphenols are thought to result from their ability to cross the blood-brain barrier and interfere with normal brain development. It is important to note that while these associations have been observed in various studies, the overall body of evidence is still evolving, and the specific mechanisms by which bisphenols may contribute to these health risks are not fully elucidated [108]. Figure 8.2 illustrates the intricate interplay between Bisphenols and the gut microbiome. Additionally, the extent of human exposure and the variability in individual responses to bisphenols make it challenging to establish definitive causal relationships [109–111]. To alleviate potential health hazards linked to bisphenol exposure, regulatory authorities have instituted measures to limit its utilization in specific items, particularly those crafted for the storage of food and beverages [112]. However, given the ubiquity of bisphenols in the environment, continued research is essential to better understand the long-term health implications and to inform public health guidelines on their use. Individuals can also take proactive measures by opting for bisphenol-free products, especially those that come into contact with food, and adopting a lifestyle that minimizes exposure to these potentially harmful substances [113].

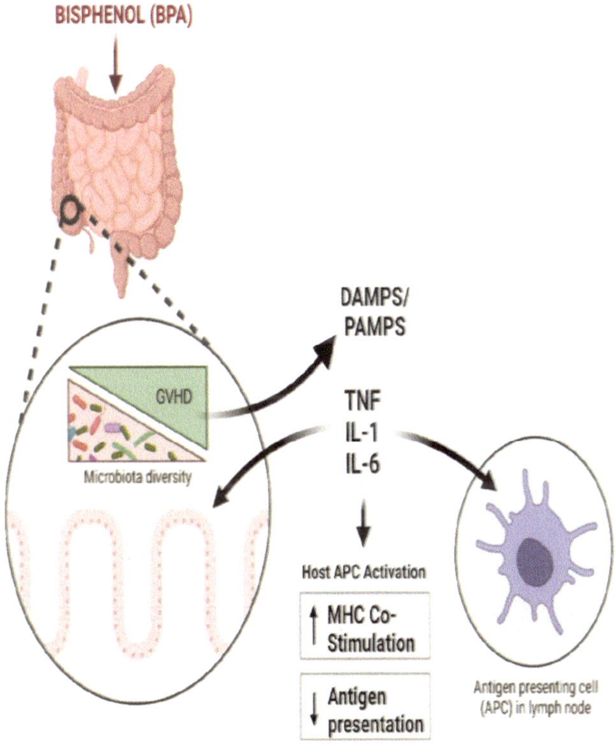

FIGURE 8.2 This image describes the intricate interplay between Bisphenols and the gut microbiome. It unravels the impact on microbiota diversity, activation of host APC in lymph nodes, portraying a comprehensive understanding of the complex dynamics at the intersection of environmental factors and the intestinal ecosystem.

8.5.2 Link between Gut Microbiome Imbalance and Health Conditions

The intricate relationship between the gut microbiome and overall health has become a focal point of scientific inquiry, shedding light on the link between gut microbiome imbalance, or dysbiosis, and various health conditions [114–116]. The gut microbiome, a dynamic community of microorganisms residing in the gastrointestinal tract, plays a crucial role in maintaining a delicate balance that contributes to the proper functioning of the immune system, metabolism, and overall physiological homeostasis [117]. "When equilibrium in the gut microbiome is disrupted, it can lead to various health issues, including IBD." IBD, encompassing disorders such as Crohn's disease and ulcerative colitis, manifests as persistent inflammation in the gastrointestinal tract [118]. Research indicates changes in the structure and variety of the gut microbiome among individuals with IBD, indicating a possible involvement of dysbiosis in the initiation or intensification of these inflammatory disorders [119]. Imbalances in specific bacterial populations and the loss of microbial diversity may contribute to the chronic inflammation observed in IBD [120, 121]. Another area of significant research interest is the connection between gut dysbiosis and metabolic disorders, including obesity and type 2 diabetes [122, 123]. The intestinal microbiota impacts energy metabolism, nutrient assimilation, and the control of adipose tissue [124]. Dysregulation has been correlated with an elevated ability to derive energy from dietary intake and foster lipid storage, thereby playing a role in obesity. Moreover, changes in the composition of gut microorganisms may induce inflammation and insulin resistance, pivotal elements in the onset of type 2 diabetes [125–128]. Psychological well-being has similarly been associated with the gut-brain axis, emphasizing the reciprocal communication between the intestinal tract and the central nervous system. Emerging research suggests a potential link between gut microbiome imbalance and conditions such as depression and anxiety [129, 130]. The gut microbiome produces neurotransmitters and communicates with the brain through the vagus nerve, influencing mood and behavior. Disruptions in this gut-brain communication, possibly stemming from dysbiosis, may contribute to the development or exacerbation of mental health disorders [131]. Furthermore, there is growing evidence of the role of the gut microbiome in immune system modulation, and dysbiosis has been implicated in autoimmune diseases [132, 133]. Conditions such as lupus, rheumatoid arthritis, and multiple sclerosis have shown associations with alterations in gut microbial composition. Dysbiosis could initiate an inadequate immune reaction, resulting in the immune system targeting the body's own tissues [134–137].

Understanding the link between gut microbiome imbalance and health conditions has led to explorations of therapeutic interventions aimed at restoring microbial balance, known as microbiome-targeted therapies. Prebiotics, probiotics, and FMT are among the approaches being investigated to address dysbiosis and potentially alleviate symptoms associated with various health conditions [138–140]. It is essential to acknowledge the complexity of these associations, as the gut microbiome is highly individualized, and the specific mechanisms by which dysbiosis contributes to health conditions remain an active area of research. Factors such as diet, lifestyle, antibiotic use, and genetic predispositions all contribute to the dynamic nature of the gut microbiome and its impact on health [141–144].

The link between gut microbiome imbalance and various health conditions underscores the importance of maintaining a healthy and diverse microbial community in the gut. Ongoing research continues to deepen our understanding of these relationships, paving the way for innovative therapeutic strategies and personalized interventions to promote gut health and overall well-being [145].

8.6 REGULATORY ASPECTS AND GUIDELINES

8.6.1 Current Regulations on Bisphenols

As concerns about the potential health risks associated with bisphenol exposure have grown, regulatory bodies worldwide have implemented measures to address the use of bisphenols in various products [146]. These regulations aim to mitigate potential health hazards and protect public health. One of the most studied and regulated bisphenols is BPA, a common component in the production of polycarbonate plastics and epoxy resins [147]. In several regions, including the European Union (EU), Canada, the United States, and Japan, there are specific regulations in place to control the use of BPA in certain products. The EU, for instance, has banned the use of BPA in the production of baby bottles since 2011 [148–150]. Additionally, the ECHA has classified BPA as a substance of very high concern due to its endocrine-disrupting properties, triggering more stringent regulations [151]. In the United States, FDA has taken steps to regulate the use of BPA in certain food and beverage containers. While BPA is still allowed in the manufacturing of some products, there have been efforts to reduce its presence, especially in items designed for use by infants and young children [152]. The FDA continues to monitor research on bisphenols and updates its regulations accordingly [153, 154]. Canada has also taken regulatory measures to address BPA exposure. In 2010, Canada became the first country to declare BPA a toxic substance, leading to the prohibition of BPA in baby bottles. The Canadian government has since continued to assess and regulate the use of bisphenols in various products [155, 156].

Japan has implemented restrictions on the use of BPA in certain food contact materials, with a focus on protecting infants and young children from potential exposure. The country's regulatory approach emphasizes the need to find alternative materials that are considered safer [157]. While regulations primarily focus on BPA, other bisphenols, such as Bisphenol S (BPS) and BPF, have gained attention as potential alternatives [158, 159]. However, there is ongoing debate and research on the safety of these substitutes, as they may share some similar chemical properties with BPA [160–162]. Regulatory bodies continue to evaluate and update guidelines based on emerging scientific evidence. It is important to note that regulations vary across regions, and the classification and restrictions on bisphenols can differ [163]. Moreover, the dynamic nature of scientific research means that regulatory bodies continuously reassess and update their guidelines as new evidence emerges. As a result, the regulatory landscape surrounding bisphenols is subject to change [164]. Current regulations on bisphenols, particularly BPA, reflect the efforts of regulatory bodies to address potential health risks associated with these compounds. The focus on protecting vulnerable populations, such as infants and young children, is evident in the restrictions on certain products [165]. Continued research and ongoing monitoring of bisphenols contribute to the evolution of regulations, ensuring that they remain reflective of the latest scientific understanding and advancements in the field [166–168].

8.6.2 Strategies to Reduce Bisphenol Exposure

Mitigating and preventing bisphenol exposure involves implementing strategies that aim to minimize the potential health risks associated with these compounds. One key approach is the promotion of consumer awareness and education regarding bisphenols and their presence in everyday products [169, 170]. Providing information on alternative materials and products that are bisphenol-free empowers individuals to make informed choices [171]. Additionally, advocating for regulatory measures to limit the use of bisphenols in specific applications, especially those involving food and beverages, is crucial for broader public health [172]. Lifestyle modifications, such as choosing glass or stainless-steel containers over plastic ones and opting for BPA-free products, can significantly reduce personal exposure [173]. Dietary habits also play a role; selecting fresh and minimally processed foods, as opposed to those packaged in materials containing bisphenols, contributes to lowering overall exposure. Moreover, supporting research and development initiatives for safer alternatives to bisphenols in manufacturing processes is essential for fostering long-term solutions [12, 174]. Collaborative efforts involving governmental bodies, industries, and the public are pivotal in developing and implementing comprehensive strategies to mitigate bisphenol exposure and proactively prevent potential health risks [175–178].

8.7 CONCLUSION AND FUTURE PERSPECTIVE

The intricate interplay between bisphenols and the gut microbiome underscores the importance of understanding the potential implications for human health. The evidence linking bisphenol exposure to alterations in gut microbial composition and function has raised concerns about the broader consequences on physiological well-being. The gut microbiome, a vital player in maintaining homeostasis within the body, can be influenced by these synthetic compounds, leading to dysbiosis and potential health risks. The diverse range of bisphenols, such as Bisphenol S (BPS), Bisphenol A (BPA), and Bisphenol F (BPF), each with distinct properties and applications, further complicates the assessment of their impact on the gut microbiome. Studies have highlighted associations between bisphenol exposure and conditions such as inflammatory bowel disease, metabolic disorders, and even neurodevelopmental issues, emphasizing the need for a comprehensive understanding of the mechanisms at play.

As we delve deeper into the future perspectives of this research area, there is a pressing need for continued investigation to unravel the complexities of the bisphenol-gut microbiome interaction. Future research should aim to elucidate the specific mechanisms by which bisphenols influence the gut microbial community and how these changes translate into health outcomes. This may involve exploring the temporal dynamics of bisphenol exposure and its sustained effects on the gut microbiome over various life stages. Additionally, investigating the potential reversibility of bisphenol-induced dysbiosis through interventions like dietary modifications or probiotic therapies is a promising avenue. Furthermore, researchers must focus on identifying biomarkers that can serve as indicators of bisphenol-induced gut microbiome alterations and associated health risks. These biomarkers could aid in early detection and intervention, offering a more personalized approach to mitigating the potential consequences of bisphenol exposure. Integrating omics technologies, such as metagenomics and metabolomics, can provide a comprehensive view of the changes occurring in the gut microbiome in response to bisphenols, enabling a more nuanced understanding of the molecular mechanisms involved. In terms of regulatory perspectives, ongoing efforts to monitor and regulate bisphenol usage in consumer products, particularly those in direct contact with food and beverages, should persist. Evidence-based adjustments to regulations, influenced by the evolving body of scientific knowledge, will be crucial for safeguarding public health. Collaborations between researchers, industry stakeholders, and regulatory agencies are essential to ensure a well-rounded approach that considers both technological advancements and health implications. The relationship between bisphenols and the gut microbiome represents a multifaceted and evolving field of study. As we progress, a holistic understanding of these interactions will contribute not only to refining our knowledge of gut microbiome dynamics but also to developing strategies to minimize potential health risks associated with bisphenol exposure. The journey ahead involves interdisciplinary collaborations, technological innovations, and a commitment to translating research findings into practical solutions for public health and environmental sustainability.

REFERENCES

1. A.G. Abela, S. Fava, Why is the incidence of type 1 diabetes increasing?, *Current Diabetes Reviews,* 17 (2021) e030521193110.
2. O. Adamovsky, J.H. Bisesi, Jr., C.J. Martyniuk, Plastics in our water: Fish microbiomes at risk?, *Comparative Biochemistry and Physiology. Part D, Genomics & Proteomics*, 39 (2021) 100834.
3. N. Akbar, R. Siddiqui, K. Sagathevan, M. Iqbal, N.A. Khan, Gut bacteria of water monitor lizard (varanus salvator) are a potential source of antibacterial compound(s), *Antibiotics*, 8 (2019) 164.
4. J. Ao, Y. Liu, W. Tang, J. Zhang, Bisphenol S exposure induces intestinal inflammation: An integrated metabolomic and transcriptomic study, *Chemosphere*, 292 (2022) 133510.

5. A.L. Arancio, E.R. Cohenour, K.D. Cole, A.R. Dominguez, J. Kadie, W.C. Maloney, C. Cilliers, S.M. Schuh, Data demonstrating distinct embryonic developmental defects induced by bisphenol a alternatives, *Data in Brief*, 25 (2019) 104091.
6. J. Balaguer-Trias, D. Deepika, M. Schuhmacher, V. Kumar, Impact of contaminants on microbiota: Linking the gut-brain axis with neurotoxicity, *International Journal of Environmental Research and Public Health*, 19 (2022) 1368.
7. K. Baralić, K. Živančević, D. Bozic, D. Đukić-Ćosić, Probiotic cultures as a potential protective strategy against the toxicity of environmentally relevant chemicals: State-of-the-art knowledge, *Food and Chemical Toxicology: An International Journal Published for the British Industrial Biological Research Association*, 172 (2023) 113582.
8. N.G. Barra, Y.H. Kwon, K.M. Morrison, G.R. Steinberg, M.G. Wade, W.I. Khan, M.M. Vijayan, J.D. Schertzer, A.C. Holloway, Increased gut serotonin production in response to bisphenol A structural analogs may contribute to their obesogenic effects, *American Journal of Physiology. Endocrinology and Metabolism*, 323 (2022) e80–e091.
9. C. Berruyer, L. Pouyet, V. Millet, F.M. Martin, A. LeGoffic, A. Canonici, S. Garcia, C. Bagnis, P. Naquet, F. Galland, Vanin-1 licenses inflammatory mediator production by gut epithelial cells and controls colitis by antagonizing peroxisome proliferator-activated receptor gamma activity, *The Journal of Experimental Medicine*, 203 (2006) 2817–2827.
10. A. Brulport, C. Lencina, M.C. Chagnon, L. Le Corre, L. Guzylack-Piriou, Transgenerational effects on intestinal inflammation status in mice perinatally exposed to bisphenol S, *Chemosphere*, 262 (2021) 128009.
11. M. Casas, M. Gascon, Prenatal exposure to endocrine-disrupting chemicals and asthma and allergic diseases, *Journal of Investigational Allergology & Clinical Immunology*, 30 (2020) 215–228.
12. K. Cerk, M. Aguilera-Gómez, Microbiota analysis for risk assessment: Evaluation of hazardous dietary substances and its potential role on the gut microbiome variability and dysbiosis, *EFSA Journal. European Food Safety Authority*, 20 (2022) e200404.
13. M.S. Hussain, G. Gupta, R. Mishra, N. Patel, S. Gupta, S.I. Alzarea, I. Kazmi, P. Kumbhar, J. Disouza, H. Dureja, N. Kukreti, S.K. Singh, K. Dua, Unlocking the secrets: Volatile Organic Compounds (VOCs) and their devastating effects on lung cancer, *Pathology, Research and Practice*, 255 (2024) 155157.
14. L. Chen, Y. Guo, C. Hu, P.K.S. Lam, J.C.W. Lam, B. Zhou, Dysbiosis of gut microbiota by chronic coexposure to titanium dioxide nanoparticles and bisphenol A: Implications for host health in zebrafish, *Environmental Pollution*, 234 (2018) 307–317.
15. P. Chen, J. Yang, B. Xiao, Y. Zhang, S. Liu, L. Zhu, Mechanisms for the impacts of graphene oxide on the developmental toxicity and endocrine disruption induced by bisphenol A on zebrafish larvae, *Journal of Hazardous Materials*, 408 (2021) 124867.
16. Z. Chen, F. Wang, D. Wen, R. Mu, Exposure to bisphenol A induced oxidative stress, cell death and impaired epithelial homeostasis in the adult Drosophila melanogaster midgut, *Ecotoxicology and Environmental Safety*, 248 (2022) 114285.
17. Z.J. Chen, X.L. Yang, H. Liu, W. Wei, K.S. Zhang, H.B. Huang, J.P. Giesy, H.L. Liu, J. Du, H.S. Wang, Bisphenol A modulates colorectal cancer protein profile and promotes the metastasis via induction of epithelial to mesenchymal transitions, *Archives of Toxicology*, 89 (2015) 1371–1381.
18. K. Chiu, G. Warner, R.A. Nowak, J.A. Flaws, W. Mei, The impact of environmental chemicals on the gut microbiome, *Toxicological Sciences: An Official Journal of the Society of Toxicology*, 176 (2020) 253–284.
19. S. Coffin, G.Y. Huang, I. Lee, D. Schlenk, Fish and seabird gut conditions enhance desorption of estrogenic chemicals from commonly-ingested plastic items, *Environmental Science & Technology*, 53 (2019) 4588–4599.
20. O. Contreras-Rodriguez, M. Solanas, R.M. Escorihuela, Dissecting ultra-processed foods and drinks: Do they have a potential to impact the brain?, *Reviews in Endocrine & Metabolic Disorders*, 23 (2022) 697–717.
21. A. Cox, Z. Bomstein, A. Jayaraman, C. Allred, The intestinal microbiota as mediators between dietary contaminants and host health, *Experimental Biology and Medicine*, 248 (2023) 2131–2150.
22. A. Czarnywojtek, M. Borowska, K. Dyrka, J. Moskal, J. Kościński, I. Krela-Kaźmierczak, A.M. Lewandowska, B. Abou Hjeily, P. Gut, K. Hoffmann, S. Van Gool, N. Sawicka-Gutaj, M. Ruchała, The influence of various endocrine disruptors on the reproductive system, *Endokrynologia Polska*, 74 (2023) 221–233.

23. A. Czarnywojtek, K. Jaz, A. Ochmańska, M. Zgorzalewicz-Stachowiak, B. Czarnocka, N. Sawicka-Gutaj, P. Ziółkowska, I. Krela-Kaźmierczak, P. Gut, E. Florek, M. Ruchała, The effect of endocrine disruptors on the reproductive system - current knowledge, *European Review for Medical and Pharmacological Sciences*, 25 (2021) 4930–4940.
24. M. Dagar, P. Kumari, A.M.W. Mirza, S. Singh, N.U. Ain, Z. Munir, T. Javed, M.F.I. Virk, S. Javed, F.H. Qizilbash, A. Kc, C. Ekhator, S.B. Bellegarde, The hidden threat: Endocrine disruptors and their impact on insulin resistance, *Cureus*, 15 (2023) e47282.
25. J.A. DeLuca, K.F. Allred, R. Menon, R. Riordan, B.R. Weeks, A. Jayaraman, C.D. Allred, Bisphenol-A alters microbiota metabolites derived from aromatic amino acids and worsens disease activity during colitis, *Experimental Biology and Medicine*, 243 (2018) 864–875.
26. G.T. Diaz-Gerevini, G. Repossi, A. Dain, M.C. Tarres, U.N. Das, A.R. Eynard, Beneficial action of resveratrol: How and why?, *Nutrition*, 32 (2016) 174–178.
27. D.R. Doerge, N.C. Twaddle, M. Vanlandingham, J.W. Fisher, Pharmacokinetics of bisphenol A in neonatal and adult Sprague-Dawley rats, *Toxicology and Applied Pharmacology*, 247 (2010) 158–165.
28. D.R. Doerge, N.C. Twaddle, K.A. Woodling, J.W. Fisher, Pharmacokinetics of bisphenol A in neonatal and adult rhesus monkeys, *Toxicology and Applied Pharmacology*, 248 (2010) 1–11.
29. D. Feng, H. Zhang, X. Jiang, J. Zou, Q. Li, H. Mai, D. Su, W. Ling, X. Feng, Bisphenol A exposure induces gut microbiota dysbiosis and consequent activation of gut-liver axis leading to hepatic steatosis in CD-1 mice, *Environmental Pollution*, 265 (2020) 114880.
30. P. Fenichel, N. Chevalier, F. Brucker-Davis, Bisphenol A: An endocrine and metabolic disruptor, *Annales D'endocrinologie*, 74 (2013) 211–220.
31. J.W. Fisher, N.C. Twaddle, M. Vanlandingham, D.R. Doerge, Pharmacokinetic modeling: prediction and evaluation of route dependent dosimetry of bisphenol A in monkeys with extrapolation to humans, *Toxicology and Applied Pharmacology*, 257 (2011) 122–136.
32. P. García-Peñarrubia, A.J. Ruiz-Alcaraz, M. Martínez-Esparza, P. Marín, F. Machado-Linde, Hypothetical roadmap towards endometriosis: prenatal endocrine-disrupting chemical pollutant exposure, anogenital distance, gut-genital microbiota and subclinical infections, *Human Reproduction Update*, 26 (2020) 214–246.
33. V. Gayrard, M.Z. Lacroix, F.C. Grandin, S.H. Collet, H. Mila, C. Viguié, C.A. Gély, B. Rabozzi, M. Bouchard, R. Léandri, P.L. Toutain, N. Picard-Hagen, Oral systemic bioavailability of Bisphenol A and Bisphenol S in pigs, *Environmental Health Perspectives*, 127 (2019) 77005.
34. C. Giommi, M. Lombó, H.R. Habibi, G. Rossi, D. Basili, S. Mangiaterra, C. Ladisa, G. Chemello, O. Carnevali, F. Maradonna, The probiotic SLAB51 as agent to counteract BPA toxicity on zebrafish gut microbiota-liver-brain axis, *The Science of the Total Environment*, 912 (2024) 169303.
35. M.V. Gomez, M. Dutta, A. Suvorov, X. Shi, H. Gu, S. Mani, J. Yue Cui, Early life exposure to environmental contaminants (BDE-47, TBBPA, and BPS) produced persistent alterations in fecal microbiome in adult male mice, *Toxicological Sciences*, 179 (2021) 14–30.
36. J. Gu, Y. Zhu, M. Guo, X. Yin, M. Liang, X. Lou, J. Chen, L. Zhou, D. Fan, L. Shi, G. Hu, G. Ji, The potential mechanism of BPF-induced neurotoxicity in adult zebrafish: Correlation between untargeted metabolomics and gut microbiota, *The Science of the Total Environment*, 839 (2022) 156221.
37. T. Hong, X. Jiang, J. Zou, J. Yang, H. Zhang, H. Mai, W. Ling, D. Feng, Hepatoprotective effect of curcumin against bisphenol A-induced hepatic steatosis via modulating gut microbiota dysbiosis and related gut-liver axis activation in CD-1 mice, *The Journal of Nutritional Biochemistry*, 109 (2022) 109103.
38. T. Hong, J. Zou, Y. He, H. Zhang, H. Liu, H. Mai, J. Yang, Z. Cao, X. Chen, J. Yao, D. Feng, Bisphenol A induced hepatic steatosis by disturbing bile acid metabolism and FXR/TGR5 signaling pathways via remodeling the gut microbiota in CD-1 mice, *The Science of the Total Environment*, 889 (2023) 164307.
39. A.B. Javurek, W.G. Spollen, S.A. Johnson, N.J. Bivens, K.H. Bromert, S.A. Givan, C.S. Rosenfeld, Effects of exposure to bisphenol A and ethinyl estradiol on the gut microbiota of parents and their offspring in a rodent model, *Gut Microbes*, 7 (2016) 471–485.
40. S. Kaur, S.J. Sarma, B.L. Marshall, Y. Liu, J.A. Kinkade, M.M. Bellamy, J. Mao, W.G. Helferich, A.K. Schenk, N.J. Bivens, Z. Lei, L.W. Sumner, J.A. Bowden, J.P. Koelmel, T. Joshi, C.S. Rosenfeld, Developmental exposure of California mice to endocrine disrupting chemicals and potential effects on the microbiome-gut-brain axis at adulthood, *Scientific Reports*, 10 (2020) 10902.
41. C. Ke, C. Meilin, M. Guangzhi, F. Yuqin, L. Lin, C. Weiting, Endocrine disruptors affect the expression of estrogen receptor genes and proteins in the white cloud mountain minnow Tanichthys albonubes (Teleostei: Cyprinidae), *Frontiers in Physiology*, 13 (2022) 1020840.

42. Z.L. Koestel, R.C. Backus, K. Tsuruta, W.G. Spollen, S.A. Johnson, A.B. Javurek, M.R. Ellersieck, C.E. Wiedmeyer, K. Kannan, J. Xue, N.J. Bivens, S.A. Givan, C.S. Rosenfeld, Bisphenol A (BPA) in the serum of pet dogs following short-term consumption of canned dog food and potential health consequences of exposure to BPA, *The Science of the Total Environment*, 579 (2017) 1804–1814.
43. J.P. Lallès, Long term effects of pre- and early postnatal nutrition and environment on the gut, *Journal of Animal Science*, 90 Suppl 4 (2012) 421–429.
44. M.F. Lemos, A.C. Esteves, B. Samyn, I. Timperman, J. van Beeumen, A. Correia, C.A. van Gestel, A.M. Soares, Protein differential expression induced by endocrine disrupting compounds in a terrestrial isopod, *Chemosphere*, 79 (2010) 570–576.
45. Z. Li, W. Wang, J. Li, S. Ru, New insight on the mechanism of eating disorder in females based on metabolic differences of bisphenol S in female and male zebrafish, *Environmental Pollution*, 317 (2023) 120820.
46. R. Thapa, G. Gupta, P. Dave, S.K. Singh, A. Raizaday, W.H. Almalki, G. Vyas, S.K. Singh, K. Dua, Y. Singh, Current update on the protective effect of epicatechin in neurodegenerative diseases, *EXCLI Journal*, 21 (2022) 897–903.
47. J.X. Liao, Y.W. Chen, M.K. Shih, Y.L. Tain, Y.T. Yeh, M.H. Chiu, S.K.C. Chang, C.Y. Hou, Resveratrol Butyrate Esters inhibit BPA-induced liver damage in male offspring rats by modulating antioxidant capacity and gut microbiota, *International Journal of Molecular Sciences*, 22 (2021) 5273.
48. R. Linares, M.F. Fernández, A. Gutiérrez, R. García-Villalba, B. Suárez, P. Zapater, J.A. Martínez-Blázquez, E. Caparrós, F.A. Tomás-Barberán, R. Francés, Endocrine disruption in Crohn's disease: Bisphenol A enhances systemic inflammatory response in patients with gut barrier translocation of dysbiotic microbiota products, *FASEB Journal: Official Publication of the Federation of American Societies for Experimental Biology*, 35 (2021) e21697.
49. R. Liu, D. Cai, X. Li, B. Liu, J. Chen, X. Jiang, H. Li, Z. Li, K. Teerds, J. Sun, W. Bai, Y. Jin, Effects of Bisphenol A on reproductive toxicity and gut microbiota dysbiosis in male rats, *Ecotoxicology and Environmental Safety*, 239 (2022) 113623.
50. R. Thapa, A. Goyal, G. Gupta, A.A. Bhat, S.K. Singh, V. Subramaniyan, S. Sharma, P. Prasher, V. Jakhmola, S.K. Singh, Recent developments in the role of protocatechuic acid in neurodegenerative disorders, *EXCLI Journal*, 22 (2023) 595.
51. R. Liu, B. Liu, L. Tian, X. Jiang, X. Li, D. Cai, J. Sun, W. Bai, Y. Jin, Exposure to Bisphenol A caused hepatoxicity and intestinal flora disorder in rats, *International Journal of Molecular Sciences*, 23 (2022) 8042.
52. Z. Liu, M. Liu, H. Wang, P. Qin, Y. Di, S. Jiang, Y. Li, L. Huang, N. Jiao, W. Yang, Glutamine attenuates bisphenol A-induced intestinal inflammation by regulating gut microbiota and TLR4-p38/MAPK-NF-κB pathway in piglets, *Ecotoxicology and Environmental Safety*, 270 (2024) 115836.
53. C.J. Lodge, S.C. Dharmage, Breastfeeding and perinatal exposure, and the risk of asthma and allergies, *Current Opinion in Allergy and Clinical Immunology,* 16 (2016) 231–236.
54. L.F. Loffredo, M.E. Coden, S. Berdnikovs, Endocrine Disruptor Bisphenol A (BPA) triggers systemic para-inflammation and is sufficient to induce airway allergic sensitization in mice, *Nutrients*, 12 (2020) 343.
55. A. López-Moreno, I. Acuña, A. Torres-Sánchez, Á. Ruiz-Moreno, K. Cerk, A. Rivas, A. Suárez, M. Monteoliva-Sánchez, M. Aguilera, Next generation probiotics for neutralizing obesogenic effects: taxa culturing searching strategies, *Nutrients*, 13 (2021) 1617.
56. A. López-Moreno, Á. Ruiz-Moreno, J. Pardo-Cacho, K. Cerk, A. Torres-Sánchez, P. Ortiz, M. Úbeda, M. Aguilera, Culturing and molecular approaches for identifying microbiota taxa impacting children's obesogenic phenotypes related to xenobiotic dietary exposure, *Nutrients*, 14 (2022) 241.
57. A. López-Moreno, A. Torres-Sánchez, I. Acuña, A. Suárez, M. Aguilera, Representative Bacillus sp. AM1 from gut microbiota harbor versatile molecular pathways for Bisphenol A biodegradation, *International Journal of Molecular Sciences*, 22 (2021) 4952.
58. N. Ma, D. Ma, X. Liu, L. Zhao, L. Ma, D. Ma, S. Dong, Bisphenol P exposure in C57BL/6 mice caused gut microbiota dysbiosis and induced intestinal barrier disruption via LPS/TLR4/NF-κB signaling pathway, *Environment International*, 175 (2023) 107949.
59. L. Maitre, O. Robinson, D. Martinez, M.B. Toledano, J. Ibarluzea, L.S. Marina, J. Sunyer, C.M. Villanueva, H.C. Keun, M. Vrijheid, M. Coen, Urine metabolic signatures of multiple environmental pollutants in pregnant women: An exposome approach, *Environmental Science & Technology*, 52 (2018) 13469–13480.

60. R. Thapa, H. Ali, O. Afzal, A.A. Bhat, W.H. Almalki, S.I. Alzarea, I. Kazmi, A.S.A. Altamimi, N. Jain, M. Pandey, Unlocking the potential of mesoporous silica nanoparticles in breast cancer treatment, *Journal of Nanoparticle Research*, 25 (2023) 169.
61. Y. Malaisé, S. Menard, C. Cartier, E. Gaultier, F. Lasserre, C. Lencina, C. Harkat, N. Geoffre, L. Lakhal, I. Castan, M. Olier, E. Houdeau, L. Guzylack-Piriou, Gut dysbiosis and impairment of immune system homeostasis in perinatally-exposed mice to Bisphenol A precede obese phenotype development, *Scientific Reports*, 7 (2017) 14472.
62. Y. Malaisé, S. Ménard, C. Cartier, C. Lencina, C. Sommer, E. Gaultier, E. Houdeau, L. Guzylack-Piriou, Consequences of bisphenol a perinatal exposure on immune responses and gut barrier function in mice, *Archives of Toxicology*, 92 (2018) 347–358.
63. W. Mao, L. Mao, F. Zhou, J. Shen, N. Zhao, H. Jin, J. Hu, Z. Hu, Influence of gut microbiota on metabolism of Bisphenol A, a major component of polycarbonate plastics, *Toxics*, 11 (2023) 340.
64. M. Marraudino, B. Bonaldo, A. Farinetti, G. Panzica, G. Ponti, S. Gotti, Metabolism disrupting chemicals and alteration of neuroendocrine circuits controlling food intake and energy metabolism, *Frontiers in Endocrinology*, 9 (2018) 766.
65. R. Thapa, O. Afzal, A.A. Bhat, A. Goyal, A.S. Alfawaz Altamimi, W.H. Almalki, S.I. Alzarea, I. Kazmi, S.K. Singh, K. Dua, New horizons in lung cancer management through ATR/CHK1 pathway modulation, *Future Medicinal Chemistry*, 15 (2023) 1807–1818.
66. C.M. McDonough, H.S. Xu, T.L. Guo, Toxicity of bisphenol analogues on the reproductive, nervous, and immune systems, and their relationships to gut microbiome and metabolism: insights from a multi-species comparison, *Critical Reviews in Toxicology*, 51 (2021) 283–300.
67. X. Mu, Z. Liu, X. Zhao, L. Chen, Q. Jia, C. Wang, T. Li, Y. Guo, J. Qiu, Y. Qian, Bisphenol analogues induced social defects and neural impairment in zebrafish, *The Science of the Total Environment,* 899 (2023) 166307.
68. R. Thapa, O. Afzal, A.S.A. Altamimi, A. Goyal, W.H. Almalki, S.I. Alzarea, I. Kazmi, V. Jakhmola, S.K. Singh, K. Dua, Galangin as an inflammatory response modulator: An updated overview and therapeutic potential, *Chemico-Biological Interactions*, 378 (2023) 110482.
69. X. Mu, Z. Liu, X. Zhao, L. Yuan, Y. Li, C. Wang, G. Xiao, J. Mu, J. Qiu, Y. Qian, Bisphenol A analogues induce neuroendocrine disruption via gut-brain regulation in zebrafish, *Environmental Science & Technology*, 58 (2024) 1022–1035.
70. X. Mu, S. Qi, H. Wang, L. Yuan, C. Wang, Y. Li, J. Qiu, Bisphenol analogues induced metabolic effects through eliciting intestinal cell heterogeneous response, *Environment International*, 165 (2022) 107287.
71. G.H. Ng, H. Xu, N. Pi, B.C. Kelly, Z. Gong, Differential GFP expression patterns induced by different heavy metals in Tg(hsp70:gfp) transgenic medaka (Oryzias latipes), *Marine Biotechnology*, 17 (2015) 317–327.
72. H. Otsuka, M. Sugimoto, S. Ikeda, S. Kume, Effects of bisphenol A administration to pregnant mice on serum Ca and intestinal Ca absorption, *Animal Science Journal = Nihon chikusan Gakkaiho*, 83 (2012) 232–237.
73. H.B. Patisaul, Endocrine disruption by dietary phyto-oestrogens: impact on dimorphic sexual systems and behaviours, *The Proceedings of the Nutrition Society*, 76 (2017) 130–144.
74. R. Thapa, O. Afzal, M. Afzal, G. Gupta, A.A. Bhat, I. Kazmi, S.I. Alzarea, S. Saleem, P. Arora, S.K. Singh, From LncRNA to metastasis: The MALAT1-EMT axis in cancer progression, *Pathology-Research and Practice*, 253 (2023) 154959.
75. H.B. Patisaul, W. Jefferson, The pros and cons of phytoestrogens, *Frontiers in neuroendocrinology*, 31 (2010) 400–419.
76. B. Peng, H. Zhao, T.P. Keerthisinghe, Y. Yu, D. Chen, Y. Huang, M. Fang, Gut microbial metabolite p-cresol alters biotransformation of bisphenol A: Enzyme competition or gene induction?, *Journal of Hazardous Materials*, 426 (2022) 128093.
77. C. Pinteur, B. Julien, N. Véga, H. Vidal, D. Naville, B. Le Magueresse-battistoni, impact of estrogen withdrawal and replacement in female mice along the intestinal tract. comparison of E2 replacement with the effect of a mixture of low dose pollutants, *International Journal of Environmental Research and Public Health*, 18 (2021) 8685.
78. R. Wadhwa, K.R. Paudel, M. Mehta, S.D. Shukla, K. Sunkara, P. Prasher, N. Panth, R. Goyal, D.K. Chellappan, G. Gupta, P.M. Hansbro, A.A.A. Aljabali, M.M. Tambuwala, K. Dua, Beyond the obvious: smoking and respiratory infection implications on Alzheimer's disease, *CNS & Neurological Disorders Drug Targets*, 19 (2020) 698–708.

79. I. Porreca, L. Ulloa-Severino, P. Almeida, D. Cuomo, A. Nardone, G. Falco, M. Mallardo, C. Ambrosino, Molecular targets of developmental exposure to bisphenol A in diabesity: a focus on endoderm-derived organs, *Obesity Reviews: An Official Journal of the International Association for the Study of Obesity*, 18 (2017) 99–108.
80. N. Singla, R. Thapa, R. Kulshrestha, A.A. Bhat, S. Gupta, M. Purohit, S.K. Singh, G. Gupta, *Introduction to Epigenetics, Targeting Epigenetics in Inflammatory Lung Diseases,* Springer, 2023, pp. 17–41.
81. L. Reddivari, D.N.R. Veeramachaneni, W.A. Walters, C. Lozupone, J. Palmer, M.K.K. Hewage, R. Bhatnagar, A. Amir, M.J. Kennett, R. Knight, J.K.P. Vanamala, Perinatal Bisphenol A exposure induces chronic inflammation in rabbit offspring via modulation of gut bacteria and their metabolites, *mSystems*, 2 (2017) 10–128.
82. M. Rönn, L. Lind, J. Örberg, J. Kullberg, S. Söderberg, A. Larsson, L. Johansson, H. Ahlström, P.M. Lind, Bisphenol A is related to circulating levels of adiponectin, leptin and ghrelin, but not to fat mass or fat distribution in humans, *Chemosphere*, 112 (2014) 42–48.
83. C.S. Rosenfeld, Xenoestrogen effects on the gut microbiome, *Current Opinion in Endocrine and Metabolic Research*, 19 (2021) 41–45.
84. P. Sharma, P. Bhagat, M.B. Mandal, T.B. Singh, Assessment of awareness regarding health hazards of plastic chemicals and their warning label among a sample population of varanasi city: a cross-sectional study, *Journal of Pharmacy & Bioallied Sciences*, 13 (2021) S1428–S1433.
85. M.S. Hussain, G. Gupta, A. Goyal, R. Thapa, W.H. Almalki, I. Kazmi, S.I. Alzarea, S. Fuloria, D.U. Meenakshi, V. Jakhmola, From nature to therapy: Luteolin's potential as an immune system modulator in inflammatory disorders, *Journal of Biochemical and Molecular Toxicology*, 37 (2023) e23482.
86. M.K. Shih, Y.L. Tain, Y.W. Chen, W.H. Hsu, Y.T. Yeh, S.K.C. Chang, J.X. Liao, C.Y. Hou, Resveratrol butyrate esters inhibit obesity caused by perinatal exposure to bisphenol a in female offspring rats, *Molecules*, 26 (2021) 4010.
87. D. Talukdar, P. Kumar, D. Sharma, V.M. Balaramnavar, O. Afzal, A.S.A. Altamimi, I. Kazmi, F.A. Al-Abbasi, S.I. Alzarea, G. Gupta, M.M. Gupta, Anticancer phytochemical-based nanoformulations: therapeutic intervention in cancer cell lines, *Journal of Environmental Pathology, Toxicology and Oncology,* 42 (2023) 79–93.
88. Y.Y. Tan, P.K. Yap, G.L. Xin Lim, M. Mehta, Y. Chan, S.W. Ng, D.N. Kapoor, P. Negi, K. Anand, S.K. Singh, N.K. Jha, L.C. Lim, T. Madheswaran, S. Satija, G. Gupta, K. Dua, D.K. Chellappan, Perspectives and advancements in the design of nanomaterials for targeted cancer theranostics, *Chemico-Biological Interactions*, 329 (2020) 109221.
89. M.J. Smith, R.J.M. Weber, M.R. Viant, Spatially mapping the baseline and Bisphenol-A exposed daphnia magna lipidome using desorption electrospray ionization-mass spectrometry, *Metabolites*, 12 (2022) 33.
90. S.Y. Son, B. Wang, H. Hur, H.H. Kim, S.U. Han, Does bisphenol-A affect alteration of gut microbiome after bariatric/metabolic surgery?: A comparative metagenomic analysis in a long-term high-fat diet induced-obesity rat model, *Annals of Surgical Treatment and Research,* 102 (2022) 342–352.
91. K. Sone, M. Hinago, A. Kitayama, J. Morokuma, N. Ueno, H. Watanabe, T. Iguchi, Effects of 17beta-estradiol, nonylphenol, and bisphenol-A on developing Xenopus laevis embryos, *General and Comparative Endocrinology*, 138 (2004) 228–236.
92. G. Gupta, M.S. Hussain, R. Thapa, R. Dahiya, D.K. Mahapatra, A.A. Bhat, N. Singla, V. Subramaniyan, S. Rawat, V. Jakhmola, *Hope on the horizon: Wharton's jelly mesenchymal stem cells in the fight against COVID-19,* Future Medicine Ltd London, UK, 2023, pp. 675–678.
93. M.M. Gupta, R. Gilhotra, D. Deopa, A.A. Bhat, R. Thapa, N. Singla, R. Kulshrestha, G. Gupta, *Epigenetics of Pulmonary Tuberculosis, Targeting Epigenetics in Inflammatory Lung Diseases,* Springer, 2023, pp. 127–144.
94. D. Song, Y. Chen, B. Wang, D. Li, C. Xu, H. Huang, S. Huang, R. Liu, Bisphenol A inhibits autophagosome-lysosome fusion and lipid droplet degradation, *Ecotoxicology and Environmental Safety*, 183 (2019) 109492.
95. S. Rohilla, M. Singh, S. Priya, W.H. Almalki, S.M. Haniffa, V. Subramaniyan, S. Fuloria, N.K. Fuloria, M. Sekar, S.K. Singh, N.K. Jha, D.K. Chellappan, P. Negi, K. Dua, G. Gupta, Exploring the mechanical perspective of a new anti-tumor agent: melatonin, *Journal of Environmental Pathology, Toxicology and Oncology,* 42 (2023) 1–16.
96. P. Sharma, M. Mehta, D.S. Dhanjal, S. Kaur, G. Gupta, H. Singh, L. Thangavelu, S. Rajeshkumar, M. Tambuwala, H.A. Bakshi, D.K. Chellappan, K. Dua, S. Satija, Emerging trends in the novel drug delivery approaches for the treatment of lung cancer, *Chemico-Biological Interactions*, 309 (2019) 108720.

97. A. Torres-Sánchez, J. Pardo-Cacho, A. López-Moreno, Á. Ruiz-Moreno, K. Cerk, M. Aguilera, Antimicrobial effects of potential probiotics of bacillus spp. isolated from human microbiota: in vitro and in silico methods, *Microorganisms*, 9 (2021) 1615.
98. M. Vacca, F.M. Calabrese, F. Loperfido, B. Maccarini, R.M. Cerbo, E. Sommella, E. Salviati, L. Voto, M. De Angelis, G. Ceccarelli, I. Di Napoli, B. Raspini, D. Porri, E. Civardi, F. Garofoli, P. Campiglia, H. Cena, R. De Giuseppe, Maternal exposure to endocrine-disrupting chemicals: analysis of their impact on infant gut microbiota composition, *Biomedicines*, 12 (2024) 234.
99. K. Wang, L. Qiu, J. Zhu, Q. Sun, W. Qu, Y. Yu, Z. Zhao, Y. Yu, G. Shao, Environmental contaminant BPA causes intestinal damage by disrupting cellular repair and injury homeostasis in vivo and in vitro, *Biomedicine & Pharmacotherapy = Biomedecine & Pharmacotherapie,* 137 (2021) 111270.
100. M.S. Hussain, G. Gupta, M. Afzal, S.M. Alqahtani, V.P. Samuel, W. Hassan Almalki, I. Kazmi, S.I. Alzarea, S. Saleem, H. Dureja, S.K. Singh, K. Dua, L. Thangavelu, Exploring the role of lncrna neat1 knockdown in regulating apoptosis across multiple cancer types: a review, *Pathology, Research and Practice*, 252 (2023) 154908.
101. W. Wang, X. Zhang, J. Qin, P. Wei, Y. Jia, J. Wang, S. Ru, Long-term bisphenol S exposure induces fat accumulation in liver of adult male zebrafish (Danio rerio) and slows yolk lipid consumption in F1 offspring, *Chemosphere*, 221 (2019) 500–510.
102. Y. Wang, M. Rui, Y. Nie, G. Lu, Influence of gastrointestinal tract on metabolism of bisphenol A as determined by in vitro simulated system, *Journal of Hazardous Materials*, 355 (2018) 111–118.
103. A.K. Ramanunny, S. Wadhwa, M. Gulati, S. Vishwas, R. Khursheed, K.R. Paudel, S. Gupta, O. Porwal, S.M. Alshahrani, N.K. Jha, D.K. Chellappan, P. Prasher, G. Gupta, J. Adams, K. Dua, D. Tewari, S.K. Singh, Journey of Alpinia galanga from kitchen spice to nutraceutical to folk medicine to nanomedicine, *Journal of Ethnopharmacology*, 291 (2022) 115144.
104. C. Rathore, C. Hemrajani, A.K. Sharma, P.K. Gupta, N.K. Jha, A.A.A. Aljabali, G. Gupta, S.K. Singh, J.C. Yang, R.P. Dwivedi, K. Dua, D.K. Chellappan, P. Negi, M.M. Tambuwala, Self-nanoemulsifying drug delivery system (SNEDDS) mediated improved oral bioavailability of thymoquinone: optimization, characterization, pharmacokinetic, and hepatotoxicity studies, *Drug Delivery and Translational Research*, 13 (2023) 292–307.
105. J. Wu, T. Zhou, H. Shen, Y. Jiang, Q. Yang, S. Su, L. Wu, X. Fan, M. Gao, Y. Wu, Y. Cheng, Y. Qi, T. Lei, Y. Xin, S. Han, X. Li, Y. Wang, Mixed probiotics modulated gut microbiota to improve spermatogenesis in bisphenol A-exposed male mice, *Ecotoxicology and Environmental Safety*, 270 (2024) 115922.
106. J. Xu, G. Huang, T.L. Guo, Developmental bisphenol A exposure modulates immune-related diseases, *Toxics*, 4 (2016) 23.
107. J. Xu, G. Huang, T. Nagy, T.L. Guo, Bisphenol A alteration of type 1 diabetes in non-obese diabetic (NOD) female mice is dependent on window of exposure, *Archives of Toxicology*, 93 (2019) 1083–1093.
108. C. Yang, Y. Tan, F. Li, H. Wang, Y. Lin, F. Lu, H. Zhao, Intestinal microecology of mice exposed to TiO(2) nanoparticles and bisphenol A, *Foods*, 11 (2022) 1696.
109. A.A. Bhat, R. Thapa, A. Goyal, V. Subramaniyan, D. Kumar, S. Gupta, S.K. Singh, K. Dua, G. Gupta, *Curcumin-based nanoformulations as an emerging therapeutic strategy for inflammatory lung diseases*, Newlands Press Ltd London, UK, 2023, pp. 583–586.
110. M. Rahman, M.Z. Ahmad, I. Kazmi, S. Akhter, M. Afzal, G. Gupta, F. Jalees Ahmed, F. Anwar, Advancement in multifunctional nanoparticles for the effective treatment of cancer, *Expert Opinion on Drug Delivery*, 9 (2012) 367–381.
111. M. Rahman, M.Z. Ahmad, I. Kazmi, S. Akhter, M. Afzal, G. Gupta, V.R. Sinha, Emergence of nanomedicine as cancer targeted magic bullets: recent development and need to address the toxicity apprehension, *Current Drug Discovery Technology*, 9 (2012) 319–329.
112. F. Yin, X. Huang, X. Lin, T.F. Chan, K.P. Lai, R. Li, Analyzing the synergistic adverse effects of BPA and its substitute, BHPF, on ulcerative colitis through comparative metabolomics, *Chemosphere*, 287 (2022) 132160.
113. H.A.H. Zaghloul, N.M. El Halfawy, Whole genome analyses of toxicants tolerance genes of Apis mellifera gut-derived Enterococcus faecium strains, *BMC Genomics*, 24 (2023) 479.
114. X. Zhan, L. Fletcher, D. Huyben, H. Cai, S. Dingle, N. Qi, L.A. Huber, B. Wang, J. Li, Choline supplementation regulates gut microbiome diversity, gut epithelial activity, and the cytokine gene expression in gilts, *Frontiers in Nutrition*, 10 (2023) 1101519.
115. M. Pandey, P.X. Wen, G.M. Ning, G.J. Xing, L.M. Wei, D. Kumar, J. Mayuren, M. Candasamy, B. Gorain, N. Jain, G. Gupta, K. Dua, Intraductal delivery of nanocarriers for ductal carcinoma in situ treatment: a strategy to enhance localized delivery, *Nanomedicine*, 17 (2022) 1871–1889.

116. P. Prasher, M. Sharma, S.K. Singh, M. Gulati, D.K. Chellappan, F. Zacconi, G. De Rubis, G. Gupta, J. Sharifi-Rad, W.C. Cho, K. Dua, Luteolin: a flavonoid with a multifaceted anticancer potential, *Cancer Cell International*, 22 (2022) 386.
117. Q. Zhang, Y. He, R. Cheng, Q. Li, Z. Qian, X. Lin, Recent advances in toxicological research and potential health impact of microplastics and nanoplastics in vivo, *Environmental Science and Pollution Research International*, 29 (2022) 40415–40448.
118. Q. Zheng, L. Cui, H. Liao, M. Junaid, Z. Li, S. Liu, D. Gao, Y. Zheng, S. Lu, J. Qiu, J. Wang, Combined exposure to polystyrene nanoplastics and bisphenol A induces hepato- and intestinal-toxicity and disturbs gut microbiota in channel catfish (Ictalurus punctatus), *The Science of the Total Environment*, 891 (2023) 164319.
119. A.A. Bhat, R. Thapa, O. Afzal, N. Agrawal, W.H. Almalki, I. Kazmi, S.I. Alzarea, A.S.A. Altamimi, P. Prasher, S.K. Singh, The pyroptotic role of Caspase-3/GSDME signalling pathway among various cancer: a review, *International Journal of Biological Macromolecules*, 242 (2023) 124832.
120. K. Zhu, Y. Zhao, Y. Yang, Y. Bai, T. Zhao, Icariin alleviates bisphenol A induced disruption of intestinal epithelial barrier by maintaining redox homeostasis in vivo and in vitro, *ACS Omega*, 5 (2020) 20399–20408.
121. M. Zhu, R. Zeng, D. Wu, Y. Li, T. Chen, A. Wang, Research progress of the effects of bisphenol analogues on the intestine and its underlying mechanisms: a review, *Environmental Research*, 243 (2023) 117891.
122. G. Bahl, M.S. Hussain, D.K. Upadhyay, M. Varma, R. Singh, S. Das, Chronic calcific pancreatitis and its association with secondary diabetes mellitus, *Journal of Endocrinology and Metabolism*, 13 (2023) 89–95.
123. G. Bahl, D.K. Upadhyay, M. Varma, R. Singh, S. Das, M.S. Hussain, Chronic calcific pancreatitis presented with secondary diabetes and diabetic ketoacidosis: a case report, *Clinical Diabetology,* 12 (2023) 209–211.
124. Committee opinion No.702: female athlete triad, *Obstetrics and Gynecology*, 129 (2017) e160–e167.
125. A.A. Bhat, G. Gupta, R. Dahiya, R. Thapa, A. Gahtori, M. Shahwan, V. Jakhmola, A. Tiwari, M. Kumar, H. Dureja, CircRNAs: Pivotal modulators of TGF-β signalling in cancer pathogenesis, *Non-Coding RNA Research*, 9 (2024) 277–287.
126. A.A. Bhat, G. Gupta, A. Goyal, R. Thapa, W.H. Almalki, I. Kazmi, S.I. Alzarea, N. Kukreti, M. Sekar, D.U. Meenakshi, Unwinding circular RNA's role in inflammatory pulmonary diseases, *Naunyn-Schmiedeberg's Archives of Pharmacology*, 397 (2023) 2567–2588.
127. A.S. More, R.R. Kumari, G. Gupta, K. Kathirvel, M.K. Lonare, R.S. Dhayagude, D. Kumar, D. Kumar, A.K. Sharma, S.K. Tandan, Effect of S-methylisothiourea in acetaminophen-induced hepatotoxicity in rat, *Naunyn-Schmiedeberg's Archives of Pharmacology*, 385 (2012) 1127–1139.
128. P.Q. Ng, L.S.C. Ling, J. Chellian, T. Madheswaran, J. Panneerselvam, A.P. Kunnath, G. Gupta, S. Satija, M. Mehta, P.M. Hansbro, T. Collet, K. Dua, D.K. Chellappan, Applications of nanocarriers as drug delivery vehicles for active phytoconstituents, *Current Pharmaceutical Design*, 26 (2020) 4580–4590.
129. M.K. Alam, F.B. Ali, R. Banik, S. Yasmin, N. Salma, Assessing the mental health condition of home-confined university level students of Bangladesh due to the COVID-19 pandemic, *Zeitschrift fur Gesundheitswissenschaften = Journal of Public Health*, 30 (2022) 1685–1692.
130. R. Alfuzaie, The link between gastrointestinal microbiome and ocular disorders, *Clinical Ophthalmology*, 17 (2023) 2133–2140.
131. S. Azam, M.E. Haque, I.S. Kim, D.K. Choi, Microglial turnover in ageing-related neurodegeneration: therapeutic avenue to intervene in disease progression, *Cells*, 10 (2021) 150.
132. A.A. Bhat, A. Goyal, R. Thapa, I. Kazmi, S.I. Alzarea, M. Singh, S. Rohilla, T.K. Saini, N. Kukreti, D.U. Meenakshi, Uncovering the complex role of interferon-gamma in suppressing type 2 immunity to cancer, *Cytokine*, 171 (2023) 156376.
133. A.A. Bhat, G. Gupta, M. Afzal, R. Thapa, H. Ali, S.M. Alqahtani, I. Kazmi, S.I. Alzarea, S. Saleem, V. Subramaniyan, Polyphenol-loaded nano-carriers for breast cancer therapy: a comprehensive review, *BioNanoScience*, 14 (2024) 4219–4237.
134. R. Bardhan, K. Heaton, M. Davis, P. Chen, D.A. Dickinson, C.T. Lungu, A cross sectional study evaluating psychosocial job stress and health risk in emergency department nurses, *International Journal of Environmental Research and Public Health*, 16 (2019) 3243.
135. J. Kaur, V. Mishra, S.K. Singh, M. Gulati, B. Kapoor, D.K. Chellappan, G. Gupta, H. Dureja, K. Anand, K. Dua, G.L. Khatik, K. Gowthamarajan, Harnessing amphiphilic polymeric micelles for diagnostic and therapeutic applications: breakthroughs and bottlenecks, *Journal of Controlled Release,* 334 (2021) 64–95.

136. M. Mehta, Deeksha, N. Sharma, M. Vyas, N. Khurana, P.K. Maurya, H. Singh, T.P. Andreoli de Jesus, H. Dureja, D.K. Chellappan, G. Gupta, R. Wadhwa, T. Collet, P.M. Hansbro, K. Dua, S. Satija, Interactions with the macrophages: an emerging targeted approach using novel drug delivery systems in respiratory diseases, *Chemico-Biological Interactions*, 304 (2019) 10–19.
137. M. Kumar, M.S. Hussain, S. Raj, R. Verma, S. Sharma, S.K. Sahu, An overview of treatment modalities and management aspects for obesity, *Current Nutrition and Food Science,* 19 (2023) 105–113.
138. A. Barrech, N. Riedel, J. Li, R.M. Herr, K. Mörtl, P. Angerer, H. Gündel, The long-term impact of a change in Effort-Reward imbalance on mental health-results from the prospective MAN-GO study, *European Journal of Public Health*, 27 (2017) 1021–1026.
139. K.F. Beckers, J.L. Sones, Maternal microbiome and the hypertensive disorder of pregnancy, preeclampsia, *American Journal of Physiology. Heart and Circulatory Physiology,* 318 (2020) H1–H10.
140. M.S. Hussain, A. Sharma, R. Kumar, Prebiotics and probiotics: a focused review of applications in respiratory disorders, *Carpathian Journal of Food Science and Technology*, 15 (2023) 183–207.
141. P. Beschoner, J. von Wietersheim, M.N. Jarczok, M. Braun, C. Schönfeldt-Lecuona, L. Jerg-Bretzke, L. Steiner, Changes in working conditions and mental health among intensive care physicians across a decade, *Frontiers in Psychiatry,* 11 (2020) 145.
142. E.M. Biagioni, L.E. May, N.T. Broskey, The impact of advanced maternal age on pregnancy and offspring health: a mechanistic role for placental angiogenic growth mediators, *Placenta*, 106 (2021) 15–21.
143. G. Gupta, S. Pathak, R. Dahiya, R. Awasthi, A. Mishra, R.K. Sharma, M. Agrawal, K. Dua, Aqueous extract of wood ear mushroom, auricularia polytricha (agaricomycetes), demonstrated antiepileptic activity against seizure induced by maximal electroshock and isoniazid in experimental animals, *International Journal of Medicinal Mushrooms*, 21 (2019) 29–35.
144. M.S. Hussain, A.S.A. Altamimi, M. Afzal, W.H. Almalki, I. Kazmi, S.I. Alzarea, G. Gupta, M. Shahwan, N. Kukreti, L.S. Wong, V. Kumarasamy, V. Subramaniyan, Kaempferol: Paving the path for advanced treatments in aging-related diseases, *Experimental Gerontology,* 188 (2024) 112389.
145. M. Boukhris, A. Hillani, F. Moroni, M.S. Annabi, F. Addad, M.H. Ribeiro, S. Mansour, X. Zhao, L.F. Ybarra, A. Abbate, L.M. Vilca, L. Azzalini, Cardiovascular implications of the COVID-19 pandemic: a global perspective, *The Canadian Journal of Cardiology*, 36 (2020) 1068–1080.
146. H.M.H. Abou El-Naga, S.A. El-Hashash, E.M. Yasen, S. Leporatti, N.A.N. Hanafy, Starch-based hydrogel nanoparticles loaded with polyphenolic compounds of moringa oleifera leaf extract have hepatoprotective activity in bisphenol A-induced animal models, *Polymers*, 14 (2022) 2846.
147. E.O. Adegoke, M.S. Rahman, M.G. Pang, Bisphenols threaten male reproductive health via testicular cells, *Frontiers in Endocrinology*, 11 (2020) 624.
148. Y. Akgul, R.C. Derk, T. Meighan, K.M. Rao, E.P. Murono, The methoxychlor metabolite, HPTE, directly inhibits the catalytic activity of cholesterol side-chain cleavage (P450scc) in cultured rat ovarian cells, *Reproductive Toxicology*, 25 (2008) 67–75.
149. G. Gupta, D.K. Chellappan, I.S. Kikuchi, T.J.A. Pinto, K. Pabreja, M. Agrawal, Y. Singh, J. Tiwari, K. Dua, Nephrotoxicity in rats exposed to paracetamol: the protective role of moralbosteroid, a steroidal glycoside, *Journal of Environmental Pathology, Toxicology and Oncology: Official Organ of the International Society for Environmental Toxicology and Cancer*, 36 (2017) 113–119.
150. G. Gupta, I. Kazmi, M. Afzal, M. Rahman, S. Saleem, M.S. Ashraf, M.J. Khusroo, K. Nazeer, S. Ahmed, M. Mujeeb, Z. Ahmed, F. Anwar, Sedative, antiepileptic and antipsychotic effects of *Viscum album* L. (Loranthaceae) in mice and rats, *Journal of Ethnopharmacology,* 141 (2012) 810–816.
151. G.J. Alekhya Sita, M. Gowthami, G. Srikanth, M.M. Krishna, K. Rama Sireesha, M. Sajjarao, K. Nagarjuna, M. Nagarjuna, G.K. Chinnaboina, A. Mishra, N. SreeHarsha, Protective role of luteolin against bisphenol A-induced renal toxicity through suppressing oxidative stress, inflammation, and upregulating Nrf2/ARE/ HO-1 pathway, *IUBMB Life*, 71 (2019) 1041–1047.
152. A.A. Bhat, O. Afzal, N. Agrawal, R. Thapa, W.H. Almalki, I. Kazmi, S.I. Alzarea, A.S.A. Altamimi, N. Kukreti, A. Chakraborty, A comprehensive review on the emerging role of long non-coding RNAs in the regulation of NF-κB signaling in inflammatory lung diseases, *International Journal of Biological Macromolecules*, 253 (2023) 126951.
153. S. Ali, G. Steinmetz, G. Montillet, M.H. Perrard, A. Loundou, P. Durand, M.R. Guichaoua, O. Prat, Exposure to low-dose bisphenol A impairs meiosis in the rat seminiferous tubule culture model: a physiotoxicogenomic approach, *PloS One*, 9 (2014) e106245.

154. B.M. Angle, R.P. Do, D. Ponzi, R.W. Stahlhut, B.E. Drury, S.C. Nagel, W.V. Welshons, C.L. Besch-Williford, P. Palanza, S. Parmigiani, F.S. vom Saal, J.A. Taylor, Metabolic disruption in male mice due to fetal exposure to low but not high doses of bisphenol A (BPA): evidence for effects on body weight, food intake, adipocytes, leptin, adiponectin, insulin and glucose regulation, *Reproductive Toxicology*, 42 (2013) 256–268.
155. N. Arnich, M.C. Canivenc-Lavier, M. Kolf-Clauw, H. Coffigny, J.P. Cravedi, K. Grob, A.C. Macherey, D. Masset, R. Maximilien, J.F. Narbonne, F. Nesslany, J. Stadler, J. Tulliez, Conclusions of the french food safety agency on the toxicity of bisphenol A, *International Journal of Hygiene and Environmental Health*, 214 (2011) 271–275.
156. E. Balcázar-López, L.H. Méndez-Lorenzo, R.A. Batista-García, U. Esquivel-Naranjo, M. Ayala, V.V. Kumar, O. Savary, H. Cabana, A. Herrera-Estrella, J.L. Folch-Mallol, Xenobiotic compounds degradation by heterologous expression of a trametes sanguineus laccase in trichoderma atroviride, *PloS One*, 11 (2016) e0147997.
157. S.A. Baluka, W.K. Rumbeiha, Bisphenol A and food safety: Lessons from developed to developing countries, *Food and Chemical Toxicology: An International Journal Published for the British Industrial Biological Research Association*, 92 (2016) 58–63.
158. V.K.R. Balaga, A. Pradhan, R. Thapa, N. Patel, R. Mishra, N. Singla, Morin: a comprehensive review on its versatile biological activity and associated therapeutic potential in treating cancers, *Pharmacological Research*, 7 (2023) 100264.
159. A.A. Bhat, O. Afzal, M. Afzal, G. Gupta, R. Thapa, H. Ali, I. Kazmi, S.I. Alzarea, S. Saleem, V.P. Samuel, MALAT1: A key regulator in lung cancer pathogenesis and therapeutic targeting, *Pathology-Research and Practice*, 253 (2023) 154991.
160. E. Baralla, V. Pasciu, M.V. Varoni, M. Nieddu, R. Demuro, M.P. Demontis, Bisphenols' occurrence in bivalves as sentinel of environmental contamination, *The Science of the Total Environment*, 785 (2021) 147263.
161. R.K. Gautam, S. Sharma, K. Sharma, G. Gupta, Evaluation of antiarthritic activity of butanol fraction of punica granatum linn. rind extract against freund's complete adjuvant-induced arthritis in rats, *Journal of Environmental Pathology, Toxicology and Oncology*, 37 (2018) 53–62.
162. G. Gupta, D.K. Chellappan, M. Agarwal, M. Ashwathanarayana, S. Nammi, K. Pabreja, K. Dua, Pharmacological evaluation of the recuperative effect of morusin against aluminium trichloride (AlC13)-induced memory impairment in rats, *Central Nervous System Agents in Medicinal Chemistry*, 17 (2017) 196–200.
163. L. Barraza, A new approach for regulating bisphenol A for the protection of the public's health, *The Journal of Law, Medicine & Ethics*, 41 Suppl 1 (2013) 9–12.
164. A. Besaratinia, The state of research and weight of evidence on the epigenetic effects of bisphenol A, *International Journal of Molecular Sciences*, 24 (2023) 7951.
165. M. Boas, K.M. Main, U. Feldt-Rasmussen, Environmental chemicals and thyroid function: an update, *Current Opinion in Endocrinology, Diabetes, and Obesity,* 16 (2009) 385–391.
166. M. Bock, K. Fetters, L. Tappert, D. Hoehne, M. Bunge, S. Tenbrock, G. Mueller, S. Gestermann, Bisphenol A in German watersheds: Part II. FlowEQ model-based characterization of sources and current and future conditions, *Integrated Environmental Assessment and Management*, 20 (2024) 226–238.
167. P. Das, G. Gupta, V. Velu, R. Awasthi, K. Dua, H. Malipeddi, Formation of struvite urinary stones and approaches towards the inhibition-A review, *Biomedicine & Pharmacotherapy = Biomedecine & Pharmacotherapie,* 96 (2017) 361–370.
168. R.K. Gautam, G. Gupta, S. Sharma, K. Hatware, K. Patil, K. Sharma, S. Goyal, D.K. Chellappan, K. Dua, Rosmarinic acid attenuates inflammation in experimentally induced arthritis in Wistar rats, using Freund's complete adjuvant, *International Journal of Rheumatic Diseases*, 22 (2019) 1247–1254.
169. J.G. Boucher, S. Ahmed, E. Atlas, Bisphenol S induces adipogenesis in primary human preadipocytes from female donors, *Endocrinology*, 157 (2016) 1397–1407.
170. S. Bucher, P. Jalili, D. Le Guillou, K. Begriche, K. Rondel, S. Martinais, D. Zalko, A. Corlu, M.A. Robin, B. Fromenty, Bisphenol a induces steatosis in HepaRG cells using a model of perinatal exposure, *Environmental Toxicology*, 32 (2017) 1024–1036.
171. M. Abubakar, E.R. Puppala, B.J. Dutta, K.C. Maharana, R. Thapa, S. Roshan, B. Tazneem, A. Khan, A.A. Bhat, Epigenetics of rhinovirus, targeting epigenetics in inflammatory lung diseases, Springer Nature Singapore, 2023, pp. 185–197.

172. F.A. Bustamante-Barrientos, M. Méndez-Ruette, A. Ortloff, P. Luz-Crawford, F.J. Rivera, C.D. Figueroa, L. Molina, L.F. Bátiz, The impact of estrogen and estrogen-like molecules in neurogenesis and neurodegeneration: beneficial or harmful?, *Frontiers in Cellular Neuroscience*, 15 (2021) 636176.
173. C. Caglayan, F.M. Kandemir, A. Ayna, C. Gür, S. Küçükler, E. Darendelioğlu, Neuroprotective effects of 18β-glycyrrhetinic acid against bisphenol A-induced neurotoxicity in rats: involvement of neuronal apoptosis, endoplasmic reticulum stress and JAK1/STAT1 signaling pathway, *Metabolic Brain Disease*, 37 (2022) 1931–1940.
174. S. Chakraborty, S. Anand, S. Coe, B. Reh, R.K. Bhandari, The PCOS-NAFLD multidisease phenotype occurred in medaka fish four generations after the removal of bisphenol A exposure, *Environmental Science & Technology*, 57 (2023) 12602–12619.
175. C. Champmartin, F. Marquet, L. Chedik, M.J. Décret, M. Aubertin, E. Ferrari, M.C. Grandclaude, F. Cosnier, Human in vitro percutaneous absorption of bisphenol S and bisphenol A: a comparative study, *Chemosphere*, 252 (2020) 126525.
176. D. Chen, K. Kannan, H. Tan, Z. Zheng, Y.L. Feng, Y. Wu, M. Widelka, Bisphenol analogues other than BPA: environmental occurrence, human exposure, and toxicity-A review, *Environmental Science & Technology*, 50 (2016) 5438–5453.
177. K.S. Alharbi, W.H. Almalki, M. Albratty, A.M. Meraya, A. Najmi, G. Vyas, S.K. Singh, K. Dua, G. Gupta, The therapeutic role of nutraceuticals targeting the Nrf2/HO-1 signaling pathway in liver cancer, *Journal of Food Biochemistry,* 46 (2022) e14357.
178. D.K. Chellappan, L.W. Yee, K.Y. Xuan, K. Kunalan, L.C. Rou, L.S. Jean, L.Y. Ying, L.X. Wie, J. Chellian, M. Mehta, S. Satija, S.K. Singh, M. Gulati, H. Dureja, M.W. Da Silva, M.M. Tambuwala, G. Gupta, K.R. Paudel, R. Wadhwa, P.M. Hansbro, K. Dua, Targeting neutrophils using novel drug delivery systems in chronic respiratory diseases, *Drug Development Research*, 81 (2020) 419–436.

9 Effect of Lead on Gut Microbiome

Yati Sharma[1], Vibhav Varshney[1], Ahsas Goyal[1], and Neetu Agrawal[1]*

[1]Institute of Pharmaceutical Research, GLA University, Mathura, U.P., India

*Corresponding Author: Ahsas Goyal.
Email id: ahsasgoyal1990@gmail.com

9.1 INTRODUCTION

Lead (Pb) is one of the unnecessary heavy metals and a common environmental contaminant that poses a significant risk to human health. Because of the many vulnerable population groups it impacts, the significant environmental damage it causes, and its exceptional ability to produce a broad variety of unfavorable effects, this universal metal specifically presents a serious danger to worldwide public health [1]. It is recognized that most urban soils have lead contamination. Research indicates that lead concentrations in urban soils have a central tendency that is higher than the ambient background value, with values ranging from 100 to 1,000 mg/kg [2]. In recent years, lead exposure in America has dramatically reduced due to the phase-out of lead in gasoline and the limitation of lead concentration in paint. However, Pb exposure in people cannot be disregarded [3]. The average blood lead content (Pb) in children was found to be greater in countries with lower and intermediate economic statuses than in the United States (0.84 μg/dl). These countries include Egypt (6 μg/dl), India (5.46 μg/dl), and China (3.71 μg/dl)[4]. Furthermore, even extremely low levels of Pb exposure (2 μg/dl) may be hazardous to one's health. Therefore, there is an urgent need for further decline [5]. Exposure to lead has a substantial impact on society since it causes many illnesses. Compared to adults, children are more susceptible to this heavy metal's notable neurotoxic effects and difficulties with cognitive development. Lead poisoning may cause problems with the nervous, hematological, circulatory, cardiac, reproductive, digestive, immunological, and renal systems [6]. According to scholarly estimations, lead exposure significantly raises mortality in the United States by 18%. Like all other dense metals, lead is known to cause general metal toxicity in terms of its effects on target organs. Pb lowers immune response, oxidative stress, inflammation, and critical metal dyshomeostasis, according to several studies [7]. A recent study revealed that the metal interferes with hormones. Unfortunately, they may not address the processes behind lead poisoning; further research is still needed to assess and mitigate the effects of the metal [8]. Given its importance in several physiological processes, the gut microbiota is gaining recognition as our "second genome" and is being investigated more extensively. Inflammation, immunology, and host metabolic regulation are responses that are closely linked to health and illness [9]. Dysbiosis, or changes in the gut microbiome, may be a mediating factor in a number of illnesses, including diabetes, intestinal disorders, cancer, obesity, cardiovascular disease, and allergies [10]. According to recent studies, a variety of environmental contaminants, including pesticides, antibacterial agents, nanomaterials, air pollutants, endocrine disruptors, fungal toxins, minute plastic particles, and dense metals, may target the gut microbiota [11]. On the other hand, there is no information on the potential involvement of dysregulated gut microorganisms in lead poisoning. Therefore, the main focus of this work has been to investigate the effects of lead exposure on gut microbiota (Figure 9.1), relevant host health, and intestinal structure and function [12]. Additionally, we calculated the data

DOI: 10.1201/9781003489221-9

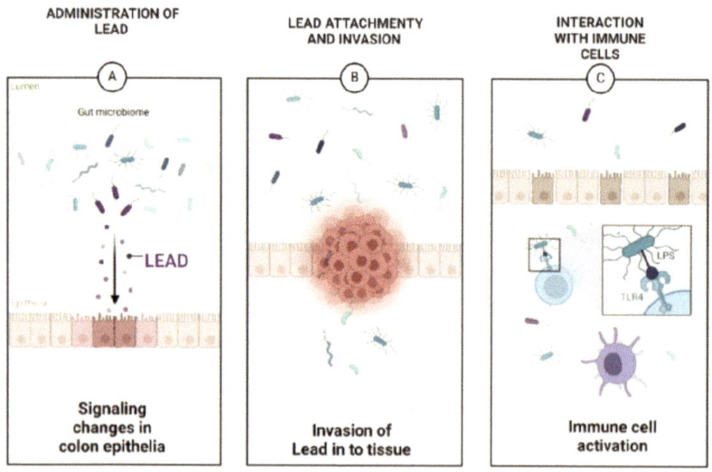

FIGURE 9.1 This figure illustrates the impact of lead exposure on the gut microbiome. Lead, a toxic heavy metal, disrupts the delicate balance of microbial communities inhabiting the gastrointestinal tract. Through mechanisms such as altered microbial composition, reduced diversity, and dysregulated metabolic functions, lead exposure adversely affects gut health.

to determine if probiotic treatment may be a helpful intervention for lead poisoning and whether the microbial population in the digestive system could be thought of as a conduit for the harmful effects of lead that travel to other organs [13].

9.2 EFFECT OF PB ON INTESTINAL MICROBIOTA

The gut microbiota is indeed an "organ" in and of itself, as has been proven. It is startling to learn that it has 150 times more genes than the human genome and is composed of more than 10^14 bacteria[14]. The two main phyla that make up the human gut microbiota are Firmicutes and Bacteroidetes. Experts cannot agree on the relevance of the F/B ratio (Firmicutes vs. Bacteroidetes) in connection to human health [15]. A growing body of research emphasizes the crucial role intestinal microbiota plays in the emergence of a range of diseases, including cancer, neurological conditions, diabetes, difficulties associated with weight, cardiovascular diseases, and gastrointestinal disorders. It is important to remember that gut flora protect the body against lead poisoning and other heavy metal toxicity [16]. A growing number of studies indicates that metabolites generated by microorganisms, such as liposaccharides (LPS), vitamins, derivatives of amino acids, SCFAs, and BA, are essential signaling molecules that connect the gut microbiota to host responses. Exposure to lead may alter the composition, diversity, and associated metabolites of the gut microbiota [17].

9.2.1 Alteration of Gut Microbiota by Pb

Short-term Pb exposure has sometimes been shown to have immediate effects on the gut flora. In an initial investigation, adult rats exposed to oral Pb for two weeks showed a large rise in the lactose-negative Escherichia coli population in the gut microbiota [18]. It is crucial to understand that Pb therapy affects the gut microbiota's phylum, order, and genus levels. The gut microbiota of zebrafish was shown to undergo significant changes after seven days of exposure to Pb (30 μg/L) [19]. These changes included a significant increase in Firmicutes and Bacteroidetes, a significant decrease in Fusobacteria and Proteobacteria, and a notable shift in microbial abundance and variety. Additional liver metabolomics studies identified changes in 41 metabolites associated with the metabolism of glucose, lipids, amino acids, and nucleotides [20]. Gk, Aco, Accl, Fas, Apo, and Dgat are among the genes linked to glycolysis and lipid metabolism that showed a consistent decrease in expression after a 14-day exposure to Pb (500 mg/kg) via water intake. Additionally, zebrafish showed a discernible increase in the phylum Alphaproteobacteria and a decrease in Gammaproteobacteria. The relative prevalence of Alteromonadales was much lower at the order level [21]. *Pseudomonas*, *Halomonadaceae*, *Arcobacter*, and *Polaribacter* showed a rise in proportional frequency in the gut microbial community at the genus level. It is interesting to note that lead-tolerant strains of *Pseudomonas* have been shown to exist in lead-contaminated soil and water, suggesting that zebrafish may have some Pb-resistant bacteria [22]. Compared to acute Pb exposure, long-term Pb exposure is less prevalent. It has been shown that prolonged exposure to lead changes the richness and diversity of bacteria, which may contribute to dysbiosis of the gut microbiota [23]. Adult mice that drank water contaminated with two parts per million lead during lactation and gestation showed an increase in cultivable anaerobes and a decrease in cultivable aerobic bacteria in their waste [24]. Exposure to Pb may have partially boosted the relative ratio of Firmicutes to Bacteroidetes (F/B), as seen by the significant increase in the proportionate predominance of Firmicutes and the opposite pattern displayed by Bacteroidetes. According to certain viewpoints, a very high F/B ratio is a key sign of obesity and lipid metabolic disease [25]. Moreover, the scientists found that early life exposure to lead changed six taxonomic groupings altogether: three genera (*Lactococcus*, *Enterorhabdus*, and *Caulobacterales*) exhibited declines, while three genera (*Desulfovibrionaceae*, *Barnesiella*, and *Clostridium* XIVb) showed increases. It has been shown that desulfovibrio converts choline to trimethylamine (TMA), which the liver then oxidizes to produce TMA N-oxide (TMAO) [26]. The buildup of TMAO and colon and cardiovascular tumors are well correlated. Research has shown that one kind of probiotic that is often present in food is Lactococcus [27]. Further correlational analysis revealed a substantial relationship between these microbial community alterations and an increase in body mass, but only in men. However, after 49 days of exposure to Pb (1,000 ppm), the F/B ratio clearly showed that the gut microbiota of Japanese quails had decreased [28]. The gut bacterial communities showed distinct genus-level characteristics, including immunological function and impaired intestinal structure, along with a notable increase in the proportionate predominance of *Bacteroides* and a decrease in *Faecalibacterium* and *Bifidobacteria* [29]. Studies with contentious F/B ratio figures have also been made public. Therefore, further research is needed to determine if the F/B ratio is a reliable indicator of the harmful consequences of lead [30]. Researchers examined the microbes found in the cecum and feces of Balb/C mice exposed to drinking water at ambient values of 100 and 500 ppm of lead for eight weeks. 16S rRNA pyrosequencing was used to identify subtle but noticeable alterations in the gut microbial population at the family and genus levels [31]. The mice given Pb had more *Lactobacillaceae* and *Erysipelotrichaceae* in their feces and cecum than the control group did, but less Lachnospiraceae [32]. The study also demonstrated a significant decrease in *Turicibacter* and an increase in *Ruminococcus* at the genus level in the intestinal microbiome of C57BL/6 mice that were given Pb (1 g/L) in their drinking water for eight weeks [33]. However, the changes in Firmicutes and Bacteroidetes did not provide any significant outcomes. It was also shown that year that Kunming mice exposed to the same duration and concentration of Pb had significant alterations in their gut flora [34]. These included alterations in the structure and function of the kidney and liver, cognitive decline, and a considerable rise in the *Lachnospiraceae*_NK4A136_group and a decrease in the *Helicobacter* population [35]. The different types of mice used in these two studies may have contributed to the lack of consistent alterations in the microorganisms [36]. Zhai's study demonstrated that alteration in the intestinal microbiome induced by Pb exposure was much more obvious in the first four weeks as compared with the latter four weeks. Additionally, the researchers hypothesized that the gut flora is going through a Pb-sensitive phase. The fact that exposure to lead dramatically reduced the population of several microbes was another interesting finding [37]. For example, after eight weeks of exposure to lead, eight basic Operational Taxonomic Units (OTUs) showed a significant reduction, while no taxa showed any significant upregulation [38, 39]. According to Monast's results, long-term lead exposure increased stomach pain and decreased the relative frequency of strains from the Ruminococcaceae family [40]. Prolonged exposure to lead also decreased the proportionate predominance of strains from the Ruminococcaceae family, which may have led to increased gastrointestinal discomfort. Notably downregulated was the amount of *Akkermansia*, which is thought to be a sign of colitis [41]. Intestinal inflammation, colitis, and other gut illnesses have previously been linked to modifications in these basic gut strains. For instance, studies have shown that some strains of *Oscillibacter* produce valeric acid and may have a negative correlation with gastrointestinal issues [42, 43]. Specific *Lachnoclostridium* strains aided in the production of secondary bile acid. Furthermore, research has shown that individuals with IBD have a significant decrease in *Ruminoclostridium* abundance [44]. Furthermore, there was a noticeable decrease in the diversity of the gut microbiome after six weeks of exposure to Pb (1 mg/L), particularly in the lactic acid bacteria (LAB) population of *Cyprinus carpio* fish. Researchers examined the fecal microbiota of mice after they were exposed to Pb (10 ppm) in their drinking water for 0, 4, and 13 weeks using 16S rRNA sequencing and metagenomic analysis [45]. As expected, significant changes were seen in the microbiome's community structures (as measured by β-diversity metrics) and α-diversity at 4 and 13 weeks. Moreover, the changes in gut flora were almost in agreement with the study results [46]. For 15 weeks, mice given modest lead concentrations (0.1 mg/L Pb) showed significant alterations in their β-diversity [47]. More specifically, the Firmicutes/Bacteroidetes (F/B) ratio decreased as a result of a drop in Firmicutes relative abundance and an increase in Bacteroidetes [48]. There was a decrease in *Dehalobacterium* strains and an increase in *Parabacteroides* strains at the genus level. Opportunistic bacteria called *Parabacteroides* are often linked to infectious diseases, including bacteremia and intra-abdominal processes [49]. Moreover, several genes linked to hepatic lipid metabolism were seen in combination with increased liver triglyceride (TG) and total cholesterol (TCH) levels. These results point to a strong link between altered gut bacteria and abnormalities in liver function [50].

Lead exposure may possibly disturb the balance of human gut microbes. A growing body of research has shown that lead exposure significantly alters the gut flora of humans, including adults and children [51]. Researchers discovered that pregnant women and schoolchildren in Tanzania had higher proportions of Succinivibrionaceae and Gamma proteobacteria in their stools when their blood lead levels were greater [52]. In a recent study, 696 adult participants in the Survey of the Health of Wisconsin (SHOW) had their urine lead concentrations and the makeup of their gut microbiota examined. Urinary lead levels were shown to be positively linked with both unique microbial β-diversity and gut microbiome richness and α-diversity [53, 54]. More precisely, a high correlation was discovered between elevated urine Pb levels and an increased proportion of Proteobacteria population, including Burkholderiales. Using 16S rRNA sequencing, we examined the gut microbiota of people living near a mining and smelting activity zone [53]. Extended contact with certain metals, including lead (Pb), changed the makeup and variety of microbes. Furthermore, the feces of the different metal exposure subjects included higher amounts of *Lachnospiraceae*, *Eubacterium eligens*, *Ruminococcaceae* UGG014, *Erysipelotrichaceae* UCG-003, *Tyzzerella* 3, *Bacteroides*, *Slackia*, *Pseudomonas*, and *Roseburia* and less *Prevotella* 9 [55]. There were a few more ways that men and women were different. For example, the microbial richness and evenness were higher in individuals who worked in the mining and smelting sectors and were exposed to metals over an extended period of time [56]. According to reports, intestinal inflammation is often linked to the groups *Eubacterium eligens*, Erysipelotrichaceae, and Lachnospiraceae. While Ruminococcaceae is mostly associated with mucosal health and is well recognized as beneficial bacteria, *Roseburia* is associated with immune system maintenance and has anti-inflammatory properties [57].

As a result, extensive research that is comprehensive indicates that exposure to Pb does have different effects on the gut flora in both human and animal models [58]. In general, most earlier

TABLE 9.1

Summary of diverse studies on lead exposure effects on gut microbiota in rats, zebrafish, mice, and Tanzanian populations. Findings include altered diversity, Firmicutes/Bacteroidetes ratio shifts, and specific bacterial changes.

Study Context	Exposure Type	Duration	Species/Subjects	Key Findings
Pb Impact on Rat Gut Microbiota	Short-term oral Pb	2 weeks	Adult rats	Increased lactose-negative E. coli; Phylum, order, genus effects.
Zebrafish Microbiota under Pb Exposure	Pb exposure (30 μg/L)	7 days	Zebrafish	Shifts in Firmicutes, Bacteroidetes; Metabolomic changes.
Long-term Pb in Mice Gestation/Lactation	Pb in water (2 ppm)	Not specified	Adult mice	Altered diversity; Firmicutes/ Bacteroidetes ratio shift.
Pb Effects on Balb/C Mice Microbiome	Pb exposure (100 and 500 ppm)	8 weeks	Balb/C mice	Changes in microbial community; Rise in specific bacteria.
Lead Impact on Tanzanian Gut Microbes	Elevated blood lead levels	Not specified	Tanzanian schoolchildren, pregnant women	Succinivibrionaceae, Gammaproteobacteria rise in feces.

research discovered a significant increase in the Firmicutes/Bacteroidetes (F/B) ratio, which was associated with either a decrease in Bacteroidetes abundance or an increase in Firmicutes predominance [59–61]. On the other hand, a few studies have yielded findings that differ from the majority, while the majority show no appreciable changes [62]. Moreover, after exposure to Pb, some important bacteria, such as *Desulfovibrio*, *Lactococcus*, *Akkermansia*, *Ruminococcus*, and *Turicibacter*, which produce SCFAs, BA, and/or other elements linked to human illness, underwent distinct transformations [63]. The reported differences in Pb effects on certain gut microorganisms may be attributed to a number of factors, including subjects, duration, intensity, and animal models [64]. Table 9.1 summarizes in vivo and vitro studies on lead exposure effects on gut microbiota.

9.2.2 Alteration of Metabolism Associated with Gut Microbiota by Pb

Microbial metabolites may interact with the host system both locally and systemically, having the potential to have profound biological impacts [65]. Consequently, changes in the relative abundance of gut microbiota have an effect on the metabolic profiles of microbes [66, 67]. Bile acids and SCFAs, two important metabolites of gut bacteria, typically bind to cellular receptors and function as signaling molecules. Specifically, bile acids may bind to G-protein-coupled receptors (GPCRs) and nuclear receptors, such as farnesin X receptor (FXR) and GPCR TGR5 [68]. On the other hand, short-chain fatty acids (SCFAs) can bind to GPCRs. Moreover, altered mucin production and secretion as well as lowered intestinal lumen pH are all possible outcomes of elevated short-chain fatty acid (SCFA) levels, especially butyric acid, which might improve intestinal barrier function [69]. Opting for signaling pathways might have profound effects on physiology. Lead exposure has been shown to affect both gut microbes and many metabolic functions (Table 9.2). Researchers examined, using multi-omics techniques, alterations in the microbiota and metabolic profile of C57 BL/6 mice exposed to Pb (10 ppm) for 4 and 13 weeks [70]. The information showed that Pb exposure significantly changed the metabolic processes of microorganisms [71–73]. Important metabolic processes were altered in particular, such as the energy metabolic pathways, oxidative

TABLE 9.2
Summary of lead's impact on microbial metabolites, metabolism, NO synthesis, and lipid pathways in mice over varying durations.

Aspect	Pb Exposure	Duration	Findings
Microbial Metabolites and Signaling	Local and systemic interaction	Not specified	Bile acids, SCFAs impact host cellular receptors.
Pb Impact on Mice Metabolism	Pb (10 ppm) exposure	4 and 13 weeks	Pb alters defense/detoxification, oxidative stress, energy pathways.
NO Synthesis and Metabolite Changes	Moderate Pb concentration (0.1 mg/L)	15 weeks	Pb exposure increases NO synthesis, alters cecum metabolites.
Lipid Metabolism and SCFA Disturbance	Pb exposure for 8 weeks	Not specified	Pb triggers lipid metabolism genes, reduces fecal SCFAs.

stress, vitamin E, bile acids, defense/detoxification mechanism, and nitrogen processing [74]. A total of 1,314 molecular features with fold changes larger than 1.5 and p-values less than 0.05 were identified. Vitamin E and bile acid concentrations were noticeably out of balance [75]. For instance, the Pb-exposed group showed a considerable reduction in cholesterol and its product coprostanol, as well as in the main biliary acids cholic acid and ursodeoxycholic acid, the secondary biliary acid deoxycholic acid, and all of these other substances. In both normal and aberrant gastrointestinal tract functions, nitric oxide (NO) is a key player [76]. Moreover, it was proposed that a 13-week Pb exposure resulting from an overexpression of the gene encoding copper-containing nitrite reductase would boost the production of NO in the gut flora. Nuclear magnetic resonance (1NMR) investigations revealed dramatically changed cecum levels for 15 metabolites in mice given a modest quantity of Pb (0.1 mg/L) in their drinking water for 15 weeks [77]. 4-Guanidinobutyrate, choline, citrate, glutamate, isobutyrate (a short-chain fatty acid, or SCFA), and lysine were all shown to be elevated [78]. On the other hand, β-galactose, phenylalanine, tyrosine, valine, isoleucine, leucine, glycine, and alanine were reduced [79]. The tricarboxylic acid cycle (TCA cycle), the metabolism of amino acids, and the host's energy metabolic pathways were all impacted by these microbial byproducts. Interestingly, modifications in the metabolites generated by microbes were consistently reflected in the host's corresponding metabolic genes [80]. For instance, the livers of mice given Pb treatment showed dose-dependent upregulation of lipid metabolism genes, including those involved in the synthesis and transit routes of newly formed fatty acids as well as the formation of triglycerides (TG) [81]. The gut microbial community of mice exposed to lead for eight weeks showed significantly lower concentrations of fecal short-chain fatty acids (SCFA) such as propionate, butyrate, and acetate [82]. Based on these observations, the authors speculated that these metabolites might also be closely related to the disruption of lipid metabolism in the Pb-exposed mice as opposed to the control group. These findings are consistent with earlier research [83–85]. All of the data suggest that exposure to lead (Pb) may cause dysbiosis in the intestinal microbiota and its metabolic byproducts, which may affect the host's metabolic processes and eventually improve the host's quality of life. These studies, however, were unable to provide a convincing justification for the particular bacteria and its byproducts [86]. It is believed that some significant bacteria, but not the whole microbiota, have an impact on host and intestine health [87]. To fully assess the development, multi-omics such as transcriptome, metabolomics, metatranscriptomics, and metagenomics may need to be used in various ways [88].

9.3 EFFECTS OF PB ON INTESTINAL BARRIER

According to reports, Pb exposure significantly affected the structure and barrier performance of the rats' small intestine. More accurately, scanning electron microscopy revealed uneven surface villi, large regions with degenerative lesions, and sometimes nonexistent enterocyte microvilli within these areas [89–91]. Figure 9.2 depicts the healthy and diseased (lead exposure) microbiota and its effect. However, a very benign idea involving intestinal permeability and barrier in Pb poisoning was developed as a result of Breton's study. Mice given lead (100 and 500 mg/L) in their drinking water for 4, 8, and 12 weeks showed colon, duodenum, and ileum histological characteristics that were identical to those of the control group [92]. More specifically, the amount of Paneth and goblet cells in the small intestine, the depth of crypts, and the size of villi were all unchanged. Additionally, Pb almost completely maintained the genes ZO-1, Foxp3, and Foxo4, which are linked to the function of the intestinal barrier in several intestinal areas [93]. The expression of genes connected to inflammation, such as Il1b, Tnf, and Ifng, as well as those linked to anti-inflammatory responses, such as Tgfb and Il-10, significantly decreased. In models derived from human cells, further research showed a decrease in transepithelial electric resistance [94]. Following eight weeks of oral Pb delivery at a high dosage (1 g/L), the intestinal barrier and permeability in mice were significantly damaged. Intestinal tight junction (TJ) mRNA expression (ZO-1, claudin-1, and occludin) was likewise downregulated, Ruminococcus prevalence increased, and Turicibacter abundance decreased [95–97]. This group examined the effects of lead poisoning on mice's colon and small intestine permeability. Mice exposed to lead exhibited a significant reduction in ZO-1, ZO-2, claudin-1, and occludin gene expression in their colon and small intestine [98]. The researchers also noticed an intriguing tendency [99]. Tight junction mRNA expression was further downregulated in mice after Pb treatment, even though their gut flora was already compromised [100]. This study suggests that commensal bacteria in the gut and stomach may compete with one another to absorb lead,

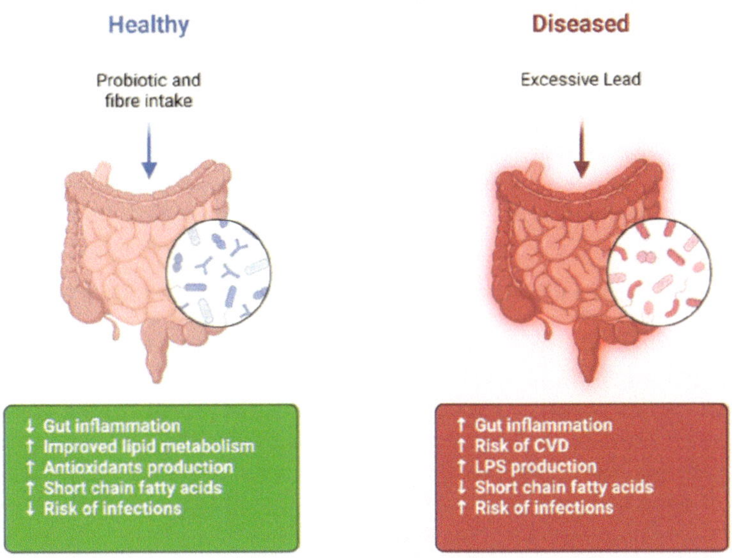

FIGURE 9.2 The image illustrates a healthy microbiome against a diseased microbiome affected by lead exposure. On the left, the healthy microbiome exhibits balanced microbial diversity and symbiosis. Conversely, the diseased microbiome on the right illustrates disruption caused by lead toxicity, evident through increased inflammation and lipopolysaccharide (LPS) levels. Lead exposure perturbs microbial equilibrium, leading to dysbiosis and heightened inflammatory responses.

reducing the bioavailability of the metal [101]. The results also showed that long-term Pb exposure had a major impact on the Japanese quail cecum's histology, causing mucosal shedding, damage to the Lieberkühn gland, and lymphocyte expansion [102]. Transmission electron microscopy was also used to examine the ultrastructural structural defects, which are defined by nucleus pyknosis, mitochondrial vacuolation, and microvilli contraction [103]. Moreover, alterations in cecum shape were linked to decreased expression levels of IFN-γ and IL-2 as well as increased levels of TNF-α, NF-κB, and IL-6. These changes were also associated with a reduction in F*aecalibacterium* and *Bifidobacteria* and an increase in *Bacteroides* [104–106].

9.4 GUT–LIVER AXIS

Numerous components present in the gut, such as bile acids (which connect the gut–liver axis), may be essential to the generation of hepatic bile, the host's intestinal health, glucose homeostasis, lipid metabolism, and heavy metal excretion. Previous studies have reported on the enterohepatic circulation several times [107]. In one study utilizing a liver cancer model, primary accumulation of BA was shown to increase the number of natural killer T cells and reduce the growth of liver tumors [108–110]. An increasing amount of data points to the gut microbiota as being essential for maintaining the balance of bile acids [111]. Long-term lead exposure has been shown to affect liver metabolism, namely lipid metabolism and the TCA cycle, and to alter gut microbes and critical genes linked to lipid metabolism [112]. Significant alterations in the composition of intestinal bacteria and the mRNA expression of genes involved in barium processing, such as Cyp8b1 in the liver and Fgf15 in the ileum, were also seen in Pb-treated mice [113]. Furthermore, the researchers noticed significant alterations in the key genes (Gk, Aco, Accl, Fas, Apo, and Dgat) connected to hepatic lipid and glycolysis metabolism, as well as in the hepatic metabolism of Pb-treated zebrafish. Additionally, changes were seen in the relative abundance of a few gut bacteria, such as *Ruminococcus*, *Bacteroides*, *Flavobacterium*, *Roseburia*, and *Alloprevotella* [114]. These bacteria have been closely associated with inflammation, sickness, and the metabolism of lipids and glycolysis. It was shown that mice given Pb (1,304 mg/kg) orally for three days showed a significant reduction in short-chain fatty acids (SCFAs), especially butyric acid [115]. Depletion of intestinal flora in a mouse model of hepatic illness has shown the ability of antibiotics to reduce liver inflammation by decreasing the liver's lipopolysaccharide transport [116]. Sections "Impact of Pb on intestinal microbiota" and "Consequences of Pb on intestinal barrier" present information indicating that changes in the quantity, diversity, or composition of the intestinal microbiota affect how fatty acids (BCFAs), triglycerides, and/or other metabolic pathways are processed, which may have an impact on the interaction between the liver and the gut [117–119].

9.5 PROBIOTICS TREATMENT AS A PROMISING PB-AGAINST STRATEGY

Numerous research on the use of lead as a therapeutic intervention to reduce lead poisoning has been reported, given that lead is one of the most dangerous dense metals [120]. Figure 9.3 depicts the beneficial effects of probiotics. EF031 and M74, two strains of Enterococcus faecium, were first tested for their ability to remove lead from water [121]. According to preliminary research, both E. faecium strains were capable of efficiently capturing lead. Variants of *E. faecium* Pb12 were the most effective lactic acid bacteria (LAB) in removing lead (Pb) from the environment. These LAB were able to absorb Pb from the fish gastrointestinal system and subsequently decrease Pb bioaccumulation in tissues and organs [122]. *Lactobacillus reuteri* P16 was isolated from the digestive system of the fish *Cyprinus carpio*, according to a recent study. It demonstrated strong resistance to bile and acidity in in vitro settings, exceptional Pb-binding ability (>15% Pb removal), and exceptional antioxidant activity (scavenging hydroxyl radicals, 42.18%). We investigated *Lactobacillus rhamnosus* GR-1 (LGR-1)'s anti-lead (Pb) effectiveness using a Caco-2 model that mimics the intestinal epithelium [123]. The findings showed that Pb caused unique agglomerations on the cell surface of LGR-1.

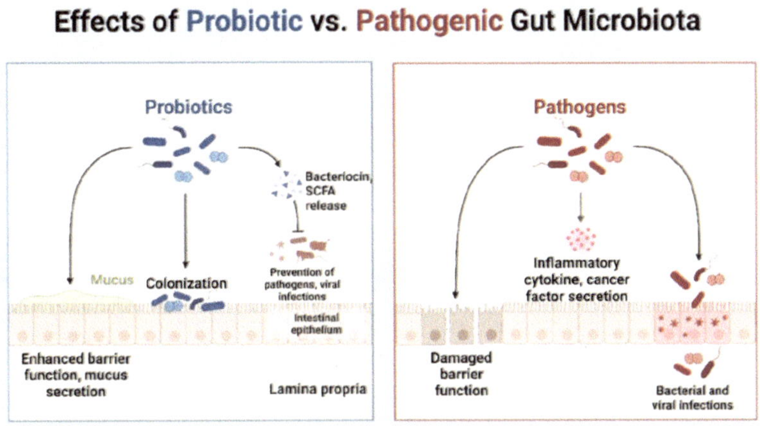

FIGURE 9.3 The image depicts the beneficial effects of probiotic intake on microbiome health through the release of bacteriocin and short-chain fatty acids (SCFA). Probiotics, represented as bacterial strains, promote the synthesis and secretion of bacteriocin, which selectively targets harmful bacteria, fostering a balanced microbial ecosystem. Additionally, probiotics enhance the production of SCFA, crucial metabolites that support gut health and immune function.

LGR-1 efficiently inhibited lead translocation across the intestinal epithelium, preventing lead from moving from the apical to the basolateral region. Probiotics and in vivo lead toxicity have also been the focus of much study in recent years [124–126]. After it was shown that *Lactobacillus reuteri* P16 was a potent Pb-removal strain, its therapeutic benefits were investigated for fish (*Cyprinus carpio*) exposed to lead in water. For six weeks, the fish were given *L. reuteri* P16 (108 CFU/g) as an extra meal and exposed to Pb in the water (1 mg/L). The outcomes suggested that this probiotic could improve growth performance, reduce tissue Pb accumulation and mortality, and partially mitigate Pb-induced alterations in the host, including oxidative stress, intestinal enzymatic functions, gut microbiota, and the expressions of heat shock proteins (HSP70 and HSP90) and proinflammatory cytokines (TNF-α and IL-1β) [127]. We isolated Leuconostoc mesenteroides, a lead-tolerant LAB strain 96, from a fermented Korean food sample. Leuconostoc mesenteroides was utilized to significantly lower the levels of glutamate pyruvate transaminase and glutamate oxaloacetate transaminase, as well as to restore some male reproductive function, including ATP content and motility of epididymal spermatozoa in mice receiving Pb [128]. Lead-intolerant intestinal microbiota (*Faecalibacterium prausnitzii*, *Oscillibacter ruminantium*, and *Akkermansia muciniphila*, especially *O. ruminantium* and *F. prausnitzii*) were given orally to mice exposed to lead in order to prevent the alteration of gut permeability. Pb removal and tight junction protein expression significantly increased as a result [129, 130]. The authors were able to claim that probiotics had the potential to be Pb-against by altering gut microbiome since they observed that colonic microbiota elevated SCFAs [131]. Additionally, a transgenic probiotic was created that significantly decreased LPS-induced deterioration in rats exposed to lead and restored normal absorption of essential metal ions [132]. Previous studies shown the ability of Lactobacillus plantarum CCFM8661 to sequester lead in vitro as well as its ability to shield mice against oxidative stress and abnormalities in their hematuria. The inclusion of *L. plantarum* CCFM8661 significantly reduced the incidence of nuclear abnormalities in red blood cells in fish peripheral blood and enhanced the innate immune state caused by lead in a follow-up investigation [133]. All of the above data point to the possibility that probiotics might prevent lead poisoning, which could come from the metal attaching directly to the surface via a quick, metabolism-independent method or by co-precipitating lead inside the gastrointestinal system [134]. Probiotics (*Lactobacillus plantarum* CCFM8661) have been shown to alter enterohepatic circulation, which increases bile flow and causes the biliary system to produce glutathione [135]. These enhancements were subsequently shown to increase biliary Pb production and fecal Pb excretion [136–138]. Moreover, the impact was considerably attenuated with the administration of an FXR agonist, and this effect was linked to target gene expression in the enterohepatic farnesoid X receptor–fibroblast growth factor (FXR-FGF15) pathway [139]. Furthermore, probiotics and other ingredients found in dietary supplements may work better as Pb-against materials; in fact, they may even help Pb-exposed mice restore their memory and learning abilities [140]. Table 9.3 summarizes probiotic interventions against lead toxicity.

9.6 CONCLUSION AND FUTURE PERPSECTIVE

According to recent research, lead (Pb) in the gastrointestinal tracts may affect the gut directly as well as indirectly. Pb poisoning primarily affects the local gut's intestinal epithelium, tight

TABLE 9.3

Summary of diverse probiotic interventions against lead toxicity, showcasing their mechanisms, host effects, and additional outcomes in brief.

Probiotic Intervention	Lead Removal	Host Effects	Additional Outcomes
Enterococcus faecium Isolates	Effective Pb capture	Reduced bioaccumulation in fish organs.	E. faecium Pb12 absorbs Pb from fish GI tract.
Lactobacillus reuteri P16	High Pb-binding	Enhanced growth, reduced tissue Pb, improved well-being.	Antioxidant activity, bile/acidity resistance, from fish Cyprinus carpio.
Lactobacillus rhamnosus GR-1	Reduces Pb translocation	Effective in reducing Pb movement through epithelium.	Induced cell-surface agglomerations in Caco-2.
L. reuteri P16 Feed	Alleviates Pb changes	Enhanced growth, reduced tissue Pb, countered oxidative stress.	Extracted from fish digestive contents; 6 weeks duration.
Leuconostoc mesenteroides	Reduces enzymes, restores fertility	Restored reproductive functionality, reduced Pb damage.	Lead-tolerant LAB strain 96; No specified duration.
Probiotic-rich Microbiota	Oral supplementation	Notable rise in Pb elimination, improved tight junctions.	Orally supplemented Faecalibacterium, Oscillibacter, Akkermansia.
Transgenic Probiotic	Restored metal absorption	Reduced lead-triggered deterioration, restored metal ions absorption.	Developed for standard absorption of metal ions; No duration.
L. plantarum CCFM8661	Sequesters Pb, enhances excretion	Improved immunity, reduced nuclear abnormalities.	Improved enterohepatic circulation, bile flow, glutathione synthesis.
FXR Agonist with Probiotics	Mitigated probiotic effect	Improved fecal Pb excretion and biliary Pb output.	Mitigated by FXR agonist; Connected to target gene expression.
Dietary Supplements	More effective	Potential aid in memory recovery for Pb-exposed mice.	Effects might be more effective as Pb-against materials; No duration.

junctions, immunological dysregulation, and inflammation. These substances may make the gut more permeable to macromolecules. Lead exposure may cause the gut microbiota, which is usually the first group of "targets" in the gastrointestinal tract, to become dysbiotic. This might harm the intestinal barrier or alter the expression of many bacterial metabolites. Moreover, disruption of the gut barrier may affect microbial diversity, structure, and metabolites, and Pb-induced changes in the intestinal microbiota may affect gut permeability. We hypothesize that lead (Pb) exposure may promote the production of lipopolysaccharides (LPS), which may result from dysbiosis of the gut microbiota and reduced intestinal integrity, which, in turn, may lead to endotoxemia as a consequence of a breached gut barrier. This is based on the changes that Pb subsequently produced to the gut permeability and intestinal flora. Furthermore, weakened gut barrier function may lead to bacterial translocation into other tissues, which can result in endotoxemia and host infections. After entering the bloodstream and passing via the enterohepatic circulation, the volatile microbial metabolites, inflammatory cytokines, and immunologic agents might affect several target organs. Numerous studies have shown how important lipopolysaccharide (LPS) is in the emergence of several illnesses, including cardiovascular disease, liver and kidney damage, and metabolic problems. By causing disruptions to human metabolism, disrupted microbial metabolites are a contributing factor to illness. Probiotics are useful detoxicants and treatments for lead poisoning, as some research studies have shown. However, most investigations examined the overall changes that Pb causes to microbial populations rather than concentrating on a specific strain. Given this, further research into the metagenome and metabonome may be necessary to fully understand a few important bacteria and how they affect the gut and other organs. The processes behind lead poisoning on the digestive tract and the consequent damage to other organs remain largely unknown, especially with regard to the modifications that lead (Pb) causes in lipopolysaccharides (LPS) and other microbial byproducts. Moreover, more research is required to determine the particular mechanisms by which probiotics reduce lead exposure.

REFERENCES

1. P.W. Andrews, A. Bharwani, K.R. Lee, M. Fox, J.A. Thomson, Jr., Is serotonin an upper or a downer? The evolution of the serotonergic system and its role in depression and the antidepressant response, *Neuroscience and Biobehavioral Reviews*, 51 (2015) 164–188.
2. H.M. Spliethoff, R.G. Mitchell, H. Shayler, L.G. Marquez-Bravo, J. Russell-Anelli, G. Ferenz, M. McBride, Estimated lead (Pb) exposures for a population of urban community gardeners, *Environmental Geochemistry and Health*, 38 (2016) 955–971.
3. D. O'Connor, D. Hou, J. Ye, Y. Zhang, Y.S. Ok, Y. Song, F. Coulon, T. Peng, L. Tian, Lead-based paint remains a major public health concern: a critical review of global production, trade, use, exposure, health risk, and implications, *Environment International*, 121 (2018) 85–101.
4. S.A. Boseila, A.A. Gabr, I.A. Hakim, Blood lead levels in Egyptian children: influence of social and environmental factors, *American Journal of Public Health*, 94 (2004) 47–49.
5. C. Gundacker, M. Forsthuber, T. Szigeti, R. Kakucs, V. Mustieles, M.F. Fernandez, E. Bengtsen, U. Vogel, K.S. Hougaard, A.T. Saber, Lead (Pb) and neurodevelopment: a review on exposure and biomarkers of effect (BDNF, HDL) and susceptibility, *International Journal of Hygiene and Environmental Health*, 238 (2021) 113855.
6. M.S. Collin, S.K. Venkatraman, N. Vijayakumar, V. Kanimozhi, S.M. Arbaaz, R.G.S. Stacey, J. Anusha, R. Choudhary, V. Lvov, G.I. Tovar, F. Senatov, S. Koppala, S. Swamiappan, Bioaccumulation of lead (Pb) and its effects on human: a review, *Journal of Hazardous Materials Advances*, 7 (2022) 100094.
7. L. Yu, H. Duan, Y. Yu, Q. Zhang, J. Zhao, H. Zhang, Q. Zhai, F. Tian, W. Chen, Dose-dependent effects of chronic lead toxicity in vivo: focusing on trace elements and gut microbiota, *Chemosphere*, 301 (2022) 134670.
8. G. Flora, D. Gupta, A. Tiwari, Toxicity of lead: a review with recent updates, *Interdisciplinary Toxicology*, 5 (2012) 47–58.
9. T.E. Adolph, L. Mayr, F. Grabherr, J. Schwärzler, H. Tilg, Pancreas-microbiota cross talk in health and disease, *Annual Review of Nutrition*, 39 (2019) 249–266.

10. M. Kumar, M.S. Hussain, S. Raj, R. Verma, S. Sharma, S.K. Sahu, An overview of treatment modalities and management aspects for obesity, *Current Nutrition and Food Science*, 19 (2023) 105–113.
11. S. Singh, P. Sharma, D.K. Sarma, M. Kumawat, R. Tiwari, V. Verma, R. Nagpal, M. Kumar, Implication of obesity and gut microbiome dysbiosis in the etiology of colorectal cancer, *Cancers*, 15 (2023) 1913.
12. W. Liu, H. Feng, S. Zheng, S. Xu, I.Y. Massey, C. Zhang, X. Wang, F. Yang, Pb toxicity on gut physiology and microbiota, *Frontiers in Physiology*, 12 (2021) 574913.
13. S. Bhattacharya, Probiotics against alleviation of lead toxicity: recent advances, *Interdisciplinary Toxicology*, 12 (2019) 89–92.
14. P.D. Cani, E. Moens de Hase, M. Van Hul, Gut microbiota and host metabolism: from proof of concept to therapeutic intervention, *Microorganisms*, 9 (2021) 1302.
15. I. Polidori, L. Marullo, C. Ialongo, F. Tomassetti, R. Colombo, F. di Gaudio, G. Calugi, G. Marrone, A. Noce, S. Bernardini, F. Broccolo, M. Pieri, Characterization of gut microbiota composition in type 2 diabetes patients: a population-based study, *International Journal of Environmental Research and Public Health,* 19 (2022) 15913.
16. O.A. Baothman, M.A. Zamzami, I. Taher, J. Abubaker, M. Abu-Farha, The role of gut microbiota in the development of obesity and diabetes, *Lipids in Health and Disease*, 15 (2016) 108.
17. J. Liu, Y. Tan, H. Cheng, D. Zhang, W. Feng, C. Peng, Functions of gut microbiota metabolites, current status and future perspectives, *Aging and Disease*, 13 (2022) 1106–1126.
18. P. Feng, Z. Ye, A. Kakade, A.K. Virk, X. Li, P. Liu, A review on gut remediation of selected environmental contaminants: possible roles of probiotics and gut microbiota, *Nutrients,* 11 (2018) 22.
19. P. Li, J. Zhang, X. Liu, L. Gan, Y. Xie, H. Zhang, J. Si, The function and the affecting factors of the zebrafish gut microbiota, *Frontiers in Microbiology*, 13 (2022) 903471.
20. H. Cheng, J. Liu, Y. Tan, W. Feng, C. Peng, Interactions between gut microbiota and berberine, a necessary procedure to understand the mechanisms of berberine, *Journal of Pharmaceutical Analysis*, 12 (2022) 541–555.
21. S.M. Ronnebaum, J.W. Joseph, O. Ilkayeva, S.C. Burgess, D. Lu, T.C. Becker, A.D. Sherry, C.B. Newgard, Chronic suppression of acetyl-CoA carboxylase 1 in beta-cells impairs insulin secretion via inhibition of glucose rather than lipid metabolism, *The Journal of Biological Chemistry*, 283 (2008) 14248–14256.
22. A. Stolz, H.J. Busse, P. Kämpfer, Pseudomonas knackmussii sp. nov, *International Journal of Systematic and Evolutionary Microbiology,* 57 (2007) 572–576.
23. L. Yu, Y. Yu, Y. Xiao, F. Tian, A. Narbad, Q. Zhai, W. Chen, Lead-induced gut injuries and the dietary protective strategies: a review, *Journal of Functional Foods*, 83 (2021) 104528.
24. S. Singh, P. Sharma, N. Pal, M. Kumawat, S. Shubham, D.K. Sarma, R.R. Tiwari, M. Kumar, R. Nagpal, Impact of environmental pollutants on gut microbiome and mental health via the gut-brain axis, *Microorganisms*, 10 (2022) 1457.
25. J. An, H. Kwon, Y.J. Kim, The firmicutes/bacteroidetes ratio as a risk factor of breast cancer, *Journal of Clinical Medicine*, 12 (2023) 2216.
26. A. Sarkar, J.Y. Yoo, S. Valeria Ozorio Dutra, K.H. Morgan, M. Groer, The association between early-life gut microbiota and long-term health and diseases, *Journal of Clinical Medicine*, 10 (2021) 459.
27. T. Ma, X. Shen, X. Shi, H.A. Sakandar, K. Quan, Y. Li, H. Jin, L.-Y. Kwok, H. Zhang, Z. Sun, Targeting gut microbiota and metabolism as the major probiotic mechanism - An evidence-based review, *Trends in Food Science & Technology*, 138 (2023) 178–198.
28. Y. Sarmiento-Andrade, R. Suárez, B. Quintero, K. Garrochamba, S.P. Chapela, Gut microbiota and obesity: new insights, *Frontiers in Nutrition*, 9 (2022) 1018212.
29. K. Hou, Z.-X. Wu, X.-Y. Chen, J.-Q. Wang, D. Zhang, C. Xiao, D. Zhu, J.B. Koya, L. Wei, J. Li, Z.-S. Chen, Microbiota in health and diseases, *Signal Transduction and Targeted Therapy*, 7 (2022) 135.
30. P. Petakh, V. Oksenych, A. Kamyshnyi, The F/B ratio as a biomarker for inflammation in COVID-19 and T2D: impact of metformin, *Biomedicine & Pharmacotherapy = Biomedecine & Pharmacotherapie*, 163 (2023) 114892.
31. A. Taibi, M. Ku, Z. Lin, G. Gargari, A. Kubant, D. Lepp, K.A. Power, S. Guglielmetti, L.U. Thompson, E.M. Comelli, Data on cecal and fecal microbiota and predicted metagenomes profiles of female mice receiving whole flaxseed or its oil and secoisolariciresinol diglucoside components, *Data in Brief*, 38 (2021) 107409.
32. V.K.R. Balaga, A. Pradhan, R. Thapa, N. Patel, R. Mishra, N. Singla, Morin: a comprehensive review on its versatile biological activity and associated therapeutic potential in treating cancers, *Pharmacological Research*, 7 (2023) 100264.

33. Z. Wang, K.L. Lam, J. Hu, S. Ge, A. Zhou, B. Zheng, S. Zeng, S. Lin, Chlorogenic acid alleviates obesity and modulates gut microbiota in high-fat-fed mice, *Food Science & Nutrition,* 7 (2019) 579–588.
34. K. Chiu, G. Warner, R.A. Nowak, J.A. Flaws, W. Mei, The impact of environmental chemicals on the gut microbiome, *Toxicological Sciences*, 176 (2020) 253–284.
35. N. Bao, F. Chen, D. Dai, The regulation of host intestinal microbiota by polyphenols in the development and prevention of chronic kidney disease, *Frontiers in Immunology*, 10 (2019) 2981.
36. D.G. Baker, Natural pathogens of laboratory mice, rats, and rabbits and their effects on research, *Clinical Microbiology Reviews*, 11 (1998) 231–266.
37. E.A. Franzosa, A. Sirota-Madi, J. Avila-Pacheco, N. Fornelos, H.J. Haiser, S. Reinker, T. Vatanen, A.B. Hall, H. Mallick, L.J. McIver, J.S. Sauk, R.G. Wilson, B.W. Stevens, J.M. Scott, K. Pierce, A.A. Deik, K. Bullock, F. Imhann, J.A. Porter, A. Zhernakova, J. Fu, R.K. Weersma, C. Wijmenga, C.B. Clish, H. Vlamakis, C. Huttenhower, R.J. Xavier, Gut microbiome structure and metabolic activity in inflammatory bowel disease, *Nature Microbiology*, 4 (2019) 293–305.
38. A.A. Bhat, O. Afzal, M. Afzal, G. Gupta, R. Thapa, H. Ali, I. Kazmi, S.I. Alzarea, S. Saleem, V.P. Samuel, MALAT1: a key regulator in lung cancer pathogenesis and therapeutic targeting, *Pathology-Research and Practice*, 253 (2023) 154991.
39. B. Wang, G. Gupta, M. Singh, K.V.S. Veerabhadrappa, A. Mishra, G.K. Chinnaboina, Pharmacological evaluation of novel flavone from morus alba in pentylenetetrazole-induced kindling and oxidative stress, *Journal of Environmental Pathology, Toxicology and Oncology,* 37 (2018) 43–52.
40. Canadian Association of Gastroenterology, Canadian Association for the Study of the Liver, Program and Abstracts from the Canadian Digestive Diseases Week™. *Canadian Journal of Gastroenterology and Hepatology*, 2016 (2016) 4792898.
41. M. Zheng, R. Han, Y. Yuan, Y. Xing, W. Zhang, Z. Sun, Y. Liu, J. Li, T. Mao, The role of Akkermansia muciniphila in inflammatory bowel disease: current knowledge and perspectives, *Frontiers in Immunology*, 13 (2022) 1089600.
42. C. Liu, S. Zhan, Z. Tian, N. Li, T. Li, D. Wu, Z. Zeng, X. Zhuang, Food additives associated with gut microbiota alterations in inflammatory bowel disease: friends or enemies?, *Nutrients*, 14 (2022) 3049.
43. R. Wadhwa, K.R. Paudel, M. Mehta, S.D. Shukla, K. Sunkara, P. Prasher, N. Panth, R. Goyal, D.K. Chellappan, G. Gupta, P.M. Hansbro, A.A.A. Aljabali, M.M. Tambuwala, K. Dua, Beyond the obvious: smoking and respiratory infection implications on Alzheimer's disease, *CNS & Neurological Disorders Drug Targets*, 19 (2020) 698–708.
44. M. Yang, Y. Gu, L. Li, T. Liu, X. Song, Y. Sun, X. Cao, B. Wang, K. Jiang, H. Cao, Bile Acid-gut microbiota axis in inflammatory bowel disease: from bench to bedside, *Nutrients*, 13 (2021) 3143.
45. A.A. Tinkov, V.A. Gritsenko, M.G. Skalnaya, S.V. Cherkasov, J. Aaseth, A.V. Skalny, Gut as a target for cadmium toxicity, *Environmental Pollution*, 235 (2018) 429–434.
46. K.V. Johnson, Gut microbiome composition and diversity are related to human personality traits, *Human Microbiome Journal*, 15 (2020) None.
47. A.A. Bhat, G. Gupta, M. Afzal, R. Thapa, H. Ali, S.M. Alqahtani, I. Kazmi, S.I. Alzarea, S. Saleem, V. Subramaniyan, Polyphenol-loaded nano-carriers for breast cancer therapy: a comprehensive review, *BioNanoScience*, 14 (2024) 4219–4237.
48. M.H. Lee, J. Kim, G.H. Kim, M.S. Kim, S.S. Yoon, Effects of Lactiplantibacillus plantarum FBT215 and prebiotics on the gut microbiota structure of mice, *Food Science and Biotechnology*, 32 (2023) 481–488.
49. J.C. Ezeji, D.K. Sarikonda, A. Hopperton, H.L. Erkkila, D.E. Cohen, S.P. Martinez, F. Cominelli, T. Kuwahara, A.E.K. Dichosa, C.E. Good, M.R. Jacobs, M. Khoretonenko, A. Veloo, A. Rodriguez-Palacios, Parabacteroides distasonis: intriguing aerotolerant gut anaerobe with emerging antimicrobial resistance and pathogenic and probiotic roles in human health, *Gut Microbes*, 13 (2021) 1922241.
50. M. Benedict, X. Zhang, Non-alcoholic fatty liver disease: an expanded review, *World Journal of Hepatology*, 9 (2017) 715–732.
51. A.A. Bhat, G. Gupta, R. Dahiya, R. Thapa, A. Gahtori, M. Shahwan, V. Jakhmola, A. Tiwari, M. Kumar, H. Dureja, CircRNAs: pivotal modulators of TGF-β signalling in cancer pathogenesis, *Non-Coding RNA Research*, 9 (2024) 277–287.
52. A.R. Sitarik, M. Arora, C. Austin, L.F. Bielak, S. Eggers, C.C. Johnson, S.V. Lynch, S. Kyun Park, K.H. Hank Wu, G.J.M. Yong, A.E. Cassidy-Bushrow, Fetal and early postnatal lead exposure measured in teeth associates with infant gut microbiota, *Environment International*, 144 (2020) 106062.
53. S. Eggers, N. Safdar, A.K. Sethi, G. Suen, P.E. Peppard, A.E. Kates, J.H. Skarlupka, M. Kanarek, K.M.C. Malecki, Urinary lead concentration and composition of the adult gut microbiota in a cross-sectional population-based sample, *Environment International*, 133 (2019) 105122.

54. S. Vishwas, S.K. Singh, M. Gulati, A. Awasthi, R. Khursheed, L. Corrie, R. Kumar, T. Collet, R. Loebenberg, O. Porwal, S. Gupta, N.K. Jha, P.K. Gupta, H.P. Devkota, D.K. Chellappan, G. Gupta, J. Adams, K. Dua, Harnessing the therapeutic potential of fisetin and its nanoparticles: journey so far and road ahead, *Chemico-Biological Interactions*, 356 (2022) 109869.
55. M. Shao, Y. Zhu, Long-term metal exposure changes gut microbiota of residents surrounding a mining and smelting area, *Scientific Reports*, 10 (2020) 4453.
56. Fatimawali, B.J. Kepel, M.A. Gani, T.E. Tallei, Comparison of bacterial community structure and diversity in traditional gold mining waste disposal site and rice field by using a metabarcoding approach, *International Journal of Microbiology*, 2020 (2020) 1858732.
57. A. Hakansson, G. Molin, Gut microbiota and inflammation, *Nutrients*, 3 (2011) 637–682.
58. P. Hemarajata, J. Versalovic, Effects of probiotics on gut microbiota: mechanisms of intestinal immunomodulation and neuromodulation, *Therapeutic Advances in Gastroenterology*, 6 (2013) 39–51.
59. A.A. Bhat, R. Gilhotra, Y. Singh, S. Sharma, T.d. Jesus Andreoli Pinto, H.G. Ferraz, S.K. Singh, K. Dua, G. Gupta, Advanced drug-delivery approaches in managing P53-mediated lung diseases remodeling, *Nanomedicine: Nanotechnology, Biology, and Medicine*, 18 (2023) 583–587.
60. R. Thapa, O. Afzal, A.S. Alfawaz Altamimi, A. Goyal, W.H. Almalki, S.I. Alzarea, I. Kazmi, V. Jakhmola, S.K. Singh, K. Dua, R. Gilhotra, G. Gupta, Galangin as an inflammatory response modulator: an updated overview and therapeutic potential, *Chemico-Biological Interactions,* 378 (2023) 110482.
61. M.B. Usman, S. Ojha, S.K. Jha, D.K. Chellappan, G. Gupta, S.K. Singh, K. Dua, S. Roychoudhury, N. Kumar, F.A. Khan, H. Dureja, V. Upadhye, F. Zacconi, P. Prasanna, K.K. Kesari, G.M. Ashraf, A. Alexiou, N.K. Jha, Biological databases and tools for neurological disorders, *Journal of Integrative Neuroscience*, 21 (2022) 41.
62. S. Stojanov, A. Berlec, B. Štrukelj, The influence of probiotics on the firmicutes/bacteroidetes ratio in the treatment of obesity and inflammatory bowel disease, *Microorganisms,* 8 (2020) 1715.
63. M.F. Maciel-Fiuza, G.C. Muller, D.M.S. Campos, P. do Socorro Silva Costa, J. Peruzzo, R.R. Bonamigo, T. Veit, F.S.L. Vianna, Role of gut microbiota in infectious and inflammatory diseases, *Frontiers in Microbiology*, 14 (2023) 1098386.
64. A.E. Mohr, R. Jäger, K.C. Carpenter, C.M. Kerksick, M. Purpura, J.R. Townsend, N.P. West, K. Black, M. Gleeson, D.B. Pyne, S.D. Wells, S.M. Arent, R.B. Kreider, B.I. Campbell, L. Bannock, J. Scheiman, C.J. Wissent, M. Pane, D.S. Kalman, J.N. Pugh, C.P. Ortega-Santos, J.A. ter Haar, P.J. Arciero, J. Antonio, The athletic gut microbiota, *Journal of the International Society of Sports Nutrition*, 17 (2020) 24.
65. A.A. Bhat, R. Thapa, A. Goyal, V. Subramaniyan, D. Kumar, S. Gupta, S.K. Singh, K. Dua, G. Gupta, *Curcumin-Based Nanoformulations as an Emerging Therapeutic Strategy for Inflammatory Lung Diseases*, Newlands Press Ltd London, UK, 2023, pp. 583–586.
66. Y. Belkaid, T.W. Hand, Role of the microbiota in immunity and inflammation, *Cell*, 157 (2014) 121–141.
67. M.S. Hussain, A. Sharma, R. Kumar, Prebiotics and probiotics: a focused review of applications in respiratory disorders, *Carpathian Journal of Food Science and Technology*, 15 (2023).
68. A. Visekruna, M. Luu, The role of short-chain fatty acids and bile acids in intestinal and liver function, inflammation, and carcinogenesis, *Frontiers in Cell and Developmental Biology*, 9 (2021) 703218.
69. P. Markowiak-Kopeć, K. Śliżewska, The effect of probiotics on the production of short-chain fatty acids by human intestinal microbiome, *Nutrients*, 12 (2020) 1107.
70. B. Gao, L. Chi, R. Mahbub, X. Bian, P. Tu, H. Ru, K. Lu, Multi-omics reveals that lead exposure disturbs gut microbiome development, key metabolites, and metabolic pathways, *Chemical Research in Toxicology*, 30 (2017) 996–1005.
71. R. Dahiya, R. Thapa, N.K. Kumawat, M. Singh, S. Jakhotiya, D. Deopa, Y. Singh, N. Singla, G. Gupta, *Epigenetic Mechanisms in Inflammation, Targeting Epigenetics in Inflammatory Lung Diseases*, Springer, 2023, pp. 43–55.
72. Y. Singh, G. Gupta, B. Shrivastava, R. Dahiya, J. Tiwari, M. Ashwathanarayana, R.K. Sharma, M. Agrawal, A. Mishra, K. Dua, Calcitonin gene-related peptide (CGRP): a novel target for Alzheimer's disease, *CNS Neuroscience & Therapeutics*, 23 (2017) 457–461.
73. V. Subramaniyan, S. Fuloria, G. Gupta, D.H. Kumar, M. Sekar, K.V. Sathasivam, K. Sudhakar, K.S. Alharbi, W.H. Al-Malki, O. Afzal, I. Kazmi, F.A. Al-Abbasi, A.S.A. Altamimi, N.K. Fuloria, A review on epidermal growth factor receptor's role in breast and non-small cell lung cancer, *Chemico-Biological Interactions*, 351 (2022) 109735.
74. V. Lobo, A. Patil, A. Phatak, N. Chandra, Free radicals, antioxidants and functional foods: impact on human health, *Pharmacognosy Reviews*, 4 (2010) 118–126.

75. K. Wegermann, C. Howe, R. Henao, Y. Wang, C.D. Guy, M.F. Abdelmalek, A.M. Diehl, C.A. Moylan, Serum bile acid, vitamin E, and serotonin metabolites are associated with future liver-related events in nonalcoholic fatty liver disease, *Hepatology Communications*, 5 (2021) 608–617.
76. T. Li, J.Y. Chiang, Regulation of bile acid and cholesterol metabolism by PPARs, P*PAR Research*, 2009 (2009) 501739.
77. R. Timkovich, M.S. Cork, Nitrogen-15 nuclear magnetic resonance investigation of nitrite reductase-substrate interaction, *Biochemistry*, 21 (1982) 3794–3797.
78. M.M. Gupta, R. Gilhotra, D. Deopa, A.A. Bhat, R. Thapa, N. Singla, R. Kulshrestha, G. Gupta, *Epigenetics of Pulmonary Tuberculosis, Targeting Epigenetics in Inflammatory Lung Diseases*, Springer, 2023, pp. 127–144.
79. Z. Li, J. Lai, P. Zhang, J. Ding, J. Jiang, C. Liu, H. Huang, H. Zhen, C. Xi, Y. Sun, L. Wu, L. Wang, X. Gao, Y. Li, Y. Fu, Z. Jie, S. Li, D. Zhang, Y. Chen, Y. Zhu, S. Lu, J. Lu, D. Wang, H. Zhou, X. Yuan, X. Li, L. Pang, M. Huang, H. Yang, W. Zhang, S. Brix, K. Kristiansen, X. Song, C. Nie, S. Hu, Multi-omics analyses of serum metabolome, gut microbiome and brain function reveal dysregulated microbiota-gut-brain axis in bipolar depression, *Molecular Psychiatry*, 27 (2022) 4123–4135.
80. A. MacLean, F. Legendre, V.D. Appanna, The tricarboxylic acid (TCA) cycle: a malleable metabolic network to counter cellular stress, *Critical Reviews in Biochemistry and Molecular Biology*, 58 (2023) 81–97.
81. M. Alves-Bezerra, D.E. Cohen, Triglyceride metabolism in the liver, *Comprehensive Physiology,* 8 (2017) 1–8.
82. A. Nogal, A.M. Valdes, C. Menni, The role of short-chain fatty acids in the interplay between gut microbiota and diet in cardio-metabolic health, *Gut Microbes*, 13 (2021) 1–24.
83. C. Zhang, T. Ma, C. Liu, D. Ma, J. Wang, M. Liu, J. Ran, X. Wang, X. Deng, PM(2.5) induced liver lipid metabolic disorders in C57BL/6J mice, *Frontiers in Endocrinology*, 14 (2023) 1212291.
84. S. Pathak, G. Gupta, L. Thangavelu, S.K. Singh, K. Dua, D.K. Chellappan, R.M. Gilhotra, Recent update on barbiturate in relation to brain disorder, *Excli Journal*, 20 (2021) 1028–1032.
85. M.A.J. Shaikh, A.S.A. Altamimi, M. Afzal, G. Gupta, N. Singla, R. Gilhotra, W.H. Almalki, I. Kazmi, S.I. Alzarea, P. Prasher, S.K. Singh, K. Dua, Unraveling the impact of miR-21 on apoptosis regulation in glioblastoma, *Pathology, Research and Practice,* 254 (2024) 155121.
86. J. Xia, C. Jin, Z. Pan, L. Sun, Z. Fu, Y. Jin, Chronic exposure to low concentrations of lead induces metabolic disorder and dysbiosis of the gut microbiota in mice, *The Science of the Total Environment*, 631–632 (2018) 439–448.
87. M.S. Hussain, G. Gupta, A. Goyal, R. Thapa, W.H. Almalki, I. Kazmi, S.I. Alzarea, S. Fuloria, D.U. Meenakshi, V. Jakhmola, From nature to therapy: Luteolin's potential as an immune system modulator in inflammatory disorders, *Journal of Biochemical and Molecular Toxicology*, 37 (2023) e23482.
88. V. Aguiar-Pulido, W. Huang, V. Suarez-Ulloa, T. Cickovski, K. Mathee, G. Narasimhan, Metagenomics, metatranscriptomics, and metabolomics approaches for microbiome analysis, *Evolutionary Bioinformatics Online*, 12 (2016) 5–16.
89. J. Martin, Scanning electron microscope studies of the small intestine of rats maintained on a low protein diet and infected with Nippostrongylus brasiliensis, *Parasitology*, 80 (1980) 39–47.
90. S.N. Pandey, G. Singh, B.C. Semwal, G. Gupta, K.S. Alharbi, W.H. Almalki, M. Albratty, A. Najmi, A.M. Meraya, Therapeutic approaches of nutraceuticals in the prevention of Alzheimer's disease, *Journal of Food Biochemistry*, 46 (2022) e14426.
91. S. Pathak, G. Gupta, R.M. Gilhotra, the role of diazepam in epigenetics: from the molecular level to clinical implications, *Advances in Mind-Body Medicine*, 35 (2021) 25–33.
92. Q. Zhai, D. Qu, S. Feng, Y. Yu, L. Yu, F. Tian, J. Zhao, H. Zhang, W. Chen, Oral supplementation of lead-intolerant intestinal microbes protects against lead (Pb) toxicity in mice, *Frontiers in Microbiology*, 10 (2019) 3161.
93. E.M. Garabedian, L.J. Roberts, M.S. McNevin, J.I. Gordon, Examining the role of Paneth cells in the small intestine by lineage ablation in transgenic mice, *The Journal of Biological Chemistry*, 272 (1997) 23729–23740.
94. J.L. Benci, L.R. Johnson, R. Choa, Y. Xu, J. Qiu, Z. Zhou, B. Xu, D. Ye, K.L. Nathanson, C.H. June, E.J. Wherry, N.R. Zhang, H. Ishwaran, M.D. Hellmann, J.D. Wolchok, T. Kambayashi, A.J. Minn, Opposing functions of interferon coordinate adaptive and innate immune responses to cancer immune checkpoint blockade, *Cell*, 178 (2019) 933–948.e914.
95. V. Volynets, A. Rings, G. Bárdos, M.J. Ostaff, J. Wehkamp, S.C. Bischoff, Intestinal barrier analysis by assessment of mucins, tight junctions, and α-defensins in healthy C57BL/6J and BALB/cJ mice, *Tissue Barriers*, 4 (2016) e1208468.

96. J. Kaur, M. Gulati, B. Kapoor, N.K. Jha, P.K. Gupta, G. Gupta, D.K. Chellappan, H.P. Devkota, P. Prasher, M.S. Ansari, F.F. Aba Alkhayl, M.F. Arshad, A. Morris, Y.E. Choonara, J. Adams, K. Dua, S.K. Singh, Advances in designing of polymeric micelles for biomedical application in brain related diseases, *Chemico-Biological Interactions*, 361 (2022) 109960.
97. S. Malhotra, N. Jain, J. Rathee, S. Kaul, U. Nagaich, M. Pandey, B. Gorain, G. Gupta, Nanoneuroscience: cutting-edge approach for disease management, *Recent Patents on Nanotechnology*, 18 (2024) 256–271.
98. L.Z. Shi, W. Zheng, Early lead exposure increases the leakage of the blood-cerebrospinal fluid barrier, in vitro, *Human & Experimental Toxicology*, 26 (2007) 159–167.
99. R. Thapa, O. Afzal, M. Afzal, G. Gupta, A.A. Bhat, I. Kazmi, S.I. Alzarea, S. Saleem, P. Arora, S.K. Singh, From LncRNA to metastasis: the MALAT1-EMT axis in cancer progression, *Pathology—Research and Practice* (2023) 154959.
100. W. Fong, Q. Li, J. Yu, Gut microbiota modulation: a novel strategy for prevention and treatment of colorectal cancer, *Oncogene*, 39 (2020) 4925–4943.
101. R. Thapa, O. Afzal, A.A. Bhat, A. Goyal, A.S. Alfawaz Altamimi, W.H. Almalki, S.I. Alzarea, I. Kazmi, S.K. Singh, K. Dua, New horizons in lung cancer management through ATR/CHK1 pathway modulation, *Future Medicinal Chemistry*, 15 (2023) 1807–1818.
102. H. Tlaskalová-Hogenová, R. Stěpánková, H. Kozáková, T. Hudcovic, L. Vannucci, L. Tučková, P. Rossmann, T. Hrnčíř, M. Kverka, Z. Zákostelská, K. Klimešová, J. Přibylová, J. Bártová, D. Sanchez, P. Fundová, D. Borovská, D. Srůtková, Z. Zídek, M. Schwarzer, P. Drastich, D.P. Funda, The role of gut microbiota (commensal bacteria) and the mucosal barrier in the pathogenesis of inflammatory and autoimmune diseases and cancer: contribution of germ-free and gnotobiotic animal models of human diseases, *Cellular & Molecular Immunology*, 8 (2011) 110–120.
103. A. Dussouchaud, J. Jacob, C. Secq, J.M. Verbavatz, M. Moras, J. Larghero, C.M. Fader, M.A. Ostuni, S.D. Lefevre, Transmission electron microscopy to follow ultrastructural modifications of erythroblasts upon ex vivo human erythropoiesis, *Frontiers in Physiology*, 12 (2021) 791691.
104. F. Cristofori, V.N. Dargenio, C. Dargenio, V.L. Miniello, M. Barone, R. Francavilla, Anti-inflammatory and immunomodulatory effects of probiotics in gut inflammation: a door to the body, *Frontiers in Immunology*, 12 (2021) 578386.
105. N.K. Jha, A. Sharma, S.K. Jha, S. Ojha, D.K. Chellappan, G. Gupta, K.K. Kesari, S. Bhardwaj, S.D. Shukla, M.M. Tambuwala, J. Ruokolainen, K. Dua, S.K. Singh, Alzheimer's disease-like perturbations in HIV-mediated neuronal dysfunctions: understanding mechanisms and developing therapeutic strategies, *Open Biology*, 10 (2020) 200286.
106. G. Kaur, N. Damodara, E. Feldstein, J. Dominguez, K.T. Huang, J.V. Ogulnick, R. Nuoman, P. Khandelwal, M. El-Ghanem, G. Gupta, S.A. Mayer, K. Amuluru, C.D. Gandhi, F. Al-Mufti, Relation between brain natriuretic peptide and delayed cerebral ischemia in patients with aneurysmalsubarachnoid hemorrhage, *Clinical Neurology and Neurosurgery*, 211 (2021) 107031.
107. A. Blesl, V. Stadlbauer, The gut-liver axis in cholestatic liver diseases, *Nutrients*, 13 (2021).
108. R. Thapa, O. Afzal, G. Kumar, A.A. Bhat, W.H. Almalki, S.I. Alzarea, I. Kazmi, A.S.A. Altamimi, V. Subramaniyan, L. Thangavelu, Unveiling the connection: long-chain non-coding RNAs and critical signaling pathways in breast cancer, *Pathology—Research and Practice*, 249 (2023) 154736.
109. M.S. Hussain, A.S.A. Altamimi, M. Afzal, W.H. Almalki, I. Kazmi, S.I. Alzarea, G. Gupta, M. Shahwan, N. Kukreti, L.S. Wong, V. Kumarasamy, V. Subramaniyan, Kaempferol: paving the path for advanced treatments in aging-related diseases, *Experimental Gerontology*, 188 (2024) 112389.
110. N.K. Jha, W.C. Chen, S. Kumar, R. Dubey, L.W. Tsai, R. Kar, S.K. Jha, P.K. Gupta, A. Sharma, R. Gundamaraju, K. Pant, S. Mani, S.K. Singh, R.B. Maccioni, T. Datta, S.K. Singh, G. Gupta, P. Prasher, K. Dua, A. Dey, C. Sharma, Y.H. Mughal, J. Ruokolainen, K.K. Kesari, S. Ojha, Molecular mechanisms of developmental pathways in neurological disorders: a pharmacological and therapeutic review, *Open Biology*, 12 (2022) 210289.
111. J.K. Xia, N. Tang, X.Y. Wu, H.Z. Ren, Deregulated bile acids may drive hepatocellular carcinoma metastasis by inducing an immunosuppressive microenvironment, *Frontiers in Oncology*, 12 (2022) 1033145.
112. R. Thapa, A. Goyal, G. Gupta, A.A. Bhat, S.K. Singh, V. Subramaniyan, S. Sharma, P. Prasher, V. Jakhmola, S.K. Singh, Recent developments in the role of protocatechuic acid in neurodegenerative disorders, *EXCLI Journal*, 22 (2023) 595.
113. X. He, Z. Qi, H. Hou, J. Gao, X.X. Zhang, Effects of chronic cadmium exposure at food limitation-relevant levels on energy metabolism in mice, *Journal of Hazardous Materials*, 388 (2020) 121791.

114. J. Xia, L. Lu, C. Jin, S. Wang, J. Zhou, Y. Ni, Z. Fu, Y. Jin, Effects of short term lead exposure on gut microbiota and hepatic metabolism in adult zebrafish, *Comparative Biochemistry and Physiology. Toxicology & Pharmacology*, 209 (2018) 1–8.
115. D. Parada Venegas, M.K. De la Fuente, G. Landskron, M.J. González, R. Quera, G. Dijkstra, H.J.M. Harmsen, K.N. Faber, M.A. Hermoso, Corrigendum: short chain fatty acids (SCFAs)-mediated gut epithelial and immune regulation and its relevance for inflammatory bowel diseases, *Frontiers in Immunology*, 10 (2019) 1486.
116. P. Prasher, M. Sharma, S.K. Singh, M. Haghi, R. McLoughlin, D.K. Chellappan, G. Gupta, K.R. Paudel, P.M. Hansbro, B.G.G. Oliver, Advances and applications of dextran-based nanomaterials targeting inflammatory respiratory diseases, *Journal of Drug Delivery Science and Technology*, 74 (2022) 103598.
117. Z. Jian, L. Zeng, T. Xu, S. Sun, S. Yan, S. Zhao, Z. Su, C. Ge, Y. Zhang, J. Jia, T. Dou, The intestinal microbiome associated with lipid metabolism and obesity in humans and animals, *Journal of Applied Microbiology*, 133 (2022) 2915–2930.
118. G. Gupta, D.K. Chellappan, M. Agarwal, M. Ashwathanarayana, S. Nammi, K. Pabreja, K. Dua, Pharmacological evaluation of the recuperative effect of morusin against aluminium trichloride (AlCl3)-induced memory impairment in rats, *Central Nervous System Agents in Medicinal Chemistry*, 17 (2017) 196–200.
119. G. Gupta, G. Singhvi, D.K. Chellappan, S. Sharma, A. Mishra, R. Dahiya, T. de Jesus Andreoli Pinto, K. Dua, Peroxisome proliferator-activated receptor gamma: promising target in glioblastoma, *Panminerva Medica*, 60 (2018) 109–116.
120. R. Thapa, G. Gupta, A.A. Bhat, W.H. Almalki, S.I. Alzarea, I. Kazmi, S. Saleem, R. Khan, N. Altwaijry, H. Dureja, A review of Glycogen Synthase Kinase-3 (GSK3) inhibitors for cancers therapies, *International Journal of Biological Macromolecules*, 253 (2023) 127375.
121. H.-C. Kim, T. Jang, H.-J. Chae, W.-J. Choi, M. Ha, B.-J. Ye, B.-G. Kim, M.-J. Jeon, S.-Y. Kim, Y.-S. Hong, Evaluation and management of lead exposure, *Annals of Occupational and Environmental Medicine*, 27 (2015) 1–9.
122. H.H. Azzaz, A.E. Kholif, H.A. Murad, E. Vargas-Bello-Pérez, A newly developed strain of Enterococcus faecium isolated from fresh dairy products to be used as a probiotic in lactating Holstein cows, *Frontiers in Veterinary Science*, 9 (2022) 989606.
123. S.S. Giri, S. Yun, J.W. Jun, H.J. Kim, S.G. Kim, J.W. Kang, S.W. Kim, S.J. Han, V. Sukumaran, S.C. Park, Therapeutic effect of intestinal autochthonous lactobacillus reuteri P16 against waterborne lead toxicity in cyprinus carpio, *Frontiers in Immunology*, 9 (2018) 1824.
124. B.A. Daisley, M. Monachese, M. Trinder, J.E. Bisanz, J.A. Chmiel, J.P. Burton, G. Reid, Immobilization of cadmium and lead by Lactobacillus rhamnosus GR-1 mitigates apical-to-basolateral heavy metal translocation in a Caco-2 model of the intestinal epithelium, *Gut Microbes*, 10 (2019) 321–333.
125. S. Bhatt, T. Devadoss, N.K. Jha, M. Baidya, G. Gupta, D.K. Chellappan, S.K. Singh, K. Dua, Targeting inflammation: a potential approach for the treatment of depression, *Metabolic Brain Disease*, 38 (2023) 45–59.
126. S. Bhatt, J. Kanoujia, S. Mohana Lakshmi, C.R. Patil, G. Gupta, D.K. Chellappan, K. Dua, Role of brain-gut-microbiota axis in depression: emerging therapeutic avenues, *CNS & Neurological Disorders Drug Targets*, 22 (2023) 276–288.
127. P. Średnicka, E. Juszczuk-Kubiak, M. Wójcicki, M. Akimowicz, M. Roszko, Probiotics as a biological detoxification tool of food chemical contamination: a review, *Food and Chemical Toxicology*, 153 (2021) 112306.
128. W.P. Chang-Liao, A. Lee, Y.H. Chiu, H.W. Chang, J.R. Liu, Isolation of a leuconostoc mesenteroides strain with anti-porcine epidemic diarrhea virus activities from kefir grains, *Frontiers in Microbiology*, 11 (2020) 1578.
129. R. Effendi, M. Anshory, H. Kalim, R.F. Dwiyana, O. Suwarsa, L.M. Pardo, T.E.C. Nijsten, H.B. Thio, Akkermansia muciniphila and faecalibacterium prausnitzii in immune-related diseases, *Microorganisms*, 10 (2022) 2382.
130. S. Bhatt, K. Anitha, D.K. Chellappan, D. Mukherjee, S. Shilpi, A. Suttee, G. Gupta, T.G. Singh, K. Dua, Targeting inflammatory signaling in obsessive compulsive disorder: a promising approach, *Metabolic Brain Disease*, 39 (2024) 335–346.
131. R. Thapa, G. Gupta, P. Dave, S.K. Singh, A. Raizaday, W.H. Almalki, G. Vyas, S.K. Singh, K. Dua, Y. Singh, Current update on the protective effect of epicatechin in neurodegenerative diseases, *EXCLI Journal*, 21 (2022) 897–903.

132. Z. Wang, L. Li, S. Wang, J. Wei, L. Qu, L. Pan, K. Xu, The role of the gut microbiota and probiotics associated with microbial metabolisms in cancer prevention and therapy, *Frontiers in Pharmacology*, 13 (2022) 1025860.
133. L. Yu, L. Zhang, H. Duan, R. Zhao, Y. Xiao, M. Guo, J. Zhao, H. Zhang, W. Chen, F. Tian, The protection of lactiplantibacillus plantarum CCFM8661 against benzopyrene-induced toxicity via regulation of the gut microbiota, *Frontiers in Immunology*, 12 (2021) 736129.
134. N.A. Abdelsalam, A.T. Ramadan, M.T. ElRakaiby, R.K. Aziz, Toxicomicrobiomics: the human microbiome vs. pharmaceutical, dietary, and environmental xenobiotics, *Frontiers in Pharmacology*, 11 (2020) 390.
135. F. Tian, Q. Zhai, J. Zhao, X. Liu, G. Wang, H. Zhang, H. Zhang, W. Chen, Lactobacillus plantarum CCFM8661 alleviates lead toxicity in mice, *Biological Trace Element Research*, 150 (2012) 264–271.
136. Y. Hamada, M. Goto, G. Nishimura, H. Nagasaki, Y. Seino, H. Kamiya, J. Nakamura, The alpha-glucosidase inhibitor miglitol increases hepatic CYP7A1 activity in association with altered short-chain fatty acid production in the gut of obese diabetic mice, *Metabolism Open,* 5 (2020) 100024.
137. K.S. Alharbi, W.H. Almalki, S.I. Alzarea, I. Kazmi, F.A. Al-Abbasi, O. Afzal, A.S.A. Altamimi, M. Albratty, A. Najmi, G. Gupta, Anaesthesia-induced changes in genomic expression leading to neurodegeneration, *CNS & Neurological Disorders Drug Targets*, 23 (2024) 411–419.
138. W.H. Almalki,s M.M. Ghoneim, S. Alshehri, S.S. Imam, I. Kazmi, G. Gupta, Sepsis triggered oxidative stress-inflammatory axis: the pathobiology of reprogramming in the normal sleep-wake cycle, *Molecular and Cellular Biochemistry*, 477 (2022) 2203–2211.
139. J.Y.L. Chiang, J.M. Ferrell, Discovery of farnesoid X receptor and its role in bile acid metabolism, *Molecular and Cellular Endocrinology*, 548 (2022) 111618.
140. Q. Zhai, L. Yang, J. Zhao, H. Zhang, F. Tian, W. Chen, Protective effects of dietary supplements containing probiotics, micronutrients, and plant extracts against lead toxicity in mice, *Frontiers in Microbiology*, 9 (2018) 2134.

10 Effect of Mercury on Gut Microbiome

Muhammad Afzal[1], Ahmed E. Altyar[2], Nehmat Ghaboura[3], Sattam Khulaif Alenezi[4], Khalid Saad Alharbi[4], Imran Kazmi[5], Mohammad Jaffar Sadiq Mantargi[1], Sana Saeed Alqarni[6], and Samiyah Alshehri[7]*

[1]Department of Pharmaceutical Sciences, Batterjee Medical College, Jeddah, Saudi Arabia
[2]Department of Pharmacy Practice, Faculty of Pharmacy, King Abdulaziz University, Jeddah, Saudi Arabia
[3]Department of Pharmacy Practice, Batterjee Medical College, Jeddah, Saudi Arabia
[4]Department of Pharmacology and Toxicology, College of Pharmacy, Qassim University, Buraydah, Al Qassim, Saudi Arabia
[5]Department of Biochemistry, Faculty of Science, King Abdulaziz University, Jeddah, Saudi Arabia
[6]Department of Clinical Laboratory Sciences, College of Applied Medical Science, King Saud University, Riyadh, Saudi Arabia
[7]Department of Pharmacology and Toxicology, College of Pharmacy, King Saud University, Riyadh, Saudi Arabia
[*]Corresponding author: Muhammad Afzal.
Email: mohmmad.afzal@bmc.edu.sa

10.1 INTRODUCTION

Mercury is known to be a ubiquitous worldwide contaminant that can take many different forms, such as elemental mercury (Hg0) and inorganic mercury (Hg+ and Hg2+). Its harmful effects on human health have been extensively researched, with a typical emphasis on how it affects important organs, including the liver, kidneys, and central nervous system [1]. These studies have shown that exposure to mercury has serious health concerns, which has led to strict legislation and public awareness efforts. The gut is a complex ecosystem that is home to billions of bacteria and is essential for immune system modulation, nutritional absorption, and digestion. Mercury was once mostly linked to organ-specific toxicity, but recent studies have started to provide insight on how it could dynamically interact with the complex microbial populations found in the digestive system [2]. Comprehending this unexplored domain is crucial, given the close relationship between the health of the gut microbiota and general well-being, which offers a nuanced view of the total effects of mercury exposure on human health.

10.1.1 Mercury Exposure and Its Varied Sources

Food contamination with mercury is a serious problem, particularly in areas where industrial activity contributes to environmental mercury pollution. Many people's go-to food is fish; however, fish may also store mercury through a process called bioaccumulation. Because eating smaller fish that already have mercury can accumulate, larger predatory fish, such as swordfish and shark, typically

DOI: 10.1201/9781003489221-10

have greater mercury levels [3]. In a similar vein, rice, a staple food for a large section of the world's population, may be grown in soil polluted with mercury, which causes the metal to accumulate in the grain. Mercury-containing medications also increase the risk of exposure, albeit less so than mercury found in food. In the past, dental amalgams and antiseptic creams, among other pharmacological and medicinal formulations, have included mercury. Because of its toxicity, the use of mercury in pharmaceuticals has decreased over time, but it is still a significant source of exposure, especially in some complementary and alternative medical procedures. Furthermore, airborne deposition of mercury on soil allows some bacteria to transform it into methylmercury, which is subsequently absorbed by plant roots, allowing mercury to enter the food chain. This makes it easier for mercury to pass from the soil to plants and then to higher-ranking animals in the food chain [4].

10.1.2 Interactions between Mercury and Gut Microbiome

Mercury can also enter the food chain by the airborne deposition of mercury on soil, where some bacteria can convert it into methylmercury, which is then absorbed by plant roots. Mercury may move more easily from the soil to plants and ultimately to higher-ranking animals in the food chain as a result. Numerous physiological functions, including as digestion, nutrition absorption, and immunological regulation, are significantly influenced by the gut microbiome, a huge population of microbes living in the digestive system [5, 6]. These microbes actively contribute to the gut's transformation of mercury compounds, which affects the dynamics of mercury exposure overall, according to recent research. Encoding proteins essential for the transformation of inorganic mercury (IHg) into methylmercury (MeHg), the hgcAB gene cluster has been identified as a major participant in mercury methylation (Figure 10.1) [7]. Comparably, the demethylation process that transforms MeHg back into IHg or elemental mercury involves a different group of genes found in the microbiome called the mer operon. These gene clusters' dynamic regulation indicates that the microbiome can control mercury speciation, which, in turn, affects the host organism's overall toxicological effect [8]. The gut microbiome is highly adaptive and sensitive to ambient mercury levels, as evidenced by the increased abundance of hgcAB and mer operon genes in response to exposure to mercury, whether in the form of IHg or in the form of MeHg. This increased expression points to a highly developed defensive mechanism or a metabolic adaptation by certain microbial communities to deal with stress brought on by mercury in the gut environment [9]. Furthermore, these interactions go beyond simple processes of methylation and demethylation. Exposure to mercury alters the makeup of the gut microbiota, which can have a domino impact on the host's health. Mercury's overall effects on gut health may be attributed to disruptions in the microbial balance, which can change immunological responses, inflammatory processes, and metabolic pathways [10].

10.1.3 Environmental Pollutants and Gut Dysfunction

Apart from mercury, gut dysfunction has been associated with environmental contaminants such as antibiotics, pesticides, and persistent organic pollutants. Comprehending this wider context is essential, as intestinal inflammation, disruption of the barrier, and changes in metabolites with regulatory roles can result from gut dysfunction [11, 12]. The way that mercury functions within this range of environmental contaminants complicates its effects on intestinal health. Mercury's harmful effects on essential organs have been well researched; however, its effects on the gut flora are still mostly unknown. Acknowledging the gut as a major target organ for mercury exposure provides new perspectives on the overall health effects of mercury poisoning [13]. This analysis has shed light on the complex relationships that exist between gut microbiota and mercury exposure, highlighting the need for more in-depth investigation to fully understand the scope of these relationships and any possible health risks [14].

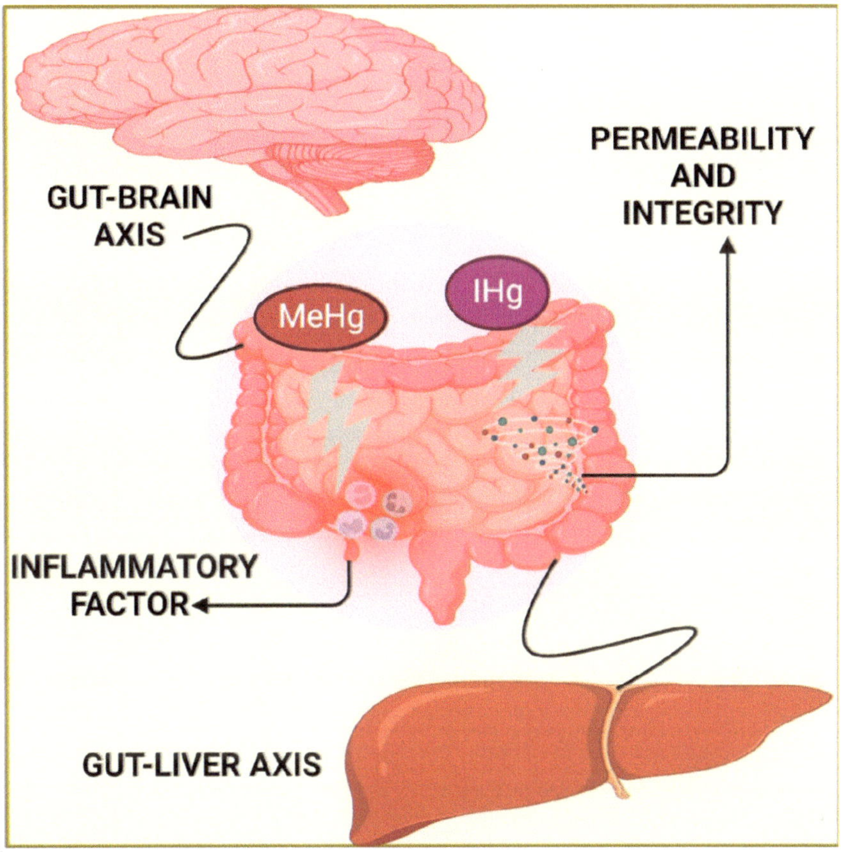

FIGURE 10.1 This image illustrates the comprehensive impact of inorganic and methylmercury on gut barriers, microbial disorders, and extraintestinal effects, emphasizing the gut's significance as a target for mercury toxicity.

10.2 MECHANISMS OF MERCURY TOXICITY

As one of the most potent environmental toxins known to science, mercury has a complex web of negative effects on human health. Its toxicity originates from a dual effect, which shows up as increased oxidative stress and focused attacks on particular organ systems. Mercury demonstrates its power in the area of oxidative stress by initiating a series of events that upset the delicate balance within the body [15]. Mercury is a strong redox-active element that promotes the production of ROS, such as hydrogen peroxide and superoxide radicals. DNA damage, protein oxidation, and lipid peroxidation result from this oxidative assault overpowering the body's natural antioxidant defense systems [16]. These molecular attacks have consequences that go beyond simple cellular damage, affecting the dynamics of proteins, the integrity of cell membranes, and the genomic stability that is encoded in DNA. Mercury also has an effect outside of cells at the same time. Its pernicious effects are confined to particular organ systems, the liver, kidneys, and central nervous system being its preferred targets [17]. As the main organ for detoxifying, the liver becomes a focal point for the buildup of mercury, which exacerbates hepatotoxicity. The buildup in the kidneys, which filter mercury, can cause nephrotoxicity and impair renal function. Mercury's neurotoxic effects are primarily felt by the central nervous system, which leads to wider neurological symptoms as well as cognitive deficits and abnormalities in motor skills [18]. This complex interaction between the induction of oxidative stress and targeted organ toxicity highlights the subtle aspects of mercury's attack on the human body and highlights the need for thorough research to fully understand its complexities and develop practical solutions for reducing mercury-related health risks [19].

10.2.1 Oxidative Damage: Unveiling Mercury's Reactive Nature

One of the main ways that mercury is hazardous to the body is by causing oxidative stress. Redox cycling can produce reactive oxygen species (ROS) such as superoxide radicals and hydrogen peroxide when elemental mercury (Hg0) and inorganic mercury (Hg+ and Hg2+) interact [20]. When these ROS levels are excessive, they overwhelm the body's antioxidant defense systems, causing lipid, protein, and DNA oxidative damage. Wide-ranging effects result from the oxidative damage caused by mercury. Unstable cell structures and membrane instability are caused by lipid peroxidation, which is the result of ROS attacking lipid membranes [21, 22]. A variety of physiological processes are impacted by protein oxidation, which modifies the structure and functionality of vital cellular proteins. Furthermore, oxidative stress-induced DNA damage can result in cellular malfunction and mutations. Comprehending the function of mercury in oxidative damage is essential because it sheds light on the direct effects on cellular constituents and advances our knowledge of mercury-induced diseases shown in many organs [23].

10.2.2 Organ-Specific Toxicity: Mercury's Varied Targets

Mercury poisoning has been primarily studied for its effects on the system, but it also has unique effects on certain organs. The principal organs affected by mercury poisoning are the liver, kidneys, and central nervous system [24].

10.2.2.1 Liver: The Detoxification Hub

After entering the circulation, mercury frequently builds up in the liver, which is an essential organ for detoxification. The liver experiences increased oxidative stress as a result of its attempts to absorb and remove mercury. Extended exposure might impair liver function and increase the risk of hepatotoxicity [25].

10.2.2.2 Kidneys: Filtration and Accumulation

When they remove mercury from the blood, the kidneys, which are in charge of this process, can sustain harm. Nephrotoxicity can result from the accumulation of inorganic mercury in renal tissues. Renal function may be compromised by an interference with the kidney's filtration and reabsorption functions [26, 27].

10.2.2.3 Central Nervous System: Neurotoxic Ramifications

Mercury poses a serious risk to the central nervous system because of its capacity to pass across the blood-brain barrier. Neurotoxicity is particularly evident in methylmercury, which impacts neurological health overall as well as motor abilities and cognitive function [28]. Various neurological indications associated with mercury exposure are partly caused by disruption of neurotransmitter pathways and activation of inflammation [29]. Essentially, the process of dissecting the toxicity of mercury entails exploring its dual effect via cellular oxidative destruction and its specific targeting of essential organs, providing a thorough understanding of the complex relationship between mercury and human physiology [30].

10.3 GUT MICROBIOTA'S INVOLVEMENT IN MERCURY PROCESSES

Examining the intricate relationship between exposure to mercury and the gut microbiota indicates a deep relationship influenced by certain gene clusters, hgcAB, and the mer operon. These gene clusters play a crucial role in controlling the path that mercury takes in the gut, mostly by means of methylation and demethylation processes. An essential component of mercury methylation is the hgcAB gene cluster, which initiates the transformation of inorganic mercury (IHg) into methylmercury (MeHg) [31]. However, the demethylation of MeHg back to IHg, or elemental mercury, is facilitated by the mer operon, a group of genes that includes merA and merB [32]. This dynamic interaction between these gene clusters and the gut microbiota highlights the complex abilities of the gut to regulate the toxicity and bioavailability of mercury compounds, while also highlighting the microbial effect on mercury transformation. Deciphering this mutualistic association is a major step toward comprehending the wider consequences of mercury exposure on human well-being [33].

10.3.1 Specific Gene Clusters: Orchestrating Mercury Transformation

Importantly, the existence of certain gene clusters – hgcAB and the mer operon – determines the gut microbiota's involvement in mercury activities. Methylmercury (MeHg) is produced when inorganic mercury (IHg) is converted to methylmercury (MeHg) thanks to the hgcAB gene cluster, which encodes proteins (Figure 10.2). Conversely, elemental mercury, or MeHg, is demethylated back into IHg by the mer operon, which is made up of the merA and merB genes [34]. These gene clusters function as microbial tools, allowing the gut microbiota to actively engage in the intricate changes that mercury undergoes in the digestive environment. These genes' increased abundance in response to exposure to mercury emphasizes the adaptability of some microbial communities and shows their important role in the gut's mercury cycle [35, 36].

10.3.2 Mercury Methylation and Demethylation: Dynamic Processes within the Gut

The mer operon and hgcAB catalyze the dynamic processes of mercury methylation and demethylation, which are essential for mercury speciation in the gut [37]. The corrinoid protein that the hgcA gene, which is part of the hgcAB cluster, codes for is necessary for Hg methylation in anaerobic bacteria. On the other hand, MerA, a crucial enzyme in the mer operon, functions as a mercuric reductase and helps to demethylate mercury [38]. These complex metabolic processes demonstrate how well the gut microbiota may control the speciation of mercury, which affects the toxicity and bioavailability of mercury compounds. These gene clusters regulate the intricate balance between methylation and demethylation activities, which determines how mercury affects the gut environment overall and may have impacts beyond the intestines [39].

10.3.3 Role of Gut Microbiota in Mercury Transformation: Beyond Methylation

Beyond simple methylation and demethylation, the gut microbiota plays a more significant role in the metabolism of mercury. The microbial populations present in the gut play a proactive role in modifying the chemical forms of mercury and impacting the organism's destiny [40, 41]. The overall toxicological results and extraintestinal consequences linked to mercury exposure are influenced by the gut microbiota's capacity to regulate speciation of mercury [42]. Furthermore, new research has started to clarify the intricate interactions between mercury-induced gut dysfunction and the gut flora, providing insight into the wider health consequences [43]. The complex role that gut bacteria plays in the transformation of mercury highlights the need for more study to understand the subtle dynamics and possible treatments that target the gut microbiome to reduce the risks of mercury-related illness [44].

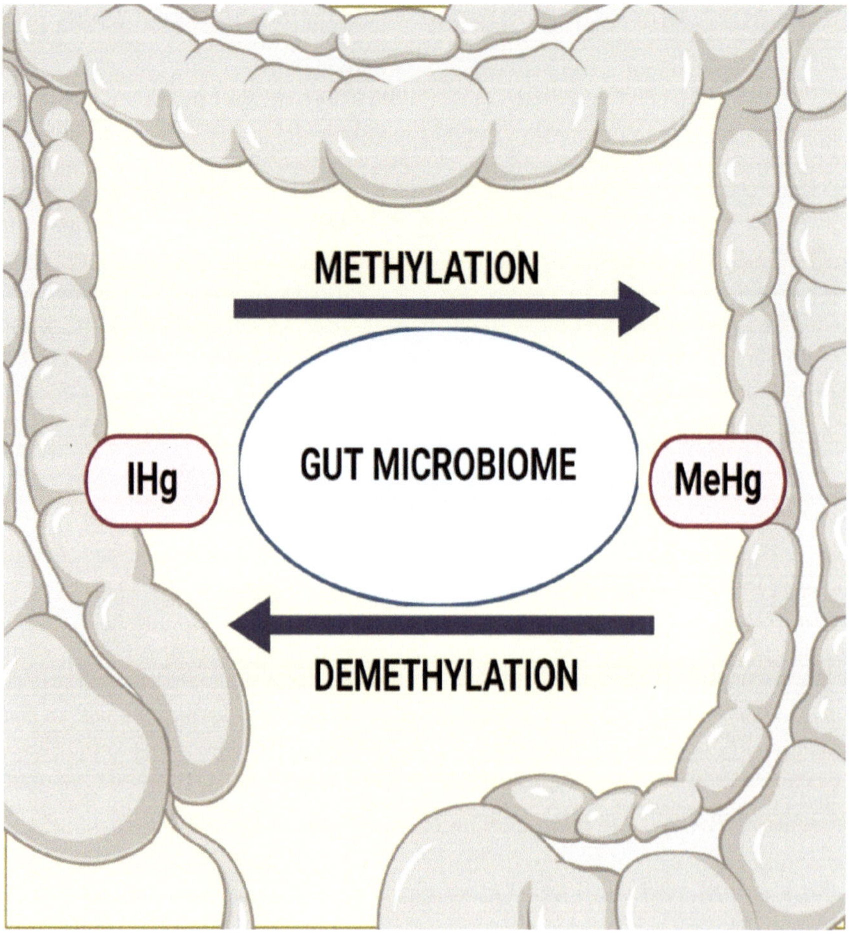

FIGURE 10.2 This image illustrates the intricate process of mercury (Hg) biotransformation in the gut, showcasing the conversion between IHg and MeHg forms. The study emphasizes the pivotal role of intestinal microbiota in Hg transformation.

10.4 GUT MICROBIOTA'S INVOLVEMENT IN MERCURY PROCESSES

Within the complex landscape of gut microbiota mercury activities, several gene clusters – most notably, hgcAB and the mer operon – emerge as important conductors. These gene clusters are important because they affect the methylation and demethylation processes, which are crucial for controlling the fate of mercury in the gut environment [45]. Mercury methylation is a complicated dance, with the hgcAB gene cluster in the core. This group, which includes the genes hgcA and hgcB, encodes the proteins needed to convert inorganic mercury (IHg) into methylmercury (MeHg), which is more hazardous and bioavailable [46]. Anaerobic bacteria in the gut undergo a mercury transformation cycle that includes this methylation process, which is mediated by the corrinoid protein encoded by hgcA [47]. On the other hand, the key to demethylating mercury back into elemental mercury or IHg is the mer operon, which includes the merA and merB genes. MerB participates in additional demethylation processes, while MerA functions as a mercuric reductase, aiding in the demethylation process [48]. The mer operon's coordinated activity guarantees the removal and detoxification of methylmercury, concluding the complex cycle of mercury transformation that the gut microbiota has been coordinating [49]. Comprehending the intricate functions of these distinct gene clusters illuminates the gut microbiota's significant impact on mercury speciation and, in turn, the general toxicological consequences linked to mercury exposure [50].

The gut microbiota's dynamic and complex interactions with mercury methylation and demethylation processes shape how mercury is metabolized in the human body. Particularly, the gut microbiota actively participates in these transformation processes by containing certain gene clusters such as the mer operon and hgcAB [51]. A major participant in mercury methylation, the hgcAB gene cluster produces the proteins needed to convert inorganic mercury (IHg) into the more hazardous methylmercury (MeHg) [52]. Enzymes such as HgcA, a corrinoid protein that is essential to anaerobic bacteria, aid in this methylation process. Conversely, the mer operon, which includes genes such as merA, controls the demethylation mechanism that transforms MeHg back into IHg, or elemental mercury [53]. The complex dance between methylation and demethylation maintains a dynamic balance by controlling the amount of hazardous methylmercury present in the gastrointestinal tract [54]. The adeptness of the gut microbiota in regulating these mechanisms underscores its crucial function in regulating the bioavailability and toxicity of mercury, offering vital perspectives on the complex correlation between mercury exposure and the gut microbiome [55]. The dynamic function that the gut microbiota plays is crucial in influencing the complicated processes of mercury transformation within the gut [56]. Leading the charge in this microbial participation are certain gene clusters that control mercury methylation and demethylation, such as hgcAB and the mer operon [57]. Table 10.1 summarizes key aspects of gut microbiota's role in mercury transformation, focusing on gene clusters hgcAB and mer operon, their functions, and toxicological implications for mercury exposure. These microbial communities actively assist in changing the chemical forms of mercury as it passes through the digestive system, influencing its eventual fate inside the host body. Recent studies demonstrate that the gut microbiota's involvement in mercury transformation goes beyond the traditional comprehension of methylation and demethylation mechanisms [58]. Mercury speciation is modulated by microbial populations in the stomach, which affects the metal's overall toxicity and bioavailability [59]. The gut microbiota's transformational potential highlights the necessity of a thorough investigation of the gut's microbial environment in order to understand the complex dynamics of changes brought on by mercury [60]. Comprehending the function of gut microbiota in mercury transformation contributes to our understanding of the complex regulatory processes of the gut and can help design focused therapies to reduce health concerns related to

TABLE 10.1

Summary of key aspects of gut microbiota's role in mercury transformation, focusing on gene clusters hgcAB and mer operon, their functions, and toxicological implications for mercury exposure.

Gene Clusters	Mercury Transformation Role	Gut Microbiota Function	Toxicological Implications
hgcAB	Converts IHg to MeHg	Active in mercury methylation process	Enhances bioavailability, increases mercury toxicity
mer operon	Controls demethylation, IHg production	Maintains dynamic balance, detoxifies MeHg	Regulates MeHg levels, mitigates mercury toxicity
Gut Microbiota	Influences mercury transformation	Actively participates, modulates mercury speciation	Shapes gut's regulatory processes, affects toxicity and bioavailability
Complex Dynamics	Gut microbiota's transformative potential	Beyond traditional methylation, demethylation roles	Highlights need for thorough investigation, aids in designing targeted therapies

mercury exposure [61]. Figure 10.2 shows the essential function of gut microbiota in Hg biotransformation, influencing Hg bioaccumulation.

10.5 ORAL EXPOSURE AND GUT–BRAIN AXIS

Recent developments in metabolomics and gene sequencing have greatly expanded our knowledge of the complex interactions between oral mercury exposure and the gut–brain axis. The intricate microbial communities found in the gut have been revealed by gene sequencing methods, including 16S rRNA sequencing and metagenomics, which have also shed light on the interactions between mercury and certain bacterial populations [62, 63]. A thorough investigation of the functional potential of the gut microbiota and its reaction to mercury exposure is made possible by this in-depth genetic research [52]. Concurrently, metabolomics – a potent technique for examining tiny molecules involved in metabolic processes – has illuminated the dynamic alterations in the biochemical landscape of the gut resulting from oral mercury intake [64]. The detection and measurement of metabolites linked to changes brought about by mercury provide a comprehensive understanding of the molecular mechanisms affecting the gut–brain axis [65]. All of these technical developments help to clarify the complex processes via which oral mercury exposure affects the gut microbiota, changes metabolic pathways, and eventually causes neurotoxicity via the gut–brain axis [66].

The complex link between gut microbiota dysfunction and oral methylmercury (MeHg) intake is an important part of the larger investigation of the gut–brain axis. Research has demonstrated how MeHg upsets the delicate balance of microbial populations in the gut [67]. MeHg is a neurotoxic form of mercury that affects gut microbiota activity and composition in addition to crossing the blood-brain barrier. Dysbiosis results from this disturbance, which takes the form of changes in microbial diversity [68, 69]. MeHg exposure has been linked to alterations in the prevalence of particular bacterial taxa, which may have an impact on microbial activities and metabolic pathways essential for gut homeostasis. MeHg-induced disruption of the gut microbiota that results has effects that extend beyond the digestive tract and increase the risk of neurotoxicity due to the complex interactions between the gut and the brain [70, 71]. The central nervous system may be affected by the inflammatory and neuroactive chemicals produced by the dysregulation of gut microbial ecosystems. Gaining knowledge about the subtleties of MeHg-induced gut microbiota disturbance in relation to the gut–brain axis is essential for comprehending the complex effects of mercury exposure on neurological and gastrointestinal health [72, 73]. Table 10.2 summarizes the intricate interplay between oral mercury exposure and the gut–brain axis, exploring metabolomics, gut microbiota responses, MeHg-induced dysbiosis, and inflammatory factors' impact on neurotoxicity. One important aspect of the complex relationship between oral exposure to mercury and the gut–brain axis is the production of inflammatory factors and the subsequent consequences they have beyond the stomach [74]. Research on the effects of mercury on gut microbiota has shown that oral exposure to this environmental toxin causes the release of components that inflame the gastrointestinal environment [75]. This inflammatory response affects distal organs, particularly the central nervous system, and is typified by the rise in pro-inflammatory mediators and cytokines [76]. The gut–brain axis is a communication channel, and changes in the gut microbiota brought on by mercury exposure might set off signaling pathways that affect neurotoxicity and neuroinflammation [77–79]. In addition to harming intestinal mucosa, the release of inflammatory factors sets off a systemic reaction that may result in neurological issues [80]. Comprehending this complex relationship offers important insights into the extraintestinal consequences of mercury exposure, highlighting the necessity for thorough studies that take the gut and the brain into account as essential elements in the range of mercury-induced health impacts [81].

TABLE 10.2
Summary of the intricate interplay between oral mercury exposure and the gut–brain axis, exploring metabolomics, gut microbiota responses, MeHg-induced dysbiosis, and inflammatory factors' impact on neurotoxicity.

Metabolomics and Gene Sequencing	Gut Microbiota Response	MeHg-Induced Dysbiosis	Inflammatory Factors and Neurotoxicity
Expands knowledge, reveals gut communities	Examines potential, illuminates interactions	Upsets microbial balance, affects diversity	Triggers systemic reactions, contributes to neurological issues
Illuminates' alterations in gut's biochemical landscape	Insights into mercury's impact on microbiota	Linked to changes in bacterial taxa, affects homeostasis	Sets off signaling pathways, affects neuroinflammation
Clarifies processes, understands gut–brain axis	Analyzes disturbance, highlights risks	Extends beyond digestive tract, increases neurotoxicity risk	Sets off signaling pathways, affects neuroinflammation
Unveils gut's intricate microbial interactions	Studies gut's response to mercury exposure	Disturbance in microbial diversity, alters homeostasis	Provokes systemic responses, impacts neurological functions
Advances knowledge, unveils gut–brain axis intricacies	Reveals microbial reactions to mercury exposure	Dysbiosis induced by MeHg, affects microbial activities	Induces inflammatory factors, poses neurological risks

10.6 OVERVIEW OF RECENT RESEARCH ON MERCURY AND GUT MICROBIOME

Mercury is a naturally occurring, extremely poisonous, non-essential element that is bad for the environment and people's health [82]. Understanding the sources and routes via which mercury cycles through the ecosystem, human exposure, and effects on the environment and human health have been the main areas of recent mercury research advancements [83]. Studies on behavior and genetics indicate that illness processes and outcomes are determined by age (at testing as well as at exposure), mercury levels, and genetic background. Additionally, recent research has demonstrated the critical role gut bacteria plays in detoxification and the biotransformation of mercury [84]. In addition to interfering with the bioavailability and toxicity of metals in humans, the gut microbiota may also influence the speciation and bioaccumulation of mercury in fish. According to recent study, it may be possible to use the bacteria found in the human stomach to lessen the body's absorption of hazardous metals like mercury [85]. However, little is known about the connection between mercury and gut flora, and further research is required to properly comprehend the part gut microbiota plays in mercury poisoning [86]. Mercury exposure is one heavy metal that significantly affects the gut flora, according to recent studies. Wetland vegetation affects mercury methylation, with vegetated sediments demonstrating a larger capacity for mercury methylation, according to research done in the St. Louis River estuary [87]. The study also showed that enhanced net methylmercury (MeHg) synthesis in unvegetated sediments absorbing leachate was accompanied by changes in microbiome structure and metagenomic alterations, indicating the role of wetland vegetation in regulating MeHg production [88]. The demethylation effectiveness of gut microbiota can be influenced by mercury exposure levels and sources, according to research on MeHg demethylation processes in humans. The work brought to light the possible contributions of certain gut microbiota to intestinal MeHg demethylation, including *Desulfovibrio a*nd Methanogens [89]. According to research on common carp, exposure to heavy metals, such as mercury, can change the makeup of the gut microbiota.

The gut microbiota of carp exposed to mercury showed a considerable increase in the number of pathogenic bacteria, including Aeromonas [90]. This finding raises the possibility of a connection between fish health, gut microbiota, and heavy metal exposure [91]. Tanzanian research looked into the possibility of probiotic-supplemented yogurt to reduce heavy metal levels, particularly mercury, in toddlers and pregnant women. The study demonstrated the potential of probiotics in reducing the negative effects of heavy metal exposure by showing that consuming probiotic yogurt protected against a further rise in mercury levels [92]. Together, these findings highlight the intricate and reciprocal relationships that exist between the gut microbiota and heavy metals, including mercury, as well as the possibility that probiotics and other therapies may be able to lessen the negative consequences of heavy metal exposure [93, 94]. An enzyme that changes more absorbable forms of mercury into less absorbable ones in the stomach is expressed by probiotic bacteria that were created for this purpose in research. In the long run, the researchers think that this may be utilized to reduce human exposure to mercury, particularly in vulnerable groups [95]. It demonstrates how exposure to mercury alters the variety and makeup of the gut microbiota and causes ferroptosis, a controlled cell death process, in the liver. The paper also shows that liver damage caused by mercury exposure can be lessened by altering the gut microbiota using antibiotics or fecal microbiota transplants [96].

10.7 CLINICAL MANIFESTATIONS

Several excellent reviews have discussed the symptoms and indicators of mercury poisoning from metallic, inorganic, and organic sources [97, 98].

10.7.1 Metallic Mercury

There have been cases of metallic mercury poisoning through the alimentary canal, albeit extremely rare. Mercury has long been believed to be innocuous in its pure metallic state; yet, this state can be maintained only when mercury is stored in a vacuum or in an atmosphere devoid of ions [99]. When metals come into touch with physiological fluids, such those in the gastrointestinal tract, mercuric ions can be slowly released. The majority of enzymes' thiol or sulphydryl groups can then react with them to reduce their activity. Every tissue protein is attracted to mercuric ions [100]. As per Sollman's account, metallic mercury was utilized in dosages ranging from 100 to 500 g to address adynamic ileus during the 18th century. While it usually went through the feces unharmed, on occasion it resulted in stomatitis, diarrhea, and even fatal poisoning. These days, the rupture of mercury-filled sacks that comprise the head of the intestinal tube is the most common method that metallic mercury damages the gastrointestinal tract [101, 102]. As reported by Birnbaum (47) in 1947, a patient with obstructive carcinoma of the rectosigmoid colon had preoperative therapy with a mercury weighted Miller-Abbott tube. After the bag burst, films revealed that the appendix contained mercury. What the surgical and pathologic examinations found in the appendix was an acute inflammatory condition [103]. In 1949, Chodoff (48) reported that a burst balloon had caused mercury to flow out of an intestinal tube after the patient's right colon was removed. Three fistulas developed close to the anastomotic site. During autopsy, granulomatous tissue containing mercury was found next to the fistulas. Lindenmuth (49) reported the formation of a fecal fistula in a patient who had a mercury bag burst following an appendectomy in 1949 [104–106]. In the lower right quadrant, mercury was detected during radiologic testing. A persistent granulomatous lesion was discovered upon removal of the fistula, and a barium enema indicated no further disease. Last but not least, Crikelair and Hiratzka (50) reported a second granulomatous reaction in an inflamed and ruptured appendix in 1953, a year after a mercury bag burst. Consequently, it is evident that there are consequences associated with having metallic mercury in the stomach [107, 108].

10.7.2 Inorganic Mercury

The gastrointestinal system is at risk from inorganic mercury salts when they are in the mercuric state rather than the mercurous form. They must also be soluble in the digestive system; the most soluble compounds are sulfide, the least soluble are iodide and chloride, and oxides and ammoniated forms have intermediate solubility [109]. The insoluble forms of mercurous salts appear to be less dangerous. When mercuric ions come into contact with other tissues, necrosis and protein precipitation occur, leading to the formation of mercuric sulfide. This explains why mercuric ions have a toxic effect [110, 111]. The local precipitation of mercuric sulfide in the capillary endothelium, which is produced by the interaction of mercuric ions with tissue proteins, is what leads to bacterial invasion of this necrotic material [112]. Most of the problems that arise after are the result of this precipitation. This process results in the mucous membrane of the upper gastrointestinal tract becoming superficially corroded or ulcerated. The astringent metallic taste is the first symptom to be noticed. This is followed by intense thirst, a burning, swollen, ashy, or pale discoloration of the tongue and throat, and stomach pain. Vomitus may show bloodstains [113–115]. Severe inorganic mercury poisoning is more prevalent and can lead to delayed renal and large intestine damage as well as ulceration, bleeding, diarrhea, and tenesmus. The most likely cause of this is the bacterial infiltration of the wounded tissue that follows the precipitation of mercuric sulfuride [116]. Large bowel symptoms may not appear for several weeks after ingesting these salts, and even then, they may be fatal [117]. The large intestine had been hemorrhaged and ulcerated, and the mucous membranes covering the upper digestive tract had been corroded and clogged, according to the autopsy. These findings are not conclusive, but they are suggestive [118, 119]. When inorganic mercury poisoning occurs, the symptoms usually include increased salivation, stomatitis, gingivitis, and a metallic taste accompanied with tremor, not the obvious gastrointestinal symptoms of more severe poisoning that have been documented in the past [120].

10.7.3 Organic Mercury

The aryl and alkyl salts all irritate mucosal membranes when taken orally, and this can lead to an acute symptom complex that includes nausea, vomiting, diarrhea, and perhaps severe tenesmus [121, 122]. Excessive fatigue, malaise, myalgias, and gastrointestinal issues may be misdiagnosed as viral disease or enteritis. In the finished product, the aryl and alkyl salts behave quite differently. Few cases of persons becoming ill after being exposed to phenylmercury salts have been reported [123, 124]. The relative long-term oral toxicities of phenylmercuric and mercuric salts have been investigated in rats. Rats given phenylmercuric salt showed increased mortality and growth impairment [125, 126]. The cecums of two rats who were given the aryl salt had an acute ulcerative, transmural process, but many other intestinal specimens were normal; the etiology of ulcerative colitis may be affected by these data. Because controlled trials, including the intake of graded doses of aryl compounds, have not been undertaken and probably will not be, studies on humans on the long-term toxicity of ingested phenylmercury are not comparable [127–129].

Conversely, the aklyl salts are quite poisonous. These drugs have the potential to produce both acute and chronic oral poisoning, with obviously neurological rather than gastrointestinal symptoms [130]. Symptoms include ataxia, difficulty speaking, sensory loss in the limbs, difficulty eating, decreased eyesight and hearing, and changes in intellectual function may appear after a latent period of weeks to months. There may also be headaches, fatigue, and difficulty focusing [131]. Severe neurologic impairment is the overall picture. On the other hand, ingesting inorganic and aryl organic substances, both acutely and over time, causes a set of symptoms that are mostly associated with the digestive system. Certain gastrointestinal symptomatology is brought on by mercurial alkyl organic compound ingestion, at least temporarily [69, 132].

10.8 METABOLISM OF THE INORGANIC AND ORGANIC SALTS INTAKE AND UPTAKE

In the small intestine and stomach, salts, both organic and inorganic, are either totally or nearly totally absorbed. Most observers believe that organic salts are better absorbed than inorganic forms of mercury, despite the fact that exact data on this matter is lacking [133]. Studies conducted on rats and humans reveal that more than 90% of the methylmercury in food is absorbed [134, 135]. Thus far, research has not been done on the process by which alkyl and aryl chemicals are transported across the intestinal wall or on the presence of a specific optimal site in the small intestine for the absorption of organic molecules [136, 137].

However, studies have been conducted on the mechanism of divalent mercuric ion absorption by the gut wall [138, 139]. The ileum and duodenum absorb the most divalent mercury, while the jejunum absorbs the least, according to research done by Sahagian and colleagues on the uptake of divalent mercury in isolated intestinal wall strips from the duodenum, jejunum, and ileum as well as in everted intestinal sacs [140, 141]. Additionally, they discovered that the mercuric ion uptake exhibited every characteristic of a passive transport mechanism. Finally, Cikrt emphasized that the duodenum did not absorb the most vitamin C12 through everted intestinal sacs; rather, ileal segments did [142, 143].

10.9 CONCENTRATION IN ORGANS, BIOTRANSFORMATION, AND EXCRETION

After ten to 15 years of oral, intramuscular, intravenous, and subcutaneous studies in humans and lab animals, we can conclude that although alkyl and aryl compounds are absorbed intact, their distribution and "biotransformation," or degradation and excretion, differ, which accounts for their varying toxicities [144, 145]. The distribution of mercury in the bodies of laboratory animals given a single dosage of phenylmercury via various routes is comparable to that of inorganic mercury; more mercury is eliminated in their feces and is concentrated in their liver and kidney. Inorganic mercury is produced when the phenylmercury salt enters the circulation and is broken down by the liver, kidney, and perhaps other tissues [146, 147]. Whether this is an active process requiring energy consumption is unknown at this time. The bulk of the mercury is then eliminated in the urine, with most of it staying in its inorganic form; 10 to 15% of the mercury is left in its unaltered, organic form after being ejected in the feces [148, 149]. A steady state is attained when intake and excretion are almost equal when intake is maintained. Finally, no phenylmercury has been found in the central nervous systems of the animals after they were given this salt [150]. Methylmercury and other alkyl salts are more widely distributed and accumulate in nearly all tissues; the largest amounts are seen in the liver, kidney, red blood cells, and brain and nervous system. Following this, methylmercury undergoes a slow degradation that eventually leaves it predominantly in an inorganic form within the kidney and liver [151, 152]. The primary excretion pathways appear to be the stomach and liver, with feces making up around half of the organic form [153, 154]. Interesting studies of the distribution and excretion of intravenous injections of methylmercury, phenylmercury, and mercuric chloride salts were carried out by Berlin and Ullberg using three different autoradiographic techniques [155, 156]. The mucosa of the liver and whole gastrointestinal tract collected mercuric chloride, with the colonic mucosa and its contents displaying the greatest concentration of action. This suggests that the excretion of mercury involves both the colon and the liver. After 16 days, the colon saw the majority of the activity [157]. Similar distribution patterns apply to methyl and phenylmercury, with the colon and its contents exhibiting the greatest activity. Unfortunately, there isn't a comparison of the three salts' levels of activity in the stomach during the trials or in the colon on day 16, when the study was canceled [158, 159].

This suggests that the excretion of mercury involves both the colon and the liver. After 16 days, the colon saw the majority of the activity. Similar distribution patterns apply to methyl and phenylmercury, with the colon and its contents exhibiting the greatest activity [160]. Unfortunately, there isn't a comparison of the three salts' levels of activity in the stomach during the trials or in the colon on day 16, when the study was canceled. Mercury was found in the cytoplasm of the colon in the deep-seated epithelial cells of the Lieberkuhn crypts, certain connective tissue cells of the lamina propria, and the capillary walls. After a single injection of mercury chloride, the mercury did not entirely disappear from the intestinal wall or mucosa; this process required 19 days [161–163]. The effects of administering radioactive methylmercury compounds orally to healthy human participants have been evaluated in three studies. Three participants were studied by Ekman et al. after methylated mercury-203 was consumed [164]. They arrived at the following conclusions: the liver was the location of major activity on whole body scanning; feces were the principal route of excretion of mercury-203; and methylmercury had a biologic half-life of 65 to 74 days, which is consistent with previous study [163, 165]. The same conclusions were achieved by Aberg and associates (84). Humans excrete methylmercury at a far slower pace than rats, according to Ekman's results. This is probably related to the greater rates of excretion and turnover of epithelial cells in rats, a cycle that happens five times faster in rats than in humans [166, 167].

10.10 CONCLUSION

To sum up, research into how mercury affects the gut microbiome shows how microbial ecosystems and environmental toxicity interact in fascinating ways. The gut, which is frequently disregarded in conventional mercury studies, turns out to be an important component in the complex dynamics of mercury exposure. Mercury speciation in the gut is regulated by methylation and demethylation processes that are coordinated by two specific gene clusters, hgcAB and the mer operon. The aforementioned discovery not only enhances our comprehension of the gut microbiota's adaptive reactions but also highlights its function in determining the bioavailability and toxicity of mercury-containing substances. The systemic consequences of gut microbiota changes are highlighted by the extraintestinal effects of oral mercury exposure, especially on the gut–brain axis. The gastrointestinal milieu releases inflammatory substances that have an influence not just on the central nervous system but also on distant organs. Acknowledging the wider consequences of gut dysfunction caused by mercury creates opportunities for a more comprehensive understanding of mercury poisoning.

Future study is desperately needed to fully understand the intricate relationship between mercury exposure and the gut flora. There is hope for reducing the dangers of mercury exposure to extraintestinal health by investigating possible therapies to modify gut flora. Furthermore, it is critical to address open-ended issues about the long-term effects of gut dysfunction caused by mercury and the implications for environmental policy and public health. The combination of cutting-edge technologies like metabolomics and metagenomics will open the door to a more sophisticated comprehension of the underlying molecular processes. Essentially, there are several facets to the impact of mercury on the gut microbiota, necessitating more research. By thoroughly dissecting the relationships that exist between gut microbiota, mercury, and overall health, we may better guide public health initiatives, regulatory actions, and interventions for a future that is healthier and more robust.

REFERENCES

1. R.M. Aaronson, H.M. Spiro, Mercury and the gut, *The American Journal of Digestive Diseases*, 18 (1973) 583–594.
2. V. Andreoli, F. Sprovieri, Genetic aspects of susceptibility to mercury toxicity: an overview, *International Journal of Environmental Research and Public Health*, 14 (2017).
3. J.J. Berzas Nevado, R.C. Rodríguez Martín-Doimeadios, F.J. Guzmán Bernardo, M. Jiménez Moreno, A.M. Herculano, J.L. do Nascimento, M.E. Crespo-López, Mercury in the Tapajós River basin, Brazilian Amazon: a review, *Environment International*, 36 (2010) 593–608.

4. T.Y. Chan, Inorganic mercury poisoning associated with skin-lightening cosmetic products, *Clinical Toxicology*, 49 (2011) 886–891.
5. S. Chandelia, V. Vohra, S. Prasad, M.K. Tiwari, K. Nallasamy, S.K. Angurana, Acute mercury poisoning in young children: when to treat?, *Indian Journal of Pediatrics*, 90 (2023) 629.
6. M.S. Hussain, A. Sharma, R. Kumar, Prebiotics and probiotics: a focused review of applications in respiratory disorders, *Carpathian Journal of Food Science and Technology*, 15 (2023) 183–207.
7. T.W. Clarkson, L. Magos, The toxicology of mercury and its chemical compounds, *Critical Reviews in Toxicology*, 36 (2006) 609–662.
8. S.A. Counter, L.H. Buchanan, Mercury exposure in children: a review, *Toxicology and Applied Pharmacology*, 198 (2004) 209–230.
9. U. Da Broi, C. Moreschi, A. Colatutto, B. Marcon, S. Zago, Medico legal aspects of self-injection of metallic mercury in cases of suicide or self-harming, *Journal of Forensic and Legal Medicine*, 50 (2017) 12–19.
10. M. Evans, Mercury poisoning, *British Medical Journal*, 1 (1962) 1458–1459.
11. K.G. Homme, J.K. Kern, B.E. Haley, D.A. Geier, P.G. King, L.K. Sykes, M.R. Geier, New science challenges old notion that mercury dental amalgam is safe, *Biometals*, 27 (2014) 19–24.
12. M.S. Hussain, G. Gupta, R. Mishra, N. Patel, S. Gupta, S.I. Alzarea, I. Kazmi, P. Kumbhar, J. Disouza, H. Dureja, N. Kukreti, S.K. Singh, K. Dua, Unlocking the secrets: Volatile Organic Compounds (VOCs) and their devastating effects on lung cancer, *Pathology, Research and Practice*, 255 (2024) 155157.
13. J.K. Kern, D.A. Geier, L.K. Sykes, B.E. Haley, M.R. Geier, The relationship between mercury and autism: a comprehensive review and discussion, *Journal of Trace ELEMENTS in Medicine and Biology,* 37 (2016) 8–24.
14. Q. Lu, Z. Liu, X. Chen, Mercury poisoning through intravenous administration: two case reports with literature review, *Medicine*, 96 (2017) e8643.
15. R. MacLehose, G. Pitt, S. Will, A. Jones, L. Duane, S. Flaherty, D. Hannant, B. Stuttard, A. Silverwood, K. Snee, V. Murray, Q. Syed, I. House, M.A. Bellis, Mercury contamination incident, *Journal of Public Health Medicine*, 23 (2001) 18–22.
16. L.C. Masur, A review of the use of mercury in historic and current ritualistic and spiritual practices, *Alternative Medicine Review: A Journal of Clinical Therapeutic*, 16 (2011) 314–320.
17. J. Mutter, J. Naumann, R. Schneider, H. Walach, B. Haley, Mercury and autism: accelerating evidence?, *Neuro Endocrinology Letters*, 26 (2005) 439–446.
18. K.L. Nuttall, Interpreting mercury in blood and urine of individual patients, *Annals of Clinical and Laboratory Science*, 34 (2004) 235–250.
19. Y. Pan, Y. Song, Broken thermometer in hand: mercury elevation caused by improper debridement, *Clinical toxicology*, 60 (2022) 654–656.
20. M. Rafati-Rahimzadeh, M. Rafati-Rahimzadeh, S. Kazemi, A.A. Moghadamnia, Current approaches of the management of mercury poisoning: need of the hour, *Daru: Journal of Faculty of Pharmacy, Tehran University of Medical Sciences*, 22 (2014) 46.
21. J.B. Rocha, M. Aschner, J.G. Dórea, S. Ceccatelli, M. Farina, L.C. Silveira, Mercury toxicity, *Journal of Biomedicine & Biotechnology*, 2012 (2012) 831890.
22. D. Saint-Phard, B. Van Dorsten, Mercury toxicity: clinical presentations in musculoskeletal medicine, *Orthopedics*, 27 (2004) 394–397; quiz 398–399.
23. H. Satoh, Occupational and environmental toxicology of mercury and its compounds, *Industrial Health*, 38 (2000) 153–164.
24. P.B. Tchounwou, W.K. Ayensu, N. Ninashvili, D. Sutton, Environmental exposure to mercury and its toxicopathologic implications for public health, *Environmental Toxicology*, 18 (2003) 149–175.
25. L. Barregård, D. Ellingsen, J. Alexander, Y. Thomassen, J. Aaseth, Mercury exposure from dental amalgam. Risk assessment and clinical evaluation, *Tidsskrift for den Norske laegeforening : tidsskrift for praktisk medicin, ny raekke*, 118 (1998) 58–62.
26. B.G. Campbell, Broadsheet number 48: Mercury, cadmium and arsenic: toxicology and laboratory investigation, *Pathology*, 31 (1999) 17–22.
27. K.B. Carman, E. Tutkun, H. Yilmaz, C. Dilber, T. Dalkiran, B. Cakir, D. Arslantas, Y. Cesaretli, S.A. Aykanat, Acute mercury poisoning among children in two provinces of Turkey, *European Journal of Pediatrics*, 172 (2013) 821–827.
28. J.B. Cavanagh, Long term persistence of mercury in the brain, *British Journal of Industrial Medicine*, 45 (1988) 649–651.
29. T.W. Clarkson, Biochemical aspects of mercury poisoning, *Journal of Occupational Medicine: Official Publication of the Industrial Medical Association*, 10 (1968) 351–355.

30. W.J. Crinnion, Environmental medicine, part three: long-term effects of chronic low-dose mercury exposure, *Alternative Medicine Review: A Journal of Clinical Therapeutic*, 5 (2000) 209–223.
31. A. El-Badry, M. Rezk, H. El-Sayed, Mercury-induced oxidative stress may adversely affect pregnancy outcome among dental staff: a cohort study, *The International Journal of Occupational and Environmental Medicine*, 9 (2018) 113–119.
32. P. Erfantalab, N. Zamani, H. Hassanian-Moghaddam, Broken thermometer in foot: a source of mercury poisoning, *Tropical Doctor*, 50 (2020) 83–84.
33. Z. Gao, X. Ying, J. Yan, J. Wang, S. Cai, C. Yan, Acute mercury vapor poisoning in a 3-month-old infant: a case report, *Clinica Chimica Acta; International Journal of Clinical Chemistry*, 465 (2017) 119–122.
34. L.R. Goldman, M.W. Shannon, Technical report: mercury in the environment: implications for pediatricians, *Pediatrics*, 108 (2001) 197–205.
35. A. Karimi, F. Moniri, A. Nasihatkon, M.J. Zarepoor, A. Alborzi, Mercury exposure among residents of a building block in Shiraz, Iran, *Environmental Research*, 88 (2002) 41–43.
36. G. Lombardi, A. Lanzirotti, C. Qualls, F. Socola, A.M. Ali, O. Appenzeller, Five hundred years of mercury exposure and adaptation, *Journal of Biomedicine & Biotechnology*, 2012 (2012) 472858.
37. M.R. Ori, J.B. Larsen, F.M. Shirazi, Mercury poisoning in a toddler from home contamination due to skin-lightening cream, *The Journal of Pediatrics*, 196 (2018) 314–317.e311.
38. S.G. Oz, M. Tozlu, S.S. Yalcin, T. Sozen, G.S. Guven, Mercury vapor inhalation and poisoning of a family, *Inhalation Toxicology*, 24 (2012) 652–658.
39. A.M. Ruha, S.C. Curry, R.D. Gerkin, K.L. Caldwell, J.D. Osterloh, P.M. Wax, Urine mercury excretion following meso-dimercaptosuccinic acid challenge in fish eaters, *Archives of Pathology & Laboratory Medicine*, 133 (2009) 87–92.
40. A.G. Abela, S. Fava, Why is the incidence of type 1 diabetes increasing?, *Current Diabetes Reviews*, 17 (2021) e030521193110.
41. M. Tufail, C. Wu, M.S. Hussain, Dietary, addictive and habitual factors, and risk of colorectal cancer, *Nutrition*, 120 (2024) 112334.
42. R. Thapa, G. Gupta, P. Dave, S.K. Singh, A. Raizaday, W.H. Almalki, G. Vyas, S.K. Singh, K. Dua, Y. Singh, Current update on the protective effect of epicatechin in neurodegenerative diseases, *EXCLI Journal*, 21 (2022) 897–903.
43. S. Assefa, G. Köhler, Intestinal microbiome and metal toxicity, *Current Opinion in Toxicology*, 19 (2020) 21–27.
44. S. Bolan, B. Seshadri, A. Kunhikrishnan, I. Grainge, N.J. Talley, N. Bolan, R. Naidu, Differential toxicity of potentially toxic elements to human gut microbes, *Chemosphere*, 303 (2022) 134958.
45. C. Chen, B.B. Zhang, A.L. Hu, H. Li, J. Liu, F. Zhang, Protective role of cinnabar and realgar in Hua-Feng-Dan against LPS plus rotenone-induced neurotoxicity and disturbance of gut microbiota in rats, *Journal of Ethnopharmacology*, 247 (2020) 112299.
46. X. Chen, J. Chi, Y. Liu, R. Du, M. Guo, W. Xu, Synthetic symbiotic bacteria reduces the toxicity of mercury ingested via contaminated food, *Food and Chemical Toxicology: An International Journal Published for the British Industrial Biological Research Association*, 178 (2023) 113937.
47. S. Gentès, A. Minet, C. Lopes, E. Tessier, C. Gassie, R. Guyoneaud, P.W. Swarzenski, P. Bustamante, M. Metian, D. Amouroux, T. Lacoue-Labarthe, In Vivo Mercury (De)Methylation Metabolism in Cephalopods under Different pCO(2) Scenarios, *Environmental Science & Technology*, 57 (2023) 5761–5770.
48. G. Guo, E. Yumvihoze, A.J. Poulain, H. Man Chan, Monomethylmercury degradation by the human gut microbiota is stimulated by protein amendments, *The Journal of Toxicological Sciences*, 43 (2018) 717–725.
49. R. Thapa, G. Gupta, A.A. Bhat, W.H. Almalki, S.I. Alzarea, I. Kazmi, S. Saleem, R. Khan, N. Altwaijry, H. Dureja, A review of Glycogen Synthase Kinase-3 (GSK3) inhibitors for cancers therapies, *International Journal of Biological Macromolecules*, 57 (2023) 127375.
50. R. Kaur, R. Rawal, Influence of heavy metal exposure on gut microbiota: Recent advances, *Journal of Biochemistry and Molecular Toxicology*, 37 (2023) e23485.
51. M.F. Khan, H. Wang, Environmental exposures and autoimmune diseases: contribution of gut microbiome, *Frontiers in Immunology*, 10 (2019) 3094.
52. R. Thapa, A. Goyal, G. Gupta, A.A. Bhat, S.K. Singh, V. Subramaniyan, S. Sharma, P. Prasher, V. Jakhmola, S.K. Singh, Recent developments in the role of protocatechuic acid in neurodegenerative disorders, *EXCLI Journal*, 22 (2023) 595.

53. X. Lin, W. Zhang, L. He, H. Xie, B. Feng, H. Zhu, J. Zhao, L. Cui, B. Li, Y.F. Li, Understanding the hepatoxicity of inorganic mercury through guts: Perturbance to gut microbiota, alteration of gut-liver axis related metabolites and damage to gut integrity, *Ecotoxicology and Environmental Safety*, 225 (2021) 112791.
54. X. Lin, J. Zhao, W. Zhang, L. He, L. Wang, H. Li, Q. Liu, L. Cui, Y. Gao, C. Chen, B. Li, Y.F. Li, Towards screening the neurotoxicity of chemicals through feces after exposure to methylmercury or inorganic mercury in rats: a combined study using gut microbiome, metabolomics and metallomics, *Journal of Hazardous Materials*, 409 (2021) 124923.
55. Y. Liu, J. Ji, W. Zhang, Y. Suo, J. Zhao, X. Lin, L. Cui, B. Li, H. Hu, C. Chen, Y.F. Li, Selenium modulated gut flora and promoted decomposition of methylmercury in methylmercury-poisoned rats, *Ecotoxicology and Environmental Safety*, 185 (2019) 109720.
56. R. Thapa, H. Ali, O. Afzal, A.A. Bhat, W.H. Almalki, S.I. Alzarea, I. Kazmi, A.S.A. Altamimi, N. Jain, M. Pandey, Unlocking the potential of mesoporous silica nanoparticles in breast cancer treatment, *Journal of Nanoparticle Research*, 25 (2023) 169.
57. N.A. Lloyd, S.E. Janssen, J.R. Reinfelder, T. Barkay, Co-selection of mercury and multiple antibiotic resistances in bacteria exposed to mercury in the fundulus heteroclitus gut microbiome, *Current Microbiology*, 73 (2016) 834–842.
58. R.C. Martín-Doimeadios, R. Mateo, M. Jiménez-Moreno, Is gastrointestinal microbiota relevant for endogenous mercury methylation in terrestrial animals?, *Environmental Research*, 152 (2017) 454–461.
59. R. Thapa, O. Afzal, G. Kumar, A.A. Bhat, W.H. Almalki, S.I. Alzarea, I. Kazmi, A.S.A. Altamimi, V. Subramaniyan, L. Thangavelu, Unveiling the connection: long-chain non-coding RNAs and critical signaling pathways in breast cancer, *Pathology—Research and Practice*, 249 (2023) 154736.
60. M.V. Miranda, F.C. González, O.S. Paredes-Godoy, M.A. Maulén, C.C. Vásquez, W.A. Díaz-Vásquez, Characterization of metal(loid)s and antibiotic resistance in bacteria of human gut microbiota from chronic kidney disease subjects, *Biological Research*, 55 (2022) 23.
61. E. Paduraru, D. Iacob, V. Rarinca, A. Rusu, R. Jijie, O.D. Ilie, A. Ciobica, M. Nicoara, B. Doroftei, Comprehensive review regarding mercury poisoning and its complex involvement in Alzheimer's disease, *International Journal of Molecular Sciences*, 23 (2022) 1992.
62. S. Pérez-Castro, G. D'Auria, M. Llambrich, S. Fernández-Barrés, M.J. Lopez-Espinosa, S. Llop, B. Regueiro, M. Bustamante, M.P. Francino, M. Vrijheid, L. Maitre, Influence of perinatal and childhood exposure to tobacco and mercury in children's gut microbiota, *Frontiers in Microbiology*, 14 (2023) 1258988.
63. M.P. Rayman, Selenium intake, status, and health: a complex relationship, *Hormones*, 19 (2020) 9–14.
64. K. Roe, Autism spectrum disorder initiation by inflammation-facilitated neurotoxin transport, *Neurochemical Research*, 47 (2022) 1150–1165.
65. K. Roe, An alternative explanation for Alzheimer's disease and Parkinson's disease initiation from specific antibiotics, gut microbiota dysbiosis and neurotoxins, *Neurochemical Research*, 47 (2022) 517–530.
66. Y. Román-Ochoa, M.C.-J. T, G.T. Choque Delgado, N. Bulut, T.R. Tejada, H.R. Yucra, A.E. Duran, B.R. Hamaker, Specific dietary fibers prevent heavy metal disruption of the human gut microbiota in vitro, *Food Research International*, 176 (2024) 113858.
67. S.E. Rothenberg, S. Keiser, N.J. Ajami, M.C. Wong, J. Gesell, J.F. Petrosino, A. Johs, The role of gut microbiota in fetal methylmercury exposure: insights from a pilot study, *Toxicology Letters*, 242 (2016) 60–67.
68. R. Thapa, O. Afzal, M. Afzal, G. Gupta, A.A. Bhat, I. Kazmi, S.I. Alzarea, S. Saleem, P. Arora, S.K. Singh, From LncRNA to metastasis: the MALAT1-EMT axis in cancer progression, *Pathology-Research and Practice*, 253 (2023) 154959.
69. R. Thapa, O. Afzal, A.S.A. Altamimi, A. Goyal, W.H. Almalki, S.I. Alzarea, I. Kazmi, V. Jakhmola, S.K. Singh, K. Dua, R. Gilhotra, G. Gupta, Galangin as an inflammatory response modulator: an updated overview and therapeutic potential, *Chemico-Biological Interactions*, 378 (2023) 110482.
70. S.E. Rothenberg, D.N. Sweitzer, B.R. Rackerby, C.E. Couch, L.A. Cohen, H.M. Broughton, S.M. Steingass, B.R. Beechler, Fecal methylmercury correlates with gut microbiota taxa in pacific walruses (Odobenus rosmarus divergens), *Frontiers in Microbiology*, 12 (2021) 648685.
71. S.E. Rothenberg, C.L. Wagner, B. Hamidi, A.V. Alekseyenko, M. Andrea Azcarate-Peril, Longitudinal changes during pregnancy in gut microbiota and methylmercury biomarkers, and reversal of microbe-exposure correlations, *Environmental Research*, 172 (2019) 700–712.
72. Y. Ruan, C. Wu, X. Guo, Z. Xu, C. Xing, H. Cao, C. Zhang, G. Hu, P. Liu, High doses of copper and mercury changed cecal microbiota in female mice, *Biological Trace Element Research*, 189 (2019) 134–144.

73. N. Seki, M. Akiyama, H. Yamakawa, K. Hase, Y. Kumagai, Y.G. Kim, Adverse effects of methylmercury on gut bacteria and accelerated accumulation of mercury in organs due to disruption of gut microbiota, *The Journal of Toxicological Sciences*, 46 (2021) 91–97.
74. A.M. Serag, M.S. Abdel-Sabour, M. El-Hadidi, M. Maged, M. Magdy, M.F. Ramadan, M.H. Refaat, Comparative 16S Metabarcoding of nile tilapia gut microbiota from the Northern Lakes of Egypt, *Applied Biochemistry and Biotechnology*, 194 (2022) 2168–2182.
75. B.V.K. Rao, A. Pradhan, R. Thapa, N. Patel, R. Mishra, N. Singla, Morin: a comprehensive review on its versatile biological activity and associated therapeutic potential in treating cancers, *Pharmacological Research-Modern Chinese Medicine*, 7 (2023) 100264.
76. S.D. Skolnick, N.H. Greig, Microbes and monoamines: potential neuropsychiatric consequences of dysbiosis, *Trends in Neurosciences*, 42 (2019) 151–163.
77. M.M. Gupta, R. Gilhotra, D. Deopa, A.A. Bhat, R. Thapa, N. Singla, R. Kulshrestha, G. Gupta, *Epigenetics of Pulmonary Tuberculosis, Targeting Epigenetics in Inflammatory Lung Diseases*, Springer, 2023, pp. 127–144.
78. M.R. Islam, A. Rauf, S. Akash, A. Kumer, M.S. Hussain, S. Akter, J.K. Gupta, L.T. Ansari, M.M.M.I. Raj, T.B. Emran, Recent perspective on the potential role of phytocompounds in the prevention of gastric cancer, *Process Biochemistry*, 135 (2023) 83–101.
79. S.K. Kaylor, I. Allen, A.D. Crim, M.L. Callihan, Calories and control: eating habits, behaviors, and motivations of Generation Z females, *Journal of American College Health*, 71 (2023) 2578–2586.
80. S. Tan, X. Xu, H. Cheng, J. Wang, X. Wang, The alteration of gut microbiome community play an important role in mercury biotransformation in largemouth bass, *Environmental Research*, 204 (2022) 112026.
81. R. Teschke, Aluminum, arsenic, beryllium, cadmium, chromium, cobalt, copper, iron, lead, mercury, molybdenum, nickel, platinum, thallium, titanium, vanadium, and zinc: Molecular Aspects in experimental liver injury, *International Journal of Molecular Sciences*, 23 (2022) 12213.
82. M. Pandey, H.X. Lang, J.S. Loh, M.C.Y. Ling, F.T.H. Long, J. Mayuren, M. Candasamy, B. Gorain, N. Jain, G. Gupta, Dendrimer platform against prostate cancer: recent update on new horizon of treatment, *Journal of Drug Delivery Science and Technology*, 85 (2023) 104589.
83. K.N. Thompson, Y. Oulhote, P. Weihe, J.E. Wilkinson, S. Ma, H. Zhong, J. Li, K. Kristiansen, C. Huttenhower, P. Grandjean, Effects of lifetime exposures to environmental contaminants on the adult gut microbiome, *Environmental Science & Technology*, 56 (2022) 16985–16995.
84. X. Tian, X. Lin, J. Zhao, L. Cui, Y. Gao, Y.L. Yu, B. Li, Y.F. Li, Gut as the target tissue of mercury and the extraintestinal effects, *Toxicology*, 484 (2023) 153396.
85. A.A. Tinkov, M. Aschner, A. Santamaria, A.R. Bogdanov, Y. Tizabi, M.B. Virgolini, J.C. Zhou, A.V. Skalny, Dissecting the role of cadmium, lead, arsenic, and mercury in non-alcoholic fatty liver disease and non-alcoholic steatohepatitis, *Environmental Research*, 238 (2023) 117134.
86. R. Dahiya, R. Thapa, N.K. Kumawat, M. Singh, S. Jakhotiya, D. Deopa, Y. Singh, N. Singla, G. Gupta, *Epigenetic Mechanisms in Inflammation, Targeting Epigenetics in Inflammatory Lung Diseases*, Springer, 2023, pp. 43–55.
87. S.E. Watson, M.A. McKinney, M. Pindo, M.J. Bull, T.C. Atwood, H.C. Hauffe, S.E. Perkins, Diet-driven mercury contamination is associated with polar bear gut microbiota, *Scientific Reports*, 11 (2021) 23372.
88. X.F. Yang, S.C. Yang, F.L. Wen, L. Feng, B. Meng, H.Y. Hu, B.L. Wang, J. Li, A.J. Poulain, P. Li, Impacts of mercury exposure levels and sources on the demethylation of methylmercury through human gut microbiota, *Bulletin of Environmental Contamination* and Toxicology, 109 (2022) 534–541.
89. B. Yin, S. Tan, K. Pan, Y. Xiao, X. Wang, Effects of diet composition on gut microbiome and mercury biotransformation in the gobyfish, *The Science of the Total Environment*, 892 (2023) 164776.
90. A.A. Bhat, R. Thapa, O. Afzal, N. Agrawal, W.H. Almalki, I. Kazmi, S.I. Alzarea, A.S.A. Altamimi, P. Prasher, S.K. Singh, The pyroptotic role of Caspase-3/GSDME signalling pathway among various cancer: a review, *International Journal of Biological Macromolecules*, 242 (2023) 124832.
91. B. Yin, S. Tan, J. Wang, K. Pan, W.X. Wang, X. Wang, Antibiotic application may raise the potential of methylmercury accumulation in fish, *The Science of the Total Environment*, 819 (2022) 152946.
92. L. Yuan, L. Wang, Z.H. Li, M.Q. Zhang, W. Shao, G.P. Sheng, Antibiotic resistance and microbiota in the gut of Chinese four major freshwater carp from retail markets, *Environmental Pollution*, 255 (2019) 113327.

93. B.B. Zhang, Y.M. Liu, A.L. Hu, S.F. Xu, L.D. Fan, M.L. Cheng, C. Li, L.X. Wei, J. Liu, HgS and Zuotai differ from HgCl(2) and methyl mercury in intestinal Hg absorption, transporter expression and gut microbiome in mice, *Toxicology and Applied Pharmacology*, 379 (2019) 114615.
94. Y. Zhang, P. Zhang, Y. Li, Gut microbiota-mediated ferroptosis contributes to mercury exposure-induced brain injury in common carp, *Metallomics: Integrated Biometal Science*, 14 (2022) mfab072.
95. Y. Zhao, C. Zhou, C. Wu, X. Guo, G. Hu, Q. Wu, Z. Xu, G. Li, H. Cao, L. Li, V. Latigo, P. Liu, S. Cheng, P. Liu, Subchronic oral mercury caused intestinal injury and changed gut microbiota in mice, *The Science of the Total Environment*, 721 (2020) 137639.
96. S. Zhou, T.P. Ko, J.W. Huang, W. Liu, Y. Zheng, S. Wu, Q. Wang, Z. Xie, Z. Liu, C.C. Chen, R.T. Guo, Structure of a gut microbial diltiazem-metabolizing enzyme suggests possible substrate binding mode, *Biochemical and Biophysical Research Communications*, 527 (2020) 799–804.
97. N. Aranda-Merino, A. Marín-Garrido, C. Román-Hidalgo, M. Ramos-Payán, N. Abril, R. Fernández-Torres, M. Bello-López, Bioavailability of flumequine and diclofenac in mice exposed to a metal-drug chemical cocktail. Evaluation of selenium protective role, *British Journal of Pharmacology*, 181 (2023) 1935–1951.
98. M. Afzal, I. Kazmi, G. Gupta, M. Rahman, V. Kimothi, F. Anwar, Preventive effect of Metformin against N-nitrosodiethylamine-initiated hepatocellular carcinoma in rats, *Saudi Pharmaceutical Journal*, 20 (2012) 365–370.
99. S. Assefa, K. Ahles, S. Bigelow, J.T. Curtis, G.A. Köhler, Lactobacilli with probiotic potential in the prairie vole (Microtus ochrogaster), *Gut Pathogens*, 7 (2015) 35.
100. J.E. Bisanz, M.K. Enos, J.R. Mwanga, J. Changalucha, J.P. Burton, G.B. Gloor, G. Reid, Randomized open-label pilot study of the influence of probiotics and the gut microbiome on toxic metal levels in Tanzanian pregnant women and school children, *mBio*, 5 (2014) e01580–01514.
101. J. Cai, B. Yin, Y. Wang, K. Pan, Y. Xiao, X. Wang, Gut microbiome play a crucial role in geographical and interspecies variations in mercury accumulation by fish, *The Science of the Total Environment*, 912 (2024) 169381.
102. K.S. Alharbi, W.H. Almalki, M. Albratty, A.M. Meraya, A. Najmi, G. Vyas, S.K. Singh, K. Dua, G. Gupta, The therapeutic role of nutraceuticals targeting the Nrf2/HO-1 signaling pathway in liver cancer, *Journal of Food and Biochemistry*, 46 (2022) e14357.
103. L.C. Conteville, J. Oliveira-Ferreira, A.C.P. Vicente, Heavy metal resistance in the Yanomami and Tunapuco microbiome, *Memorias do Instituto Oswaldo Cruz*, 118 (2023) e230086.
104. E. Davis, K.M. Bakulski, J.M. Goodrich, K.E. Peterson, M.L. Marazita, B. Foxman, Low levels of salivary metals, oral microbiome composition and dental decay, *Scientific Reports*, 10 (2020) 14640.
105. S.E. George, R. Devereux, J. James, Y. Wan, G.L. Diamond, K.D. Bradham, D.J. Thomas, Dietary lead modulates the mouse intestinal microbiome: subacute exposure to lead acetate and lead contaminated soil, *Ecotoxicology and Environmental Safety*, 249 (2023) 114430.
106. M. Altamish, R. Dahiya, A.K. Singh, A. Mishra, A.A.A. Aljabali, S. Satija, M. Mehta, H. Dureja, P. Prasher, P. Negi, D.N. Kapoor, R. Goyal, M.M. Tambuwala, D.K. Chellappan, K. Dua, G. Gupta, Role of the serine/threonine kinase 11 (STK11) or liver kinase B1 (LKB1) gene in Peutz-Jeghers syndrome, *Critical Reviews in Eukaryotic Gene Expression*, 30 (2020) 245–252.
107. E. Gorokhova, A.L. Soerensen, N.H. Motwani, Mercury-methylating bacteria are associated with copepods: a proof-of-principle survey in the Baltic Sea, *PloS One*, 15 (2020) e0230310.
108. L. Jin, H. Wu, G. Li, S. Yang, R. Wei, Y. Huang, P. Penttinen, W. Deng, J. Chen, X. Han, C. Li, L. Hu, T. Li, H. Zhang, K. Zhao, L. Zou, Gastrointestinal microbiome, resistance genes, and risk assessment of heavy metals in wild giant pandas, *The Science of the Total Environment*, 899 (2023) 165671.
109. O.V. Karnachuk, A.V. Beletsky, A.L. Rakitin, O.P. Ikkert, M.R. Avakyan, V.S. Zyusman, A. Napilov, A.V. Mardanov, N.V. Ravin, Antibiotic-resistant desulfovibrio produces H(2)S from supplements for animal farming, *Microorganisms*, 11 (2023) 838.
110. A.A. Bhat, G. Gupta, R. Dahiya, R. Thapa, A. Gahtori, M. Shahwan, V. Jakhmola, A. Tiwari, M. Kumar, H. Dureja, CircRNAs: pivotal modulators of TGF-β signalling in cancer pathogenesis, *Non-Coding RNA Research*, 9 (2024) 277–287.
111. A.A. Bhat, G. Gupta, A. Goyal, R. Thapa, W.H. Almalki, I. Kazmi, S.I. Alzarea, N. Kukreti, M. Sekar, D.U. Meenakshi, Unwinding circular RNA's role in inflammatory pulmonary diseases, *Naunyn-Schmiedeberg's Archives of Pharmacology*, 397 (2023) 2567–2588.
112. A. Lapanje, D. Drobne, N. Nolde, J. Valant, B. Muscet, V. Leser, M. Rupnik, Long-term Hg pollution induced Hg tolerance in the terrestrial isopod Porcellio scaber (Isopoda, Crustacea), *Environmental Pollution*, 153 (2008) 537–547.

113. A. Lapanje, M. Rupnik, D. Drobne, Gut bacterial community structure (Porcellio scaber, Isopoda, Crustacea) as a measure of community level response to long-term and short-term metal pollution, *Environmental Toxicology and Chemistry*, 26 (2007) 755–763.
114. H.E. Laue, Y. Moroishi, B.P. Jackson, T.J. Palys, J.C. Madan, M.R. Karagas, Nutrient-toxic element mixtures and the early postnatal gut microbiome in a United States longitudinal birth cohort, *Environment International*, 138 (2020) 105613.
115. K. Anand, N.S. Abdul, T. Ghazi, M. Ramesh, G. Gupta, M.M. Tambuwala, H. Dureja, S.K. Singh, D.K. Chellappan, K. Dua, B. Pandi, M. Saravanan, A.A. Chuturgoon, Induction of caspase-mediated apoptosis in HepG2 liver carcinoma cells using mutagen-antioxidant conjugated self-assembled novel carbazole nanoparticles and in silico modeling studies, *ACS Omega*, 6 (2021) 265–277.
116. A.A. Bhat, G. Gupta, M. Afzal, R. Thapa, H. Ali, S.M. Alqahtani, I. Kazmi, S.I. Alzarea, S. Saleem, V. Subramaniyan, Polyphenol-loaded nano-carriers for breast cancer therapy: a comprehensive review, *BioNanoScience*, 14 (2024) 4219–4237.
117. N.A. Lloyd, S. Nazaret, T. Barkay, Whole genome sequences to assess the link between antibiotic and metal resistance in three coastal marine bacteria isolated from the mummichog gastrointestinal tract, *Marine Pollution Bulletin*, 135 (2018) 514–520.
118. S. Mukhi, B. Dhanashree, R.M. Srikantiah, P. Manjrekar, S. Harish, Evaluation of minimum inhibitory concentration of heavy metals contained in packaging material digest on prominent gut microbiota, *International Journal of Food Science*, 2023 (2023) 3840795.
119. M. Nagano, M. Fujimura, Y. Tada, Y. Seko, Dietary fructooligosaccharides reduce mercury levels in the brain of mice exposed to methylmercury, *Biological & Pharmaceutical Bulletin*, 44 (2021) 522–527.
120. K.M. Nielsen, Y. Zhang, T.E. Curran, J.T. Magnuson, B.J. Venables, K.E. Durrer, M.S. Allen, A.P. Roberts, Alterations to the intestinal microbiome and metabolome of pimephales promelas and mus musculus following exposure to dietary methylmercury, *Environmental Science & Technology*, 52 (2018) 8774–8784.
121. G. Pelletier, Y.L. Feng, K. Leingartner, P. Black, Co-administration of a Rhododendron tomentosum extract does not affect mercury tissue concentrations and excretion rate in methylmercury-treated adult male rats, *BMC Research Notes*, 12 (2019) 369.
122. D.K. Chellappan, S. Ganasen, S. Batumalai, M. Candasamy, P. Krishnappa, K. Dua, J. Chellian, G. Gupta, The protective action of the aqueous extract of auricularia polytricha in paracetamol induced hepatotoxicity in rats, *Recent Patents on Drug Delivery & Formulation*, 10 (2016) 72–76.
123. A.A. Bhat, O. Afzal, N. Agrawal, R. Thapa, W.H. Almalki, I. Kazmi, S.I. Alzarea, A.S.A. Altamimi, N. Kukreti, A. Chakraborty, A comprehensive review on the emerging role of long non-coding RNAs in the regulation of NF-κB signaling in inflammatory lung diseases, *International Journal of Biological Macromolecules*, 253 (2023) 126951.
124. A.A. Bhat, A. Goyal, R. Thapa, I. Kazmi, S.I. Alzarea, M. Singh, S. Rohilla, T.K. Saini, N. Kukreti, D.U. Meenakshi, Uncovering the complex role of interferon-gamma in suppressing type 2 immunity to cancer, *Cytokine*, 171 (2023) 156376.
125. Q. Pope, M.D. Rand, Variation in methylmercury metabolism and elimination in humans: physiological pharmacokinetic modeling highlights the role of gut biotransformation, skeletal muscle, and hair, *Toxicological Sciences*, 180 (2021) 26–37.
126. X. Shang, B. Wang, Q. Sun, Y. Zhang, Y. Lu, S. Liu, Y. Li, Selenium-enriched Bacillus subtilis reduces the effects of mercury-induced on inflammation and intestinal microbes in carp (*Cyprinus carpio* var. specularis), *Fish Physiology and Biochemistry*, 48 (2022) 215–226.
127. Y. Shen, H.E. Laue, M.J. Shrubsole, H. Wu, T.R. Bloomquist, A. Larouche, K. Zhao, F. Gao, A. Boivin, D. Prada, D.J. Hunting, V. Gillet, L. Takser, A.A. Baccarelli, Associations of childhood and perinatal blood metals with children's gut microbiomes in a canadian gestation cohort, *Environmental Health Perspectives*, 130 (2022) 17007.
128. A. Simon, J. Colom, S. Mazhar, E. Khokhlova, J. Deaton, K. Rea, Bacillus megaterium Renuspore(®) as a potential probiotic for gut health and detoxification of unwanted dietary contaminants, *Frontiers in Microbiology*, 14 (2023) 1125616.
129. G. Gupta, W.H. Al-Malki, I. Kazmi, L. Thangavelu, P.K. Gupta, N.K. Jha, P. Prasher, S.K. Singh, K. Dua, The role of HGF/MET in liver cancer, *Future Medicinal Chemistry*, 13 (2021) 1829–1832.
130. M. Trinder, J.E. Bisanz, J.P. Burton, G. Reid, Probiotic lactobacilli: a potential prophylactic treatment for reducing pesticide absorption in humans and wildlife, *Beneficial microbes*, 6 (2015) 841–847.
131. Y. Zhang, T. Chen, Y. Zhang, Q. Hu, X. Wang, H. Chang, J.H. Mao, A.M. Snijders, Y. Xia, Contribution of trace element exposure to gestational diabetes mellitus through disturbing the gut microbiome, *Environment International*, 153 (2021) 106520.

132. C. Zhou, P. Xu, C. Huang, G. Liu, S. Chen, G. Hu, G. Li, P. Liu, X. Guo, Effects of subchronic exposure of mercuric chloride on intestinal histology and microbiota in the cecum of chicken, *Ecotoxicology and Environmental Safety*, 188 (2020) 109920.
133. J.A. Acosta, A. Abbaspour, G.R. Martínez, S. Martínez-Martínez, R. Zornoza, M. Gabarrón, A. Faz, Phytoremediation of mine tailings with Atriplex halimus and organic/inorganic amendments: a five-year field case study, *Chemosphere*, 204 (2018) 71–78.
134. V.K.R. Balaga, A. Pradhan, R. Thapa, N. Patel, R. Mishra, N. Singla, Morin: a comprehensive review on its versatile biological activity and associated therapeutic Potential in treating cancers, *Pharmacological Research*, 7 (2023) 100264.
135. G. Gupta, A.A. Bhat, A. Goyal, N. Singla, S. Gupta, S. Sharma, S. Bhatt, K. Dua, Exploring ACSL4/LPCAT3/ALOX15 and SLC7A11/GPX4/NFE2L2 as potential targets in ferroptosis-based cancer therapy, *Future Science* (2023) 1209–1212.
136. M.S. Ahmad, M. Ashraf, Essential roles and hazardous effects of nickel in plants, *Reviews of Environmental Contamination and Toxicology*, 214 (2011) 125–167.
137. M.A. Arshad, H.M. Ebeid, F.U. Hassan, Revisiting the effects of different dietary sources of selenium on the health and performance of dairy animals: a review, *Biological Trace Element Research*, 199 (2021) 3319–3337.
138. A. Bartos, J. Sikora, Bioinorganic modulators of ferroptosis: a review of recent findings, *International Journal of Molecular Sciences*, 24 (2023) 3634.
139. G. Gupta, G. Krishna, D.K. Chellappan, K.S. Gubbiyappa, M. Candasamy, K. Dua, Protective effect of pioglitazone, a PPARγ agonist against acetaminophen-induced hepatotoxicity in rats, *Molecular Cell and Biochemistry*, 393 (2014) 223–228.
140. J. Bełtowski, Hydrogen sulfide in pharmacology and medicine--an update, *Pharmacological Reports*, 67 (2015) 647–658.
141. V.P. Samuel, G. Gupta, R. Dahiya, D.A. Jain, A. Mishra, K. Dua, Current update on preclinical and clinical studies of resveratrol, a naturally occurring phenolic compound, *Critical Reviews in Eukaryotic Gene Expression*, 29 (2019) 529–537.
142. S. Bocanegra-Rodríguez, N. Jornet-Martínez, C. Molins-Legua, P. Campíns-Falcó, Delivering inorganic and organic reagents and enzymes from zein and developing optical sensors, *Analytical Chemistry*, 90 (2018) 8501–8508.
143. I. Boulogne, P. Petit, L. Desfontaines, G. Durambur, C. Deborde, C. Mirande-Ney, Q. Arnaudin, C. Plasson, J. Grivotte, C. Chamot, S. Bernard, G. Loranger-Merciris, Biological and chemical characterization of musa paradisiaca leachate, *Biology,* 12 (2023) 1326.
144. J.L. Boyer, Bile formation and secretion, *Comprehensive Physiology*, 3 (2013) 1035–1078.
145. G.A. Brooks, N.A. Martin, Cerebral metabolism following traumatic brain injury: new discoveries with implications for treatment, *Frontiers in Neuroscience*, 8 (2014) 408.
146. A. Carocci, N. Rovito, M.S. Sinicropi, G. Genchi, Mercury toxicity and neurodegenerative effects, *Reviews of Environmental Contamination and Toxicology*, 229 (2014) 1–18.
147. I.O. Costa, N.S. Rios, P.J.M. Lima, L.R.B. Gonçalves, Synthesis of organic-inorganic hybrid nanoflowers of lipases from Candida antarctica type B (CALB) and Thermomyces lanuginosus (TLL): improvement of thermal stability and reusability, *Enzyme and Microbial Technology*, 163 (2023) 110167.
148. C. Coudray, M. Rambeau, C. Feillet-Coudray, E. Gueux, J.C. Tressol, A. Mazur, Y. Rayssiguier, Study of magnesium bioavailability from ten organic and inorganic Mg salts in Mg-depleted rats using a stable isotope approach, *Magnesium Research*, 18 (2005) 215–223.
149. A. Reshma, T. Tamilanban, V. Chitra, V. Subramaniyan, G. Gupta, N.K. Fuloria, M. Sekar, S. Fuloria, R. Sahu, J. Narayanan, S. Chakravarthy, S. Selvaraj, Anti-obesity effects of olivetol in adult zebrafish model induced by short-term high-fat diet, *Scientific Reports*, 13 (2023) 18449.
150. R. Eivazzadeh-Keihan, M. Ghafori Gorab, H. Aghamirza Moghim Aliabadi, M. Mahdavi, A.R. Akbarzadeh, A. Maleki, H. Ghafuri, Novel magnetic organic-inorganic hybrids based on aromatic polyamides and ZnFe(2)O(4) nanoparticles with biological activity, *Scientific Reports*, 11 (2021) 20310.
151. E.V. Fedorova, A.V. Buriakina, N.M. Vorob'eva, N.I. Baranova, [The vanadium compounds: chemistry, synthesis, insulinomimetic properties], *Biomeditsinskaia Khimiia*, 60 (2014) 416–429.
152. E.L. Forker, Mechanisms of hepatic bile formation, *Annual Review of Physiology*, 39 (1977) 323–347.
153. M. Abubakar, E.R. Puppala, B.J. Dutta, K.C. Maharana, R. Thapa, S. Roshan, B. Tazneem, A. Khan, A.A. Bhat, *Epigenetics of Rhinovirus, Targeting Epigenetics in Inflammatory Lung Diseases*, Springer Nature Singapore, 2023, pp. 185–197.

154. I. Kazmi, F.A. Al-Abbasi, M. Afzal, H.N. Altayb, M.S. Nadeem, G. Gupta, Formulation and evaluation of kaempferol loaded nanoparticles against experimentally induced hepatocellular carcinoma: in vitro and in vivo studies, *Pharmaceutics*, 13 (2021) 2086.
155. S. Haldar, S. Samanta, R. Banarjee, B. Sharma, T.K. Ghosh, Glucose tolerance and serum concentrations of hormones and metabolites in goats (Capra hircus) fed diets supplemented with inorganic and organic chromium salts, *Animal*, 1 (2007) 347–356.
156. E. Hikita, M. Arai, S. Tanaka, K. Onda, H. Utsumi, B. Yuan, H. Toyoda, T. Hirano, Effects of inorganic and organic arsenic compounds on growth and apoptosis of human T-lymphoblastoid leukemia cells, *Anticancer Research*, 31 (2011) 4169–4178.
157. S.G. Holt, E.R. Smith, Fetuin-A-containing calciprotein particles in mineral trafficking and vascular disease, *Nephrology, Dialysis, Transplantation*, 31 (2016) 1583–1587.
158. E.A. Jaramillo, M.S. Ferreira Santos, A.C. Noell, M.F. Mora, Capillary electrophoresis method for analysis of inorganic and organic anions related to habitability and the search for life, *Electrophoresis*, 42 (2021) 1956–1964.
159. A.B. Omer, H.N. Altayb, F.A. Al-Abbasi, G. Gupta, M.M. Ahmed, A.M. Alghamdi, S.I. Alzarea, N. Sayyed, M.S. Nadeem, I. Kazmi, Acemannan ameliorates STZ-activated diabetes by attenuating high glucose via inhibiting inflammatory cytokines and apoptosis pathway, *International Journal of Biological Macromolecules*, 253 (2023) 127127.
160. P. Jurtshuk, Jr., Bacterial metabolism, in: S. Baron (Ed.) *Medical Microbiology,* University of Texas Medical Branch at Galveston. Copyright © 1996, *The University of Texas Medical Branch at Galveston*, Galveston (TX), 1996.
161. Z. Khurshid, M.S. Zafar, R.S. Khan, S. Najeeb, P.D. Slowey, I.U. Rehman, Role of salivary biomarkers in oral cancer detection, *Advances in Clinical Chemistry*, 86 (2018) 23–70.
162. D. Kiejza, U. Kotowska, W. Polińska, J. Karpińska, Peracids - New oxidants in advanced oxidation processes: The use of peracetic acid, peroxymonosulfate, and persulfate salts in the removal of organic micropollutants of emerging concern - A review, *The Science of the Total Environment*, 790 (2021) 148195.
163. I. Kazmi, F.A. Al-Abbasi, M. Afzal, M.S. Nadeem, H.N. Altayb, G. Gupta, Phosphodiesterase-4 inhibitor roflumilast-mediated protective effect in sepsis-induced late-phase event of acute kidney injury: a narrative review, *Pharmaceuticals*, 15 (2022).
164. J.H. Kim, J. Lee, H.B. Lee, J.H. Shin, E.K. Kim, Water-retentive and anti-inflammatory properties of organic and inorganic substances from Korean sea mud, *Natural Product Communications*, 5 (2010) 395–398.
165. N.S. Kudryasheva, Bioluminescence and exogenous compounds: physico-chemical basis for bioluminescent assay, *Journal of Photochemistry and Photobiology. B, Biology*, 83 (2006) 77–86.
166. B. Mion, L. Ogilvie, B. Van Winters, J.F.W. Spricigo, S. Anan, M. Duplessis, B.W. McBride, S.J. LeBlanc, M.A. Steele, E.S. Ribeiro, Effects of replacing inorganic salts of trace minerals with organic trace minerals in the pre- and postpartum diets on mineral status, antioxidant biomarkers, and health of dairy cows, *Journal of Animal Science*, 101 (2023).
167. B. Mion, B. Van Winters, K. King, J.F.W. Spricigo, L. Ogilvie, L. Guan, T.J. DeVries, B.W. McBride, S.J. LeBlanc, M.A. Steele, E.S. Ribeiro, Effects of replacing inorganic salts of trace minerals with organic trace minerals in pre- and postpartum diets on feeding behavior, rumen fermentation, and performance of dairy cows, *Journal of Dairy Science,* 105 (2022) 6693–6709.

11 Effect of Cadmium on Gut Microbiome

Rahamat Unissa Syed[1], Mohammed Khaled Bin Break[2], Taghreed Mohammad Saad Alhaidan[3,4], and Fulvia di Pillo[5]*

[1]Department of Pharmaceutics, College of Pharmacy, University of Ha'il, Ha'il, Saudi Arabia
[2]Department of Pharmaceutical Chemistry, College of Pharmacy, University of Ha'il, Ha'il, Saudi Arabia
[3]Department of Public Health, College of Public Health and Health Informatics, University of Ha'il, Ha'il, Saudi Arabia
[4]Scientific office manager in Advanz Pharma, Riyadh, Saudi Arabia
[5]Research Scientist, Centre for Medicinal Chemistry (CQMED), SP, Brazil
[*]Corresponding Author: Rahamat Unissa Syed.
Email: ru.syed@outlook.com

11.1 INTRODUCTION

Cadmium, an environmental contaminant with far-reaching health implications, has become a global concern. This heavy metal is omnipresent in various environmental compartments, notably sediments, stream water, and topsoil, posing significant risks to human health [1]. Regions such as Europe grapple with elevated cadmium levels, with countries like Germany and Slovakia displaying concerning concentrations in sediments and stream water, while others, including Greece, Italy, France, Austria, and Ireland, exhibit heightened levels in topsoil [2]. The average weekly consumption of cadmium for adult Europeans, standing at 2.5 mg/kg, underscores the ubiquity of this exposure. Similarly, China faces substantial cadmium exposure, while countries like Bangladesh and Sri Lanka contend with heightened levels of cadmium in rice [3].

11.1.1 Cadmium Exposure and Health Consequences

Despite some studies indicating a decrease in cadmium exposure in specific regions, further reduction remains imperative due to the profound socioeconomic consequences associated with its exposure [4]. Cadmium has been implicated in a spectrum of severe health conditions, including cancer, diabetes mellitus, cardiovascular disorders, chronic renal disease, osteoporosis, liver disease, and adverse neurodevelopmental outcomes (Figure 11.1). Shockingly, high levels of cadmium exposure are estimated to cause a substantial 17% increase in mortality, underscoring the urgency of addressing this environmental challenge [5]. Historically, the understanding of cadmium's health effects has revolved around comprehensive pathways elucidating the range of detrimental outcomes linked to exposure. Critical metal dyshomeostasis, oxidative stress, inflammation, endoplasmic reticulum stress, and genomic instability have all been documented as consequences of cadmium exposure. Moreover, cadmium's role as an endocrine disruptor adds complexity to its effects, necessitating in-depth research into the underlying processes for effective evaluation and mitigation strategies [6]. Recognizing the pivotal role of the gut microbiome in various physiological processes introduces another layer of complexity to the study of cadmium's impact on human health [7]. The

 DOI: 10.1201/9781003489221-11

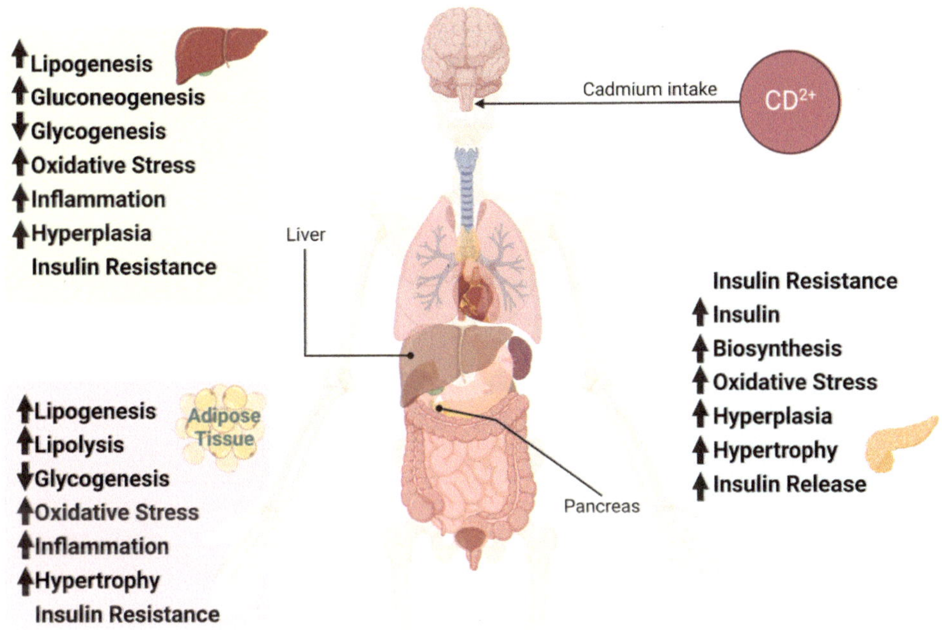

FIGURE 11.1 This image illustrates the cellular response to cadmium toxicity in the pancreas, liver, and adipose tissues, emphasizing its impact on metabolism, cell cycle regulation, oxidative stress, and inflammation.

gut microbiome, crucial for immune regulation and mineral metabolism, is susceptible to disruptions that may arise from exposure to environmental contaminants [8]. Previous research has implicated antibiotics, herbicides, heavy metals, and persistent organic pollutants in altering the composition and function of the gut microbiota. However, the existing body of reviews and data falls short in conclusively supporting the hypothesis that gut dysbiosis mediates cadmium poisoning [9]. Understanding how cadmium interacts with the delicate balance of the gut microbiome is crucial for unraveling potential mechanisms that underlie health outcomes. From alterations in microbial diversity to disruptions in metabolic functions, exploring the impact on the gut microbiota is paramount [10]. The intricate relationship between cadmium and the gut microbiome raises questions about the potential direct effects on intestinal health [11]. Investigating this aspect is essential for delineating the specific pathways through which cadmium may exert its toxic effects on the gastrointestinal system. This comprehensive exploration into the effect of cadmium on the gut microbiome represents a critical step toward understanding the complex interactions between environmental contaminants and human health [12].

11.2 CADMIUM: SOURCES AND EXPOSURE

Cadmium, a toxic heavy metal, pervades the environment through various natural and anthropogenic sources, posing potential threats to human health. Understanding the diverse origins and exposure routes is crucial for assessing the risks associated with cadmium contamination [13].

11.2.1 Natural Sources of Cadmium

Cadmium is present in the Earth's crust and enters the environment through natural geological processes. Its occurrence is primarily associated with zinc, and both metals often coexist in mineral deposits. Weathering and erosion of rocks release cadmium into the soil, where it can persist for extended periods [14]. Additionally, volcanic activity contributes to atmospheric cadmium levels. Plants, as primary producers, can accumulate cadmium from the soil, leading to its bioavailability in the food chain. Thus, natural sources, though intrinsic to the Earth's composition, set the stage for potential human exposure [15].

11.2.2 Human Exposure Routes

Understanding how cadmium enters the human body is essential for evaluating the associated health risks. Human exposure to cadmium occurs through multiple routes, with ingestion, inhalation, and dermal contact being the primary pathways. Food is a significant source of cadmium exposure for the general population [16]. Plants absorb cadmium from soil, and it bioaccumulates in various crops, especially leafy vegetables, grains, and root vegetables. Additionally, seafood, including shellfish and certain fish species, may contain elevated cadmium levels. Tobacco smoke represents another notable source, as tobacco plants readily absorb and accumulate cadmium from the soil [17]. Inhalation of cadmium occurs primarily in occupational settings, such as mining, smelting, and manufacturing processes. Workers in these industries may inhale cadmium-containing dust and fumes. Furthermore, cadmium present in ambient air from industrial activities can contribute to general population exposure through respiratory uptake [18]. While not a primary route, dermal contact with cadmium-containing products or contaminated soil can contribute to exposure. This is particularly relevant in occupational settings where direct skin contact with cadmium may occur [19].

11.2.3 Occupational and Environmental Sources

Certain occupational settings pose a higher risk of cadmium exposure. Workers involved in industries like battery manufacturing, electroplating, and welding are at an increased risk due to the potential release of cadmium-containing particles and fumes. Miners and smelters, dealing with ores containing cadmium, face elevated exposure levels [20]. Proper occupational safety measures are crucial in minimizing the risk for workers in these environments. Anthropogenic activities contribute significantly to environmental cadmium contamination. Industrial discharges, agricultural runoff, and improper disposal of cadmium-containing waste can lead to soil and water contamination [21]. Once in the environment, cadmium can persist and accumulate, further exacerbating the risk of exposure through various pathways, including food chain contamination. Cadmium exposure is a complex interplay of natural sources and anthropogenic activities. Understanding the diverse routes through which cadmium enters the human body is essential for implementing effective risk management strategies and regulatory measures [22]. Whether through the food we consume, the air we breathe, or the occupational environments we work in, being cognizant of cadmium exposure sources is paramount for safeguarding human health [23].

11.3 MECHANISMS OF CADMIUM TOXICITY

Cadmium, a heavy metal with well-documented toxicity, exerts its adverse effects through intricate cellular and molecular mechanisms. Understanding how cadmium interferes at these fundamental levels is crucial for unraveling the complexity of its toxicological impact [24]. At the cellular level, cadmium disrupts critical metal homeostasis, leading to dysregulation of essential trace elements like zinc and copper [25]. This disruption occurs primarily through competitive binding, where cadmium competes with these essential metals for binding sites on transport proteins [26]. The imbalance in metal homeostasis can compromise cellular functions and contribute to the manifestation of various health issues. Molecularly, cadmium induces oxidative stress, a condition characterized by an imbalance between the production of reactive oxygen species (ROS) and the body's ability to neutralize them. Cadmium stimulates the generation of ROS, such as superoxide radicals and hydrogen peroxide, which can damage cellular components like lipids, proteins, and DNA [27]. This oxidative assault can lead to cellular dysfunction and trigger downstream signaling pathways associated with inflammation and apoptosis. Furthermore, cadmium induces endoplasmic reticulum (ER) stress, a condition where the protein-folding capacity of the ER is overwhelmed. Prolonged exposure to cadmium disrupts proper protein folding, leading to the accumulation of misfolded proteins and triggering the unfolded protein response (UPR) [28]. The UPR aims to restore ER homeostasis but, when overwhelmed, can contribute to cellular damage and apoptosis. Genomic instability is another facet of cadmium toxicity, impacting the integrity of the cell's genetic material. Cadmium has been linked to DNA damage and mutations, potentially contributing to the development of cancer and other adverse health outcomes [29]. The metal's ability to interfere with DNA repair mechanisms further exacerbates its genotoxic effects. Moreover, cadmium functions as an endocrine disruptor, interfering with hormonal regulation. It can mimic the action of hormones, binding to hormone receptors and disrupting normal endocrine function [30]. This disruption can have far-reaching consequences, particularly in reproductive and developmental processes. The cellular and molecular impact of cadmium toxicity is characterized by disruptions in metal homeostasis, oxidative stress, ER stress, genomic instability, and endocrine disruption. These intricate mechanisms collectively contribute to the adverse health outcomes associated with cadmium exposure, spanning a range of physiological systems and highlighting the multifaceted nature of its toxicological effects [31].

11.3.1 Interactions with Gut Microbiome Components

Cadmium toxicity extends beyond direct cellular interactions, intricately involving the dynamic ecosystem of the gut microbiome. The gut, host to trillions of microorganisms, plays a pivotal role in maintaining homeostasis and influencing overall health [32]. Cadmium, once ingested, embarks on a journey through the gastrointestinal tract, where it encounters the diverse microbial community residing in the gut. The first point of contact is often the mucosal lining, where cadmium may interact with mucin-producing goblet cells [33]. These interactions can compromise the integrity of the gut barrier, facilitating the translocation of cadmium into deeper layers of the intestinal tissue. Moreover, cadmium's ability to induce oxidative stress can disrupt the delicate balance of the gut microbiota by promoting the growth of certain bacteria while suppressing others [34]. Cadmium also exhibits a penchant for binding to sulfhydryl groups, crucial components of enzymes and proteins. This sulfhydryl binding capability can directly impact the structure and function of microbial proteins within the gut. Enzymes involved in essential microbial processes may become targets for cadmium-induced inhibition, leading to disruptions in metabolic pathways and nutrient cycling [35]. Furthermore, cadmium's influence extends to the immune-modulating functions of the gut microbiome. The delicate interplay between the microbiota and the immune system is crucial for maintaining a balanced and responsive defense against pathogens [36]. Cadmium-induced alterations in microbial composition may trigger immune responses, contributing to inflammation and potentially exacerbating gastrointestinal conditions. The intricate nature of cadmium's interactions with gut microbiome components underscores the need for a holistic understanding of its mechanisms of toxicity [37]. Research in this realm not only unveils the nuanced relationships between environmental contaminants and the microbiome but also opens avenues for targeted interventions to mitigate the adverse health effects of cadmium exposure. As we delve deeper into this intricate interplay, unraveling the specific ways in which cadmium influences the gut microbiome

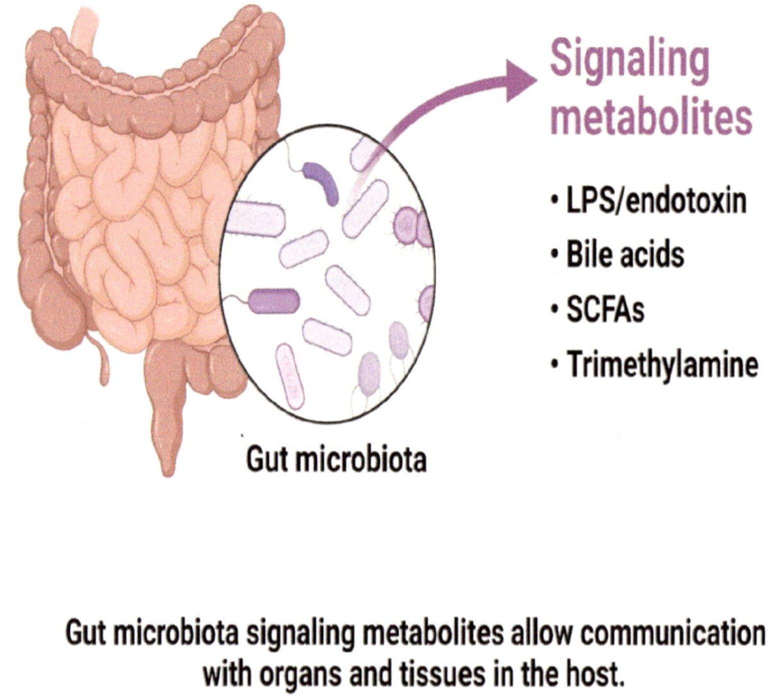

FIGURE 11.2 This image illustrates the systemic impact of gut microbiota metabolites, influencing distant organs and regulating gut microbiota composition through host-mediated release of anti-bacterial materials into the gut lumen.

components holds promise for advancing our knowledge and developing strategies to preserve gut health in the face of environmental challenges [38]. Figure 11.2 depicts the systemic impact of gut microbiota metabolites.

11.4 CADMIUM AND THE GUT MICROBIOME

Recognizing the pivotal role of the gut microbiome in various physiological processes introduces another layer of complexity to the study of cadmium's impact on human health [39]. The gut microbiome, crucial for immune regulation and mineral metabolism, is susceptible to disruptions that may arise from exposure to environmental contaminants [40]. Previous research has implicated antibiotics, herbicides, heavy metals, and persistent organic pollutants in altering the composition and function of the gut microbiota. However, the existing body of reviews and data falls short in conclusively supporting the hypothesis that gut dysbiosis mediates cadmium poisoning [41]. The gut microbiota is thought to be a key player in mediating the harmful effects of heavy metals, such as cadmium. When germ-free mice were treated with Cd, the amount of Cd in their kidney, liver, blood, and spleen increased significantly, and their duodenum and colon expressed MT1 and MT2, suggesting that the gut microbiota may play a protective role against Cd toxicity [42]. Studies on laboratory animals of various ages have shown the effects of short-term (and subchronic) and long-term (and chronic) Cd exposure on the physiology of the gut microbiome (Figure 11.3). Adult rats exposed to cadmium for 42 days showed a significant decrease in both aerobic and anaerobic lactic bacteria in their digestive tracts. Furthermore, Cd caused particular alterations in the gut microbiota at the genus and family levels [43]. In mice exposed to Cd in drinking water for 45 days, previous research showed a substantial dose-dependent decrease in the intestinal populations of B. cereus, Lactobacillus spp., Clostridium spp., and E. coli. Additionally, it has been demonstrated that an eight-week exposure to Cd increased the relative abundance of Actinobacteria in the mouse caecum in a dose-dependent manner, but alterations in Bacteroidetes and Firmicutes did not reach statistical significance [44]. The relative numbers of *Coriobacteriaceae* and *Lactobacillaceae* increased at the family level of caecal flora in a dose-dependent manner, whereas the percentage of *Lachnospiraceae* declined. In mice exposed to Cd, the alterations in fecal microorganisms were not as noticeable [45]. Additionally, they have shown a notable rise in the relative abundance of the *Odoribacter* and *Alistipes* genera, while the adult male C57black/6 mice exposed to the same amount of metal showed a considerable reduction in the amount of Cd-induced Mollicutes and unclassified Ruminococcaceae [46]. It is noteworthy that during the initial four weeks of exposure, as opposed to the last four weeks, Cd-induced changes in microbiota are more noticeable. Interestingly, since germ-free animals were less likely to exhibit a proinflammatory response to Cd exposure, it is noteworthy that the observed NF-kB activation and proinflammatory cytokine production was at least partially reliant on Cd-induced modification in gut microbiota [47]. Long after the chemical was removed, the gut's identified alterations remained. Notably, Cd-unexposed animals that had their gut flora transferred to them had

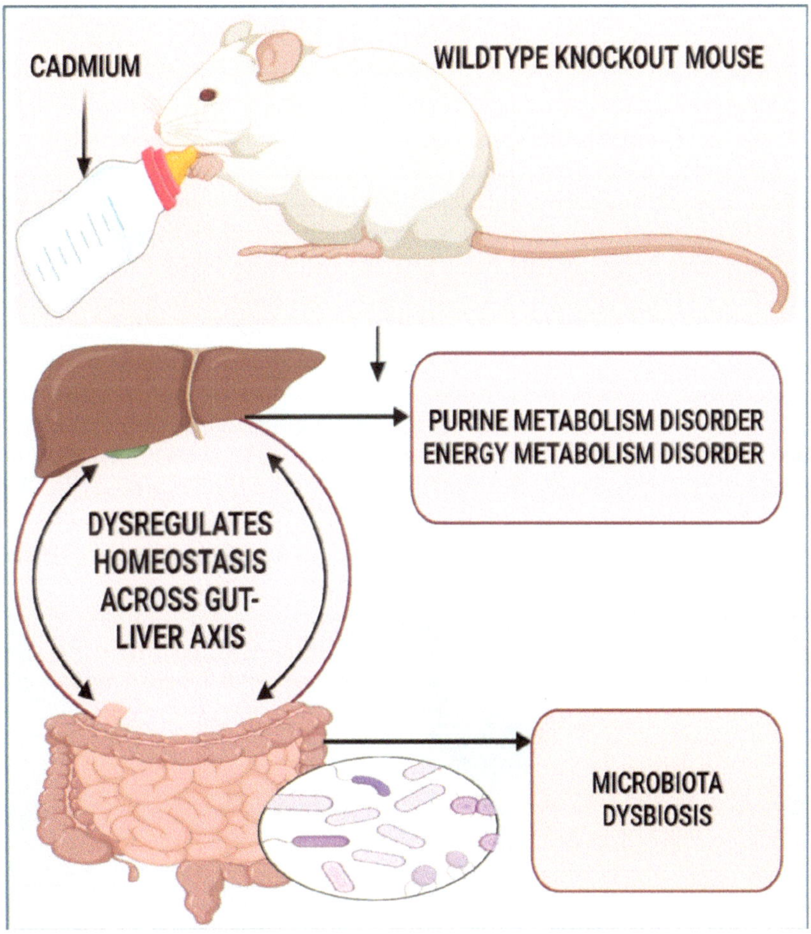

FIGURE 11.3 This image illustrates the toxicological impact of chronic low-dose Cadmium (Cd) exposure on liver purine metabolism, energy metabolism, and gut microbiota, contributing to metabolic diseases.

comparable inflammatory and allergy reactions, but germ-free mice showed no alterations at all [48]. Overall, their results are consistent with the previous discovery that mice exposed to Cd for 28 days had enhanced intestinal permeability to macromolecules. They showed that beginning after three weeks of exposure, the administration of escalating dosages of Cd dramatically reduced intestine bacterial populations and the ratio of Firmicutes to Bacteroidetes [49]. While the population of lactobacilli showed a substantial dose-dependent drop only after three weeks of exposure, the number of Bifidobacteria also started to decline after one week of exposure to Cd. Increased TNFa synthesis and altered genes related to the bacterial metabolism of short-chain fatty acids (SCFAs) also accompanied these modifications [50]. Likewise, there was a notable drop in cell viability when Cd was added to the cultured fecal microbiota in vitro. On the other hand, a further investigation revealed a noteworthy reduction in the quantity of Bacteroides following an eight-week administration of Cd. It has also been demonstrated that Cd exposure lowers the amount of SCFAs in the fecal pellet [51].

In addition to affecting gut biodiversity, short-term Cd therapy also has an impact on the functioning of the gut microbiome [52]. Specifically, it has been shown that exposure to Cd causes a substantial change in the bacterial proteome beginning at 15 minutes. Based on these results, the authors hypothesized that three temporal response types to CD exposure – sustained tolerance, adaptation, and resistance – may be connected to the distinctive changes in protein expression in the gut microbiota [53]. Wang and Crowly also showed that Cd altered global gene expression in an earlier in vitro investigation. Notably, exposure to Cd for 0, 5, 15, and 25 minutes increased the expression of genes that may be involved in the production or transport of lipopolysaccharides, in accordance with the effects on energy metabolism, transport, cell cycle, and binding protein genes [54]. These findings are really significant since they show how Cd affects gut microbiota almost immediately. Low-dose Cd exposure during infancy also had a major impact on the amount and variety of the gut microbiota. Specifically, male (but not female) mice exposed to Cd at eight weeks of age showed a notable rise in *Bacteroides* populations and a corresponding drop in Firmicutes populations [55]. At the genus level, populations of *Bifidobacterium* and *Prevotella* showed the greatest decline. It's also noteworthy that animals exposed to cadmium had a noticeably larger population of *Sphingomonas* that accumulate cadmium. It's interesting to note that the altered gut flora of mice given Cd was linked to higher levels of obesity [56]. The observed increase in adipose tissue mass accumulation following the transfer of fecal flora from Cd-exposed to Cd-unexposed mice validated this effect of Cd-induced modification of the microbiome. Furthermore, the administration of Gram negative targeted antibiotics prior to the mice's Cd-induced rise in obesity was avoided. Another study showed that five-week-old C57BL/6 male mice exposed to Cd showed a substantial drop in caecal Firmicutes, but not in Bacteroidetes or g-proteobacteria [57]. These modifications were linked to alterations in the liver's energy balance genes, which may be associated with changes in fatty acid production and transport, as well as triglyceride synthesis. These changes, potentially triggered by LPS, resulted in an increased expression of proinflammatory genes. Therefore, because of increasing Bacteroidetes abundance and/or decreased Firmicutes populations, most studies show a considerable rise in the Bacteroidetes-to-Firmicutes ratio [58]. The inconsistencies that currently exist regarding the impact of lead exposure on specific bacterial populations might be related to various exposure regimens (dose, duration, life span), or specific lead species [59]. Table 11.1 summarizes cadmium's interaction with the gut microbiota.

11.4.1 Gut Permeability, Inflammation, and Endotoxemia

Evidence indicates that Cd exposure significantly diminishes the viability of intestinal epithelial cells and disrupts paracellular permeability by compromising tight junctions. Subsequent investigations unveiled an uneven distribution of tight junction proteins in HT-29 cell monolayers and a notable reduction in ZO-1, ZO-2, occludin, and claudin-1 mRNA expression in the colon and jejunum of Cd-exposed animals [60]. These alterations correlate with heightened intestinal permeability, leading to increased LPS levels in the bloodstream. In a Caco-2 cell culture, Cd exposure markedly disrupted tight junction permeability, causing intercellular junction damage and upregulation

TABLE 11.1
Outline of diverse facets of cadmium's interaction with the gut microbiota, encompassing protective roles, compositional alterations, and immediate responses, shedding light on potential health outcomes and gender-specific impacts.

Factors and Impacts on Gut Microbiota	Experimental Insights	Alterations in Microbial Composition	Health Outcomes
Role of Gut Microbiota in Cadmium's Impact	Environmental Disruptions and Gut Dysbiosis	Germfree Mice's Protective Role	Protective effects on Cd toxicity
Short-Term and Chronic Cd Exposure Effects	Laboratory Animal Studies	Altered Microbial Diversity	Changes in Cd-induced gut flora
NF-kB Activation and Persistent Gut Alterations	Post-Exposure Gut Changes	Transfer of Gut Flora Effects	Long-lasting effects of Cd on gut
Temporal Response Types to Cd Exposure	Bacterial Proteome Alterations at 15 Minutes	Global Gene Expression Changes	Immediate effects on gut microbiota
Cd Exposure during Infancy and Gender-Specific Effects	Cadmium Accumulation in Sphingomonas	Obesity Link with Altered Gut Flora	Cd's influence on obesity via gut changes
Liver Gene Expression Changes and Cd-Induced Ratios	Inconsistencies in Lead Exposure Effects	Effects of Caecal Microbiota and Liver Genes on Cd	Cd's complex effects on gut microbiota

of genes associated with oxidative stress, heat shock protein, and caspase 3 [61]. These findings validate prior research indicating a substantial increase in paracellular permeability in the Caco-2 layer, accompanied by elevated HSP70 levels and the presence of necrotic lesions [62]. Moreover, Cd exposure resulted in elevated serum levels of lipopolysaccharide-binding protein (LPS) in mice in a dose-dependent manner [63]. This exposure also upregulated pro-inflammatory genes in the liver, including IL-1b, TNFa, and IL-6, along with LPS target genes like TLR4, MD2, and CD14. The study proposes that changes in tight junctions and subsequent disruptions in gut dysbiosis are driven by increased production of the prehaptoglobin protein zonulin [64, 65]. Extended Cd exposure altered intestinal histology, manifesting as thick, short villi with fusion and necrotic regions. The gut exhibited reduced lactobacilli count and elevated levels of TNFa, IFN-y, IL-1b, and IL-17, contributing to altered gut morphology [66]. Mesenteric lymph nodes displayed Cd-induced alterations, including proliferation and increased cellularity, accompanied by elevated levels of IFN-y, IL-17, and IL-10 at both protein and mRNA levels. Subsequent research revealed strain-specific differences, more pronounced in Dark Agouti rats than in the Albino Oxford strain [67, 68].

Treatment with CdC12 caused significant damage to villi and inflammatory cell infiltration into the proximal intestine's lamina propria in mice. These effects were linked to increased macrophage inflammatory protein-2 (MIP-2) mRNA production. While a Caco-2 cell model suggested Cd-induced activation of NF-kB via I-kBa degradation contributing to the gut's proinflammatory response, histological examination in mice did not reveal notable intestinal lesions [69]. Interactions between Cd and LPS toxicity were evident, with simultaneous exposure leading to greater liver damage through oxidative and nitrosative stress and an inflammatory response compared to control or single-agent exposure [70, 71]. Cd and LPS co-exposure also influenced nitric oxide formation and displaced essential metals (Cu, Zn) by Cd. Remarkably, there was a significant correlation between blood Cd levels and the frequency of metabolic syndrome when measuring high levels of Cd exposure [72, 73]. Additional research suggested that Cd could impede typical LPS signaling in cells, inhibiting the NF-kB pathway in macrophages and compromising their inflammatory response to LPS, potentially reducing resistance to infectious agents. Oxidative stress induced by Cd may contribute to the deregulation of reactivity in murine macrophages [74]. Furthermore, in mice with colitis, exposure to Cd demonstrated a dose-dependent reduction in proinflammatory cytokine expression. Most studies to date indicate that Cd may obstruct LPS signaling, elevate LPS levels, and induce endotoxemia by increasing gut permeability [75, 76]. However, the nature of the relationship between LPS and Cd may exhibit both antagonistic and synergistic effects, depending on dosage and target mechanisms [77, 78].

11.5 PROBIOTICS AS A THERAPEUTIC STRATEGY IN CD EXPOSURE

Probiotics could exert a significant protective influence, as numerous inquiries have demonstrated, supporting the pivotal role of gut dysbiosis in cadmium (Cd) poisoning. Consumption of meals fortified with *B. coagulans*, *L. plantarum*, and inulin has been proven to mitigate liver and kidney damage induced by Cd. Specifically, in comparison to Cd-exposed rats without treatment, a meal rich in probiotics markedly diminished the Cd accumulation in the liver and kidney, evidenced by decreased blood levels of ALT, AST, creatinine, and carbamide [79]. The abundance of intestinal microorganisms, particularly lactic bacteria, increased in Cd-exposed rats receiving probiotics, aligning with the observed alterations. Correspondingly, in Cd-exposed colonic (HT-29) cells, the prebiotic Lactobacillus plantarum significantly attenuated the Cd-induced inflammatory response, reducing TNFa, IL-1b, IL-6, and IL-8 production [80]. Moreover, *L. plantarum* therapy in Cd-exposed rats mitigated endotoxemia, intestinal permeability, and tight junction abnormalities [81–83]. Furthermore, it was demonstrated that *L. plantarum* bolstered the antioxidant response in the kidney and liver, diminishing Cd-induced damage [84]. These findings led researchers to propose that probiotics, in contrast to antioxidant therapy, might be a more efficacious treatment for Cd intoxication. Notably, administering *L. plantarum* intraperitoneally to Cd-exposed animals did not reduce Cd accumulation in organs, although it enhanced antioxidant activity and protected tissues from harm. The Cd-induced decrease in lactobacilli and bifidobacilli populations was reversed by administering a probiotic containing *L. rhamnosus*, *L. acidophilus*, and *B. longum*, resulting in a substantial reduction in Cd genotoxicity, as evidenced by reduced DNA damage in hepatocytes [85]. Probiotics may exert a preventive effect against mercury exposure by directly binding to metal ions in the colon, thereby reducing their bioaccessibility. A significant correlation has been demonstrated between the decrease in blood metal concentration, fecal Cd levels, and the increase in fecal lactobacilli content [86–88]. Preincubation of Cd(NO3)2 with Lactobacillus kefir considerably diminished its toxicity in HepG2 colonies in an in vitro model. It is suggested that probiotics primarily target tight junction disruption and intestinal inflammation caused by Cd in their preventive activity against Cd poisoning. The addition of zinc supplements to probiotics may augment their protective effects [89].

11.6 CURRENT REGULATIONS ON CADMIUM EXPOSURE

Recognizing the significant health risks associated with cadmium exposure, regulatory bodies worldwide have implemented measures to control and limit human exposure to this toxic heavy metal. The regulations encompass various aspects, including permissible levels in food, occupational settings, and environmental discharge [90]. These standards aim to safeguard public health and the environment, acknowledging the diverse routes through which cadmium can enter the human body [91]. In the realm of food safety, regulatory agencies such as the Food and Drug Administration (FDA) in the United States and the European Food Safety Authority (EFSA) have established maximum allowable limits for cadmium in various food products. These limits are based on comprehensive risk assessments that consider factors such as dietary habits, bioavailability, and potential health effects [92–94]. For instance, in the European Union, the maximum allowable limit for cadmium in certain vegetables is set at 0.05 mg/kg, reflecting a cautious approach to protect consumers from excessive cadmium intake. Occupational exposure to cadmium is also subject to stringent regulations [95]. Organizations like the Occupational Safety and Health Administration (OSHA) in the United States and the European Agency for Safety and Health at Work (EU-OSHA) have established permissible exposure limits (PELs) to protect workers from the adverse effects of cadmium in various industries [96]. These limits specify the maximum allowable concentration of airborne cadmium in workplace environments, mitigating the risk for those employed in occupations where cadmium exposure is likely, such as battery manufacturing, smelting, and welding. Environmental regulations address the discharge of cadmium into air, water, and soil, recognizing its potential to persist and accumulate in the ecosystem [76, 97, 98]. The Environmental Protection Agency (EPA) in the United States and the European Chemicals Agency (ECHA) in the European Union (EU) set standards for cadmium emissions and concentrations in ambient air and water bodies [99–102]. Efforts to reduce industrial discharges and promote responsible waste management contribute to minimizing environmental contamination and subsequent human exposure. Despite these regulations, challenges persist in ensuring comprehensive protection. Monitoring, enforcement, and continuous reassessment of regulatory standards remain essential to address emerging scientific findings and evolving understanding of cadmium toxicity [103, 104]. Collaborative efforts between regulatory bodies, industry stakeholders, and the scientific community are crucial for refining and adapting regulations to effectively mitigate the risks associated with cadmium exposure, safeguarding both human health and the environment [32, 105–107].

11.6.1 Environmental Standards and Monitoring: Regulatory Perspectives on Cadmium

Regulating cadmium exposure is a critical aspect of safeguarding public health and environmental integrity. Various regulatory bodies worldwide have established environmental standards and monitoring protocols to mitigate the risks associated with cadmium contamination. International organizations, national agencies, and regional bodies have set stringent standards for cadmium levels in air, water, soil, and food [108]. These standards aim to limit cadmium concentrations to levels deemed safe for human health and the environment. For example, the World Health Organization (WHO) has established guidelines for acceptable levels of cadmium in drinking water, while the EU has defined maximum permissible levels for cadmium in various food products [109]. These standards serve as benchmarks for regulatory compliance and provide a basis for assessing the extent of environmental contamination. Monitoring cadmium levels in the environment is a multifaceted endeavor that involves routine surveillance, data collection, and risk assessment. Regulatory agencies employ a range of monitoring techniques to track cadmium concentrations and assess potential exposure pathways [110, 111]. These techniques include air and water sampling, soil analysis, and monitoring of cadmium levels in food products. Continuous monitoring allows regulatory bodies to identify trends, hotspots of contamination, and potential sources of exposure [112–114].

In addition to environmental monitoring, regulatory frameworks often include measures to protect occupational health. Workers in industries associated with cadmium, such as mining, smelting, and manufacturing, may face elevated exposure risks. Occupational safety guidelines and monitoring programs aim to ensure that workplace exposure remains within acceptable limits, protecting the health and well-being of those engaged in cadmium-associated activities [115]. Regulatory perspectives on cadmium also encompass comprehensive risk assessments to evaluate the potential impact on human health and the environment. Risk management strategies are then implemented based on these assessments. This may include setting limits on industrial emissions, enforcing proper waste disposal practices, and regulating the use of cadmium-containing products [116]. As scientific understanding evolves, regulatory bodies continuously reassess and update standards to align with the latest research findings. Moreover, international cooperation and information sharing contribute to a collective effort to address the challenges posed by cadmium contamination on a global scale. By combining robust environmental standards with effective monitoring programs, regulatory authorities play a pivotal role in mitigating the adverse effects of cadmium, ensuring a healthier and more sustainable future [117–119].

11.6.2 Technological Advances in Studying Cadmium–Gut Microbiome Interaction

The intricate relationship between cadmium exposure and the gut microbiome demands innovative research methodologies. Recent technological advances have revolutionized the study of this interaction, providing researchers with tools to delve deeper into the molecular and microbial nuances [120].

Metagenomic sequencing has emerged as a powerful tool for deciphering the complex structure of the gut microbiome. This technique allows researchers to analyze the collective genetic material of the microbial community without the need for individual isolates [121]. In the context of cadmium exposure, metagenomics enables the identification and quantification of microbial species, shedding light on how the composition of the gut microbiome changes in response to cadmium. It provides a comprehensive snapshot of the microbial diversity and potential functional alterations induced by cadmium toxicity [122–124]. Metabolomics explores the unique fingerprint of small molecules – metabolites – within a biological system. Advancements in mass spectrometry and nuclear magnetic resonance spectroscopy enable researchers to analyze the metabolic profiles of the gut microbiome under cadmium exposure [125]. This approach unveils alterations in microbial metabolic pathways, offering insights into how cadmium influences the functional dynamics of the microbiome. Metabolomics aids in identifying specific metabolites associated with cadmium toxicity, paving the way for a more nuanced understanding of the mechanistic links between cadmium exposure and gut microbial function [126]. Culturomics involves cultivating and characterizing previously uncultured microorganisms from complex samples. This technique is particularly valuable in isolating and studying specific bacteria within the gut microbiome that may respond uniquely to cadmium exposure. By growing these bacteria in laboratory settings, researchers can explore their behavior, resilience, and potential mitigating factors against cadmium toxicity. Culturomics complements other high-throughput techniques, providing a bridge between culture-dependent and culture-independent methods [127, 128].

The revolutionary CRISPR-Cas9 gene-editing technology allows researchers to manipulate the genetic material of microorganisms with unprecedented precision. In the context of cadmium–gut microbiome interaction studies, CRISPR-Cas9 can be employed to engineer specific microbial strains, elucidating the role of individual genes in responding to cadmium exposure. This targeted approach facilitates the identification of key genes involved in microbial resilience or susceptibility to cadmium, offering insights into potential therapeutic targets [129]. Microfluidic devices, often referred to as microbiome-on-a-chip platforms, mimic the complex microenvironments of the human gut. These platforms integrate advanced imaging, microsensors, and real-time monitoring capabilities, allowing researchers to observe the dynamic interactions between cadmium and the gut microbiome in a controlled setting. Microbiome-on-a-chip systems offer a more physiologically relevant approach, bridging the gap between in vitro experiments and in vivo complexity [48, 49]. The amalgamation of these cutting-edge technologies propels the study of cadmium–gut microbiome interaction into a new era of precision and depth. These tools not only unravel the intricacies of microbial responses to cadmium but also open avenues for developing targeted interventions and therapeutic strategies to mitigate the adverse effects of cadmium exposure on gut health [130–132].

11.7 CONCLUSION AND FUTURE PERSPECTIVE

The current knowledge regarding the impact of Cadmium (Cd) on gut physiology reveals a multifaceted relationship with far-reaching consequences. Exposure to Cd demonstrates a dual effect on the gut microbiota, inducing significant alterations in bacterial populations, particularly in the Bacteroidetes to Firmicutes ratio. This shift triggers an increase in lipopolysaccharide (LPS) production and disrupts the metabolic activity of the gut microbiome from the earliest moments of exposure. The repercussions of Cd exposure extend beyond the gut wall, leading to cell damage, inflammation, and a crucial breakdown of tight junctions. The compromised barrier function results in increased permeability, leading to endotoxemia and systemic inflammation, compounded by heightened LPS generation. Studies suggest that Cd-induced intestinal permeability may enhance bacterial translocation into tissues, exacerbating endotoxemia and compromising the immune response, potentially increasing susceptibility to infections. The Cd-induced decrease in bacteriophage lytic activity further raises concerns about the potential lowering of viral resistance, contributing to increased vulnerability to infections. This observation aligns with findings that Cd-exposed mice exhibit heightened susceptibility to infections, highlighting the broader impact of Cd on immune responses. Target organs, significantly affected by both reduced microbial metabolic activity and systemic inflammation induced by LPS, manifest various disorders. Altered hemopoiesis, inflammatory bowel disorders, atherosclerosis, alcoholic steatohepatitis, and non-alcoholic fatty liver disease have all been linked to poor gut microbiota metabolism. Furthermore, inflammatory conditions like obesity and diabetes mellitus are associated with LPS, underlining the systemic implications of Cd exposure [133–135]. The integral role of altered gut physiology in mediating the harmful effects of Cd is underscored by the observed efficacy of probiotics in mitigating Cd toxicity. Probiotic therapy, coupled with chelation, antioxidant, and anti-inflammatory interventions, emerges as a potential strategy for preventing Cd poisoning. This holistic approach acknowledges the interconnectedness of gut health and systemic well-being.

Looking forward, further research is imperative to unravel the intricate dynamics of Cd exposure on gut microbiota. Investigating biodiversity, symbiotic relationships, metabolic activities, and the impact on the intestinal epithelium will provide a deeper understanding of Cd-induced alterations. This knowledge will not only contribute to refining therapeutic approaches for Cd toxicity but may also unveil insights into the role of altered gut physiology in the pathophysiology of various disorders associated with Cd exposure. As we delve into these complexities, the potential for probiotic therapy to emerge as a key player in Cd toxicity prevention opens exciting avenues for future research and clinical applications.

REFERENCES

1. W. Ali, Y. Ma, J. Zhu, H. Zou, Z. Liu, Mechanisms of cadmium-induced testicular injury: a risk to male fertility, *Cells*, 11 (2022) 3601.
2. M.A. Arruebarrena, C.T. Hawe, Y.M. Lee, R.C. Branco, Mechanisms of cadmium neurotoxicity, *International Journal of Molecular Sciences*, 24 (2023) 16558.
3. D. Diaz, F. Ujueta, G. Mansur, G.A. Lamas, A. Navas-Acien, I.A. Arenas, Low-level cadmium exposure and atherosclerosis, *Current Environmental Health Reports*, 8 (2021) 42–53.
4. E. Freisinger, M. Vašák, Cadmium in metallothioneins, *Metal Ions in Life Sciences*, 11 (2013) 339–371.
5. G. Genchi, M.S. Sinicropi, G. Lauria, A. Carocci, A. Catalano, The effects of cadmium toxicity, *International Journal of Environmental Research and Public Health*, 17 (2020) 3782.
6. H.X. Geng, L. Wang, Cadmium: toxic effects on placental and embryonic development, *Environmental Toxicology and Pharmacology*, 67 (2019) 102–107.
7. M. Abubakar, E.R. Puppala, B.J. Dutta, K.C. Maharana, R. Thapa, S. Roshan, B. Tazneem, A. Khan, A.A. Bhat, *Epigenetics of Rhinovirus, Targeting Epigenetics in Inflammatory Lung Diseases*, Springer Nature Singapore, Singapore, 2023, pp. 185–197.
8. F.U. Haider, C. Liqun, J.A. Coulter, S.A. Cheema, J. Wu, R. Zhang, M. Wenjun, M. Farooq, Cadmium toxicity in plants: impacts and remediation strategies, *Ecotoxicology and Environmental Safety*, 211 (2021) 111887.
9. H. Küpper, B. Leitenmaier, Cadmium-accumulating plants, *Metal Ions in Life Sciences*, 11 (2013) 373–393.
10. V.K.R. Balaga, A. Pradhan, R. Thapa, N. Patel, R. Mishra, N. Singla, Morin: a comprehensive review on its versatile biological activity and associated therapeutic potential in treating cancers, *Pharmacological Research*, 7 (2023) 100264.
11. Y. Ma, Q. Su, C. Yue, H. Zou, J. Zhu, H. Zhao, R. Song, Z. Liu, The effect of oxidative stress-induced autophagy by cadmium exposure in kidney, liver, and bone damage, and neurotoxicity, *International Journal of Molecular Sciences*, 23 (2022) 13491.

12. M. Nordberg, G.F. Nordberg, Metallothionein and cadmium toxicology-historical review and commentary, *Biomolecules*, 12 (2022) 360.
13. M. Peana, A. Pelucelli, C.T. Chasapis, S.P. Perlepes, V. Bekiari, S. Medici, M.A. Zoroddu, Biological effects of human exposure to environmental cadmium, *Biomolecules*, 13 (2022) 36.
14. A. Popov Aleksandrov, I. Mirkov, D. Tucovic, J. Kulas, M. Ninkov, M. Kataranovski, Cadmium and immunologically-mediated homeostasis of anatomical barrier tissues, *Toxicology Letters*, 337 (2021) 38–45.
15. P. Richter, O. Faroon, R.S. Pappas, Cadmium and cadmium/zinc ratios and tobacco-related morbidities, *International Journal of Environmental Research and Public Health*, 14 (2017) 1154.
16. S. Satarug, S.H. Garrett, M.A. Sens, D.A. Sens, Cadmium, environmental exposure, and health outcomes, *Environmental Health Perspectives*, 118 (2010) 182–190.
17. H.R. Schaefer, S. Dennis, S. Fitzpatrick, Cadmium: mitigation strategies to reduce dietary exposure, *Journal of Food Science*, 85 (2020) 260–267.
18. M.P. Waalkes, Cadmium carcinogenesis, *Mutation Research*, 533 (2003) 107–120.
19. R. Wang, P. Sang, Y. Guo, P. Jin, Y. Cheng, H. Yu, Y. Xie, W. Yao, H. Qian, Cadmium in food: Source, distribution and removal, *Food Chemistry*, 405 (2023) 134666.
20. M. Webb, Cadmium, *British Medical Bulletin*, 31 (1975) 246–250.
21. L.J. Yan, D.C. Allen, Cadmium-induced kidney injury: oxidative damage as a unifying mechanism, *Biomolecules*, 11 (2021) 1575.
22. F.J. Zhao, X.Y. Huang, Cadmium phytoremediation: call rice CAL1, *Molecular Plant*, 11 (2018) 640–642.
23. J.K. Bhardwaj, P. Bikal, S.N. Sachdeva, Cadmium as an ovarian toxicant: a review, *Journal of Applied Toxicology*, 44 (2024) 129–147.
24. S. Bhattacharya, The role of probiotics in the amelioration of cadmium toxicity, *Biological Trace Element Research*, 197 (2020) 440–444.
25. A.A. Bhat, O. Afzal, N. Agrawal, R. Thapa, W.H. Almalki, I. Kazmi, S.I. Alzarea, A.S.A. Altamimi, N. Kukreti, A. Chakraborty, A comprehensive review on the emerging role of long non-coding RNAs in the regulation of NF-κB signaling in inflammatory lung diseases, *International Journal of Biological Macromolecules*, 253 (2023) 126951.
26. G. Cannino, E. Ferruggia, C. Luparello, A.M. Rinaldi, Cadmium and mitochondria, *Mitochondrion*, 9 (2009) 377–384.
27. M. Chatterjee, A. Kortenkamp, Cadmium exposures and deteriorations of cognitive abilities: estimation of a reference dose for mixture risk assessments based on a systematic review and confidence rating, *Environmental Health: A Global Access Science Source*, 21 (2022) 69.
28. X.X. Chen, Y.M. Xu, A.T.Y. Lau, Metabolic effects of long-term cadmium exposure: an overview, *Environmental Science and Pollution Research International*, 29 (2022) 89874–89888.
29. Q. Gu, C. Wang, Q. Xiao, Z. Chen, Y. Han, Melatonin confers plant cadmium tolerance: an update, *International Journal of Molecular Sciences*, 22 (2021) 11704.
30. B. Hussain, M.N. Ashraf, R. Shafeeq Ur, A. Abbas, J. Li, M. Farooq, Cadmium stress in paddy fields: effects of soil conditions and remediation strategies, *The Science of the Total Environment*, 754 (2021) 142188.
31. N.K. Isaev, E.V. Stelmashook, E.E. Genrikhs, G.E. Onishchenko, Interaction between mitophagy, cadmium and zinc, *Journal of Trace Elements in Medicine and Biology: Organ of the Society for Minerals and Trace Elements (GMS)*, 79 (2023) 127230.
32. S. Kumar, A. Sharma, Cadmium toxicity: effects on human reproduction and fertility, *Reviews on Environmental Health*, 34 (2019) 327–338.
33. C.T. Kunioka, M.C. Manso, M. Carvalho, Association between environmental cadmium exposure and osteoporosis risk in postmenopausal women: a systematic review and meta-analysis, *International Journal of Environmental Research and Public Health*, 20 (2022) 485.
34. G. Li, L. Yan, X. Chen, S.S. Lam, J. Rinklebe, Q. Yu, Y. Yang, W. Peng, C. Sonne, Phytoremediation of cadmium from soil, air and water, *Chemosphere*, 320 (2023) 138058.
35. H. Li, N. Luo, Y.W. Li, Q.Y. Cai, H.Y. Li, C.H. Mo, M.H. Wong, Cadmium in rice: Transport mechanisms, influencing factors, and minimizing measures, *Environmental Pollution*, 224 (2017) 622–630.
36. T.N. Lucon, A.T. Costa, P. Galvão, M.G.P. Leite, Cadmium behavior in a karst environment hydrological cycle, *Environmental Science and Pollution Research International*, 27 (2020) 8965–8979.
37. Y. Ma, D. Ran, X. Shi, H. Zhao, Z. Liu, Cadmium toxicity: a role in bone cell function and teeth development, *The Science of the Total Environment*, 769 (2021) 144646.

38. A.C. Martins, A.A.D. Santos, A. Lopes, A.V. Skalny, M. Aschner, A.A. Tinkov, M.M.B. Paoliello, Endothelial dysfunction induced by cadmium and mercury and its relationship to hypertension, *Current Hypertension Reviews*, 17 (2021) 14–26.
39. M.S. Hussain, A. Sharma, R. Kumar, Prebiotics and probiotics: a focused review of applications in respiratory disorders, *Carpathian Journal of Food Science and Technology*, 15 (2023) 183–207.
40. R. Nagaraju, R. Kalahasthi, R. Balachandar, B.S. Bagepally, Cadmium exposure and DNA damage (genotoxicity): a systematic review and meta-analysis, *Critical Reviews in Toxicology,* 52 (2022) 786–798.
41. M. Ozturk, M. Metin, V. Altay, L. De Filippis, B.T. Ünal, A. Khursheed, A. Gul, M. Hasanuzzaman, K. Nahar, T. Kawano, P.G. Caparrós, Molecular biology of cadmium toxicity in saccharomyces cerevisiae, *Biological Trace Element Research*, 199 (2021) 4832–4846.
42. V. Rapisarda, E. Miozzi, C. Loreto, S. Matera, C. Fenga, R. Avola, C. Ledda, Cadmium exposure and prostate cancer: insights, mechanisms and perspectives, *Frontiers in Bioscience*, 23 (2018) 1687–1700.
43. D. Reyes-Hinojosa, C.A. Lozada-Pérez, Y. Zamudio Cuevas, A. López-Reyes, G. Martínez-Nava, J. Fernández-Torres, A. Olivos-Meza, C. Landa-Solis, M.C. Gutiérrez-Ruiz, E. Rojas Del Castillo, K. Martínez-Flores, Toxicity of cadmium in musculoskeletal diseases, *Environmental Toxicology and Pharmacology*, 72 (2019) 103219.
44. M. Rezapour, H.A. Rezapour, M. Chegeni, N. Khanjani, Exposure to cadmium and head and neck cancers: a meta-analysis of observational studies, *Reviews on Environmental Health*, 36 (2021) 577–584.
45. A.A. Bhat, A. Goyal, R. Thapa, I. Kazmi, S.I. Alzarea, M. Singh, S. Rohilla, T.K. Saini, N. Kukreti, D.U. Meenakshi, Uncovering the complex role of interferon-gamma in suppressing type 2 immunity to cancer, *Cytokine*, 171 (2023) 156376.
46. J. Song, Q. Song, D. Wang, Y. Liu, Mitigation strategies for excessive cadmium in rice, *Comprehensive Reviews in Food Science and Food Safety*, 22 (2023) 3847–3869.
47. V. Souza-Arroyo, J.J. Fabián, L. Bucio-Ortiz, R.U. Miranda-Labra, L.E. Gomez-Quiroz, M.C. Gutiérrez-Ruiz, The mechanism of the cadmium-induced toxicity and cellular response in the liver, *Toxicology*, 480 (2022) 153339.
48. A. Tuffour, A. Adebayiga Kosiba, F. Addai Peprah, J. Gu, Y. Zhou, H. Shi, Cadmium-induced stress: a close look at the relationship between autophagy and apoptosis, *Toxicological Sciences*, 194 (2023) 1–12.
49. A. Turner, Cadmium pigments in consumer products and their health risks, *The Science of the Total Environment*, 657 (2019) 1409–1418.
50. H.T. Yu, J. Zhen, J.Y. Leng, L. Cai, H.L. Ji, B.B. Keller, Zinc as a countermeasure for cadmium toxicity, *Acta Pharmacologica SINICA*, 42 (2021) 340–346.
51. M.M. Ardestani, C.A. van Gestel, Sorption and pH determine the long-term partitioning of cadmium in natural soils, *Environmental Science and Pollution Research International*, 23 (2016) 18492–18501.
52. S. Asaf, R. Jan, M.A. Khan, Lubna, A.L. Khan, S. Asif, S. Bilal, W. Ahmad, M. Waqas, K.M. Kim, A. Al-Harrasi, I.J. Lee, Unraveling the mutualistic interaction between endophytic Curvularia lunata CSL1 and tomato to mitigate cadmium (Cd) toxicity via transcriptomic insights, *The Science of the Total Environment*, 861 (2023) 160542.
53. S.O. Asagba, G.E. Eriyamremu, M.E. Igberaese, Bioaccumulation of cadmium and its biochemical effect on selected tissues of the catfish (*Clarias gariepinus*), *Fish Physiology and Biochemistry*, 34 (2008) 61–69.
54. L.B. Bradley, M. Jacob, E.E. Jacobs, D.R. Sanadi, Uncoupling of oxidative phosphorylation by cadmium ion, *The Journal of Biological Chemistry*, 223 (1956) 147–156.
55. O.Y. Chan, S.C. Poh, H.S. Lee, K.T. Tan, S.F. Kwok, Respiratory function in cadmium battery workers--a follow-up study, *Annals of the Academy of Medicine*, 17 (1988) 283–287.
56. H. Doshi, A. Ray, I.L. Kothari, Biosorption of cadmium by live and dead Spirulina: IR spectroscopic, kinetics, and SEM studies, *Current Microbiology*, 54 (2007) 213–218.
57. D.W. Engel, B.A. Fowler, Factors influencing cadmium accumulation and its toxicity to marine organisms, *Environmental Health Perspectives*, 28 (1979) 81–88.
58. A.A. Bhat, G. Gupta, R. Dahiya, R. Thapa, A. Gahtori, M. Shahwan, V. Jakhmola, A. Tiwari, M. Kumar, H. Dureja, CircRNAs: pivotal modulators of TGF-β signalling in cancer pathogenesis, *Non-Coding RNA Research*, 9 (2024) 277–287.
59. L. Gervais, Y. Lacasse, J. Brodeur, A. P'an, Presence of cadmium in the saliva of adult male workers, *Toxicology Letters*, 8 (1981) 63–66.

60. C.T. He, Y.H. Zhou, Y.Y. Huang, H.L. Fu, X.S. Wang, F.Y. Gong, X. Tan, Z.Y. Yang, Different Proteomic Processes Related to the Cultivar-dependent cadmium accumulation of Amaranthus gangeticus, *Journal of Agricultural and Food Chemistry*, 66 (2018) 1085–1095.
61. T.J. Horner, R.B. Lee, G.M. Henderson, R.E. Rickaby, Nonspecific uptake and homeostasis drive the oceanic cadmium cycle, *Proceedings of the National Academy of Sciences of the United States of America*, 110 (2013) 2500–2505.
62. T.J. Horner, R.B. Lee, G.M. Henderson, R.E. Rickaby, Reply to morel: cadmium as a micronutrient and macrotoxin in the oceans, *Proceedings of the National Academy of Sciences of the United States of America*, 110 (2013) E1878.
63. A.A. Bhat, R. Thapa, O. Afzal, N. Agrawal, W.H. Almalki, I. Kazmi, S.I. Alzarea, A.S.A. Altamimi, P. Prasher, S.K. Singh, The pyroptotic role of Caspase-3/GSDME signalling pathway among various cancer: a review, *International Journal of Biological Macromolecules*, 242 (2023) 124832.
64. W. Huang, Z. Bai, J. Jiao, H. Yuan, Z. Bao, S. Chen, M. Ding, Z. Liang, Distribution and chemical forms of cadmium in Coptis chinensis Franch. determined by laser ablation ICP-MS, cell fractionation, and sequential extraction, *Ecotoxicology and Environmental Safety*, 171 (2019) 894–903.
65. E.G. Hughes, Biological monitoring of cadmium workers, *Lancet*, 2 (1984) 1467–1468.
66. R. Dahiya, R. Thapa, N.K. Kumawat, M. Singh, S. Jakhotiya, D. Deopa, Y. Singh, N. Singla, G. Gupta, *Epigenetic Mechanisms in Inflammation, Targeting Epigenetics in Inflammatory Lung Diseases*, Springer, 2023, pp. 43–55.
67. M. Inouhe, M. Hiyama, H. Tohoyama, M. Joho, T. Murayama, Cadmium-binding protein in a cadmium-resistant strain of Saccharomyces cerevisiae, *Biochimica et Biophysica Acta*, 993 (1989) 51–55.
68. R. Thapa, O. Afzal, A.A. Bhat, A. Goyal, A.S. Alfawaz Altamimi, W.H. Almalki, S.I. Alzarea, I. Kazmi, S.K. Singh, K. Dua, L. Thangavelu, G. Gupta, New horizons in lung cancer management through ATR/CHK1 pathway modulation, *Future Medicinal Chemistry,* 15 (2023) 1807–1818.
69. Z. Li, H. Yuan, X. Hu, Cadmium-resistance in growing Rhodotorula sp. Y11, *Bioresource Technology*, 99 (2008) 1339–1344.
70. B. Chaudhary, P. Kumar, P. Arya, D. Singla, V. Kumar, D. Kumar, R. S, S. Wadhwa, M. Gulati, S.K. Singh, K. Dua, G. Gupta, M.M. Gupta, Recent developments in the study of the microenvironment of cancer and drug delivery, *Curr Drug Metab*, 23 (2023) 1027–1053.
71. M.S. Hussain, G. Gupta, A. Goyal, R. Thapa, W.H. Almalki, I. Kazmi, S.I. Alzarea, S. Fuloria, D.U. Meenakshi, V. Jakhmola, From nature to therapy: Luteolin's potential as an immune system modulator in inflammatory disorders, *Journal of Biochemical and Molecular Toxicology*, 37 (2023) e23482.
72. J. Luo, S. Qi, L. Peng, X. Xie, Phytoremediation potential of cadmium-contaminated soil by Eucalyptus globulus under different coppice systems, *Bulletin of Environmental Contamination and Toxicology*, 94 (2015) 321–325.
73. R. Thapa, O. Afzal, G. Gupta, A.A. Bhat, W.H. Almalki, S.I. Alzarea, I. Kazmi, A.S.A. Altamimi, V. Subramaniyan, L. Thangavelu, S.K. Singh, K. Dua, Unveiling the connection: Long-chain non-coding RNAs and critical signaling pathways in breast cancer, *Pathology, Research and Practice*, 249 (2023) 154736.
74. J.A. Marigomez, M.P. Ireland, Accumulation, distribution and loss of cadmium in the marine prosobranch *Littorina littorea* (L.), *The Science of the Total Environment*, 78 (1989) 1–12.
75. N. Singla, R. Thapa, R. Kulshrestha, A.A. Bhat, S. Gupta, M. Purohit, S.K. Singh, G. Gupta, *Introduction to Epigenetics, Targeting Epigenetics in Inflammatory Lung Diseases,* Springer, 2023, pp. 17–41.
76. R. Thapa, G. Gupta, A.A. Bhat, W.H. Almalki, S.I. Alzarea, I. Kazmi, S. Saleem, R. Khan, N. Altwaijry, H. Dureja, S.K. Singh, K. Dua, A review of Glycogen Synthase Kinase-3 (GSK3) inhibitors for cancers therapies, *International Journal of Biological Macromolecules*, 253 (2023) 127375.
77. M.D. Mashitah, Y. Yus Azila, S. Bhatia, Biosorption of cadmium (II) ions by immobilized cells of Pycnoporus sanguineus from aqueous solution, *Bioresource Technology*, 99 (2008) 4742–4748.
78. M.A.J. Shaikh, K.S. Alharbi, W.H. Almalki, S.S. Imam, M. Albratty, A.M. Meraya, S.I. Alzarea, I. Kazmi, F.A. Al-Abbasi, O. Afzal, A.S.A. Altamimi, Y. Singh, S.K. Singh, K. Dua, G. Gupta, Sodium alginate based drug delivery in management of breast cancer, *Carbohydr Polym*, 292 (2022) 119689.
79. W.D. Morgan, New ways of measuring cadmium in man, Nature, 282 (1979) 673–674.
80. R. Thapa, O. Afzal, A.S.A. Altamimi, A. Goyal, W.H. Almalki, S.I. Alzarea, I. Kazmi, V. Jakhmola, S.K. Singh, K. Dua, Galangin as an inflammatory response modulator: An updated overview and therapeutic potential, *Chemico-Biological Interactions*, 378 (2023) 110482.
81. J. Parizek, Interactions between selenium compounds and those of mercury or cadmium, *Environmental Health Perspectives*, 25 (1978) 53–55.

82. S. Rohilla, M. Singh, S. Priya, W.H. Almalki, S.M. Haniffa, V. Subramaniyan, S. Fuloria, N.K. Fuloria, M. Sekar, S.K. Singh, N.K. Jha, D.K. Chellappan, P. Negi, K. Dua, G. Gupta, Exploring the mechanical perspective of a new anti-tumor agent: melatonin, *Journal of Environmental Pathology, Toxicology and Oncology,* 42 (2023) 1–16.
83. V.P. Samuel, R. Dahiya, Y. Singh, G. Gupta, S.K. Sah, S.K. Gubbiyappa, D.K. Chellappan, K. Dua, Metformin: a salutary candidate for colorectal cancer treatment in patients with diabetes, *Journal of Environmental Pathology, Toxicology and Oncology*, 38 (2019) 133–141.
84. J.S. Park, S. Chung, I.S. Park, Y. Kim, C. Koh, I.S. Lee, Purification and characterization of metallothionein-like cadmium binding protein from Asian periwinkle *Littorina brevicula*, Comparative biochemistry and physiology, *Toxicology & Pharmacology: CBP,* 131 (2002) 425–431.
85. J. Pu, K. Fukushi, Measurement of cadmium ion in the presence of metal-binding biopolymers in aqueous sample, *The Scientific World Journal*, 2013 (2013) 270478.
86. K. Sasaki, S. Oguma, T. Glass, Y. Namiki, H. Sugiyama, N. Ohmura, D.A. Blake, Simple method to reduce interference from excess magnesium in cadmium immunoassays, *Journal of Agricultural and Food Chemistry*, 56 (2008) 7613–7616.
87. R. Pradhan, G. Singhvi, S.K. Dubey, G. Gupta, K. Dua, MAPK pathway: a potential target for the treatment of non-small-cell lung carcinoma, *Future Medicinal Chemistry*, 11 (2019) 793–795.
88. S. Rohilla, M. Singh, S.I. Alzarea, W.H. Almalki, F.A. Al-Abbasi, I. Kazmi, O. Afzal, A.S.A. Altamimi, S.K. Singh, D.K. Chellappan, K. Dua, G. Gupta, Recent developments and challenges in molecular-targeted therapy of non-small-cell lung cancer, *Journal of Environmental Pathology, Toxicology and Oncology: Official Organ of the International Society for Environmental Toxicology and Cancer*, 42 (2023) 27–50.
89. N. Semerci, F. Ceçen, Importance of cadmium speciation in nitrification inhibition, *Journal of Hazardous Materials*, 147 (2007) 503–512.
90. B.O. Anyanwu, O.E. Orisakwe, Current mechanistic perspectives on male reproductive toxicity induced by heavy metals, Journal of environmental science and health. Part C, *Toxicology and Carcinogenesis*, 38 (2020) 204–244.
91. R. Thapa, O. Afzal, G. Kumar, A.A. Bhat, W.H. Almalki, S.I. Alzarea, I. Kazmi, A.S.A. Altamimi, V. Subramaniyan, L. Thangavelu, Unveiling the connection: long-chain non-coding RNAs and critical signaling pathways in breast cancer, *Pathology-Research and Practice*, 249 (2023) 154736.
92. A.C. Bosch, B. O'Neill, G.O. Sigge, S.E. Kerwath, L.C. Hoffman, Heavy metals in marine fish meat and consumer health: a review, *Journal of the Science of Food and Agriculture*, 96 (2016) 32–48.
93. M.S. Hussain, O. Afzal, G. Gupta, A.S.A. Altamimi, W.H. Almalki, S.I. Alzarea, I. Kazmi, N.K. Fuloria, M. Sekar, D.U. Meenakshi, L. Thangavelu, A. Sharma, Long non-coding RNAs in lung cancer: unraveling the molecular modulators of MAPK signaling, *Pathology, Research and Practice*, 249 (2023) 154738.
94. M.S. Hussain, G. Gupta, M. Afzal, S.M. Alqahtani, V.P. Samuel, W. Hassan Almalki, I. Kazmi, S.I. Alzarea, S. Saleem, H. Dureja, S.K. Singh, K. Dua, L. Thangavelu, Exploring the role of lncrna neat1 knockdown in regulating apoptosis across multiple cancer types: a review, *Pathology, Research and Practice*, 252 (2023) 154908.
95. J.L. Dempsey, J.Y. Cui, Long non-coding RNAs: a novel paradigm for toxicology, *Toxicological Sciences: An Official Journal of the Society of Toxicology*, 155 (2017) 3–21.
96. E.N. Dover, N.Y. Patel, M. Stýblo, Impact of in vitro heavy metal exposure on pancreatic β-cell function, *Toxicology Letters*, 299 (2018) 137–144.
97. R. Thapa, A. Goyal, G. Gupta, A.A. Bhat, S.K. Singh, V. Subramaniyan, S. Sharma, P. Prasher, V. Jakhmola, S.K. Singh, Recent developments in the role of protocatechuic acid in neurodegenerative disorders, *EXCLI Journal*, 22 (2023) 595.
98. R. Thapa, G. Gupta, P. Dave, S.K. Singh, A. Raizaday, W.H. Almalki, G. Vyas, S.K. Singh, K. Dua, Y. Singh, Current update on the protective effect of epicatechin in neurodegenerative diseases, *EXCLI Journal*, 21 (2022) 897–903.
99. M. Guney, G.J. Zagury, Heavy metals in toys and low-cost jewelry: critical review of U.S. and Canadian legislations and recommendations for testing, *Environmental Science & Technology*, 46 (2012) 4265–4274.
100. J. Gzyl, Assessment of Polish population exposure to lead and cadmium with special emphasis to the Katowice Province on the basis of metal concentrations in environmental compartments, *Central European Journal of Public Health*, 5 (1997) 93–96.
101. G. Gupta, T. de Jesus Andreoli Pinto, D.K. Chellappan, A. Mishra, H. Malipeddi, K. Dua, A clinical update on metformin and lung cancer in diabetic patients, *Panminerva Medica*, 60 (2018) 70–75.

102. G. Gupta, G. Singhvi, D.K. Chellappan, S. Sharma, A. Mishra, R. Dahiya, T. de Jesus Andreoli Pinto, K. Dua, Peroxisome proliferator-activated receptor gamma: promising target in glioblastoma, *Panminerva Medica*, 60 (2018) 109–116.
103. S.H. Kim, Y.J. Jeong, M.C. Pyo, K.W. Lee, Combined effects of cadmium and ochratoxin A on intestinal barrier dysfunction in human Caco-2 cells and pig small intestinal epithelial cells, *Mycotoxin Research*, 38 (2022) 253–264.
104. Y. Komoike, H. Inamura, M. Matsuoka, Effects of salubrinal on cadmium-induced apoptosis in HK-2 human renal proximal tubular cells, *Archives of Toxicology*, 86 (2012) 37–44.
105. U. Kukongviriyapan, K. Apaijit, V. Kukongviriyapan, Oxidative stress and cardiovascular dysfunction associated with cadmium exposure: beneficial effects of curcumin and tetrahydrocurcumin, *The Tohoku Journal of Experimental Medicine*, 239 (2016) 25–38.
106. G. Gupta, A.A. Bhat, A. Goyal, N. Singla, S. Gupta, S. Sharma, S. Bhatt, K. Dua, Exploring ACSL4/LPCAT3/ALOX15 and SLC7A11/GPX4/NFE2L2 as potential targets in ferroptosis-based cancer therapy, *Future Medicinal Chemistry*, 15 (2023) 1209–1212.
107. G. Gupta, D.K. Chellappan, T. de Jesus Andreoli Pinto, P.M. Hansbro, M. Bebawy, K. Dua, Tumor suppressor role of miR-503, *Panminerva Medica*, 60 (2018) 17–24.
108. Y. Li, K. He, L. Cao, X. Tang, R. Gou, T. Luo, S. Xiao, Z. Chen, T. Li, J. Qin, Z. Zhang, J. Cai, Association between plasma cadmium and renal stone prevalence in adults in rural areas of Guangxi, China: a case-control study, *BMC Nephrology*, 23 (2022) 323.
109. Y. Li, X. Shi, J. Xu, X. Huang, J. Feng, Y. Huang, K. Liu, F. Yu, Proteomics-based analysis on the stress response mechanism of Bidens pilosa L. under cadmium exposure, *Journal of Hazardous Materials*, 462 (2024) 132761.
110. Y. Liang, Z. Pan, M. Zhu, R. Gao, Y. Wang, Y. Cheng, N. Zhang, Exposure to essential and non-essential trace elements and risks of congenital heart defects: a narrative review, *Frontiers in Nutrition*, 10 (2023) 1121826.
111. N. Lombaert, M. Gilles, V. Verougstraete, Cadmium monitoring at the workplace: effectiveness of a combination of air- and biomonitoring, *Toxics*, 11 (2023) 354.
112. B. Lv, H.L. Yang, Y.D. Peng, J. Wang, Z. Zeng, N. Li, Y.E. Tang, Z. Wang, Q.S. Song, Cadmium exposure alters expression of protective enzymes and protein processing genes in venom glands of the wolf spider Pardosa pseudoannulata, *Environmental Pollution*, 268 (2021) 115847.
113. G. Gupta, W.H. Al-Malki, I. Kazmi, L. Thangavelu, P.K. Gupta, N.K. Jha, P. Prasher, S.K. Singh, K. Dua, The role of HGF/MET in liver cancer, *Future Medicinal Chemistry*, 13 (2021) 1829–1832.
114. G. Gupta, M. Bebawy, T.J.A. Pinto, D.K. Chellappan, A. Mishra, K. Dua, Role of the tristetraprolin (Zinc Finger Protein 36 Homolog) gene in cancer, *Critical Reviews in Eukaryotic Gene Expression*, 28 (2018) 217–221.
115. L. Ma, J. Mo, Y. Chen, L. Li, L. Xie, X. Chen, X. Li, Y. Wang, Z. Lin, R.S. Ge, In utero cadmium and dibutyl phthalate combination exposure worsens the defects of fetal testis in rats, *Environmental Pollution*, 265 (2020) 114842.
116. E.M. Martin, R.C. Fry, Environmental influences on the epigenome: exposure- associated DNA methylation in human populations, *Annual Review of Public Health*, 39 (2018) 309–333.
117. N.K. McRae, S. Gaw, C.N. Glover, Effects of waterborne cadmium on metabolic rate, oxidative stress, and ion regulation in the freshwater fish, inanga (Galaxias maculatus), *Aquatic Toxicology*, 194 (2018) 1–9.
118. A.A. Bhat, G. Gupta, R. Dahiya, R. Thapa, A. Gahtori, M. Shahwan, V. Jakhmola, A. Tiwari, M. Kumar, H. Dureja, S.K. Singh, K. Dua, V. Kumarasamy, V. Subramaniyan, CircRNAs: Pivotal modulators of TGF-β signalling in cancer pathogenesis, *Noncoding RNA Res*, 9 (2024) 277–287.
119. A.A. Bhat, R. Thapa, O. Afzal, N. Agrawal, W.H. Almalki, I. Kazmi, S.I. Alzarea, A.S.A. Altamimi, P. Prasher, S.K. Singh, K. Dua, G. Gupta, The pyroptotic role of Caspase-3/GSDME signalling pathway among various cancer: a review, *International Journal of Biological Macromolecules*, 242 (2023) 124832.
120. L.H. Oshiquiri, K.R.A. Dos Santos, S.A. Ferreira Junior, A.S. Steindorff, J.R. Barbosa Filho, T.M. Mota, C.J. Ulhoa, R.C. Georg, Trichoderma harzianum transcriptome in response to cadmium exposure, *Fungal Genetics and Biology: FG & B*, 134 (2020) 103281.
121. S. Satarug, M.R. Moore, Emerging roles of cadmium and heme oxygenase in type-2 diabetes and cancer susceptibility, *The Tohoku Journal of Experimental Medicine*, 228 (2012) 267–288.
122. G. Schwalfenberg, I. Rodushkin, S.J. Genuis, Heavy metal contamination of prenatal vitamins, *Toxicology Reports*, 5 (2018) 390–395.

123. A.A. Bhat, A. Goyal, R. Thapa, W.H. Almalki, I. Kazmi, S.I. Alzarea, M. Singh, S. Rohilla, T.K. Saini, N. Kukreti, D.U. Meenakshi, N.K. Fuloria, M. Sekar, G. Gupta, Uncovering the complex role of interferon-gamma in suppressing type 2 immunity to cancer, *Cytokine*, 171 (2023) 156376.
124. A.A. Bhat, G. Gupta, K.S. Alharbi, O. Afzal, A.S.A. Altamimi, W.H. Almalki, I. Kazmi, F.A. Al-Abbasi, S.I. Alzarea, D.K. Chellappan, S.K. Singh, R. MacLoughlin, B.G. Oliver, K. Dua, Polysaccharide-based nanomedicines targeting lung cancer, *Pharmaceutics*, 14 (2022) 2788.
125. Y. Shinkai, A. Masuda, M. Akiyama, M. Xian, Y. Kumagai, Cadmium-mediated activation of the HSP90/HSF1 pathway regulated by reactive persulfides/polysulfides, *Toxicological Sciences: An Official Journal of the Society of Toxicology*, 156 (2017) 412–421.
126. S.B. Silva, T.F.R. Ruiz, F.C.A. Dos Santos, S.R. Taboga, P.S.L. Vilamaior, Impacts of heavy metal exposure on the prostate of murine models: mechanisms of toxicity, *Reproductive toxicology*, 120 (2023) 108448.
127. M.R. Smith, J. Fernandes, Y.M. Go, D.P. Jones, Redox dynamics of manganese as a mitochondrial life-death switch, *Biochem Biophys Res Commun*, 482 (2017) 388–398.
128. A.A. Bhat, O. Afzal, M. Afzal, G. Gupta, R. Thapa, H. Ali, W. Hassan Almalki, I. Kazmi, S.I. Alzarea, S. Saleem, V.P. Samuel, S.K. Gubbiyappa, V. Subramaniyan, MALAT1: A key regulator in lung cancer pathogenesis and therapeutic targeting, *Pathology, Research and Practice*, 253 (2024) 154991.
129. E.T. Snow, Metal carcinogenesis: mechanistic implications, *Pharmacology & Therapeutics*, 53 (1992) 31–65.
130. A. Vanlaeys, G. Fouquet, P. Kischel, F. Hague, S. Pasco-Brassart, T. Lefebvre, P. Rybarczyk, I. Dhennin-Duthille, B. Brassart, H. Ouadid-Ahidouch, M. Gautier, Cadmium exposure enhances cell migration and invasion through modulated TRPM7 channel expression, *Archives of Toxicology*, 94 (2020) 735–747.
131. K.S. Alharbi, N.K. Fuloria, S. Fuloria, S.B. Rahman, W.H. Al-Malki, M.A. Javed Shaikh, L. Thangavelu, S.K. Singh, V.S. Rama Raju Allam, N.K. Jha, D.K. Chellappan, K. Dua, G. Gupta, Nuclear factor-kappa B and its role in inflammatory lung disease, *Chemico-Biological Interactions*, 345 (2021) 109568.
132. K.S. Alharbi, M.A. Javed Shaikh, O. Afzal, A.S. Alfawaz Altamimi, W.H. Almalki, S.I. Alzarea, I. Kazmi, F.A. Al-Abbasi, S.K. Singh, K. Dua, G. Gupta, An overview of epithelial growth factor receptor (EGFR) inhibitors in cancer therapy, *Chemico-Biological Interactions*, 366 (2022) 110108.
133. M. Kumar, M.S. Hussain, S. Raj, R. Verma, S. Sharma, S.K. Sahu, An overview of treatment modalities and management aspects for obesity, *Current Nutrition and Food Science*, 19 (2023) 105–113.
134. K.S. Alharbi, W.H. Almalki, M. Albratty, A.M. Meraya, A. Najmi, G. Vyas, S.K. Singh, K. Dua, G. Gupta, The therapeutic role of nutraceuticals targeting the Nrf2/HO-1 signaling pathway in liver cancer, *J Food Biochem*, 46 (2022) e14357.
135. K.S. Alharbi, W.H. Almalki, H.A. Makeen, M. Albratty, A.M. Meraya, R. Nagraik, A. Sharma, D. Kumar, D.K. Chellappan, S.K. Singh, K. Dua, G. Gupta, Role of medicinal plant-derived Nutraceuticals as a potential target for the treatment of breast cancer, *J Food Biochem*, 46 (2022) e14387.

12 The Effect of Arsenic on the Gut Microbiome

Benu Chaudhary[1], Preeti Arya[1], Parveen Kumar[2], Arvind Kumar[3], and Madan Mohan Gupta[2,4]*

[1]Shri Ram College of Pharmacy, Ramba, Karnal, Haryana, India
[2]NIMS Institute of Pharmacy, NIMS University Rajasthan, Jaipur, Rajasthan, India
[3]Department of Chemistry, Faculty of Science and Technology, The University of the West Indies, St. Augustine, Trinidad and Tobago
[4]School of Pharmacy, Faculty of Medical Sciences, The University of the West Indies, St. Augustine, Trinidad and Tobago
*Corresponding Author: Madan Mohan Gupta.
Email: mmgupta1432@gmail.com

12.1 INTRODUCTION

Arsenic is one of the most prevalent and hazardous environmental pollutants globally. It is omnipresent in the environment and mostly comes from the erosion of the earth's surface. Consequently, it is necessarily linked to food and drink and soils and particles in air [1]. Human activities can also contribute significantly to contamination, particularly in regions where mining and industrial use of arsenic are prevalent [2]. High amounts of compounds consisting of arsenic are acutely hazardous to most cellular life forms. Additionally, long-lived species such as humans can experience chronic, low-level exposure to arsenic, which is also linked to disease. The International Agency for Research on Cancer (IARC) has categorized arsenic or inorganic arsenic compounds as carcinogens group I, based on substantial evidence indicating that substances are capable of causing skin, lung, and bladder cancers [3]. Arsenic exposure has been associated with several noncancerous conditions such as metabolic dysregulation, skin lesions, diabetes mellitus, pregnancy difficulties, cardiovascular disease, and neurological manifestations [4]. Crucially, the occurrence of these diseases in populations that are exposed to arsenic varies significantly among people, even when considering factors such as heredity and unique ways in which the body processes arsenic [5]. The utilization of genome-wide association research and the development of sophisticated analytical tools have greatly enhanced our comprehension of the significance of human genetic factors in determining the process of arsenical metabolism, which, in turn, influences the chemical transformation of arsenic in cells from mammals [6]. Nevertheless, the extent of human genetic diversity does not sufficiently account for the degree to which diseases manifest in populations that are exposed, hence allowing for the possibility of identifying other significant explanatory factors [7, 8]. A microbiome refers to the collective population of microorganisms that inhabit a specific habitat and encompasses all of their physical, metabolic, and ecological functions [9]. The microbiome of humans consists of a wide range of taxonomically varied microorganisms, including viruses, bacteria, fungus, archaea, and micro-eukaryotes, The scientific community now recognizes that the interactions among the host cells, the microbiome, and the environment that is abiotic have substantial effects on human wellness and illness [10]. The primary focus is on the functioning of the microbiota in the gastrointestinal tract of humans, also known as the gut. From

DOI: 10.1201/9781003489221-12

an ecological standpoint, the human stomach can be considered as an ecosystem that performs similar activities to those of a forest or ocean "biome." Metabolism, which refers to the biochemical processes carried out by microorganisms and human cells in the gastrointestinal tract, is a vital function that has the potential to directly and indirectly affect human well-being. Given that all living things must address the issue of arsenic to reduce its potential harm, it is not unexpected that the gut microbiota can metabolize arsenicals, therefore altering its toxicity in the tissues of the host. Conversely, if there is arsenic exposure that leads to the death of specific components of the microbiome in the gut, their roles will be eradicated, potentially impacting the condition of the host inadvertently. Arsenic (As) is a harmful molecule that is present in nearly all environments and can create several chemical forms with varied levels of movement, toxicity, and accessibility [11]. The toxicities of several arsenic species include dmethylarsenite, arsenite, monomethylarsenite, arsenate, monomethylarsenate, and dimethylarsenate [12]. Water is the primary source of arsenic exposure, containing arsenic in the form of both As(III) and As(V). The rate of bioavailability of inorganic arsenic (iAs) in drinking water is close to 100% [13]. Internationally, a significant number of individuals consume water with hazardous levels of arsenic, posing a threat to human well-being. This is observed in countries including Chile, Argentina, El Salvador, and Mexico [14]. Being exposed to the environment has been associated with a range of cancers and other illnesses, such as kidney failure, diabetes, or cerebrovascular disease [15, 16]. The gastrointestinal tract of humans harbors a minimum of 1,014 bacteria, consisting of over 1,000 distinct species [17]. The gastrointestinal tract is the primary location where microorganisms metabolize foreign substances that are consumed. The gut exhibits a significantly higher level of microbe metabolism compared to other regions of the body, making it a crucial area of focus for investigating xenobiotic metabolism and toxicity [18]. Research indicates that the gut microbiota possesses an impact in the metabolism of arsenic [19, 20]. The microbiota present in the gastrointestinal tract of rats convert inorganic arsenic (iAs(V)) into inorganic arsenic (iAs(III)), as well as various forms of methylated or dimethylated oxyarsenicals. These microbial activities also have an impact on the metabolic reactions involving dimethylarsinic acid (DMA V), such as methylation along with thiolation. This has been demonstrated in studies referenced as [21–23]. In humans, the gut microbiota facilitates the processes of reducing, methylating, and thiolating iAs(V) when As is consumed by food (such as rice) or soil [24–26]. However, there is still a paucity of thorough studies on the alteration of As by human gut flora. The main types of bacteria found in the human gut microbiota are primarily from the bacteroidetes, phyla Firmicutes and proteobacteria and actinomycetes. There is also a wide range of bacterial genera, such as *lactobacillus*, *clostridium*, *enterobacter*, *bifidobacterium*, and *bacteroides* [27]. These bacterial taxa play crucial roles in determining the fate of arsenic in the human gut. As reduction, resistance, and methylation have all been demonstrated by bacteria, including those in the genera *Clostridium* as well as *Bacteroides*. The family *Lactobacillus* naturally resides in the intestines of humans and has garnered interest due to its beneficial impact on human well-being [28]. A previous investigation has demonstrated that the genus *Lactobacillus* is capable of adsorbing iAs and mitigating its toxicity inside the human gastrointestinal environment [29]. While it has been documented that genera of bacteria in the human gut can alter arsenic through biotransformation, the particular strains responsible for this process have yet to be identified. Therefore, it is necessary to examine the process of biological transformation of As by particular strains. The study aimed to evaluate the abilities of 21 different strains of intestinal bacteria from seven genera to absorb and transform inorganic arsenic (iAs), as well as their resistance to arsenic. This assessment was conducted to identify the specific bacterial strains that are involved in the metabolism of arsenic, considering the main sources of exposure for humans, namely food and water that is consumed. The findings of this study may offer fresh data on how gut bacteria control the levels and types of arsenic (As) in order to impact human health. The different historical occurrences of arsenic contamination are shown in Table 12.1.

TABLE 12.1
Milestones in the Historical Timeline of Arsenic.

Sr No.	Years	Historical Timeline of Arsenic Contamination	References
1.	1786	The solution referred to as Fowler's solution containing 1% of potassium arsenite.	[30]
2.	1836	A novel method, known as the Marsh Test, has been devised for the purpose of detecting arsenic.	[31]
3.	1842	Dimethylarsinic acid has been identified in the environment as well. In the United States which is composed of copper acetoarsenite, is employed as an insecticide.	[32]
4.	1867	Hutchison suggests that arsenic is a carcinogen for the human skin. While Salvarsan is employed as a chemotherapeutic drug.	[33]
5.	1887	A new type of antilewisite, known as British antilewisite, has been created.	[34]
6.	1940s	Tseng et al. have published their research findings on the occurrence of skin cancer in a population from Taiwan that has been exposed to arsenic.	[35]
7.	1942	The Environmental Protection Agency (EPA) has implemented an interim regulation for drinking water at a concentration of 50 micrograms per liter (ug/l).	[36]
8.	1968	The World Health Organization (WHO) proposes a drinking water standard of 10 micrograms per liter (ug/l).	[37]
9.	1975	The solution referred to as Fowler's solution containing 1% of potassium arsenite.	[38]
10.	1995	Dimethylarsinic acid is identified as a carcinogen that promotes tumor growth in four different organs of rats.	[39]
11	1998	Dimethylarsinic acid is a known carcinogen that causes cancer in the urinary bladder of rats.	[40]
12.	2002	Arsenic methyltransferase with a +3 oxidation state, obtained from the cytosol of rat liver.	[41]
13.	2010	Inorganic arsenic is classified as a total carcinogen. This has led to the adoption of more stringent arsenic regulations for rice products in different nations, as there is growing apprehension about the presence of arsenic in foods made from rice.	[41]
14.	2011	Dimethylarsinic acid is identified as a carcinogen that promotes tumor growth in four different organs of rats.	[43]
15.	2013	The identification of extensive arsenic pollution in groundwater across the United States has resulted in heightened surveillance and public consciousness.	[44]
16.	2014	Introducing arsenic-free wood treatments as a way to address environmental and health issues.	[45]
17.	2015	Enhanced emphasis on tackling arsenic contamination in privately owned wells, specifically in rural regions, including educational initiatives and water quality assessment initiatives.	[46]
18.	2017	The dissemination of research establishing a connection between arsenic exposure and negative pregnancy outcomes, hence increasing knowledge regarding the possible hazards to the health of both the mother and the fetus.	[47]
19.	2022	Ongoing investigation on the molecular processes behind the harmful effects of arsenic and possible treatment approaches.	[48]
20.	Present	International efforts and cooperative endeavors aimed at tackling issues associated to arsenic, such as raising public awareness, advocating for governmental changes, and establishing research alliances.	

12.2 INTERACTIONS BETWEEN ARSENIC AND THE MICROBIOME

Understanding of arsenic-microbe associations is mostly derived from the field study of environmental microbiology, particularly in ecosystems that include soil and subsurface. In these environments, microbial metabolisms have an important role in determining the forms, movement, and harmful effects of arsenic. The ideas employed in environmental microbiology research of arsenic can be effectively utilized to comprehend the interactions between arsenic and the human microbiome. For instance, the interactions between arsenic and the microbiome might result in three broad and occasionally overlapping consequences: absence of any detectable impact, disturbance of taxonomic organization and function of gut microbiome, and modification of the pharmacologic and/or toxicological characteristics of the hazardous substances [49]. Several studies contend that host metabolism is equally, if not more, significant than microbial metabolism in relation to the biotransformation as well as toxicity of arsenic [50]. There is compelling data indicating that the microbiome plays a crucial role in determining exposure consequences. The modification of the form and function of the gut microbiome is frequently a component of the body's natural response to both physical and perceived dangers. The connections between the structure and function of the microbiome have been identified in numerous syndromes, human diseases, and behaviors. The ability to restore the altered structure of microbiomes relies on the specific microbiome's members that are impacted and their potential for recovery. If they are unable to do so, a situation of "dysbiosis" may continue to exist, which might lead to harmful health consequences. For instance, the disturbance of the gut microbiota caused by a hazardous substance might affect the regular process of the host's absorption, processing, and elimination of nutrients from the diet, and this can, in turn, impact the absorption of the poisonous substance. Likewise, the elimination of harmful substances can disrupt the preservation of the protective layer of cells in the gut, the control of the body's inflammatory reactions, and the production or reutilization of essential substances and compounds that play a role in the body's response to toxins. Further examination of this form of contact is addressed in the following sections. Pathologic results resulting from the modification of microbiome structure-function by chemical toxicants are contrasted with outcomes induced by the microbiome's transformation of chemical toxicants, which, in turn, modifies their physicochemical qualities. However, in other instances, they occur either indirectly or non-enzymatically when the byproducts of the metabolism of microbes chemically interact with the hazardous compounds. In addition, many biotransformations can have little impact, while others can improve the situation, and some can greatly enhance toxicity [51].

12.3 METABOLISM OF ARSENIC IN HUMANS

The main method of systemic detoxification among humans and a large number of other living entities is the methylation of iAs that approach the body through food and water. The excretion of methylated arsenicals becomes more efficient, resulting in improved removal from the body [52]. Effective arsenic methylation is linked to favorable long-term illness outcomes [53]. The precise process by which mammals methylate arsenic, including the specific reactants involved and the order in which the products are formed, is still a subject of debate and has been thoroughly examined in previous studies [54–55]. In summarizing, iAs is absorbed into the bloodstream through both transcellular and paracellular transport mechanisms, as explained in reference [56]. Inside cells, iAs undergoes reduction and oxidative methylation processes, resulting nearly entirely in the formation of four methylated compounds: monomethylarsonous, monomethylarsonic, dimethylarsinous and dimethylarsinic acids. Methylation is facilitated by the enzyme arsenic with oxidation state of 3^+ methyltransferase, which utilizes S-adenosylmethionine (SAM) for a source of methyl groups. This process is aided by natural reducing agents like glutathione (GSH) and thioredoxin (Trx) a [57–58]. The latest supported crystallographic evidence has shown that intermediates which are pentavalent in nature are probably decreased and still attached to AS3MT. This implies that the more harmful trivalent arsenicals could be the final results of methylation, instead of the fewer harmful pentavalent molecules [59]. Although GSH plays a crucial role as a reducing agent in the methylation process, it is worth noting that antimony-GSH-conjugates are additionally transported and expelled from cells. Additionally, urine and bile samples from animals have been found to contain arsenic tri-glutathione and monomethyl arsenite di-glutathione [60]. These findings indicate that conjugation has a significant impact in the chemical breakdown of arsenic, particularly in its movement inside human tissue. Furthermore, the body has the capability to produce other arsenicals that contain sulfur, in addition to arsenic-GSH conjugates. The research conducted in both laboratory (in vitro) and in animals (in vivo) has shown that exposure to arsenic leads to the synthesis of thioarsenicals [61]. Thioarsenicals are regularly found in human urine as well as in the urine and feces of models of animals [21, 62]. Arsenic can be thiolate in vitro by both human and rat red blood cells [63], however it is still unknown what the main mechanism(s) and toxicological effects of arsenic thiolation within the body are. However, pentavalent monothiolated arsenic compounds such as monomethyl and dimethyl monothioarsonic acid exhibit cytotoxicity that is more comparable to the trivalent arsenicals than to the non-thiolated counterparts [64]. Several distinct chemical routes have been suggested for the creation of thioarsenicals, which have undergone comprehensive examination. Typically, these mechanisms do not include enzymes and instead rely on the relationship between methylated arsenicals and sulfide ions or bonded sulfane sulfur. This interaction leads to the replacement of a number of arsenic-oxygen bonds with similar arsenic-sulfur bonds [54]. Hydrogen sulfide (H2S) serves as a crucial signaling molecule in the body and is a significant producer of sulfide ions. Hepatocytes are well recognized as the primary producer of biogenic H_2S, indicating their substantial involvement in thiolation. Nevertheless, this theory has not been explored by experimentation.

12.3.1 Metabolism of Arsenic by Bacteria

The study of bacterial arsenic processing has primarily focused on investigating microbial "resistance" to the harmful effects of arsenic, such as its ability to cause cell death. While metabolism along with resistance commonly neutralize arsenicals, it is important to note that not all resistance routes entail the metabolic conversion of arsenic. Ars operons, which are clusters of resistance genes, were initially identified in the plasmids obtained from bacteria *Escherichia coli* and *Staphylococcus aureus* [65, 66]. Subsequently, bacterial genes were discovered and examined in several medically significant pathogens, such as *Yersinia strains Listeria monocytogenes* and *Campylobacter jejuni,* [67]. The ars operon were identified in human gut symbionts such as *Bacillus subtilis* [68] and an obligate anaerobe, *Bacteroides vulgatus* [69]. A comprehensive analysis has been conducted to examine the molecular distribution, functions, and development of the bacterial resistance to arsenic [70–72]. Researchers present a comprehensive analysis of the typical methods by which bacteria develop resistance to arsenic and metabolize it. Its emphasis is on those mechanisms that have been observed or anticipated to occur in the environment of the intestines, as depicted in Figure 12.1.

Bacteria-encoded enzymes have the ability to change both inorganic and organic arsenicals through oxidation, reduction, and methylation as well as demethylation processes. These transformations occur in conjunction with certain substrates. Bacterial metabolites, shown by a dotted box and represented by green text, may have a significant impact in the process of transforming arsenic compounds in the gastrointestinal tract. Arrows are used to indicate the beginning and end molecules in this biotransformation. Although bacteria are recognized as the catalysts for these events, there is currently no evidence to support the occurrence of demethylation and oxidation in bacteria residing in the gut of mammals. Additionally, the direct demonstration of thiolation resulting from bacterial hydrogen sulfide (H_2S) synthesis is also lacking. The ultimate destiny of arsenic throughout the GIT. Figure 12.2 depicts the arsenic in biomass determining by gut.

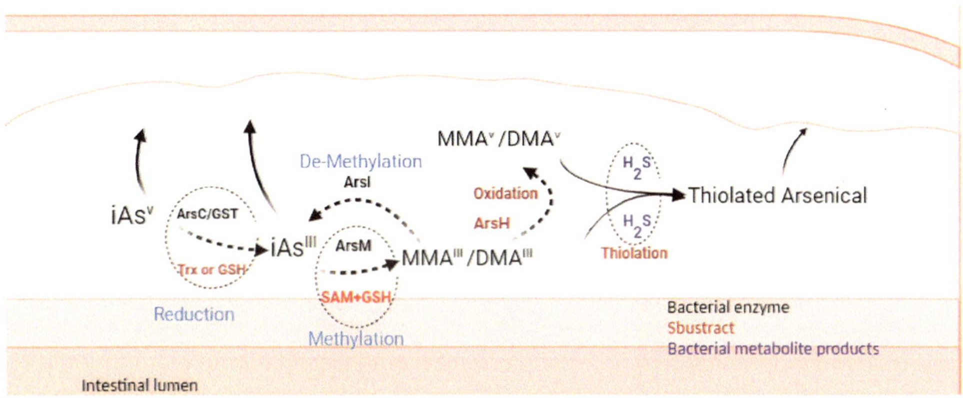

FIGURE 12.1 Interactions between arsenic and the microbiome in the gut of mammals.

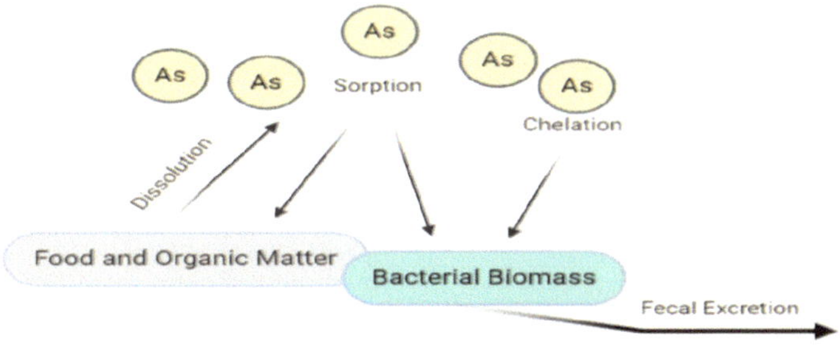

FIGURE 12.2 Arsenic in biomass determining by chemical composition of the gut. The routes have similarities to the source-sink processes observed in the natural world.

12.3.2 The Microbiome Modifies the Level of Arsenic Exposure

The concept that gut microbiota play a role in determining health outcomes after being exposed to arsenic has been established for over a century. In 1917, Puntoni documented that spore-producing microorganisms obtained from human excrement were capable of generating a strong garlic scent when grown in a solution of cacodyl arsenic [73]. At that point in time, it was observed that the smell was frequently present in individuals who were orally consuming therapeutic arsenic substances, but less frequently in those who were receiving it subcutaneously. This led him to hypothesize that the microbiome of the gut was chemically modifying the arsenic. In a different investigation, Higginbottom and Challenger (1935) recognized trimethyl arsine gas as a comparable gaseous arsenical generated by Scopulariopsis brevicaulis (previously Penicillium brevicaule) [74]. A recent study demonstrated that *E. coli* bacteria found in the cecal secretions of rats have the ability to metabolize DMAsV, resulting in the production of TMAVO and an undetermined arsenical compound [75]. Apart from conducting experiments on bacteria in pure cultures, researchers have also investigated arsenic absorption in microbiome members that have been tested experimentally in the laboratory (ex vivo). The microbiome in mice showed a strong ability to enzymatically reduce and methylate iAs, as evidenced by incubating the intestinal contents of the rodents [76, 77]. Moreover, two separate investigations have shown that human microbiomes decreased and methylated inorganic arsenic (iAsV) in a simulated human gut surroundings, resulting in the production of both toxic and relatively less toxic methylated arsenicals [26]. Remarkably, iAsV was converted to iAsIII even in autoclave sterilized experimental controls, indicating that non-enzymatic mechanisms may play a role in this conversion. Although the aforementioned research offers compelling evidence of the microbiome's ability to affect arsenic toxicity, neither of them empirically assessed the overall influence of the microbiome on exposure to the host. The study conducted by researchers demonstrated that administering antibiotics to mice before exposing them to arsenic resulted in a notable decrease in the amount of arsenic excreted in their feces. Additionally, it led to an increase in the buildup of arsenic in the liver and lung tissues of the mice. They hypothesized that the reduction of the microbial biomass for the gut caused by antibiotic exposure played a role in facilitating the excretion of ingested arsenic through feces. Experiments conducted by other researchers have shown that iAs is adsorbed onto extracellular polymeric components of Gram-positive bacterial isolates, not Gram-negative ones. This finding suggests that microbial arsenic buildup in the gut may occur through this method. Our research provided evidence that supported this process. Specifically, we discovered that the Gram-positive bacterium i.e. *Faecalibacterium prausnitzii*, offered protection to that of gnotobiotic AS3MT-KO mice when they were exposed. In contrast, the Gram-negative bacterium, the *E. coli* did not provide the same level of protection [78].

12.3.3 Arsenic Toxicity Decreases by Microbiome

Within living organisms, the microbiome's "phenotypes" have been associated with changes in the proportions of methylated or inorganic arsenicals within the host [79]. Nevertheless, there is a scarcity of research that has explicitly established a direct correlation between modifications, changes, or the absence of the microbiome and the well-being of the host. Researchers conducted an experiment using AS3MT-KO mice which had been humanized (meaning they received a human fecal transplant) and were germ free. The purpose of the experiment was to assess how the heterogeneity in the microbiome between individuals affects the outcome of the disease, specifically mortality. This study is referenced as [80]. Each cohort of humanized mice corresponded to a distinct human donor, while each had significantly distinct microbiome compositions. Researchers discovered that only a small number of bacteria consistently showed a positive effect (i.e., longer survival) across different groups of humans, and these bacteria belonged to some of the most prevalent species found in the human gut. For instance, two members of the *Blautia* genus, together with representatives from the *Lachnospiraceae*, *Ruminococcus*, and *Faecalibacterium* families, showed a strong correlation with survival across different humanized species. However, it is worth noting that these bacteria groupings are both widespread and diverse. While these taxa were consistently related among participants, it is plausible that significant variations at the strain level might potentially undermine or eliminate the statistical relationships. Put simply, the bacteria that are helpful to one person may not be helpful to another person since the classification of bacteria does not always accurately determine their function. For example, the first three *E. coli* genomes that were sequenced only had 40% of their protein-coding regions in common [81]. In any case, we clearly demonstrated that a preserved microbiome substantially prolonged the time until death caused by arsenic in A3mt-KO mice, when compared to groups of mice that were germ free or treated with antibiotics. Consequently, there is compelling data indicating that the microbiota shields the host from the harmful effects of arsenic [80]. The search is currently underway to identify the particular members along with metabolic pathways of the donor microbiome that are most advantageous in providing protection against arsenicosis. These findings might potentially be used to develop innovative therapeutic and preventative approaches for arsenicosis [81, 82].

12.4 IMPACT OF A REALISTIC CONCENTRATION OF ARSENIC UPON THE GUT MICROBIOME, INCLUDING ITS FUNCTIONAL METAGENOME

Dysbiosis in the microbiome of the gut is triggered by numerous environmental variables, leading to a range of human disorders. Earlier, we showed that arsenic modifies the makeup of the gut microbiota. Nevertheless, the extent to which arsenic affects the gut microbiome's functionality has not been sufficiently evaluated, especially at quantities that are significant to the environment. This work employed 16S rRNA sequencing or metagenomics sequencing techniques to examine the impact of a 13-week treatment to 100 ppb arsenic on the makeup and functional capabilities of the gut microbiome in mice. Exposure to arsenic caused changes in both the alpha and the beta diversities, in addition to the composition profile, of the microbiota in the intestines. The metagenomics results indicated considerable alterations in the levels of genes related to carbohydrate metabolism, particularly in fermentation of pyruvate, fatty acid synthesis, and starch utilization. Furthermore, the gut microbiota of arsenic-exposed mice exhibited a large increase in lipopolysaccharide manufacturing genes, numerous responses to stress genes, and repair of DNA genes. The genes responsible for the synthesis or modification of other vitamins, such as folic acid or vitamins B12, B6, and K2, were also found in higher abundance in mice treated with arsenic. Furthermore, the expression of genes associated with resistance to multiple drugs and conjugative transposon proteins exhibited a significant increase following arsenic treatment. Ultimately, we have shown that exposure to arsenic, even at a level that is appropriate to the environment, not only disrupted the overall makeup of the microbiome in the gut but also significantly changed certain crucial bacterial functioning pathways. Our findings indicate that exposure to a realistic amount of arsenic (100 ppb) can disrupt the typical makeup and functioning systems of the gut microbiome in mice, which is important to the environment. The expression of energy metabolism genes was modified, and there was a widespread upregulation of genes associated with LPS production, oxidative stress, and DNA repair. Furthermore, arsenic exposure led to an increase in the expression of genes responsible for producing conjugative transposon proteins, elements of the efflux multidrug framework, and the manufacture of several vitamins. These findings may offer novel insights into the impact of arsenic upon the gut microbiome, particularly at dosages that are relevant to the environment. Given that the main objective of this investigation is to analyze and describe the impact of arsenic affecting the gut microbiome, our forthcoming research will explore the ways in which the disturbance caused by arsenic in the gut microbiota affects the overall health of the host. Additionally, it would be highly enlightening to investigate the impact of the gut microbiome on the process of arsenic processing and the accumulation of arsenic within the host's tissues [83].

Arsenic exposure has a significant impact on a vast number of people globally and has been linked to several diseases such as bladder, skin, lung, and liver cancers as well as diabetes and cardiovascular issues. Furthermore, significant variations exist in the vulnerability to diseases caused by arsenic, which are often linked to distinct arsenic metabolism pathways. Various fundamental mechanisms, including genetic polymorphisms as well as epigenetics, are being recommended, as these factors significantly influence an individual's ability to metabolize arsenic. Within this framework, the gut microbiome's function in directly metabolizing arsenic and initiating systemic reactions in several organs suggests that disturbances in the gut microbial populations may impact the range of metabolized arsenic compounds and consequent toxicological consequences. This study utilized an animal model in which the gut microbiome was modified through bacterial infection. Researchers employed 16S rRNA gene sequencing and inductively linked plasma mass spectrometry–associated arsenic speciation to investigate how disruptions in the gut microbiota impact the conversion of arsenic. The metagenomics sequencing identified that bacterial infection had a major impact on the structure of the gut microbiome in C57BL/6 mice. This, in turn, led to changes in the types of arsenic metabolites found in urine, specifically affecting the presence of inorganic arsenic species and methylation and thiolated arsenic. The data provided compelling evidence that variations in gut microbiota phenotypes had a major impact on arsenic metabolic processes, such as reduction, thiolation, and methylation. These findings enhance our comprehension of the interaction between infectious diseases and environmental exposure. They also offer fresh insights into the composition for the gut microbiome as a potential risk factor for individual vulnerability to environmental toxins [22].

12.5 THE MICROBIOME OF THE GUT AND ARSENIC EXPOSURE PHYSIOLOGICAL INTERACTION

There is a significant lack of comprehension of the impact of interactions between exposure to the environment with the gut flora on human health. The human digestive tracts microbiome exerts a significant impact on human health due to its crucial involvement in various host-related activities. Inorganic arsenic (iAs) exposure is a major global public health concern and has been associated with various diseases, such as cancer, metabolic problems, cardiovascular diseases, and compromised immunological functioning. Increasing data suggests that imbalanced gut microorganisms play a key role in the development of many disorders, highlighting the potential importance of the microbiome of the gut in arsenic-associated diseases. Nevertheless, there is a significant lack of research on the functional relationship between the gut microbiome with arsenic exposure, with the exception of the authors' latest investigations. Their main proposition is that the interaction between the gut flora and arsenic influences the toxicity of inorganic arsenic (iAs). The aforementioned theory has been developed based on promising preliminary data generated in the laboratory of the Principal Investigator. This theory will be tested by seeking three specific objectives, aiming to determine the impact of iAs on the microbiome of the gut and its function, as well as to clarify the role of the gut microbiome structure in mediating arsenic toxic effects. The application stems from the fact that the relationship between gut microbiome as well as exposure is still widely studied, despite the microbiome's broad and significant contributions to human health. The suggested research by the investigators is very unique and signify a significant change in toxicology. These center on disruptions in the gut microbiome as a new mechanism for exposure-induced disorders and an unknown risk factor that contributes to specific susceptibility. This application is innovative because it applies integrated bioanalytical with computational technologies to examine the interplay between the gut microbiota and exposures at a systems level, which has not been done before. The results of this study are anticipated to establish a basis for future research endeavors that seek to enhance our comprehension of novel pathways by which exposure to iAs contributes to or worsens human diseases, specifically focusing on the influence of the microbiome [84].

12.5.1 Arsenic Exposure's Impacts on the Path and Function of the Gut Microbiota in Relation to Sex

The gut microbiota plays a significant role in various areas of human health. Nevertheless, it can be easily disrupted by environmental toxins, such as arsenic. The interaction between the gut microbiome, host, and xenobiotics is a highly intricate and continuous process. Earlier, researchers have shown that the characteristics of the gut microbiome, which are influenced by the genetics of the host as well as bacterial infection, impact how the body responds to exposure to arsenic. The influence of the host's gender on the composition of the gut microbiome prompts an inquiry into whether gender affects the microbiome's reaction to exposure. To investigate this, we employed 16S rRNA sequencing along with metagenomics sequencing to assess the alterations in the microbiome of the gut and its corresponding functional metagenome for both female as well as male C57/BL6 mice. The findings of the investigation clearly indicated that exposure to arsenic disrupted the development and functioning of the gut microbiome in a manner that varied between sexes [85].

12.5.2 Serum Metabolomics Shows Gut Microbiome Perturbation Mediates Arsenic Exposure-Induced Metabolic Disruption in Mice

The issue of arsenic pollution in drinking water has been considered a global health problem for many years. Arsenic exposure, aside from being well acknowledged as a carcinogen, has also been associated with diabetes, neurological impacts, and cardiovascular ailments. Emerging data suggests that the gut microbiota plays a crucial role in influencing the onset of illnesses. The goal of this study is to examine the impact of changes in the gut microbiome upon diseases caused by arsenic exposure. To do this, they will combine a metabolomics method that uses mass spectrometry using an animal model that has been intentionally infected with bacteria to change its gut microbiota. An analysis of serum metabolic profiles has shown that the disturbance of the gut microbiome as well as exposure to arsenic caused significant alterations in different metabolic processes in the body. These pathways include phospholipids, fatty acid metabolism, cholesterols, sphingolipids, and tryptophan metabolism. These changes were either absent or less pronounced when the gut microbiome remained unaffected. To summarize, our work indicates that disturbing the gut microbiome can worsen or trigger metabolic problems caused by exposure to arsenic [86].

12.5.3 Long-Term Exposure to Arsenic Disrupts the Balance of Microorganisms in the Gut and the Regulation of Bile Acids in Mice

Exposure to arsenic can disrupt the composition and metabolic activities of the gut microbiota. They subjected C57BL/6 mice to a concentration of 1 part per million (ppm) of arsenic in their drinking water. The purpose of this study was to examine the impact of arsenic exposure on the equilibrium of bile acids, which are essential microbiome-regulated signaling compounds involved in microbiome-host interactions. The results of the study indicate that arsenic exposure caused distinct alterations in primary bile acids, specifically those that are not conjugated, and consistently reduced secondary bile acids in both the serum and the liver. The ratio of Bacteroidetes to Firmicutes was correlated with the concentration of bile acids in the bloodstream. The present research provides evidence that the imbalance in gut microbiota caused by arsenic exposure may contribute to the disruption of bile acid regulation affected by arsenic [87].

12.5.4 Relationship between the Gut Microbiota and Diseases Caused by Arsenic Exposure – iAs Investigating the Metabolism of Inorganic Arsenic in Mice

In recent times, numerous investigations have commenced to unveil the interplay between iAs and the gut flora. Empirical data demonstrates that the composition of gut bacteria can have an impact on the process of iAs biotransformation and the likelihood of developing certain diseases. The microbiome of the gut has the ability to metabolize iAs directly, while it can also have an indirect role in iAs metabolism via the host by modifying iAs absorption, cofactors, and genes associated with iAs metabolism. Various factors, including the way in which the gut microbiome affects the metabolism of iAs, as well as the impact of iAs on microbial metabolites, can contribute to varying risks of disease. Arsenic (iAs) is a highly prevalent poisonous metalloid in the environment, and its toxicity has emerged as a significant global health concern. Upon entering the human body, iAs undergoes many metabolic processes, such as demethylation, methylation, reduction, and thiolation and oxidation. The toxicity of arsenic is determined by several species, such as trivalent and pentavalent forms, as well as inorganic and organic forms. The poisoning of aresenic is mostly induced by the consumption of polluted drinking water and food. Prolonged exposure to arsenic can lead to the development of numerous ailments. Consequently, it is crucial to investigate the metabolism of inorganic arsenic (iAs) in order to comprehend the dangers of diseases related with iAs [88].

12.6 HOST GENETICALLY DETERMINED GUT MICROBIOME PHENOTYPES IMPACT ON ARSENIC METABOLISM

A significant amount of evidence shows there are substantial variations among individuals in their vulnerability to diseases caused by arsenic exposure. These variances are often linked to distinct ways in which arsenic is processed in the body. Within this particular environment, the gut microbiome plays a direct role in metabolizing arsenic and initiating systemic reactions in many organs. This suggests that the characteristics of the gut microbiota may influence the range of arsenic species that are metabolized. Nevertheless, the precise mechanism by which host genetics as well as the gut microbiome combine to influence the biotransformation of arsenic is still not fully understood. We employ a comprehensive method that combines 16S rRNA gene sequencing with HPLC-ICP-MS arsenic speciation to show that the absence of the IL-10 gene causes a notable alteration in the composition of the gut microbiome, resulting in a major impact on arsenic metabolism [81].

12.7 CHALLENGES

The terms "pharmacomicrobiomics" have been introduced in recent research due to the growing interest in the microbiome's role in metabolizing foreign substances and its role on the physiological system. These nascent disciplines offer a structure for integrating microbiome research within quantitative risk assessment, public health even includes precision medicine. Although there are still many uncertainties, the evidence presented in this review indicates that the microbiome has a substantial impact on reducing the host's exposure to arsenic. Additional investigation of these functions is necessary to ascertain if the changes are a result of direct metabolism of arsenic and other indirect mechanisms by microbes. The primary focus of mechanistic research should be on determining the right dosage, conducting relevant experimentation, and showing direct cause-and-effect relationships with medical conditions in the host. The ability to replicate and apply findings from health research, particularly in relation to the microbiome, is a persistent difficulty. The heterogeneity in microbiome composition among different animal vivariums might lead to unanticipated fluctuations in experimental outcomes and a dearth of consistency in findings. This not only underscores the imperative to build improved standardizations in models of animals and emphasizes the necessity for additional exploration of natural microbiome variance among populations and individuals utilization of germ-free and gnotobiotic models, as well as the transplantation of human microbiomes, are highly effective methods for evaluating the impacts of both natural and artificial variations in living organisms. Proposals are being made to improve the standardization of experimental designs, sample collection, processing of data, and the integration of microbiome record with other targeted/non-targeted datasets. Further efforts are required to establish animal models that accurately simulate chronic arsenic toxicity, as prolonged exposure poses the most significant risk to human health. Until now, only a limited number of epidemiological research have examined the relationships between arsenic and the microbiome, and even fewer have explored health of the as the variable. The single-timepoint monitoring studies have the ability to detect potential connections with the makeup or activity of the microbiome, but they are restricted to making correlative inferences. Conducting prospective cohort studies that monitor oral or fecal communities over a period of time could provide valuable insights into the specific relationships between microbiome-related disorders caused by arsenic. It is crucial to acknowledge that the microbiome undergoes significant changes at the strain level. Both cohort and case-control studies might include data on the occurrence of enteric infection or antibiotic usage in populations who were exposed to arsenic [89]. This information would be valuable in assessing the possible impact of these well-known modifications of the microbiome on the long-term risk of arsenic-related diseases. The incorporation of microbiota analysis into empirical case reports and mechanistic investigations should enhance the clinical study on the utilization of ATO in the medical treatment of APL. Based on the findings from animal models, it is reasonable to believe that the use of antibiotics or disruptions to the microbiome may affect the speed and effectiveness of orally delivered ATO. The existing knowledge regarding the effect of the microbiome on the absorption and elimination of arsenic may or may not be applicable to the administration of arsenic trioxide (ATO), as animal models of the toxicity of arsenic often involve different chemical sources that contain arsenic.

12.8 CONCLUSION

The field of arsenic toxicity experienced a significant change in perspective due to the advancement of technology and procedures that enabled quick and precise identification of different forms of arsenic compounds. The way we understand microbial interactions in human health has also evolved as a result of advances in DNA sequencing as well as meta-omics technologies. We have examined and condensed recent findings about the impact of the microbiome on the processes of arsenic excretion, toxicity, and metabolism. Additionally, we have explored the effects of disruptions in the microbiome on exposure to arsenic in models of humans and animals. The microbiome possesses the capacity to significantly modify the way mice metabolize arsenic and influence the consequences of diseases. Further investigation is required to measure the microbiological breakdown of arsenic within a living organism and to determine the fundamental factors that affect the absorption, breakdown, and elimination of arsenic by the host. The current constraint in animal research is in the scarcity of suitable animal models for studying diseases caused by arsenic, particularly in relation to the long-term health consequences of chronic arsenic exposure, which is considered a major peril to human well-being. Continued investigation and utilization of germ-free and gnotobiotic model organisms could aid in the identification of causal connections between arsenic and the microbiome, which are relevant to clinical practices and therapies.

REFERENCES

1. Nordstrom DK. Worldwide occurrences of arsenic in ground water. *Science*. 2002 Jun 21;296(5576):2143–5.
2. Williams M. Arsenic in mine waters: an international study. *Environmental Geology*. 2001 Jan;40:267–78.
3. Herceg Z, Lambert MP, van Veldhoven K, Demetriou C, Vineis P, Smith MT, Straif K, Wild CP. Towards incorporating epigenetic mechanisms into carcinogen identification and evaluation. *Carcinogenesis*. 2013 Sep 1;34(9):1955–67.
4. Schuhmacher–Wolz U, Dieter HH, Klein D, Schneider K. Oral exposure to inorganic arsenic: evaluation of its carcinogenic and non-carcinogenic effects. *Critical Reviews in Toxicology*. 2009 Apr 1;39(4):271–98.
5. Cohen SM, Arnold LL, Beck BD, Lewis AS, Eldan M. Evaluation of the carcinogenicity of inorganic arsenic. *Critical Reviews in Toxicology*. 2013 Oct 1;43(9):711–52.
6. Antonelli R, Shao K, Thomas DJ, Sams II R, Cowden J. AS3MT, GSTO, and PNP polymorphisms: impact on arsenic methylation and implications for disease susceptibility. *Environmental Research*. 2014 Jul 1;132:156–67.
7. Gribble MO, Voruganti VS, Cole SA, Haack K, Balakrishnan P, Laston SL, Tellez-Plaza M, Francesconi KA, Goessler W, Umans JG, Thomas DC. Linkage analysis of urine arsenic species patterns in the strong heart family study. *Toxicological Sciences*. 2015 Nov 1;148(1):89–100.
8. Tellez-Plaza M, Gribble MO, Voruganti VS, Francesconi KA, Goessler W, Umans JG, Silbergeld EK, Guallar E, Franceschini N, North KE, Kao WH. Heritability and preliminary genome-wide linkage analysis of arsenic metabolites in urine. *Environmental Health Perspectives*. 2013 Mar;121(3):345–51.
9. Whipps JM, Lewis K, Cooke RC. Mycoparasitism and plant disease control. *Fungi in Biological Control Systems*. 1988;1988:161–87.
10. Clemente JC, Ursell LK, Parfrey LW, Knight R. The impact of the gut microbiota on human health: an integrative view. *Cell*. 2012 Mar 16;148(6):1258–70.
11. Wang P, Sun G, Jia Y, Meharg AA, Zhu Y. A review on completing arsenic biogeochemical cycle: microbial volatilization of arsines in environment. *Journal of Environmental Sciences*. 2014 Feb 1;26(2):371–81.
12. Yu H, Liu S, Li M, Wu B. Influence of diet, vitamin, tea, trace elements and exogenous antioxidants on arsenic metabolism and toxicity. *Environmental Geochemistry and Health*. 2016 Apr;38:339–51.
13. Luvonga C, Rimmer CA, Yu LL, Lee SB. Analytical methodologies for the determination of organoarsenicals in edible marine species: a review. *Journal of Agricultural and Food Chemistry*. 2020 Jan 30;68(7):1910–34.
14. Kumar R, Patel M, Singh P, Bundschuh J, Pittman Jr CU, Trakal L, Mohan D. Emerging technologies for arsenic removal from drinking water in rural and peri-urban areas: Methods, experience from, and options for Latin America. *Science of the Total Environment*. 2019 Dec 1;694:133427.
15. Jomova K, Jenisova Z, Feszterova M, Baros S, Liska J, Hudecova D, Rhodes CJ, Valko M. Arsenic: toxicity, oxidative stress and human disease. *Journal of Applied Toxicology*. 2011 Mar;31(2):95–107.

16. Hussain MS, Gupta G, Mishra R, Patel N, Gupta S, Alzarea SI, Kazmi I, Kumbhar P, Disouza J, Dureja H, Kukreti N. Unlocking the secrets: Volatile Organic Compounds (VOCs) and their devastating effects on lung cancer. *Pathology-Research and Practice*. 2024 Jan 26; 255:155157.
17. Thursby E, Juge N. Introduction to the human gut microbiota. *Biochemical Journal*. 2017 Jun 1;474(11):1823–36.
18. Diaz-Bone RA, Van de Wiele T. Biotransformation of metal (loid) s by intestinal microorganisms. *Pure and Applied Chemistry*. 2010 Jan 25;82(2):409–27.
19. Coryell M, Roggenbeck BA, Walk ST. The human gut microbiome's influence on arsenic toxicity. *Current Pharmacology Reports*. 2019 Dec;5:491–504.
20. Van de Wiele T, Du Laing G, Calatayud M. Arsenic from food: biotransformations and risk assessment. *Current Opinion in Food Science*. 2015 Dec 1;6:1–6.
21. Kubachka KM, Kohan MC, Herbin-Davis K, Creed JT, Thomas DJ. Exploring the in vitro formation of trimethylarsine sulfide from dimethylthioarsinic acid in anaerobic microflora of mouse cecum using HPLC–ICP-MS and HPLC–ESI-MS. *Toxicology and Applied Pharmacology*. 2009 Sep 1;239(2):137–43.
22. Lu K, Cable PH, Abo RP, Ru H, Graffam ME, Schlieper KA, Parry NM, Levine S, Bodnar WM, Wishnok JS, Styblo M. Gut microbiome perturbations induced by bacterial infection affect arsenic biotransformation. *Chemical Research in Toxicology*. 2013 Dec 16;26(12):1893–903.
23. Pinyayev TS, Kohan MJ, Herbin-Davis K, Creed JT, Thomas DJ. Preabsorptive metabolism of sodium arsenate by anaerobic microbiota of mouse cecum forms a variety of methylated and thiolated arsenicals. *Chemical Research in Toxicology*. 2011 Apr 18;24(4):475–7.
24. Alava P, Du Laing G, Tack F, De Ryck T, Van De Wiele T. Westernized diets lower arsenic gastrointestinal bioaccessibility but increase microbial arsenic speciation changes in the colon. *Chemosphere*. 2015 Jan 1;119:757–62.
25. Alava P, Tack F, Du Laing G, Van de Wiele T. Arsenic undergoes significant speciation changes upon incubation of contaminated rice with human colon micro biota. *Journal of Hazardous Materials*. 2013 Nov 15;262:1237–44.
26. Van de Wiele T, Gallawa CM, Kubachk KM, Creed JT, Basta N, Dayton EA, Whitacre S, Laing GD, Bradham K. Arsenic metabolism by human gut microbiota upon in vitro digestion of contaminated soils. *Environmental Health Perspectives*. 2010 Jul;118(7):1004–9.
27. Martinez-Guryn K, Leone V, Chang EB. Regional diversity of the gastrointestinal microbiome. *Cell Host & Microbe*. 2019 Sep 11;26(3):314–24.
28. Pereira DI, Gibson GR. Cholesterol assimilation by lactic acid bacteria and bifidobacteria isolated from the human gut. *Applied and Environmental Microbiology*. 2002 Sep;68(9):4689–93.
29. de Matuoka e Chiocchetti G, Monedero V, Zúñiga M, Vélez D, Devesa V. In vitro evaluation of the protective role of Lactobacillus StrainsAgainst inorganic arsenic toxicity. *Probiotics and Antimicrobial Proteins*. 2020 Dec;12:1484–91.
30. Lander JJ, Stanley RJ, Sumner HW, Boswell DC, Aach RD. Angiosarcoma of the liver associated with Fowler's solution (potassium arsenite). *Gastroenterology*. 1975 Jun 1;68(6):1582–6.
31. Sanger CR. The quantitative determination of arsenic, by the Berzelius-Marsh process, especially as applied to the analysis of wall papers and fabrics. In *Proceedings of the American Academy of Arts and Sciences* 1890 May 1 (Vol. 26, pp. 24–45). American Academy of Arts & Sciences.Nriagu JO. Arsenic poisoning through the ages. *Environmental Chemistry of Arsenic*. 2002;1:1–26.
32. Nriagu JO. Arsenic poisoning through the ages. *Environmental Chemistry of Arsenic*. 2002;1:1–26.
33. Mego M, Holec V, Drgona L, Hainova K, Ciernikova S, Zajac V. Probiotic bacteria in cancer patients undergoing chemotherapy and radiation therapy. *Complementary Therapies in Medicine*. 2013 Dec 1;21(6):712–23.
34. Hughes MF. History of arsenic as a poison and a medicinal agent. *Arsenic: Exposure Sources, Health Risks, and Mechanisms of Toxicity*. 2015 Oct 16:1–22.
35. Wu MM. DNA repair and arsenical skin cancer: A case-control study in a blackfoot disease endemic area in Taiwan. The Johns Hopkins University; 1997 (vol. 1, pp. 9730811).
36. Tiemann M. Arsenic in drinking water: Regulatory developments and issues. Water Pollution Issues and Developments. 2008:103.Pip E. Urban drinking water quality: a survey of selected literature. *Institute of Urban Studies*; 1993 Jan 1; 1.
37. Dheer R, Patterson J, Dudash M, Stachler EN, Bibby KJ, Stolz DB, Shiva S, Wang Z, Hazen SL, Barchowsky A, Stolz JF. Arsenic induces structural and compositional colonic microbiome change and promotes host nitrogen and amino acid metabolism. *Toxicology and Applied Pharmacology*. 2015 Dec 15;289(3):397–408.

38. Gibaud S, Jaouen G. Arsenic-based drugs: from fowler's solution to modern anticancer chemotherapy. In *Medicinal Organometallic Chemistry* (pp. 1–20). Berlin, Heidelberg: Springer; 2010.
39. Yamamoto S, Konishi Y, Matsuda T, Murai T, Shibata MA, Matsui-Yuasa I, Otani S, Kuroda K, Endo G, Fukushima S. Cancer induction by an organic arsenic compound, dimethylarsinic acid (cacodylic acid), in F344/DuCrj rats after pretreatment with five carcinogens. *Cancer Research.* 1995 Mar 15;55(6):1271–6.
40. Wang A, Wolf DC, Sen B, Knapp GW, Holladay SD, Huckle WR, Caceci T, Robertson JL. Dimethylarsinic acid in drinking water changed the morphology of urinary bladder but not the expression of DNA repair genes of bladder transitional epithelium in F344 rats. *Toxicologic Pathology.* 2009 Jun;37(4):425–37.
41. Li J, Waters SB, Drobna Z, Devesa V, Styblo M, Thomas DJ. Arsenic (+ 3 oxidation state) methyltransferase and the inorganic arsenic methylation phenotype. *Toxicology and applied pharmacology.* 2005 Apr 15;204(2):164–9.
42. LaGrega MD, Buckingham PL, Evans JC. *Hazardous Waste Management.* Long Grove, Illinois: Waveland Press; 2010 Jul 30.
43. Kitchin KT, Sun H. *Arsenic's Interactions with Macromolecules and Its Relationship to Carcinogenesis.* West Sussex, UK: John Wiley & Sons; 2011.
44. Carlin DJ, Naujokas MF, Bradham KD, Cowden J, Heacock M, Henry HF, Lee JS, Thomas DJ, Thompson C, Tokar EJ, Waalkes MP. Arsenic and environmental health: state of the science and future research opportunities. *Environmental Health Perspectives.* 2016 Jul;124(7):890–9.
45. Basu A, Saha D, Saha R, Ghosh T, Saha B. A review on sources, toxicity and remediation technologies for removing arsenic from drinking water. *Research on Chemical Intermediates.* 2014 Feb;40:447–85.
46. Kulkarni H, Shah M, Shankar PV. Shaping the contours of groundwater governance in India. *Journal of Hydrology: Regional Studies.* 2015 Sep 1;4:172–92.
47. Kumar S, Sharma S, Thaker R. Occupational, environmental, and lifestyle factors and their contribution to preterm birth–an overview. *Indian Journal of Occupational and Environmental Medicine.* 2017 Jan;21(1):9.
48. Pawar A, Singh S, Ramamurthy PC, Anil AG, Shehata N, Dhanjal DS, Naik TS, Parihar P, Prasad R, Singh J. Toxicity, Environmental monitoring and removal strategies of arsenic. *International Journal of Environmental Research.* 2022 Oct;16(5):66.
49. Naranmandura H, Rehman K, Le XC, Thomas DJ. Formation of methylated oxyarsenicals and thioarsenicals in wild-type and arsenic (+ 3 oxidation state) methyltransferase knockout mice exposed to arsenate. *Analytical and Bioanalytical Chemistry.* 2013 Feb;405:1885–91.
50. Bu N, Wang HY, Hao WH, Liu X, Xu S, Wu B, Anan Y, Ogra Y, Lou YJ, Naranmandura H. Generation of thioarsenicals is dependent on the enterohepatic circulation in rats. *Metallomics.* 2011 Oct;3(10):1064–73.
51. Németi B, Anderson ME, Gregus Z. Glutathione synthetase promotes the reduction of arsenate via arsenolysis of glutathione. *Biochimie.* 2012 Jun 1;94(6):1327–33.
52. Cohen SM, Arnold LL, Eldan M, Lewis AS, Beck BD. Methylated arsenicals: the implications of metabolism and carcinogenicity studies in rodents to human risk assessment. *Critical Reviews in Toxicology.* 2006 Jan 1;36(2):99–133.
53. Gamboa-Loira B, Cebrian ME, Franco-Marina F, Lopez-Carrillo L. Arsenic metabolism and cancer risk: a meta-analysis. *Environmental Research.* 2017 Jul 1;156:551–8.
54. Fan C, Liu G, Long Y, Rosen B, Cai Y. Thiolation in arsenic metabolism: a chemical perspective. *Metallomics.* 2018;10(10):1368–82.
55. Cullen WR. Chemical mechanism of arsenic biomethylation. Chemical research in toxicology. 2014 Apr 21;27(4):457–61.
56. Roggenbeck BA, Banerjee M, Leslie EM. Cellular arsenic transport pathways in mammals. *Journal of Environmental Sciences.* 2016 Nov 1;49:38–58.
57. Dheeman DS, Packianathan C, Pillai JK, Rosen BP. Pathway of human AS3MT arsenic methylation. *Chemical Research in Toxicology.* 2014 Nov 17;27(11):1979–89.
58. Roggenbeck BA, Leslie EM, Walk ST, Schmidt EE. Redox metabolism of ingested arsenic: Integrated activities of microbiome and host on toxicological outcomes. *Current Opinion in Toxicology.* 2019 Feb 1;13:90–8.
59. Packianathan C, Kandavelu P, Rosen BP. The structure of an As (III) S-adenosylmethionine methyltransferase with 3-coordinately bound As (III) depicts the first step in catalysis. *Biochemistry.* 2018 Jun 12;57(28):4083–92.
60. Leslie EM. Arsenic–glutathione conjugate transport by the human multidrug resistance proteins (MRPs/ABCCs). *Journal of Inorganic Biochemistry.* 2012 Mar 1;108:141–9.

61. Sun Y, Liu G, Cai Y. Thiolated arsenicals in arsenic metabolism: Occurrence, formation, and biological implications. *Journal of Environmental Sciences*. 2016 Nov 1;49:59–73.
62. Raml R, Rumpler A, Goessler W, Vahter M, Li L, Ochi T, Francesconi KA. Thio-dimethylarsinate is a common metabolite in urine samples from arsenic-exposed women in Bangladesh. *Toxicology and Applied Pharmacology*. 2007 Aug 1;222(3):374–80.
63. Naranmandura H, Suzuki KT. Formation of dimethylthioarsenicals in red blood cells. *Toxicology and Applied Pharmacology*. 2008 Mar 15;227(3):390–9.
64. Moe B, Peng H, Lu X, Chen B, Chen LW, Gabos S, Li XF, Le XC. Comparative cytotoxicity of fourteen trivalent and pentavalent arsenic species determined using real-time cell sensing. *Journal of Environmental Sciences*. 2016 Nov 1;49:113–24.
65. Chen CM, Mobley HL, Rosen BP. Separate resistances to arsenate and arsenite (antimonate) encoded by the arsenical resistance operon of R factor R773. *Journal of Bacteriology*. 1985 Feb;161(2):758–63.
66. Silver S, Budd K, Leahy KM, Shaw WV, Hammond D, Novick RP, Willsky GR, Malamy MH, Rosenberg H. Inducible plasmid-determined resistance to arsenate, arsenite, and antimony (III) in Escherichia coli and Staphylococcus aureus. *Journal of Bacteriology*. 1981 Jun;146(3):983–96.
67. Ben Fekih I, Zhang C, Li YP, Zhao Y, Alwathnani HA, Saquib Q, Rensing C, Cervantes C. Distribution of arsenic resistance genes in prokaryotes. *Frontiers in Microbiology*. 2018 Oct 23;9:2473.
68. Sato T, Kobayashi Y. The ars operon in the skin element of Bacillus subtilis confers resistance to arsenate and arsenite. *Journal of Bacteriology*. 1998 Apr 1;180(7):1655–61.
69. Li J, Mandal G, Rosen BP. Expression of arsenic resistance genes in the obligate anaerobe Bacteroides vulgatus ATCC 8482, a gut microbiome bacterium. *Anaerobe*. 2016 Jun 1;39:117–23.
70. Argudín MA, Hoefer A, Butaye P. Heavy metal resistance in bacteria from animals. *Research in Veterinary Science*. 2019 Feb 1;122:132–47.
71. Koppel N, Maini Rekdal V, Balskus EP. Chemical transformation of xenobiotics by the human gut microbiota. *Science*. 2017 Jun 23;356(6344):eaag2770.
72. Garbinski LD, Rosen BP, Chen J. Pathways of arsenic uptake and efflux. *Environment International*. 2019 May 1;126:585–97.
73. Coryell MP. *Investigating Arsenic-Microbiome Interactions in the Gut Using Murine Models* (Doctoral dissertation, Montana State University).
74. Challenger F, Higginbottom C. The production of trimethylarsine by Penicillium brevicaule (Scopulariopsis brevicaulis). *Biochemical Journal*. 1935 Jul;29(7):1757.
75. Kuroda K, Yoshida K, Yasukawa A, Wanibuchi H, Fukushima S, Endo G. Enteric bacteria may play a role in mammalian arsenic metabolism. *Applied Organometallic Chemistry*. 2001 Jun;15(6):548–52.
76. Rowland IR, Davies MJ. In vitro metabolism of inorganic arsenic by the gastro-intestinal microflora of the rat. *Journal of Applied Toxicology*. 1981 Oct;1(5):278–83.
77. Hall LL, George SE, Kohan MJ, Styblo M, Thomas DJ. In vitro methylation of inorganic arsenic in mouse intestinal cecum. *Toxicology and Applied Pharmacology*. 1997 Nov 1;147(1):101–9.
78. Coryell M, McAlpine M, Pinkham NV, McDermott TR, Walk ST. The gut microbiome is required for full protection against acute arsenic toxicity in mouse models. *Nature Communications*. 2018 Dec 21;9(1):5424.
79. Lu K, Mahbub R, Cable PH, Ru H, Parry NM, Bodnar WM, Wishnok JS, Styblo M, Swenberg JA, Fox JG, Tannenbaum SR. Gut microbiome phenotypes driven by host genetics affect arsenic metabolism. *Chemical Research in Toxicology*. 2014 Feb 17;27(2):172–4.
80. Gaulke CA, Rolshoven J, Wong CP, Hudson LG, Ho E, Sharpton TJ. Marginal zinc deficiency and environmentally relevant concentrations of arsenic elicit combined effects on the gut microbiome. *Msphere*. 2018 Dec 26;3(6):110–28.
81. Welch RA, Burland V, Plunkett III G, Redford P, Roesch P, Rasko D, Buckles EL, Liou SR, Boutin A, Hackett J, Stroud D. Extensive mosaic structure revealed by the complete genome sequence of uropathogenic Escherichia coli. *Proceedings of the National Academy of Sciences*. 2002 Dec 24;99(26):17020–4.
82. Chi L, Xue J, Tu P, Lai Y, Ru H, Lu K. Gut microbiome disruption altered the biotransformation and liver toxicity of arsenic in mice. *Archives of Toxicology*. 2019 Jan 21;93:25–35.
83. Chi L, Bian X, Gao B, Tu P, Ru H, Lu K. The effects of an environmentally relevant level of arsenic on the gut microbiome and its functional metagenome. *Toxicological Sciences*. 2017 Dec 1;160(2):193–204.
84. Rahaman MS, Rahman MM, Mise N, Sikder MT, Ichihara G, Uddin MK, Kurasaki M, Ichihara S. Environmental arsenic exposure and its contribution to human diseases, toxicity mechanism and management. *Environmental Pollution*. 2021 Nov 15;289:117940.
85. Chi L, Bian X, Gao B, Ru H, Tu P, Lu K. Sex-specific effects of arsenic exposure on the trajectory and function of the gut microbiome. *Chemical Research in Toxicology*. 2016 Jun 20;29(6):949–51.

86. Xue J, Lai Y, Chi L, Tu P, Leng J, Liu CW, Ru H, Lu K. Serum metabolomics reveals that gut microbiome perturbation mediates metabolic disruption induced by arsenic exposure in mice. *Journal of Proteome Research.* 2019 Jan 10;18(3):1006–18.
87. Yang Y, Chi L, Liu CW, Hsiao YC, Lu K. Chronic arsenic exposure perturbs gut microbiota and bile acid homeostasis in mice. *Chemical Research in Toxicology.* 2023 Jun 9;36(7):1037–1043.
88. Yang Y, Chi L, Lai Y, Hsiao YC, Ru H, Lu K. The gut microbiome and arsenic-induced disease—iAs metabolism in mice. *Current Environmental Health Reports.* 2021 Jun;8:89–97.
89. Gao B, Chi L, Mahbub R, Bian X, Tu P, Ru H, Lu K. Multi-omics reveals that lead exposure disturbs gut microbiome development, key metabolites, and metabolic pathways. *Chemical Research in Toxicology.* 2017 Apr 17;30(4):996–1005.

13 Future Prospects and Challenges

*Md Sadique Hussain[1], Somya Sharma[1], Riya Mishra[2], Asif Ahmad Bhat[2], Riya Thapa[2], Raihan Siddiqui[2], Rahul Sharma[1], Rashi Kulshrestha[2], Neeraj Patel[2], and Gaurav Gupta[3,4]**

[1]School of Pharmaceutical Sciences, Jaipur National University, Jaipur, Rajasthan, India
[2]School of Pharmacy, Suresh Gyan Vihar University, Jaipur, Rajasthan, India
[3]Centre of Medical and Bio-allied Health Sciences Research, Ajman University, Ajman, United Arab Emirates
[4]Centre for Research Impact & Outcome, Chitkara College of Pharmacy, Chitkara University, Rajpura, Punjab, India
*Corresponding Author: Gaurav Gupta.
Email: gauravpharma25@gmail.com

13.1 INTRODUCTION

13.1.1 Brief Overview of the Gut Microbiome and Environmental Toxicants

The term "gut microbiome" (GM) refers to the diverse array of microorganisms inhabiting the human gastrointestinal tract (GIT). These microorganisms perform a crucial function in maintaining a mutually beneficial relationship with their host, contributing significantly to digestion, immune function, and overall well-being. Various factors, including developmental, lifestyle, and environmental impacts, distinguish the makeup and activity of the GM [1–3]. Recent studies have demonstrated that environmental toxicants (ETs) can also exert an influence on the GM, potentially leading to health complications. For instance, research has indicated that exposure to polycyclic aromatic hydrocarbons (PAHs) can induce variations in the microbial makeup associated with the gut of fish, resulting in changes in feeding behavior and heightened susceptibility to disease [4, 5]. Furthermore, exposure to organophosphate insecticides such as diazinon has been found to disrupt the community structure and functional metagenome of the GM in mice, with more pronounced effects observed in male subjects [6, 7]. These findings underscore the importance of comprehending the impact of ETs on the GM and the potential ramifications for human health.

Several prevalent ETs have the potential to influence the GM. Arsenic (As) is classified as a class 1 carcinogen that can adversely affect the GM, resulting in GI disorders, disruptions in immune function, and alterations in the growth patterns of bacterial species commonly found in the GIT [8]. Trichloroethylene (TCE) is an industrial byproduct and ETs and TCE, when ingested, can induce modifications in the GM, which may have relevance to idiopathic Parkinson's disease (PD) [9]. Exposure to PAHs can instigate shifts in the microbial communities associated with the gut of fish, potentially leading to alterations in foraging behavior and an increased susceptibility to disease [9]. Exposure to organophosphate insecticides, such as diazinon, has been demonstrated to disrupt the community structure and functional metagenome of the GM in mice, with more pronounced effects observed in male subjects [10]. Moreover, alterations in diet and environmental conditions can impact the composition and abundance of GM in animals, as observed in the common kestrel,

 DOI: 10.1201/9781003489221-13

where differences in GM before and after captivity have been documented [11]. These toxicants have the capacity to disrupt the GM, subsequently affecting host health and potentially contributing to the development of various diseases and conditions.

13.1.2 Importance of Understanding Their Impact on Human Health

Comprehending the influence of ETs on the GM holds paramount importance for multiple facets of human health, encompassing the digestive system, immune system, mental health, obesity, and cancer. The GM serves as a cornerstone in the establishment and operation of the human immune system, termed the GM-immune system axis (GMISA). This system governs immune reaction, facilitating the preservation of vital aspects of host-microorganism symbiosis through a response that regulates responses in the most advantageous manner [12, 13]. Moreover, the GM is intricately linked to mental health via the gut-brain axis, emphasizing a microbiome-gut-brain axis (MGBA). Various neurocognitive and mental health–related conditions have been associated with alterations in the GM, prompting investigations into the potential impact of medicinal plants on mental disorders and GM [14].

Environmental factors have the capacity to modulate the GMISA positively or negatively, exerting repercussions on the body's overall health status. For instance, research has explored the influence of ETs on the gut-brain axis in PD, as well as the function of probiotics and postbiotics in regulating the GMISA in pediatric populations [12]. The GM also plays a role in obesity, with variations in the makeup of the intestinal microbiome observed in response to various obesity therapies. Probiotics, prebiotics, and synbiotics have been studied for their potential to influence the release of hormones, neurotransmitters, and inflammatory factors, thereby mitigating the stimuli of food consumption that contribute to weight gain [15, 16]. In the realm of cancer, the GM has been implicated in the onset and progression of certain cancers, as well as in the response to cancer treatments [17]. Understanding the repercussions of ETs on the GM can facilitate the development of targeted therapies and preventive strategies for diverse health conditions.

13.1.3 Objectives of the Chapter

The chapter intends to investigate the evolving landscape of research concerning the intricate interplay between the GM and ETs, with a particular emphasis on its implications for human health. It aims to evaluate and anticipate the trajectory of research in elucidating how ETs influence the structure and function of the GM, and subsequently impact human health. This endeavor entails exploring potential avenues for inquiry and technological advancements that might shape future research pursuits. It also involves examining the current obstacles and limitations in thoroughly investigating the influence of ETs on the GM and human wellness. This encompasses addressing knowledge gaps, technological constraints, and methodological challenges that impede progress in this domain.

Furthermore, the chapter delves into the therapeutic potential of interventions targeting the GM-environment axis to alleviate the detrimental effects of ETs on human health. This exploration entails investigating strategies for tailored therapies and interventions based on a nuanced understanding of these interactions. It also involves identifying gaps in current knowledge and proposing future research directions to tackle unresolved questions and further enhance our understanding of the GM and ETs within the context of human health. This includes exploring interdisciplinary collaborations and innovative research approaches to confront the intricate challenges in this field.

13.2 ADVANCES IN GUT MICROBIOME RESEARCH

13.2.1 Sequencing Technologies

Recent progress in sequencing technologies for the GM has emphasized multi-omic approaches, integrating diverse high-throughput sequencing methods to investigate host-microbe interactions in conditions like inflammatory bowel disease (IBD) and colorectal cancer (CRC). These methodologies encompass 16S rRNA sequencing, shotgun metagenomics, meta-transcriptomics, metabolomics, and metaproteomics [18]. Machine learning algorithms have also been utilized to predict and compare both complete and partial pathogenic proteins from genomic and metagenomic datasets, offering potential for early-stage disease detection, particularly in CRC [19, 20].

The adoption of emerging technologies in GM research has been motivated by the necessity to progress beyond phylogenetic descriptions and delve into mechanistic techniques. Techniques such as metatranscriptomics, metabolomics, culturomics, and synthetic biology are employed with the objective of providing clinicians a robust framework for studying the GM and expediting the integration of these methods into clinical settings. The GM's association with a diverse array of diseases and its dynamic nature present significant opportunities for uncovering mechanisms that influence human physiology. In recent years, there has been a notable surge in GM studies, facilitated by the rapid evolution of high-throughput sequencing methods [21–23].

13.2.2 Insights from Metagenomics and Metabolomics Studies

In the realm of GM research, exploring associations with diseases stands as a significant focal point. The available literature demonstrates that GM exhibits connections with a broad spectrum of diseases, encompassing IBD, non-alcoholic fatty liver disease (NAFLD), chronic kidney disease (CKD), and others [24–26]. Studies utilizing metagenomics and metabolomics have served an essential part in clarifying the precise functionalities of GM within the context of functional host-microbe interactions [27].

For instance, an investigation into the GM and metabolome among IBD patients revealed that mesenchymal stem cell-derived exosomes (MSC-Ex) ameliorated colitis by modulating the gut metagenomics and metabolomics profile, augmenting colonic FXR, and enhancing the composition of GM [28]. Another study employed LIVE modeling to amalgamate various types of GM multi-omics data, thereby reducing the number of feature correlations and facilitating the prioritization of biological associations among microbes, metabolites, enzymes, and IBD status [29]. In essence, GM has been implicated in a diverse array of diseases, and metagenomics and metabolomics studies are indispensable for comprehending the specific functionalities of GM and their implications for host health.

13.3 EMERGING TRENDS IN ENVIRONMENTAL TOXICOLOGY

Current directions in environmental toxicology encompass diverse areas, reflecting evolving methodologies and research foci. These include the employment of lipidomics to investigate how ETs influence lipid metabolism, the exploration of prophylactic strategies to counter organophosphate toxicity, the examination of microplastic pollution and its potential health implications, the utilization of plant nanobionics and nanobiosensors in toxicology applications, the adoption of zebrafish bioassays for aquatic ecotoxicology, and the scrutiny of PCBs and their associated health impacts [30–35]. In parallel, there is a concerted effort to enhance the statistical reporting in environmental toxicology research, aiming to improve transparency and facilitate the interpretation of study outcomes [36].

13.3.1 Novel Approaches for Assessing Exposure to Environmental Toxicants

Innovative strategies for evaluating exposure to ETs include:

13.3.1.1 Wearable Passive Samplers

These devices enable the evaluation of organic chemical exposure by measuring chemical concentrations in the surroundings [37]. A study conducted in South Africa employed wearable passive candidates in conjunction with high-resolution mass spectrometry (HR-MS), revealing that children in low- and middle-income nations frequently face elevated chemical exposure levels and are particularly vulnerable to the health ramifications of contaminants in the air. The investigation identified 637 environmental exposures, encompassing 50 airborne compounds of concern, such as combustion byproducts, dyes, fragrances, organophosphates, pesticides, and plasticizers. Noteworthy, wristbands detected biocides like p,p'-DDT, p,p'-DDD, p,p'-DDE, propoxur, piperonyl butoxide, and triclosan. Exposure levels exhibited seasonal variations, with 27% of the identified chemicals displaying distinctions between wet and dry seasons [38, 39]. Another study, employing wearable passive samplers and gas chromatography HR-MS, identified 50 chemical compounds of concern in South African children. This investigation aimed to mitigate the adverse health impacts experienced by vulnerable populations due to environmental exposures [40]. These studies underscore the efficacy of wearable passive samplers in evaluating exposure to an extensive array of ETs, offering invaluable insights into the intricacies and fluctuations of chemical exposures at a molecular scale.

13.3.1.2 Geographic Information System (GIS)

GIS aids in assessing pesticide exposure by analyzing the proximity of residential areas to agricultural regions and the pesticide usage in those areas [41]. A study carried out in Italy employed GIS to evaluate the risk of amyotrophic lateral sclerosis (ALS) stemming from passive residential exposure to pesticides. The study compared two methodologies: individual questionnaire-based assessment and GIS-based assessment. The GIS-based approach involved calculating the proportion of rural land use within a 100m radius around each subject's residence using cover maps from various periods. The odds ratio for passive residential pesticide exposure was 1.05 and 1.13 based on GIS assessment, in contrast to 1.67 from the questionnaire-based evaluation [41]. In another study, a novel method termed Wastewater-Based Epidemiology (WBE) was suggested to use GIS to measure community consumption of organophosphates, pyrethroids, and triazines. This technique entailed measuring human urinary metabolites of pesticides found in influent wastewater collected from treatment facilities for wastewater in European urban areas. The outcomes unveiled spatial disparities in pesticide exposure within Italy and Europe, with seasonal fluctuations noted in human pyrethroid intake [42]. These investigations underscore the efficacy of GIS in evaluating exposure to ETs, furnishing valuable insights into the spatial dispersion and potential health hazards linked with pesticide exposure.

13.3.1.3 Exposure to Heavy Metals (HMs)

A systematic review evaluates health hazards linked with HMs in oil and gas waste, like cadmium, lead, and mercury, and devises mitigation strategies [43]. An assessment of essential elements and HMs in food analyzed essential elements and toxic HMs in food samples using methods like inductively coupled plasma-mass spectrometry (ICP-MS) to ascertain food consumption safety [44]. A thorough examination of the pollution status and concurrent health risk assessment concerning human exposure to specific HMs in road dust across diverse global cities was conducted [45]. An investigation assessing the non-carcinogenic health risk associated with HMs (As, Cd, Cr, Cu, Mn, Pb, and Zn) present in indoor dust within residences situated near an e-waste dismantling area in South China was completed [46]. A South African case study scrutinized heavy metal levels in surface and groundwater, pinpointing sources, and appraising the risk of human exposure through the drinking water route [47]. A study determined the concentrations of essential elements and toxic HMs in rice samples subjected to diverse processing methods, including washing, soaking, and cooking, utilizing ICP-MS [44]. These investigations underscore the significance of appraising HMs as ETs, given their potential to pose substantial health risks to humans, particularly when exposure transpires through various avenues like road dust, residential indoor dust, and drinking water. The utilization of sophisticated analytical techniques such as ICP-MS enables precise detection and quantification of HMs in diverse environmental samples, a pivotal aspect for risk assessment and safeguarding public health.

13.3.1.4 Human Inhalation Exposure to PAHs

This method gauges the general population's exposure to carcinogenic, teratogenic, and mutagenic PAHs by collecting environmental pollution concentrations and modeling exposure patterns [48]. Research has emphasized the importance of evaluating human inhalation exposure to PAHs due to their carcinogenic, teratogenic, and mutagenic characteristics. A study summarized various assessment methods for PAH exposure, highlighting the necessity of determining the exposure levels among the general population. These methods encompass collecting environmental pollution concentrations via active or passive samplers, analyzing time-activity patterns to capture exposure durations and behaviors, utilizing modeling for exposure estimation, and investigating microenvironmental exposure. Accurately quantifying exposure levels is pivotal for establishing exposure-risk associations and safeguarding public health [48]. PAHs permeate inside humans through diverse pathways, including eating of tainted food, potable water, and exposure to smoke from cigarettes, vehicle emissions, and tainted air in homes and surroundings. Dietary ingestion of PAHs emerges as a significant concern of human exposure and significantly contributes to GIT cancer development. Grasping the involvement of diet, smoking, and other variables in GIT cancers is imperative for gauging the risks linked to PAH exposure [49]. These investigations underscore the criticality of evaluating human inhalation exposure to PAHs, considering their detrimental health impacts and potential hazards to public health. Comprehending the origins and routes of exposure is essential for formulating strategies to alleviate these risks and safeguard human health.

13.4 BIOLOGICAL PATHWAYS ORCHESTRATED BY THE GUT MICROBIOTA FOLLOWING EXPOSURE TO ENVIRONMENTAL CHEMICALS

There are diverse processes through which GM can potentially influence, either directly or indirectly, the interplay between ETs and GM altering health (Figure 13.1). Numerous investigations suggest that exposure to ETs has the potential to alter the composition and function of microbes. Alternatively, the microbial processing of chemicals might exert positive or negative effects on the host, either via metabolic byproducts or by influencing the toxicity of the compounds. These procedures encompass both direct metabolic alterations and secondary alterations, including the deconjugation of host-generated metabolites, control of permeation of epithelial cells, and modulation of crucial metabolic pathways. It is noteworthy that the primary mechanism can intersect with other indirect mechanisms mediated downstream by GM.

For instance, GM may initially metabolically alter ETs directly. A study in mice demonstrated that absorption of As from ingesting water could significantly disrupt GM and disturb both bile acid and indole concentrations [50]. Additionally, chemical reactions involving the biotransformation of toxicants by GM are conceivable, including reduction and hydrolysis reactions [51]. Indeed, information from both preclinical and clinical microbiome investigations suggested that GM can modify the toxicity of toxicants, such as HMs, through various mechanisms [52]. For example, in humans, methyltransferase activity catalyzes the complete methylation of As to dimethyl arsenic. However, research utilizing human bacteria has shown that following decline and methylation, it can be transformed into intermediary variations, a few of which might be more harmful than the initial metalloid [53].

Secondary transformations may involve the disruption of GM epithelial barrier. Several studies have indicated that alterations in intestinal permeability, akin to those observed with stress, infection, and dietary changes, can occur following gnotobiotic nurturing, antibiotic treatment, or probiotic

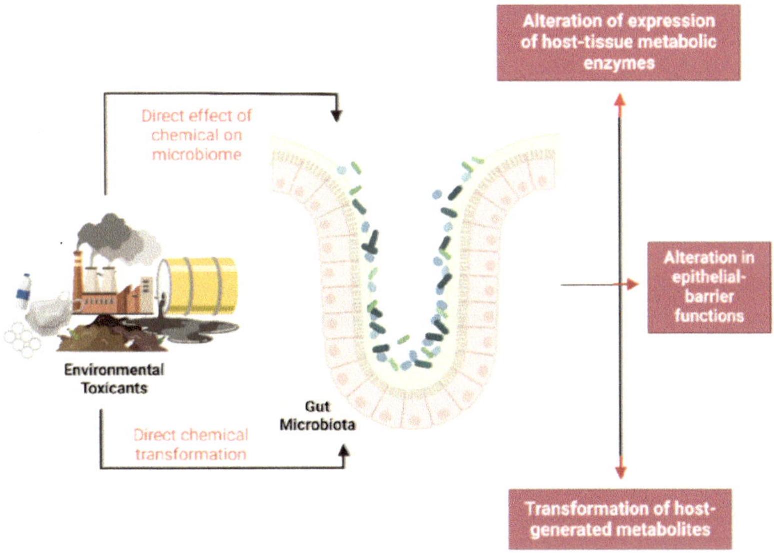

FIGURE 13.1 The primary mechanisms through which the microbiome regulates the effects of toxicants.

administration [54, 55]. In a rat investigation, the spread of a novel bacterium species named *Akk. muciniphila* resulted in higher levels of some tight-junction proteins linked to decreased intestine absorptive ability [54, 56]. In a further investigation employing an experimental model of autism, it was shown that administering the Bacteroides fragilis species led to increased gene expression of two claudin proteins (8 and 15) associated with impaired GI barrier stability [57]. Zhao et al. used metabolomics investigation on mice urine to study the adverse consequences of Chlorpyrifos. They found that it caused prolonged intestinal inflammation and affected the flexibility of GI cells [58]. Recent RNA-seq investigations of intestinal and hepatic epithelia have shown confirmation of GM's involvement in modulating chemical processes of metabolism [59, 60]. The first evaluation of liver gene expression in untreated mice compared to bacteria-free animals showed substantial variations in the expression of genes related to hepatic chemical breakdown [60]. CYP3 expression significantly decreased in the absence of bacteria but returned to normal values after colonization by microbes, similar to that of control [60, 61]. These findings are significant as the modulation of CYP3a activity occurs via the Pregnane X pathway, an essential receptor known to be one of the primary biochemical regulators between GM and the host [62].

13.5 INTEGRATION OF MULTI-OMICS APPROACHES

The GM exerts influence on all organ systems and physiological aspects. As researchers increasingly investigate scantily inhabited microbiome ecosystems with low volume within ostensibly "sterile" organs like the skin, lungs, reproductive organs, and bile ducts, their discoveries often face scrutiny. These studies are frequently confronted with significant incorrect outcomes caused by potential interference and execution errors. Thus, it is typical to investigate the microbiome to internally validate microbial data, especially in investigations involving low-biomass patient samples. This validation typically employs experimental methods such as electron microscopy and microbiome multi-omics approaches like culturomics, metatranscriptomics, and metabolomics.

Mishra et al. examined microbial makeup in human fetal matter during the second trimester of gestation employing 16S rRNA sequencing, the supplied more microbiological data to verify the existence of these microorganisms' using methods like culturomics, electron microscopy, and in vitro experimental validation experiments. This groundbreaking research underscores the crucial function for microbial contact in early-life immunological stimulation, where T cells are activated by GM found in the second trimester of pregnancy. This microbial multi-omics exhibition provides a more profound insight into the revealed microbial ecosystems [63].

Multi-omics also empowers microbiota experts to elucidate the significance of microbial populations within GM by validating metagenomic readouts, identifying expressed genes, and translating them into proteins (Figure 13.2). In a study by Schirmer et al. on IBD metatranscriptomics, meta-transcriptional profiles were utilized to gain insight into GM community interactions, including IBD-specific transcriptional functioning. Moreover, microbial transcriptional programs are believed to be more sensitive to external stimuli that might not be captured at the DNA level, such as variations in oxygen availability and inflammation. However, interpreting fecal metatranscriptomics readouts faces limitations due to variations in subject-specific transit times and possible bias in screening wherein feces specimens contain only organisms whose RNA can be extracted and is not broken down [64].

The impact of the microbiome extends beyond the immediate surroundings, notably through microbiome metabolites, which are categories of small molecules generated or altered by GM and play crucial roles in regulating host-microbial relations. Franzosa et al. researched IBD microbiota metabolomics, where they found particular microbial compounds linked to IBD by analyzing the correlation between fecal metabolomics and fecal metagenomics. Individuals with IBD showed increased concentrations of bile acids and sphingolipids, as well as reduced concentrations of triacylglycerols and tetrapyrroles [65, 66]. These metabolites of bacteria have the potential to serve as innovative indicators for individualized medical treatment. IBD people who have GM and are able to transform primary bile acids to secondary ones may have a higher chance of responding positively to anti-cytokine biological treatment [67]. Individuals with Crohn's disease with higher levels of butyrate-producing microorganisms are more prone to have remission after using anti-integrin medication [68–70].

As mentioned earlier, a significant group of microbial metabolites comprises short-chain fatty acids (SCFAs), which result from the fermentation of carbohydrates by GM in the colon [71]. SCFAs are essential gut elements with varied immune regulatory functions [72, 73]. Due to their probable

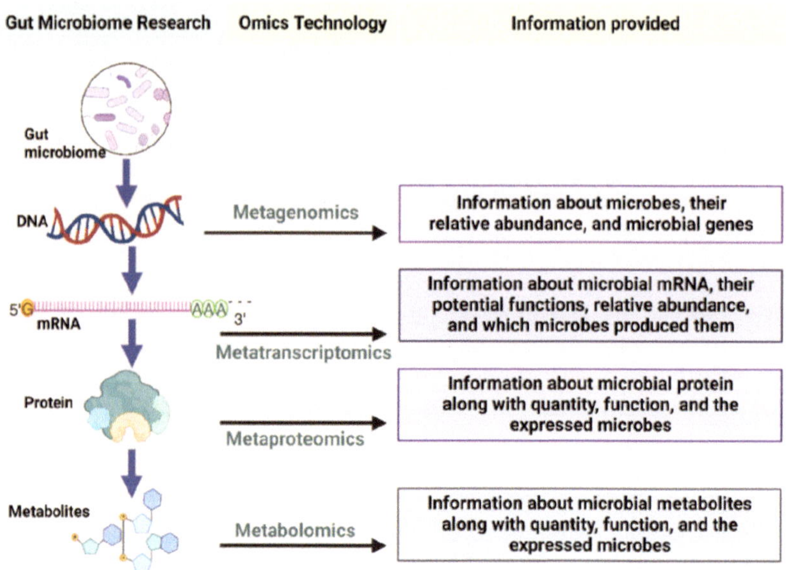

FIGURE 13.2 Multi-omics technology and information they provide.

anti-inflammatory properties, coupled with the observed decline of SCFAs and SCFA-mediating microbes in IBD individuals, SCFAs are seen as favorable treatment options for managing IBD clinically [74, 75]. Consequently, there have been encouraging albeit limited results in animal models and a variation of IBD individual responding to augmentation of SCFA-mediating microbes. Nonetheless, increasingly rigorous clinical trials are needed to assess the security and efficacy of these procedures.

IgA-Seq has demonstrated enhanced accuracy in identifying pathogenic members of GM driving inflammatory responses in IBD. This innovative method integrates current sequencing techniques with prior understanding of human biology, enabling researchers to target bacteria interacting with the immune system, specifically those tagged with IgA. Shapiro et al. used IgA-SEQ in the Ocean State Crohn's and Colitis Area Database to pinpoint several possible bacterial factors in IBD by focusing on IgA covering. A diminished IgA coating of Oscillospira was associated with a poorer prognosis, with a considerable proportion of individuals exhibiting low IgA coating requiring subsequent surgical resection. While reduced IgA coating in IBD individuals identifies bacteria linked to elevated inflammation, it is crucial to recognize that IgA coating does not solely denote pathogenesis but is also vital for the physiological host defense system. Thus, while analyzing possible IgA-targeted bacteria strains, caution is warranted as these microbes may regulate disease pathogenesis while also playing essential roles in normal immunoregulatory responses, potentially exhibiting protective effects against intestinal inflammation [76]. IgA-SEQ provides a new way to combine bacterial genomics with various other biological study techniques, offering a more comprehensive understanding of host-microbiome connections in both healthy and diseased conditions.

13.5.1 Future Directions in Multi-Omics Research for Studying Gut Microbiome and Environmental Toxicants

Future directions in multi-omics research for investigating GM and ETs involve integrating data from various biomolecular levels, including DNA, RNA, proteins, metabolites, and epigenetic marks, to gain a comprehensive understanding of biological systems and their interactions [77–79]. Analyzing individual cells will enhance our comprehension of the heterogeneity and intricacy of biological systems, shedding light on how ETs affect GM. Mapping gene expression patterns within tissues in their spatial context will further elucidate the impact of ETs on GM and its interactions with host tissues. Integrating multi-omics data with computational models will enable the prediction of biological system behavior and the identification of potential intervention targets. These tools will be instrumental in studying the effects of ETs on GM and its interactions with host tissues, facilitating the development of personalized therapies and predictive models [80]. Additionally, investigating the effects of ETs on GM and its interactions with host tissues will aid in identifying potential biomarkers and therapeutic targets [81–83].

Multi-omics research focused on investigating GM and ETs aims to integrate diverse biological data, providing a comprehensive insight into the intricate interactions among GM, host, and environmental elements. This holistic approach facilitates the discovery of potential biomarkers and the elucidation of mechanisms related to diseases such as IBD [29], CRC [84], and PD [85, 86]. The applications of multi-omics research in this domain involves the identification of microbial functions associated with diseases and their linkage to specific taxa, development of frameworks for reconstructing extensive microbial communities [87]. It includes exploration of the "gut-to-brain" hypothesis by employing metabolomic-combined 16S rRNA gene sequencing methods [86]; investigation of the interplay between GM and CRC [84]; evaluation of the therapeutic potential of compounds, such as the resveratrol hydroxypropyl-β-cyclodextrin inclusion complex, for PD treatment through the MGBA [85]. These applications are designed to advance our comprehension of the intricate relationships between GM, host, and environmental factors. Such insights hold the potential to foster the creation of more potent methods for various illness avoidance, detection, and therapy.

Leveraging multi-omics data will also contribute to the development of innovative and effective therapies by unraveling proteome complexity [77]. Coupling genomic approaches with the environmental chemistry of anthropogenic organic matter will allow for the assessment of microbial pollutant degradation and the understanding of organic pollutant effects on microbial community structure and function [88]. Efforts will also focus on enhancing seed oil content and quality, targeting specific fatty acid components to meet human health needs or industrial requirements [89]. Investigating the regulatory impact and mechanism of *Clostridium* butyricum-derived extracellular vesicles (CbEVs) in ulcerative colitis will be explored as a novel agent for colitis treatment [90]. Moreover, exploring the role of GM in neurodegenerative diseases via the MGBA will identify potential therapeutic targets and interventions [91]. Lastly, applying multi-omics approaches to tea breeding will provide insights into the effects of environmental factors on plant growth and development [92–94].

13.6 MICROBIOME-BASED THERAPEUTICS AND INTERVENTIONS

The rapid expansion of microbiome insights gained through the methodologies outlined earlier has solidified the significance of the microbiome in understanding health and illnesses. Moreover, the microbiome holds promising potential for modification, offering untapped avenues to leverage mechanisms impacting human physiology to promote health and prevent illness. Interventions aimed at the microbiome can be broadly categorized as either untargeted, aiming for overall enhancement in microbial makeup, or targeted, seeking specific alterations in metabolism-related GM (Figure 13.3). The advantages and disadvantages of microbiome-based interventions are listed in Table 13.1.

Typical instances of non-specific microbiome-focused interventions encompass antibiotics, fecal microbiota transplant (FMT), prebiotics, probiotics, postbiotics, and dietary adjustments. FMT involves transferring fecal material from healthy donors devoid of identifiable pathogens into the gut of host's lacking healthy GM, proving notably effective in treating recurrent *Clostridioides*

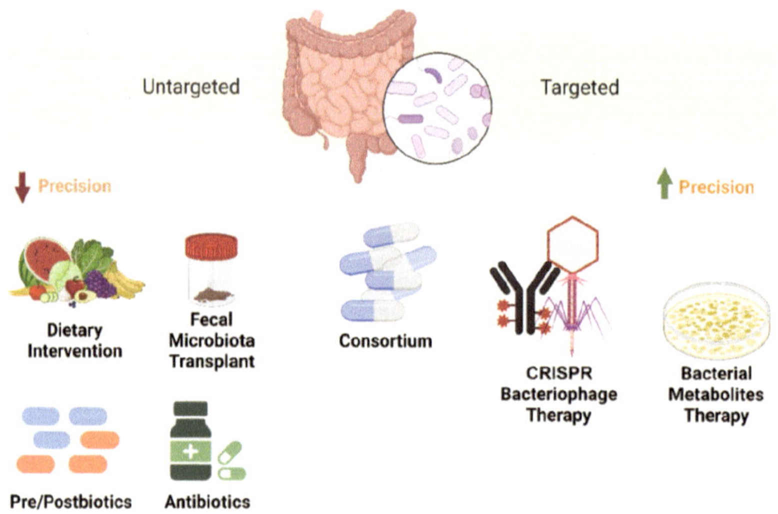

FIGURE 13.3 The diagram illustrates the present pathway for microbiome-targeted therapeutics, spanning from broad untargeted approaches to focused interventions.

TABLE 13.1
The advantages and disadvantages for microbiome-based therapeutics.

Therapeutic	Advantages	Disadvantages
Dietary interventions	Secure. Susceptible to manipulation.	Varied elements in each meal item. Inadequate dosage for efficacy in therapy. Transient remedial reaction.
Prebiotics	Secure. Food elements that are conveniently provided.	Relying on the colonization of a particular microorganism. Reliant on the GM. Treatment reaction is transient. Possible negative reactions (e.g., bloating).
Probiotics	Secure. Quickly accessible as a standardized mixture.	Not specific to a particular condition or individual. Relying on certain microorganism colonization. Reliant on gut ecosystem. Treatment reaction is transient. Regulation does not mandate survival.
Synbiotics	Comparatively secure. Comprises every element required for effectiveness.	The treatment reaction is short-term. Need a certain GM habitat. Possible negative reactions (e.g., after taking antibiotics).
Antibiotics	Secure. Inexpensive. Authorized drug. Current framework for regulation.	Possible unintended consequences such as susceptibility to antibiotics and impairment of colonization tolerance. Restricted to disturbance of the microbiome.
Phage therapy	Exceptionally precise.	Restricted to disturbance of the microbiome. Targets need precise cultivation. Novel treatment.
FMT	Comprises all microorganisms and essential nutrients. Proven effective for treating *Clostridioides difficile*.	Diversity among donors. Demands extensive testing. Effectiveness observed only in certain situations. Certain administrative tasks are expensive. Failure to standardize balance.
Live biotherapeutics	Authorized for particular conditions.	Demands to preserve viable bacteria. Possible negative long-term health consequences. Challenging to establish a causal link.
Microbiome mimetics	Dependent on the status of the microbiome at the time.	Development on developing mimetics is scarce.

FMT, fecal microbiota transplant; GM, gut microbiome; NEC, necrotizing enterocolitis.

difficile infections, boasting over 90% effectiveness (Figure 13.4) [95]. Despite its broad-spectrum nature, FMT trials for various conditions, including ulcerative colitis [96–98], immune checkpoint inhibitor-related anticancer response [99, 100], and type 2 diabetes mellitus [101, 102], have shown significant efficacy. While FMT has demonstrated a low incidence of serious adverse events, its safe implementation relies heavily on standardized, highly specialist labs for stool analysis [103]. These facilities necessitate expertise in proper spicemens collection, processing, and storage as well as rigorous donor health screening. Rigorously conducted clinical FMT trials hold promise for discovering the precise bacteria or molecules that contribute to the curative advantages of microbiota to produce tailored bacteriotherapy for an environmentally friendly and safer treatment method.

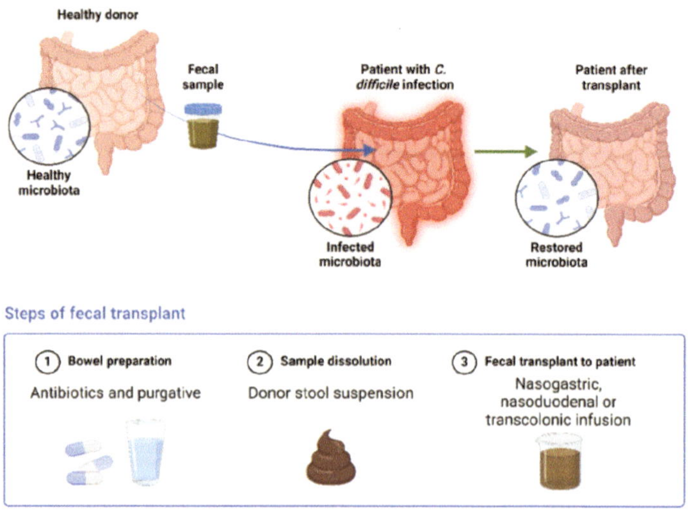

FIGURE 13.4 Fecal microbiota transplant.

A notable targeted microbial-based intervention is exemplified by Ho et al., who engineered *Escherichia coli* (*E. coli*) *Nissle* (*E. Nissle*) into a probiotic capable of adhering to CRC cell surfaces to secrete myrosinases. When this bioengineered *E. Nissle* strain was administered with broccoli extract, it achieved over 95% suppression of CRC cell lines [104]. Phage therapy is a different developing microbial-based treatment, where bacteriophages show promise for treating life-threatening antibiotic-resistant infections. This potential was demonstrated in the treatment of an individual with disseminated resistant *Acinetobacter baumannii* infection [105–107]. However, challenges linked to phage therapy include the potential for bacterial tolerance emergence against phages, diminished function due to immune responses, and the labor-intensive process of isolating and formulating specific phages for clinical use [108].

The advancement of microbiome-driven treatments aimed at detoxifying ETs presents significant potential in safeguarding public health and alleviating the adverse impacts of ETs. Persistent research endeavors are essential to address the current obstacles, laying the groundwork for a future in which leveraging the capabilities of GM can successfully counteract ETs and safeguard human health.

13.7 EXPLORING NOVEL TOOLS FOR MICROBIOME ANALYSIS

Amplicon sequencing emerged as one of the initial methods to delve into the function of GM in reaction to ETs [109]. However, the latest advancements in next-gen sequencing technology represent the most cutting-edge tolls for conducting high-resolution microbiota investigations. With decreasing average costs, the untargeted approach of shotgun metagenome sequencing has become more widely accessible, facilitating the sequencing of millions of reads per sample. Furthermore, recent developments in high-throughput methods have enabled the comprehensive detection of extended sequences from metagenomic sequencing to RNA metatranscriptomes [110]. These technological breakthroughs have paved the way for exploring the functional aspects of GM concerning the ETs, thereby expanding the domain of genomic toxicology into a broader metagenomic toxicology investigation domain [111–113].

Recently, significant progress has also been achieved in metabolomic and metaproteomic practices. These state-of-the-art methodologies offer enhanced precision in quantifying and identifying cellular proteins along with their post-translational modifications [114]. Moreover, they furnish valuable insights into the characteristics of the microbiome and the connection of microbial communities with the person's surrounding. Research into metabolomics holds the potential to develop biomarkers indicative of ETs [115, 116].

13.8 CURRENT CHALLENGES AND FUTURE DIRECTIONS

Gut bacteria exhibit diverse properties and can metabolize ETs in varied ways, either amplifying or reducing their toxicity to the host. Conversely, ETs can also impact both the diversity and the makeup of GM. Current studies suggest that the GM are intricately involved in the toxicity of ETs. However, despite identifying certain robust connections between changes in GM and ETs, many discrepancies in investigations persist. Zebrafish and drosophila are examples of common experimental models used in investigations [117–119]. There is a need for more germ-free models (e.g., fish) and humanized animal models with more defined and stable GM communities to minimize variability in GM measurements and experiments across studies [120–122]. Furthermore, further studies involving humans will need close oversight over nutrition and other variables that might greatly disrupt GM [123].

A current concern is the complexity of measuring the combined effects of several substances at once. The pollutants are widespread and can interact jointly as a combination [124–126]. Several investigations have focused on analyzing individual ETs to comprehend their impact on GM and human wellness, but this method fails to provide an exhaustive overview of the adverse health consequences caused by the various harmful ETs people are exposed to simultaneously. Furthermore, it is uncertain how GM responds when exposed to numerous harmful stimuli (such as toxins) together with one or more helpful substances (such as micronutrients). Evaluating the collective impact of several toxins or contaminants as a whole combination, instead of analyzing individually, may more correctly evaluate and pinpoint potential hazards and connections, eventually solving real-world issues related to public health. Furthermore, whole-genome sequencing systems provide a hurdle because of the complex and ever-changing relationships and networks across various 'omics variables [116]. A more thorough strategy for this evaluation is necessary. Lu et al. [50] demonstrated an integrated approach by showing substantial influence on both GM diversity and metabolite synthesis after introducing As into the mouse gut ecosystem. Zhang et al. performed research on disinfection byproducts in drinking water, showing the interconnectedness of various 'omics and the benefits of an integrated strategy for more insightful results [127–129].

Additional research is required to combine all 'omics knowledge thoroughly to understand the genetic makeup's reaction to ET exposure and how it influences physical traits. Numerous epidemiological investigations have found differences in the early-life GM diversity that do not exist in the subsequent years of existence, indicating that this phase may be an essential one when people are more susceptible to changes in their surroundings and GM [130]. ETs at this crucial stage, such as food, illnesses, and antibiotic treatments, might disturb the ideal development of GM. However, it also provides a chance for specific treatments such as nutrition, probiotics, and prebiotics to support the proper development and growth of GM [131–133]. It is essential to take into account how a remedy or disturbance might have different effects depending on the situation, regardless of whether it is in a healthy adult or a growing newborn.

Although currently constrained, a therapeutic approach based on metabolites holds significant promise. Over the next five years, it will be imperative for microbiome researchers to incorporate metabolomic profiling of the microbial ecosystem into their standard toolkit. This expansion will enable the community to delineate functional signatures for disease states previously linked only to changes in composition and metagenomics. Concurrently, the discovery of novel sensors for microbial metabolites will provide new insights into how the host assesses the activity of the microbial ecosystem, with direct implications for regulating inflammation versus maintaining homeostasis. The identification of these sensors will ultimately facilitate targeted manipulation of downstream signaling pathways in cases where the host exhibits an exaggerated response to the GM. Metabolite-based therapies offer a direct and actionable approach to addressing the impacts of dysbiosis on the host. The era of metabolite research in GM science has already commenced, and concerted endeavors may lead to the development of clinical interventions for dysbiosis-related disorders [97, 134–136].

13.9 CONCLUSION

Based on the literature provided, it becomes evident that the interaction between GM and ETs carries significant implications for human health. It underscores the crucial role of GM in various physiological functions, including digestion, immune regulation, and mental health, while highlighting the disruptive impact of ETs on its composition and function. Furthermore, this chapter emphasizes the importance of understanding these interactions across multiple facets of human health, spanning from the digestive system to mental health, obesity, and cancer. The modulation of GMISA and the gut-brain axis by environmental factors highlights the complexity of these interactions and their influence on disease onset and progression. Advancements in sequencing technologies, metagenomics, and metabolomics, as well as emerging trends in environmental toxicology, offer valuable insights into the mechanisms underlying these interactions. Additionally, innovative approaches for assessing exposure to ETs, such as wearable passive samplers and GIS, provide valuable tools for understanding spatial distribution and health implications. Both conclusions also discuss the therapeutic potential of interventions targeting the GM-environment axis to mitigate adverse health effects. Tailored therapies and preventive strategies based on a nuanced understanding of these interactions hold promise for improving human health outcomes.

Future research endeavors should focus on enhancing our understanding of GM and its complex relationship with ETs, ultimately leading to improved strategies for promoting human health and well-being. The integration of multi-omics approaches, along with sophisticated experimental models and comprehensive data analysis, will be instrumental in advancing our understanding of GM's role in ETs metabolism and toxicity. This knowledge will facilitate the development of more effective strategies for disease prevention, diagnosis, and treatment, ultimately benefiting global public health.

REFERENCES

1. L. Blackmer-Raynolds, T.R. Sampson, Overview of the gut microbiome, *Seminars in Neurology*, 43 (2023) 518–529.
2. K. Sangiuolo, E. Cheng, A. Terala, F. Dubrosa, R.L. Milanaik, The gut microbiome: an overview of current trends and risks for paediatric populations, *Current Opinion in Pediatrics,* 34 (2022) 634–642.
3. M.S. Hussain, P. Sharma, D.S. Dhanjal, N. Khurana, M. Vyas, N. Sharma, M. Mehta, M.M. Tambuwala, S. Satija, S.S. Sohal, B.G.G. Oliver, H.S. Sharma, Nanotechnology based advanced therapeutic strategies for targeting interleukins in chronic respiratory diseases, *Chemico-Biological Interactions*, 348 (2021) 109637.
4. M. Wigren, *The Sub-Chronic Effects of Polycyclic Aromatic Hydrocarbons on the Sheepshead Minnow (Cyrpinodon variegatus) Gut-Microbiome and Foraging Behavio*r, Department of Forestry and Natural Resources, Purdue University, West Lafayette, IN, 2020.
5. S. Santos-Paulo, S.P. Costello, S.C. Forster, S.P. Travis, R.V. Bryant, The gut microbiota as a therapeutic target for obesity: a scoping review, *Nutrition Research Reviews*, 35 (2022) 207–220.
6. B. Gao, X. Bian, R. Mahbub, K. Lu, Sex-specific effects of organophosphate diazinon on the gut microbiome and its metabolic functions, *Environmental Health Perspectives*, 125 (2017) 198–206.
7. M.S. Hussain, G. Gupta, R. Mishra, N. Patel, S. Gupta, S.I. Alzarea, I. Kazmi, P. Kumbhar, J. Disouza, H. Dureja, N. Kukreti, S.K. Singh, K. Dua, Unlocking the secrets: Volatile Organic Compounds (VOCs) and their devastating effects on lung cancer, *Pathology, Research and Practice*, 255 (2024) 155157.
8. A. Jain, S. Jain, R. Jain, S. Thakur, S. Jain, Gut microbiome and health assessment due to arsenic toxicity: a review, *International Journal of Life science and Pharma Research*, 11 (2021) L52–L60.

9. N.M. Ilieva, Z.D. Wallen, B.R. De Miranda, Oral ingestion of the environmental toxicant trichloroethylene in rats induces alterations in the gut microbiome: Relevance to idiopathic Parkinson's disease, *Toxicology and Applied Pharmacology*, 451 (2022) 116176.
10. M.L. Smith, J.B. Wade, J. Wolstenholme, J.S. Bajaj, *Gut Microbiome-Brain-Cirrhosis Axis*, Hepatology, Baltimore, MD (2023).
11. K. Zhang, X. Wang, X. Gong, J. Sui, Gut microbiome differences in rescued common kestrels (*Falco tinnunculus*) before and after captivity, *Frontiers in Microbiology*, 13 (2022) 858592.
12. L. Carucci, S. Coppola, A. Luzzetti, V. Giglio, J. Vanderhoof, R. Berni Canani, The role of probiotics and postbiotics in modulating the gut microbiome-immune system axis in the pediatric age, *Minerva Pediatrics*, 73 (2021) 115–127.
13. M.S. Hussain, G. Gupta, M. Afzal, S.M. Alqahtani, V.P. Samuel, W. Hassan Almalki, I. Kazmi, S.I. Alzarea, S. Saleem, H. Dureja, S.K. Singh, K. Dua, L. Thangavelu, Exploring the role of lncrna neat1 knockdown in regulating apoptosis across multiple cancer types: A review, *Pathology, Research and Practice*, 252 (2023) 154908.
14. E.M. Pferschy-Wenzig, M.R. Pausan, K. Ardjomand-Woelkart, S. Röck, R.M. Ammar, O. Kelber, C. Moissl-Eichinger, R. Bauer, Medicinal plants and their impact on the gut microbiome in mental health: a systematic review, *Nutrients*, 14 (2022) 2111.
15. J. Noor, A. Chaudhry, S. Batool, R. Noor, G. Fatima, Exploring the impact of the gut microbiome on obesity and weight loss: a review article, *Cureus*, 15 (2023) e40948.
16. M.S. Hussain, A. Sharma, R. Kumar, Prebiotics and probiotics: a focused review of applications in respiratory disorders, *Carpathian Journal of Food Science and Technology*, 15 (2023) 183–207.
17. E. Bezirtzoglou, E. Stavropoulou, K. Kantartzi, C. Tsigalou, C. Voidarou, G. Mitropoulou, I. Prapa, V. Santarmaki, V. Kompoura, A.E. Yanni, M. Antoniadou, T. Varzakas, Y. Kourkoutas, Maintaining digestive health in diabetes: the role of the gut microbiome and the challenge of functional foods, *Microorganisms*, 9 (2021) 516.
18. C. Sauceda, C. Bayne, K. Sudqi, A. Gonzalez, P.S. Dulai, R. Knight, D.J. Gonzalez, C.G. Gonzalez, Stool multi-omics for the study of host-microbe interactions in inflammatory bowel disease, *Gut Microbes*, 14 (2022) 2154092.
19. A. Gupta, Functional *Insights into Genomes and Metagenomes through Integrative Omics and Machine Learning Approaches*, Indian Institute of Science Education and Research Bhopal, Bhopal, India, 2018.
20. M. Tufail, C. Wu, M.S. Hussain, Dietary, addictive and habitual factors, and risk of colorectal cancer, *Nutrition*, 120 (2024) 112334.
21. W.T. Kwa, S. Sundarajoo, K.Y. Toh, J. Lee, Application of emerging technologies for gut microbiome research, *Singapore Medical Journal*, 64 (2023) 45–52.
22. Y. Wang, Y. Zhou, X. Xiao, J. Zheng, H. Zhou, Metaproteomics: A strategy to study the taxonomy and functionality of the gut microbiota, *Journal of Proteomics*, 219 (2020) 103737.
23. A.J. Denton, D.A. Godur, J. Mittal, N.B. Bencie, R. Mittal, A.A. Eshraghi, Recent advancements in understanding the gut microbiome and the inner ear axis, *Otolaryngologic Clinics of North America*, 55 (2022) 1125–1137.
24. J. Mayneris-Perxachs, M. Cardellini, L. Hoyles, J. Latorre, F. Davato, J.M. Moreno-Navarrete, M. Arnoriaga-Rodríguez, M. Serino, J. Abbott, R.H. Barton, J. Puig, X. Fernández-Real, W. Ricart, C. Tomlinson, M. Woodbridge, P. Gentileschi, S.A. Butcher, E. Holmes, J.K. Nicholson, V. Pérez-Brocal, A. Moya, D.M. Clain, R. Burcelin, M.E. Dumas, M. Federici, J.M. Fernández-Real, Iron status influences non-alcoholic fatty liver disease in obesity through the gut microbiome, *Microbiome,* 9 (2021) 104.
25. L. Zhu, X. Jian, B. Zhou, R. Liu, M. Muñoz, W. Sun, L. Xie, X. Chen, C. Peng, M. Maurer, J. Li, Gut microbiota facilitate chronic spontaneous urticaria, *Nature Communications,* 15 (2024) 112.
26. H. Wang, A. Ainiwaer, Y. Song, L. Qin, A. Peng, H. Bao, H. Qin, Perturbed gut microbiome and fecal and serum metabolomes are associated with chronic kidney disease severity, *Microbiome,* 11 (2023) 3.
27. C.S. Reigstad, P.C. Kashyap, Beyond phylotyping: understanding the impact of gut microbiota on host biology, *Neurogastroenterology and Motility*, 25 (2013) 358–372.
28. D.K.W. Ocansey, Z. Zhang, X. Xu, L. Liu, S. Amoah, X. Chen, B. Wang, X. Zhang, F. Mao, Mesenchymal stem cell-derived exosome mitigates colitis via the modulation of the gut metagenomics-metabolomics-farnesoid X receptor axis, *Biomaterials Science*, 10 (2022) 4822–4836.
29. J.E. Munoz, D.K. Brubaker, Latent interacting variable-effects modeling of gut microbiome multi-omics in inflammatory bowel disease, *bioRxiv*, (2022) 2022.2007. 2008.499280.

30. J.J. Aristizabal-Henao, A. Ahmadireskety, E.K. Griffin, B.F. Da Silva, J.A. Bowden, Lipidomics and environmental toxicology: recent trends, *Current Opinion in Environmental Science and Health*, 15 (2020) 26–31.
31. A. Ranjan, A. Chauhan, T. Jindal, Toxicology of organophosphate and recent trends in prophylactic approaches, *Toxicology New Frontiers in Environmental*, (2022) 103–123.
32. Y. Jiang, Recent progress and research trends in microplastic pollution and the potential health risks, *Bulletin of Environmental Contamination and Toxicology*, 107 (2021) 575–576.
33. M.H. Ansari, S. Lavhale, R.M. Kalunke, P.L. Srivastava, V. Pandit, S. Gade, S. Yadav, P. Laux, A. Luch, D. Gemmati, Recent advances in plant nanobionics and nanobiosensors for toxicology applications, *Current Nanoscience*, 16 (2020) 27–41.
34. I. Magyary, Recent advances and future trends in zebrafish bioassays for aquatic ecotoxicology, *Ecocycles*, 4 (2018) 12–18.
35. D.O. Carpenter, Polychlorinated biphenyls (PCBs): routes of exposure and effects on human health, *Reviews on Environmental Health*, 21 (2006) 1–24.
36. T. Bosker, J.F. Mudge, K.R. Munkittrick, Statistical reporting deficiencies in environmental toxicology, *Environmental Toxicology and Chemistry*, 32 (2013) 1737–1739.
37. J.O. Okeme, J.P. Koelmel, E. Johnson, E.Z. Lin, D. Gao, K.J.G. Pollitt, Wearable passive samplers for assessing environmental exposure to organic chemicals: current approaches and future directions, *Current Environmental Health Reports*, 10 (2023) 84–98.
38. J.P. Koelmel, E.Z. Lin, K. DeLay, A.J. Williams, Y. Zhou, R. Bornman, M. Obida, J. Chevrier, K. Pollitt, Assessing the external exposome of South African children using wearable passive samplers and high-resolution mass spectrometry, *Research Square*, (2021).
39. J.P. Koelmel, E.Z. Lin, K. DeLay, A.J. Williams, Y. Zhou, R. Bornman, M. Obida, J. Chevrier, K.J. Godri Pollitt, Assessing the external exposome using wearable passive samplers and high-resolution mass spectrometry among south african children participating in the VHEMBE study, *Environmental Science & Technology*, 56 (2022) 2191–2203.
40. J.P. Koelmel, E.Z. Lin, K. Delay, A.J. Williams, Y. Zhou, R. Bornman, M. Obida, J. Chevrier, K.J. Godri Pollitt, 50 chemical exposures of concern discovered using wearable passive samplers and gas chromatography high-resolution mass spectrometry in South African children, *ISEE Conference Abstracts*, 2021.
41. T. Filippini, C. Malagoli, S. Costanzini, F. Violi, S. Cilloni, S. Teggi, M. Fiore, M. Ferrante, M. Vinceti, 9 Risk of amyotrophic lateral sclerosis and passive residential exposure to pesticides: comparison of questionnaire-based with gis-based exposure assessment methods, *Occupational and Environmental Medicine*, 75 (2018) A28–A29.
42. N. Rousis, *Identification and Screening of Biomarkers of Human Exposure to Environmental and Food Toxicants in Sewage*, Open University (United Kingdom), Milan, Italy 2017.
43. B. Samaila, Z.M. Kalgo, B. Maidamma, Exposure to heavy metals in oil and gas wastes: A systematic review on health hazards assessment and mitigation in Nigeria, *Advanced Research in Medical and Health Sciences*, 1 (2023) 22–32.
44. H.F. Al-Harbi, A.M. Al-Mohaimeed, M.F. El-Tohamy, Assessment of essential elements and heavy metals in Saudi Arabian rice samples underwent various processing methods, *Open Chemistry*, 21 (2023) 20220328.
45. A. Shahab, Z. Hui, S. Rad, H. Xiao, J. Siddique, L.L. Huang, H. Ullah, A. Rashid, M.R. Taha, N. Zada, A comprehensive review on pollution status and associated health risk assessment of human exposure to selected heavy metals in road dust across different cities of the world, *Environmental Geochemistry and Health*, 45 (2023) 585–606.
46. X.C. Chen, X.H. Zhu, B.G. Lin, L.Z. Li, Z.L. Yu, M.D. Xiang, Y.J. Yu, [Children's non-carcinogenic health risk assessment of heavy metals exposure to residential indoor dust around an e-waste dismantling area in South China], *Zhonghua yu fang yi xue za zhi [Chinese Journal of Preventive Medicine]*, 53 (2019) 360–364.
47. V. Elumalai, K. Brindha, E. Lakshmanan, Human exposure risk assessment due to heavy metals in groundwater by pollution index and multivariate statistical methods: a case study from South Africa, *Water*, 9 (2017) 234.
48. X. Zhang, L. Yang, H. Zhang, W. Xing, Y. Wang, P. Bai, L. Zhang, K. Hayakawa, A. Toriba, Y. Wei, N. Tang, Assessing approaches of human inhalation exposure to polycyclic aromatic hydrocarbons: a review, *International Journal of Environmental Research and Public Health*, 18 (2021) 3124.

49. D.L. Diggs, A.C. Huderson, K.L. Harris, J.N. Myers, L.D. Banks, P.V. Rekhadevi, M.S. Niaz, A. Ramesh, Polycyclic aromatic hydrocarbons and digestive tract cancers: a perspective, Journal of environmental science and health. *Part C, Environmental Carcinogenesis & Ecotoxicology Reviews*, 29 (2011) 324–357.
50. K. Lu, R.P. Abo, K.A. Schlieper, M.E. Graffam, S. Levine, J.S. Wishnok, J.A. Swenberg, S.R. Tannenbaum, J.G. Fox, Arsenic exposure perturbs the gut microbiome and its metabolic profile in mice: an integrated metagenomics and metabolomics analysis, *Environmental Health Perspectives*, 122 (2014) 284–291.
51. S.P. Claus, H. Guillou, S. Ellero-Simatos, The gut microbiota: a major player in the toxicity of environmental pollutants?, *NPJ Biofilms and Microbiomes*, 2 (2016) 16003.
52. R.A. Diaz-Bone, T.R. van de Wiele, Biovolatilization of metal(loid)s by intestinal microorganisms in the simulator of the human intestinal microbial ecosystem, *Environmental Science & Technology*, 43 (2009) 5249–5256.
53. T. Van de Wiele, C.M. Gallawa, K.M. Kubachka, J.T. Creed, N. Basta, E.A. Dayton, S. Whitacre, G. Du Laing, K. Bradham, Arsenic metabolism by human gut microbiota upon in vitro digestion of contaminated soils, *Environmental Health Perspectives*, 118 (2010) 1004–1009.
54. A. Everard, C. Belzer, L. Geurts, J.P. Ouwerkerk, C. Druart, L.B. Bindels, Y. Guiot, M. Derrien, G.G. Muccioli, N.M. Delzenne, W.M. de Vos, P.D. Cani, Cross-talk between Akkermansia muciniphila and intestinal epithelium controls diet-induced obesity, *Proceedings of the National Academy of Sciences of the United States of America*, 110 (2013) 9066–9071.
55. S. Leclercq, S. Matamoros, P.D. Cani, A.M. Neyrinck, F. Jamar, P. Stärkel, K. Windey, V. Tremaroli, F. Bäckhed, K. Verbeke, P. de Timary, N.M. Delzenne, Intestinal permeability, gut-bacterial dysbiosis, and behavioral markers of alcohol-dependence severity, *Proceedings of the National Academy of Sciences of the United States of America*, 111 (2014) E4485–4493.
56. H. Plovier, A. Everard, C. Druart, C. Depommier, M. Van Hul, L. Geurts, J. Chilloux, N. Ottman, T. Duparc, L. Lichtenstein, A. Myridakis, N.M. Delzenne, J. Klievink, A. Bhattacharjee, K.C. van der Ark, S. Aalvink, L.O. Martinez, M.E. Dumas, D. Maiter, A. Loumaye, M.P. Hermans, J.P. Thissen, C. Belzer, W.M. de Vos, P.D. Cani, A purified membrane protein from Akkermansia muciniphila or the pasteurized bacterium improves metabolism in obese and diabetic mice, *Nature Medicine*, 23 (2017) 107–113.
57. E.Y. Hsiao, S.W. McBride, S. Hsien, G. Sharon, E.R. Hyde, T. McCue, J.A. Codelli, J. Chow, S.E. Reisman, J.F. Petrosino, P.H. Patterson, S.K. Mazmanian, Microbiota modulate behavioral and physiological abnormalities associated with neurodevelopmental disorders, *Cell*, 155 (2013) 1451–1463.
58. Y. Zhao, Y. Zhang, G. Wang, R. Han, X. Xie, Effects of chlorpyrifos on the gut microbiome and urine metabolome in mouse (Mus musculus), Chemosphere, 153 (2016) 287–293.
59. C.Y. Li, H.J. Renaud, C.D. Klaassen, J.Y. Cui, Age-specific regulation of drug-processing genes in mouse liver by ligands of xenobiotic-sensing transcription factors, *Drug Metabolism and Disposition: The Biological Fate of Chemicals,* 44 (2016) 1038–1049.
60. F.P. Selwyn, S.L. Cheng, C.D. Klaassen, J.Y. Cui, Regulation of hepatic drug-metabolizing enzymes in germ-free mice by conventionalization and probiotics, *Drug Metabolism and Disposition: The Biological Fate of Chemicals*, 44 (2016) 262–274.
61. National Academies of Sciences, Engineering, and Medicine, Division on Earth and Life Studies, Board on Life Sciences, Board on Environmental Studies and Toxicology, Committee on Advancing Understanding of the Implications of Environmental-Chemical Interactions with the Human Microbiome. *The National Academies Collection: Reports Funded by National Institutes of Health. Environmental Chemicals, the Human Microbiome, and Health Risk: A Research Strategy*, National Academies Press, Washington, DC, 2017.
62. B. Björkholm, C.M. Bok, A. Lundin, J. Rafter, M.L. Hibberd, S. Pettersson, Intestinal microbiota regulate xenobiotic metabolism in the liver, *PLoS One*, 4 (2009) e6958.
63. A. Mishra, G.C. Lai, L.J. Yao, T.T. Aung, N. Shental, A. Rotter-Maskowitz, E. Shepherdson, G.S.N. Singh, R. Pai, A. Shanti, R.M.M. Wong, A. Lee, C. Khyriem, C.A. Dutertre, S. Chakarov, K.G. Srinivasan, N.B. Shadan, X.M. Zhang, S. Khalilnezhad, F. Cottier, A.S.M. Tan, G. Low, P. Chen, Y. Fan, P.X. Hor, A.K.M. Lee, M. Choolani, D. Vermijlen, A. Sharma, G. Fuks, R. Straussman, N. Pavelka, B. Malleret, N. McGovern, S. Albani, J.K.Y. Chan, F. Ginhoux, Microbial exposure during early human development primes fetal immune cells, *Cell*, 184 (2021) 3394–3409.e3320.

64. M. Schirmer, E.A. Franzosa, J. Lloyd-Price, L.J. McIver, R. Schwager, T.W. Poon, A.N. Ananthakrishnan, E. Andrews, G. Barron, K. Lake, M. Prasad, J. Sauk, B. Stevens, R.G. Wilson, J. Braun, L.A. Denson, S. Kugathasan, D.P.B. McGovern, H. Vlamakis, R.J. Xavier, C. Huttenhower, Dynamics of metatranscription in the inflammatory bowel disease gut microbiome, *Nature Microbiology*, 3 (2018) 337–346.
65. E.A. Franzosa, A. Sirota-Madi, J. Avila-Pacheco, N. Fornelos, H.J. Haiser, S. Reinker, T. Vatanen, A.B. Hall, H. Mallick, L.J. McIver, J.S. Sauk, R.G. Wilson, B.W. Stevens, J.M. Scott, K. Pierce, A.A. Deik, K. Bullock, F. Imhann, J.A. Porter, A. Zhernakova, J. Fu, R.K. Weersma, C. Wijmenga, C.B. Clish, H. Vlamakis, C. Huttenhower, R.J. Xavier, Gut microbiome structure and metabolic activity in inflammatory bowel disease, *Nature Microbiology*, 4 (2019) 293–305.
66. R. Thapa, G. Gupta, A.A. Bhat, W.H. Almalki, S.I. Alzarea, I. Kazmi, S. Saleem, R. Khan, N. Altwaijry, H. Dureja, S.K. Singh, K. Dua, A review of Glycogen Synthase Kinase-3 (GSK3) inhibitors for cancers therapies, *International Journal of Biological Macromolecules*, 253 (2023) 127375.
67. J.W.J. Lee, D. Plichta, L. Hogstrom, N.Z. Borren, H. Lau, S.M. Gregory, W. Tan, H. Khalili, C. Clish, H. Vlamakis, R.J. Xavier, A.N. Ananthakrishnan, Multi-omics reveal microbial determinants impacting responses to biologic therapies in inflammatory bowel disease, *Cell Host & Microbe*, 29 (2021) 1294–1304.e1294.
68. A.N. Ananthakrishnan, C. Luo, V. Yajnik, H. Khalili, J.J. Garber, B.W. Stevens, T. Cleland, R.J. Xavier, Gut microbiome function predicts response to anti-integrin biologic therapy in inflammatory bowel diseases, *Cell Host & Microbe*, 21 (2017) 603–610.e603.
69. X.N. Tew, N.J. Xin Lau, D.K. Chellappan, T. Madheswaran, F. Zeeshan, M.M. Tambuwala, A.A. Aljabali, S.R. Balusamy, H. Perumalsamy, G. Gupta, B.G. Oliver, A. Hsu, P. Wark, K. Reddy, R. Wadhwa, P.M. Hansbro, K. Dua, Immunological axis of berberine in managing inflammation underlying chronic respiratory inflammatory diseases, *Chemico-Biological Interactions*, 317 (2020) 108947.
70. R. Thapa, O. Afzal, A.S. Alfawaz Altamimi, A. Goyal, W.H. Almalki, S.I. Alzarea, I. Kazmi, V. Jakhmola, S.K. Singh, K. Dua, R. Gilhotra, G. Gupta, Galangin as an inflammatory response modulator: an updated overview and therapeutic potential, *Chemico-Biological Interactions*, 378 (2023) 110482.
71. J.M. Wong, R. de Souza, C.W. Kendall, A. Emam, D.J. Jenkins, Colonic health: fermentation and short chain fatty acids, *Journal of Clinical Gastroenterology*, 40 (2006) 235–243.
72. J.R. Davie, Inhibition of histone deacetylase activity by butyrate, *The Journal of Nutrition,* 133 (2003) 2485s–2493s.
73. N. Arpaia, C. Campbell, X. Fan, S. Dikiy, J. van der Veeken, P. deRoos, H. Liu, J.R. Cross, K. Pfeffer, P.J. Coffer, A.Y. Rudensky, Metabolites produced by commensal bacteria promote peripheral regulatory T-cell generation, *Nature*, 504 (2013) 451–455.
74. S. Deleu, K. Machiels, J. Raes, K. Verbeke, S. Vermeire, Short chain fatty acids and its producing organisms: an overlooked therapy for IBD?, *EBioMedicine*, 66 (2021) 103293.
75. D. Parada Venegas, M.K. De la Fuente, G. Landskron, M.J. González, R. Quera, G. Dijkstra, H.J.M. Harmsen, K.N. Faber, M.A. Hermoso, Short Chain Fatty Acids (SCFAs)–mediated gut epithelial and immune regulation and its relevance for inflammatory bowel diseases, *Frontiers in Immunology* , 10 (2019) 277.
76. J.M. Shapiro, M.R. de Zoete, N.W. Palm, Y. Laenen, R. Bright, M. Mallette, K. Bu, A.A. Bielecka, F. Xu, A. Hurtado-Lorenzo, S.A. Shah, J.H. Cho, N.S. LeLeiko, B.E. Sands, R.A. Flavell, J.C. Clemente, Immunoglobulin A targets a unique subset of the microbiota in inflammatory bowel disease, *Cell Host & Microbe*, 29 (2021) 83–93.e83.
77. T. Ivanisevic, R.N. Sewduth, Multi-omics integration for the design of novel therapies and the identification of novel biomarkers, *Proteomes*, 11 (2023) 34.
78. M. Purohit, G. Gupta, O. Afzal, A.S.A. Altamimi, S.I. Alzarea, I. Kazmi, W.H. Almalki, M. Gulati, I.P. Kaur, S.K. Singh, K. Dua, Janus kinase/signal transducers and activator of transcription (JAK/STAT) and its role in Lung inflammatory disease, *Chemico-Biological Interactions*, 371 (2023) 110334.
79. N. Solanki, G. Gupta, D.K. Chellappan, S.K. Singh, M. Gulati, K.R. Paudel, P.M. Hansbro, K. Dua, S. Bhan, M. Saini, H. Dureja, Boswellic Acids: a critical appraisal of their therapeutic and nutritional benefits in chronic inflammatory diseases, *Endocr Metab Immune Disord Drug Targets*, 24 (2024) 116–129.
80. H.B.L. Tallaksen, E.B. Johannsen, J. Just, M.H. Viuff, C.H. Gravholt, A. Skakkebæk, The multi-omic landscape of sex chromosome abnormalities: current status and future directions, *Endocrine Connections*, 12 (2023) e23001112.
81. B. Gao, L. Chi, R. Mahbub, X. Bian, P. Tu, H. Ru, K. Lu, Multi-omics reveals that lead exposure disturbs gut microbiome development, key metabolites, and metabolic pathways, *Chemical Research in Toxicology*, 30 (2017) 996–1005.

82. K.R. Paudel, V. Patel, S. Vishwas, S. Gupta, S. Sharma, Y. Chan, N.K. Jha, J. Shrestha, M. Imran, N. Panth, S.D. Shukla, S.K. Jha, H.P. Devkota, M.E. Warkiani, S.K. Singh, M.K. Ali, G. Gupta, D.K. Chellappan, P.M. Hansbro, K. Dua, Nutraceuticals and COVID-19: A mechanistic approach toward attenuating the disease complications, *Journal of Food Biochemistry*, 46 (2022) e14445.
83. P. Prasher, M. Sharma, S.K. Singh, M. Gulati, N.K. Jha, P.K. Gupta, G. Gupta, D.K. Chellappan, F. Zacconi, T. de Jesus Andreoli Pinto, Y. Chan, G. Liu, K.R. Paudel, P.M. Hansbro, B.G. George Oliver, K. Dua, Targeting mucus barrier in respiratory diseases by chemically modified advanced delivery systems, *Chemico-Biological Interactions*, 365 (2022) 110048.
84. A.J. Wang, D. Song, Y.M. Hong, N.N. Liu, Multi-omics insights into the interplay between gut microbiota and colorectal cancer in the "microworld" age, *Molecular Omics*, 19 (2023) 283–296.
85. X. Sun, S. Feng, B. Qin, J. Ye, L. Xie, J. Gui, M. Sang, Integrated multi-omics profiling highlights the benefits of resveratrol hydroxypropyl-β-cyclodextrin inclusion complex for A53T transgenic mice through the microbiota-gut-brain axis, *Food & Function*, 15 (2024) 1489–1512.
86. W. Wang, G. Zhu, Y. Wang, W. Li, S. Yi, K. Wang, L. Fan, J. Tang, R. Chen, Multi-omics integration in mice with parkinson's disease and the intervention effect of cyanidin-3-O-glucoside, *Frontiers in Aging Neuroscience*, 14 (2022) 877078.
87. T.M. Sirangelo, Human gut microbiome analysis and multi-omics approach, *International Journal Of Pharmaceutical and Bio-Medical Science*, 7 (2018) 52–57.
88. J. Dachs, M. Vila-Costa, Toward a multi-omics-based single-cell environmental chemistry and toxicology, *Environmental Science & Technology*, 56 (2022) 10550–10552.
89. H. Li, R. Che, J. Zhu, X. Yang, J. Li, A.R. Fernie, J. Yan, Multi-omics-driven advances in the understanding of triacylglycerol biosynthesis in oil seeds, *The Plant Journal: For Cell and Molecular Biology*, 117 (2024) 999–1017.
90. L. Ma, W. Lyu, Y. Song, K. Chen, L. Lv, H. Yang, W. Wang, Y. Xiao, Anti-inflammatory effect of clostridium butyricum-derived extracellular vesicles in ulcerative colitis: impact on host micrornas expressions and gut microbiome profiles, *Molecular Nutrition & Food Research*, 67 (2023) e2200884.
91. X. Liu, Y. Liu, J. Liu, H. Zhang, C. Shan, Y. Guo, X. Gong, M. Cui, X. Li, M. Tang, Correlation between the gut microbiome and neurodegenerative diseases: a review of metagenomics evidence, *Neural Regeneration Research*, 19 (2024) 833–845.
92. H. Li, K. Song, X. Zhang, D. Wang, S. Dong, Y. Liu, L. Yang, Application of multi-perspectives in tea breeding and the main directions, *International Journal of Molecular Sciences*, 24 (2023) 12643.
93. I. Kazmi, F.A. Al-Abbasi, M. Afzal, M.S. Nadeem, H.N. Altayb, G. Gupta, Phosphodiesterase-4 inhibitor roflumilast-mediated protective effect in sepsis-induced late-phase event of acute kidney injury: a narrative review, *Pharmaceuticals*, 15 (2022) 899.
94. M. Mehta, D.S. Dhanjal, K.R. Paudel, B. Singh, G. Gupta, S. Rajeshkumar, L. Thangavelu, M.M. Tambuwala, H.A. Bakshi, D.K. Chellappan, P. Pandey, H. Dureja, N.B. Charbe, S.K. Singh, S.D. Shukla, S. Nammi, A.A. Aljabali, P.R. Wich, P.M. Hansbro, S. Satija, K. Dua, Cellular signalling pathways mediating the pathogenesis of chronic inflammatory respiratory diseases: an update, *Inflammopharmacology*, 28 (2020) 795–817.
95. C.Y. Lai, J. Sung, F. Cheng, W. Tang, S.H. Wong, P.K.S. Chan, M.A. Kamm, J.J.Y. Sung, G. Kaplan, F.K.L. Chan, S.C. Ng, Systematic review with meta-analysis: review of donor features, procedures and outcomes in 168 clinical studies of faecal microbiota transplantation, *Alimentary Pharmacology & Therapeutics*, 49 (2019) 354–363.
96. N. Narula, Z. Kassam, Y. Yuan, J.F. Colombel, C. Ponsioen, W. Reinisch, P. Moayyedi, Systematic review and meta-analysis: fecal microbiota transplantation for treatment of active ulcerative colitis, *Inflammatory Bowel Diseases*, 23 (2017) 1702–1709.
97. M.S. Hussain, G. Gupta, A. Goyal, R. Thapa, W.H. Almalki, I. Kazmi, S.I. Alzarea, S. Fuloria, D.U. Meenakshi, V. Jakhmola, M. Pandey, S.K. Singh, K. Dua, From nature to therapy: Luteolin's potential as an immune system modulator in inflammatory disorders, *Journal of Biochemical and Molecular Toxicology*, 37 (2023) e23482.
98. M.S. Hussain, G. Gupta, V.P. Samuel, W.H. Almalki, I. Kazmi, S.I. Alzarea, S. Saleem, R. Khan, N. Altwaijry, S. Patel, A. Patel, S.K. Singh, K. Dua, Immunopathology of herpes simplex virus-associated neuroinflammation: unveiling the mysteries, *Reviews in Medical Virology*, 34 (2024) e2491.
99. E.N. Baruch, I. Youngster, G. Ben-Betzalel, R. Ortenberg, A. Lahat, L. Katz, K. Adler, D. Dick-Necula, S. Raskin, N. Bloch, D. Rotin, L. Anafi, C. Avivi, J. Melnichenko, Y. Steinberg-Silman, R. Mamtani, H. Harati, N. Asher, R. Shapira-Frommer, T. Brosh-Nissimov, Y. Eshet, S. Ben-Simon, O. Ziv, M.A.W.

Khan, M. Amit, N.J. Ajami, I. Barshack, J. Schachter, J.A. Wargo, O. Koren, G. Markel, B. Boursi, Fecal microbiota transplant promotes response in immunotherapy-refractory melanoma patients, *Science*, 371 (2021) 602–609.

100. D. Davar, A.K. Dzutsev, J.A. McCulloch, R.R. Rodrigues, J.M. Chauvin, R.M. Morrison, R.N. Deblasio, C. Menna, Q. Ding, O. Pagliano, B. Zidi, S. Zhang, J.H. Badger, M. Vetizou, A.M. Cole, M.R. Fernandes, S. Prescott, R.G.F. Costa, A.K. Balaji, A. Morgun, I. Vujkovic-Cvijin, H. Wang, A.A. Borhani, M.B. Schwartz, H.M. Dubner, S.J. Ernst, A. Rose, Y.G. Najjar, Y. Belkaid, J.M. Kirkwood, G. Trinchieri, H.M. Zarour, Fecal microbiota transplant overcomes resistance to anti-PD-1 therapy in melanoma patients, *Science*, 371 (2021) 595–602.
101. S.C. Ng, Z. Xu, J.W.Y. Mak, K. Yang, Q. Liu, T. Zuo, W. Tang, L. Lau, R.N. Lui, S.H. Wong, Y.K. Tse, A.Y.L. Li, K. Cheung, J.Y.L. Ching, V.W.S. Wong, A.P.S. Kong, R.C.W. Ma, E.Y.K. Chow, S.K.H. Wong, I.C.H. Ho, P.K.S. Chan, F.K.L. Chan, Microbiota engraftment after faecal microbiota transplantation in obese subjects with type 2 diabetes: a 24–week, double-blind, randomised controlled trial, *Gut*, 71 (2022) 716–723.
102. G. Bahl, D.K. Upadhyay, M. Varma, R. Singh, S. Das, M.S. Hussain, Chronic calcific pancreatitis presented with secondary diabetes and diabetic ketoacidosis: a case report, *Clinical Diabetology*, 12 (2023) 209–211.
103. G. Cammarota, G. Ianiro, H. Tilg, M. Rajilić-Stojanović, P. Kump, R. Satokari, H. Sokol, P. Arkkila, C. Pintus, A. Hart, J. Segal, M. Aloi, L. Masucci, A. Molinaro, F. Scaldaferri, G. Gasbarrini, A. Lopez-Sanroman, A. Link, P. de Groot, W.M. de Vos, C. Högenauer, P. Malfertheiner, E. Mattila, T. Milosavljević, M. Nieuwdorp, M. Sanguinetti, M. Simren, A. Gasbarrini, European consensus conference on faecal microbiota transplantation in clinical practice, *Gut*, 66 (2017) 569–580.
104. C.L. Ho, H.Q. Tan, K.J. Chua, A. Kang, K.H. Lim, K.L. Ling, W.S. Yew, Y.S. Lee, J.P. Thiery, M.W. Chang, Engineered commensal microbes for diet-mediated colorectal-cancer chemoprevention, *Nature Biomedical Engineering*, 2 (2018) 27–37.
105. R.T. Schooley, B. Biswas, J.J. Gill, A. Hernandez-Morales, J. Lancaster, L. Lessor, J.J. Barr, S.L. Reed, F. Rohwer, S. Benler, A.M. Segall, R. Taplitz, D.M. Smith, K. Kerr, M. Kumaraswamy, V. Nizet, L. Lin, M.D. McCauley, S.A. Strathdee, C.A. Benson, R.K. Pope, B.M. Leroux, A.C. Picel, A.J. Mateczun, K.E. Cilwa, J.M. Regeimbal, L.A. Estrella, D.M. Wolfe, M.S. Henry, J. Quinones, S. Salka, K.A. Bishop-Lilly, R. Young, T. Hamilton, Development and use of personalized bacteriophage-based therapeutic cocktails to treat a patient with a disseminated resistant acinetobacter baumannii infection, *Antimicrobial Agents and Chemotherapy*, 61 (2017) 10–128.
106. D.D. Clarence, K.R. Paudel, B. Manandhar, S.K. Singh, H.P. Devkota, J. Panneerselvam, V. Gupta, N. Chitranshi, N. Verma, S. Saad, G. Gupta, P.M. Hansbro, B.G. Oliver, T. Madheswaran, K. Dua, D.K. Chellappan, Unravelling the therapeutic potential of nano-delivered functional foods in chronic respiratory diseases, *Nutrients*, 14 (2022) 3828.
107. M.S. Hussain, O. Afzal, G. Gupta, A.S.A. Altamimi, W.H. Almalki, S.I. Alzarea, I. Kazmi, N. Kukreti, S. Gupta, K. Sulakhiya, S.K. Singh, K. Dua, Probing the links: Long non-coding RNAs and NF-κB signalling in atherosclerosis, *Pathology, Research and Practice,* 249 (2023) 154773.
108. N. Principi, E. Silvestri, S. Esposito, Advantages and Limitations of Bacteriophages for the Treatment of Bacterial Infections, *Frontiers in Pharmacology*, 10 (2019) 513.
109. M. Hamady, R. Knight, Microbial community profiling for human microbiome projects: Tools, techniques, and challenges, *Genome Research*, 19 (2009) 1141–1152.
110. E.A. Franzosa, X.C. Morgan, N. Segata, L. Waldron, J. Reyes, A.M. Earl, G. Giannoukos, M.R. Boylan, D. Ciulla, D. Gevers, J. Izard, W.S. Garrett, A.T. Chan, C. Huttenhower, Relating the metatranscriptome and metagenome of the human gut, *Proceedings of the National Academy of Sciences of the United States of America,* 111 (2014) E2329–E2338.
111. R. Mesnage, M.N. Antoniou, D. Tsoukalas, G.N. Goulielmos, A. Tsatsakis, Gut microbiome metagenomics to understand how xenobiotics impact human health, *Current Opinion in Toxicology*, 11 (2018) 51–58.
112. D.K. Chellappan, K.R. Paudel, N.W. Tan, K.S. Cheong, S.S.Q. Khoo, S.M. Seow, J. Chellian, M. Candasamy, V.K. Patel, P. Arora, P.K. Singh, S.K. Singh, G. Gupta, B.G. Oliver, P.M. Hansbro, K. Dua, Targeting the mitochondria in chronic respiratory diseases, *Mitochondrion*, 67 (2022) 15–37.
113. D.K. Chellappan, L.W. Yee, K.Y. Xuan, K. Kunalan, L.C. Rou, L.S. Jean, L.Y. Ying, L.X. Wie, J. Chellian, M. Mehta, S. Satija, S.K. Singh, M. Gulati, H. Dureja, M.W. Da Silva, M.M. Tambuwala, G. Gupta, K.R. Paudel, R. Wadhwa, P.M. Hansbro, K. Dua, Targeting neutrophils using novel drug delivery systems in chronic respiratory diseases, *Drug Development Research*, 81 (2020) 419–436.

114. Y. Soufi, B. Soufi, Mass spectrometry-based bacterial proteomics: focus on dermatologic microbial pathogens, *Frontiers in Microbiology*, 7 (2016) 181.
115. V. Aguiar-Pulido, W. Huang, V. Suarez-Ulloa, T. Cickovski, K. Mathee, G. Narasimhan, Metagenomics, metatranscriptomics, and metabolomics approaches for microbiome analysis, *Evolutionary Bioinformatics Online*, 12 (2016) 5–16.
116. E.A. Franzosa, T. Hsu, A. Sirota-Madi, A. Shafquat, G. Abu-Ali, X.C. Morgan, C. Huttenhower, Sequencing and beyond: integrating molecular 'omics' for microbial community profiling, *Nature Reviews Microbiology*, 13 (2015) 360–372.
117. P. Tu, L. Chi, W. Bodnar, Z. Zhang, B. Gao, X. Bian, J. Stewart, R. Fry, K. Lu, Gut Microbiome toxicity: connecting the environment and gut microbiome-associated diseases, *Toxics*, 8 (2020) 19.
118. A.A. Bhat, G. Gupta, A. Goyal, R. Thapa, W.H. Almalki, I. Kazmi, S.I. Alzarea, N. Kukreti, M. Sekar, D.U. Meenakshi, S.K. Singh, R. MacLoughlin, K. Dua, Unwinding circular RNA's role in inflammatory pulmonary diseases, *Naunyn-Schmiedeberg's Archives of Pharmacology,* 397 (2023) 2567–2588.
119. S. Bhatt, T. Devadoss, N.K. Jha, M. Baidya, G. Gupta, D.K. Chellappan, S.K. Singh, K. Dua, Targeting inflammation: a potential approach for the treatment of depression, *Metabolic Brain Disease*, 38 (2023) 45–59.
120. H. Tlaskalova-Hogenova, L. Vannucci, K. Klimesova, R. Stepankova, J. Krizan, M. Kverka, Microbiome and colorectal carcinoma: insights from germ-free and conventional animal models, *Cancer Journal*, 20 (2014) 217–224.
121. K. Chiu, G. Warner, R.A. Nowak, J.A. Flaws, W. Mei, The impact of environmental chemicals on the gut microbiome, *Toxicological Sciences: An Official Journal of the Society of Toxicology*, 176 (2020) 253–284.
122. A.E. Douglas, Simple animal models for microbiome research, *Nature Reviews. Microbiology*, 17 (2019) 764–775.
123. L.A. David, C.F. Maurice, R.N. Carmody, D.B. Gootenberg, J.E. Button, B.E. Wolfe, A.V. Ling, A.S. Devlin, Y. Varma, M.A. Fischbach, S.B. Biddinger, R.J. Dutton, P.J. Turnbaugh, Diet rapidly and reproducibly alters the human gut microbiome, *Nature*, 505 (2014) 559–563.
124. A.F. Hernandez, A. Buha, C. Constantin, D.R. Wallace, D. Sarigiannis, M. Neagu, B. Antonijevic, A.W. Hayes, M.F. Wilks, A. Tsatsakis, Critical assessment and integration of separate lines of evidence for risk assessment of chemical mixtures, *Archives of Toxicology,* 93 (2019) 2741–2757.
125. S. Ashique, S. Kumar, A. Hussain, N. Mishra, A. Garg, B.H.J. Gowda, A. Farid, G. Gupta, K. Dua, F. Taghizadeh-Hesary, A narrative review on the role of magnesium in immune regulation, inflammation, infectious diseases, and cancer, *Journal of Health, Population, and Nutrition*, 42 (2023) 74.
126. S. Ashique, E. Sirohi, S. Kumar, M. Rihan, N. Mishra, S. Bhatt, R.K. Gautam, S.K. Singh, G. Gupta, D.K. Chellappan, K. Dua, Aducanumab in Alzheimer's disease: a critical update, *Current Medicinal Chemistry*, 31 (2023) 5004–5026.
127. Y. Zhang, F. Zhao, Y. Deng, Y. Zhao, H. Ren, Metagenomic and metabolomic analysis of the toxic effects of trichloroacetamide-induced gut microbiome and urine metabolome perturbations in mice, *Journal of Proteome Research*, 14 (2015) 1752–1761.
128. W.H. Almalki, M.M. Ghoneim, S. Alshehri, S.S. Imam, I. Kazmi, G. Gupta, Sepsis triggered oxidative stress-inflammatory axis: the pathobiology of reprogramming in the normal sleep-wake cycle, *Molecular and Cellular Biochemistry*, 477 (2022) 2203–2211.
129. S. Ashique, G. De Rubis, E. Sirohi, N. Mishra, M. Rihan, A. Garg, R.J. Reyes, B. Manandhar, S. Bhatt, N.K. Jha, T.G. Singh, G. Gupta, S.K. Singh, D.K. Chellappan, K.R. Paudel, P.M. Hansbro, B.G. Oliver, K. Dua, Short chain fatty acids: fundamental mediators of the gut-lung axis and their involvement in pulmonary diseases, *Chemico-Biological Interactions*, 368 (2022) 110231.
130. M. Tanaka, J. Nakayama, Development of the gut microbiota in infancy and its impact on health in later life, *Allergology International: Official Journal of the Japanese Society of Allergology*, 66 (2017) 515–522.
131. H. Sbihi, R.C. Boutin, C. Cutler, M. Suen, B.B. Finlay, S.E. Turvey, Thinking bigger: how early-life environmental exposures shape the gut microbiome and influence the development of asthma and allergic disease, *Allergy*, 74 (2019) 2103–2115.
132. V. Allam, D.K. Chellappan, N.K. Jha, M.D. Shastri, G. Gupta, S.D. Shukla, S.K. Singh, K. Sunkara, N. Chitranshi, V. Gupta, P.R. Wich, R. MacLoughlin, B.G.G. Oliver, S. Wernersson, G. Pejler, K. Dua, Treatment of chronic airway diseases using nutraceuticals: mechanistic insight, *Critical Reviews in Food Science and Nutrition*, 62 (2022) 7576–7590.

133. V. Allam, K.R. Paudel, G. Gupta, S.K. Singh, S. Vishwas, M. Gulati, S. Gupta, M. Chaitanya, N.K. Jha, P.K. Gupta, V.K. Patel, G. Liu, M.A. Kamal, P.M. Hansbro, B.G.G. Oliver, D.K. Chellappan, K. Dua, Nutraceuticals and mitochondrial oxidative stress: bridging the gap in the management of bronchial asthma, *Environmental Science and Pollution Research International*, 29 (2022) 62733–62754.
134. A.C. Wong, M. Levy, New approaches to microbiome-based therapies, *mSystems*, 4 (2019) 10–128.
135. T. Aggarwal, R. Wadhwa, R. Gupta, K.R. Paudel, T. Collet, D.K. Chellappan, G. Gupta, H. Perumalsamy, M. Mehta, S. Satija, P.M. Hansbro, K. Dua, P.K. Maurya, MicroRNas as biomarker for breast cancer, *Endocrine, Metabolic & Immune Disorders—Drug Targets*, 20 (2020) 1597–1610.
136. K.S. Alharbi, N.K. Fuloria, S. Fuloria, S.B. Rahman, W.H. Al-Malki, M.A. Javed Shaikh, L. Thangavelu, S.K. Singh, V.S. Rama Raju Allam, N.K. Jha, D.K. Chellappan, K. Dua, G. Gupta, Nuclear factor-kappa B and its role in inflammatory lung disease, *Chemico-Biological Interactions*, 345 (2021) 109568.

Index